The Classical Electromagnetic Field

The Classical Electromagnetic Field

and special relativity

Richard A.W. Bradford

principia publications unlimited

This edition first published May 2026

A catalogue record for this book is available from the British Library

Cover design by rawb

https://principiapublications.wordpress.com/

ISBN: 978-1-8380216-5-8

The Shorter Contents (chapter titles)

The Longer Contents (section & subsection titles)

Nomenclature, Acronyms and Notation

Acronyms and Definitions	
AC	Alternating Current
AVR	Automatic Voltage Regulator
BJT	Bipolar Junction Transistor
bps	bits per second (hence kbps, Mbps, Gbps)
CCGT	Combined Cycle Gas Turbines
CEGB	Central Electricity Generating Board
CME	Coronal Mass Ejection
DC	Direct Current
DFIG	Doubly Fed Induction Generator
e/m	electromagnetic
emf	electromotive force
EPR	European Pressurised Reactor
FET	Field Effect Transistor
flux	Denotes the amount of a substance flowing through unit area in unit time. Given a vector field $\bar{J} = \rho \bar{v}$ where ρ is the density of some substance (i.e., amount of the substance per unit volume) and $\bar{v}$ is its velocity, the interpretation of "flux" is that, for any small element of vector area, $d\bar{S}$, it is the amount of the substance which flows through that area in unit time, specifically $\bar{J} \cdot d\bar{S}$.
flux linkage	(Usually of magnetic flux) the flux through a specified closed loop (usually an electrical circuit) where each loop of a coil with N turns counts separately so that the total flux is N times the flux through a single loop.
HE, EH	Modes of e/m transmission in waveguides (esp. optic fibres)
iff	if and only if
impedance	ratio of (complex) voltage to current
inertial frame	an inertial frame of reference is a coordinate system moving with constant and uniform velocity wrt another inertial frame (the implicit assumption is the existence of a certain absolute frame which acts as the ultimate reference, a Newtonian position addressed further only in general relativity)
ITER	International Thermonuclear Experimental Reactor
IT	Information Technology
JFET	Junction-Field Effect Transistor
keV	kiloelectron volts

LCR	An electrical circuit consisting of an inductor (L), a capacitor (C), and a resistor (R) in series.
LHC	Large Hadron Collider
LIH	Linear, Isotropic and Homogeneous
MeV	megaelectron-volts
MKSA	metres, kilograms, seconds, amps (SI base units used here)
MOSFET	Metal Oxide Semiconductor Field Effect Transistor
NMI	Nuclear Magnetic Imaging (medical device)
PFC	Power Factor Correction
PMSG	Permanent Magnet Synchronous Generator
PWR	Pressurised Water Reactor
reactance	the imaginary part of impedance
rms	root mean square, $\sqrt{\langle X^2 \rangle}$
SAR	Specific Absorption Rate (non-ionising e/m radiation)
SI	Système International (international system of units)
TE	Transverse Electric (of e/m waves)
TM	Transverse Magnetic (of e/m waves)
VARs	Volts Amps Reactive
wrt	with respect to
Algebraic Quantities Different symbols may be used on occasion if necessary, and any symbol may take on other meanings in specific contexts	
$\bar{A}$	magnetic vector potential (a 3-vector)
A_μ	electromagnetic 4-vector potential
$\bar{B}$	$= \mu_0 \mu_r \bar{H}$, magnetic flux density (a variant magnetic field)
c	velocity of light in vacuum
C	capacitance
χ_e	$= \varepsilon_r - 1$, electric susceptibility
χ_m	$= \mu_r - 1$, magnetic susceptibility
$\bar{D}$	$= \varepsilon_0 \varepsilon_r \bar{E}$, electric displacement (a variant form of electric field)
δ	skin depth in AC (and equals ζ)
e	The quantum of charge (charge of the proton) or the Naperian constant 2.718281828, depending on context
$\bar{E}$	electric field (strength)
ε_0	permittivity of the vacuum
ε_r	the relative permittivity (or dielectric constant)
φ	magnetic flux
Φ	electric scalar potential (in a field)

ϕ	azimuthal angle in spherical polars
g_e	g-ratio of the electron
g_p	g-ratio of the proton
g_n	g-ratio of the neutron
G	gravitational constant
h	Planck's constant,
$\hbar$	Planck's reduced constant, $h/2\pi$
$\bar{H}$	magnetic field (strength)
I	current (electric)
$\bar{J}$	current density (strictly the charge flux)
κ	conductivity (electrical)
L	inductance (self or mutual), or angular momentum
$\bar{L}$	vectorial angular momentum
λ	line charge per unit length
m_e	electron mass
m_p	proton mass
m_n	neutron mass
$\bar{m}$	magnetic dipole moment
$\bar{M}$	magnetisation (magnetic dipole moment per unit volume)
μ_0	permeability of the vacuum
μ_r	relative permeability
μ_B	Bohr magneton
μ_N	nuclear magneton
n	real part of the refractive index, see §11.8
$\bar{N}$	Poynting vector (e/m radiant energy flux)
$\mathcal{N}$	complex refractive index, see §11.8
ω	angular frequency (radians per second)
Ω	complex valued angular frequency
$\bar{p}$	electric dipole moment
$\bar{P}$	polarisation (electric dipole moment per unit volume)
r	radial coordinate in either cylindrical or spherical polars, or the magnitude of some arbitrary position vector, $\bar{r}$
R	resistance
ρ	charge density (per unit volume)
ds	line element (units m), scalar
$d\bar{s}$	line element (units m), vector
dS	surface element (units m^2), scalar
$d\bar{S}$	surface element (units m^2), vector

σ	surface charge per unit area
$\bar{\varsigma}$	surface current 'density' (per unit length)
τ	Ambiguous, varies with context, e.g., As $d\tau$: 3D spatial volume element (units m^3). In relativity, τ may be the proper time, or, in Lorentz transformations sometimes $\tau = ct$. τ may also be used as a time constant in $e^{-t/\tau}$, e.g., in discharging capacitors or inductors
θ	polar angle in either cylindrical or spherical polars
u_E	energy density of an electric field
u_M	energy density of a magnetic field
V	electric scalar potential (in circuit theory)
ξ	imaginary (absorptive) part of the refractive index, see §11.8
ζ	penetration depth (of e/m wave into a medium)
Operators	
$\times$	vector cross product
$\cdot$	inner (or dot) product of vectors
$\wedge$	wedge product
∂_ξ $=_{,\xi}$	differential operator wrt the coordinate indicated by ξ and the comma notation for the same, e.g., $\partial_\xi v_i \equiv v_{i,\xi} \equiv \frac{\partial v_i}{\partial \xi}$. If ξ is an integer, 0, 1, 2 or 3, the implied derivative variable is the Cartesian or Minkowskian coordinate $x_0, x_1, x_2, x_3 \equiv ct, x, y, z$
∂^2	the d'Alembertian operator, $\partial_\mu \partial^\mu = \partial_0^2 - \nabla^2$
$c\partial_0 = \partial_t$	because $x^0 = ct$
$\boxdot$	the d'Alembertian operator in a medium $\left(\frac{1}{v^2}\partial_t^2 - \nabla^2\right)$ where $v = \frac{c}{\sqrt{\mu_r \varepsilon_r}} = \frac{1}{\sqrt{\mu_r \varepsilon_r \mu_0 \varepsilon_0}}$
d	exterior derivative

The Preface and Important Conventions

Perennial Preface skippers should not do so because this Preface includes key text conventions of which the reader needs to be aware.

There are already many texts on classical electromagnetism at the undergraduate level. Why another? To any author of physics texts, attempting one on classical electromagnetism is by way of being a rite of passage. Maxwell's equations, the indisputable heart of classical electromagnetism, still represent a pinnacle of achievement in physics theory. The subject remains the archetypal example of a classical field theory. There is, therefore, an element of homage involved in authoring such a text. To be blunt, such a book is really written for the author's own satisfaction. Actually, they always are. Despite that, I hope you find it beneficial.

Having given you the hard sell, let me emphasise that this book includes some material which is not so commonly found in texts. Actually, the book starts in that way with a discussion of objectivity in physics, and science generally. This is not usually encountered in physics texts, being consigned to specialist books on the philosophy of science. But there is a reason for its inclusion, namely that it is emphasised throughout that we must deal with objects whose existence and nature are independent of how or by whom they are observed or measured. Immediately, then, it is established that we are in the world of classical realism. Of quantum mechanics we say nothing here.

The next chapter reinforces this perspective with an elementary account of vectors, stressing that they are objects in their own right not merely sets of numbers. This sets the tone to be adopted throughout. Vector notation is used in the bulk of the book, consistent with the practice that became common after Maxwell's death in the last decade or two of the 19th century. The use of vectors is itself an indication of the objective nature of the electric and magnetic fields, illustrating that being an "object" does not necessarily imply being composed of lumpen stuff. Rather it conveys existence independent of observers or arbitrary coordinate systems.

It may be helpful to give some indication of the subjects covered which may feature less often in other texts. The first is the detailed analysis of the Feynman disc paradox in chapter 8. This is a lovely illustration of the necessary reality of the momentum of an electromagnetic field – even when the fields are static, remarkably. Feynman included this gedanken experiment in his famous lecture series. He did not present an analysis of it, but this is required to complete the picture. The same chapter also explains why the magnetic force acting between a pair of moving charges appears to violate Newton's 3rd law – but does not when the field momentum is included.

Chapter 11's focus on reflection, refraction and such things as total internal reflection and Brewster's angle may seem rather prosaic. But the purpose of the chapter is to show how all these phenomena, which are familiar from geometrical optics, are actually the result of Maxwell's equations and their compatible boundary conditions. If this were not the case, the claim that light is an electromagnetic wave could hardly be maintained.

Chapter 12 is largely devoted to the safety of electromagnetic radiation. We are now marinaded in myriad electromagnetic fields so the issue is apposite and sometimes causes popular unease (generally ill-informed). Human safety in the context of exposure to such radiation is not normally considered the province of theoretical physics. Perhaps my personal background inclines me not to ignore it.

Chapter 16 proves Earnshaw's Theorem and goes on to discuss various types of magnetic levitation. While these topics have probably become standard fare in modern texts, the chapter serves to address a reasonable confusion between Earnshaw's Theorem, which appears to forbid magnetic levitation, and the fact that we have all seen videos of levitating frogs.

The skin effect, considered in chapter 17, has long been standard fare in texts. But the linking of it to the operation of domestic induction hobs is not necessarily covered in other texts. The matter is subtle because the significance of using pans of ferromagnetic material is not merely because it enhances the H-field within the pan's base. It also involves the skin effect.

Chapter 18 on circuit electricity is mainly for the purpose of emphasising that the methodology of circuit analysis arises as a cut-down accounting for the effects of the electromagnetic field. Conventional circuit analysis is essential in practice, but it is worth being mindful that what is really happening lies in the electromagnetic fields which mostly reside outside the conducting wires and circuit components.

That lesson reaches its climax with what is perhaps the best example of a crucial topic that is generally not mentioned at all: energy does not flow in wires. Chapters 19 and 20 provide detailed solutions of the Maxwell equations relating to geometrically simple conducting circuits to drive home this point. You will never look at overhead transmission lines the same way again. Previously, you may have committed the folly of thinking that perhaps 2 GW of power was travelling inside the metal conductors in the cables. Not so. The power is travelling through the air in the space between the cables in the form of an electromagnetic wave. Power flows through the field around wires and cables, not within the conductors which bear the current.

Chapter 21 addresses a subject more likely to be found in electrical engineering texts than physics texts, namely large-scale electricity generation and transmission. It is undoubtedly my personal background in that industry which encourages its inclusion. The matter is highly topical, even controversial, due to the current drive for ever more wind and solar sources. Consequently, this chapter makes observations which relate to policy and hence politics. This includes the challenges to grid stability posed by large proportions of inverter-based sources with no intrinsic inertia.

Conducting boundary waveguides have always been a staple of electromagnetism texts. But older texts often said little about dielectric waveguides. This will have changed dramatically in more recent texts because of the ubiquity of optic fibre communications. The theory of dielectric waveguides is therefore addressed in some detail. In particular, it is explained how the phenomenon of the "single mode fibre" arises and its significance explained. While this is not a text on IT communication technology some observations are made of relevance to data transmission rates. In particular, why does your WiFi speed vary around your house? We understand that the electric and magnetic fields suffer attenuation, i.e., a reduction in their amplitude. But how does this translate into digital units, e.g., Mbps (mega-bits per second)? This is explained in chapter 23 in simple terms.

The emphasis from the start on the importance of objects – as the entities to which objectivity applies - stands us in good stead when introducing relativity in chapter 26, which is the pre-eminent theory of objectivity. Rather than saying that "everything is relative", relativity theory is based solidly on the view that we are dealing with objects of reality independent of ourselves – though our measurements will indeed be dependent on, for example, our state of motion and our choice of an arbitrary coordinate frame. Relativity theory is a formalised procedure for decontaminating our measurements of their subjective aspects to leave their objective core. Only the latter can be expected to appear in the laws of physics.

On the subject of relativity, the topic is introduced by deriving the Lorentz transform based solely on the equivalence of inertial observers. It is easy to lose sight of whence the Lorentz transform originates physically after being emersed in the world of 4-vectors and tensors for a long time. However, those entities are not optional and emphatically continue the theme that we are studying objects of reality. This is rewarded by the formal simplification of Maxwell's equations that result from the use of tensors, chapter 27, which became standard after Einstein – especially after the 1916 general relativity paper. This sets the scene for chapter 28 which introduces the energy-momentum tensor, the formulation of the conservation of energy and momentum in tensorial form, and the link between spacetime symmetries and these conservation laws via Noether's Theorem.

Unusually, a chapter is included on how Maxwell's equations can be written in terms of quaternions (or biquaternions, to be more precise). This results in the most compact expression of those key equations, reduced rather remarkably to just one equation: $DF + J = 0$. Although Maxwell did use quaternions to a rather restricted degree in his Treatise on electromagnetism, this compactness of notation was unknown to him, as far as I am aware.

The book closes with an elementary account of the exterior derivative, exterior algebra and the Cartan-Stokes theorem. This has become the standard toolset in general relativity since the masterful Gravitation by Misner, Thorne and Wheeler in 1970 but is less commonly encountered in texts devoted purely to electromagnetism. It's main benefit here, in addition to further emphasis on objectivity, is the Cartan-Stokes theorem which is the general theorem of which Green's theorem, Stokes's theorem and Gauss's theorem are merely special cases. Moreover, it extends them to arbitrary dimensions and so consolidates integrals encountered in the 4D Minkowski spacetime of special relativity.

Notation, Conventions, Abbreviations, Finding Things

The book has no Index. To find things your first port of call should be the shorter or the longer Contents (above). Failing that, a word search on the pdf may be used. This is surely a more powerful method than a traditional Index. The pdf of this book is available for free download on http://rickbradford.co.uk. In due course I hope to have a facility to "ask my book a question" on ChatGPT.

References are listed at the end of each chapter in which they are cited. Most are html linked, for which you will need the pdf. References are indicated by square brackets, such as [ch.x], where ch is the chapter number and x is the reference number in that chapter. However, there are only a few references, many chapters having none. This is a pedagogic work, not a work of deep scholarship.

Citations to other parts of the book are either explicitly by reference to a specific chapter or by reference to a specific section or equation. Sections and subsections are indicated by the section mark, §, for example, §5.4.8. In contrast, (5.4.8) refers to the equation of that number. Do not confuse the two. Equations are indicated by round brackets, (ch.s.x), where ch is the chapter number, s is the section number within that chapter, and x is the sequential equation number within that section.

MKSA units are used throughout. These are the SI approved units. Unfortunately, electromagnetism is plagued by differing conventions on units. Rather than attempting to define all commonly used units I have opted to stick rigorously with just MKSA which I think is less confusing. This means that formulae commonly include factors like $\mu_0/4\pi$ or $1/4\pi\varepsilon_0$ and often powers of c, the velocity of light. (Only rarely do I switch to putting $c = 1$, the convention so beloved by particle physicists).

The Einstein summation convention is used throughout. The convention will be that Latin indices take only spatial values x, y, z whereas Greek indices take spacetime values t, x, y, z. Hence, for example, $a_i b_i$ is shorthand for $a_x b_x + a_y b_y + a_z b_z$ whereas $a_\mu b^\mu$ is shorthand for $a_t b^t + a_x b^x + a_y b^y + a_z b^z$.

Vectors in 3D space are denoted by a bar over the symbol, e.g., $\bar{a}$. Hence, the dot product is $\bar{a} \cdot \bar{b} = a_i b_i$, assuming Cartesian coordinates. The notation for p-forms need not trouble the reader until the last chapter.

In tensorial expressions I use index 0 interchangeably with t except that $x^0 = ct$ so that $\frac{\partial}{\partial x^0} = \partial_0 = \frac{1}{c}\frac{\partial}{\partial t} = \frac{1}{c}\partial_t$ so that ∂_0 is dimensionally consistent with ∂_i. On occasion I also use $\tau \equiv x^0 = ct$.

The Minkowski metric is time-like.

Bracket notation is used to denote matrices. For example (M) denotes the whole matrix whose components might be written M_{ij} or perhaps $M_{\mu\nu}$.

Herein the term "wavenumber" is used as synonymous with the magnitude of the wavevector, $\bar{k}$, and hence means what some authors refer to as "angular wavenumber", i.e., $2\pi/\lambda$ where λ is the wavelength.

"log" is always natural when unqualified. When base 10 or base 2 is meant this is indicated explicitly, i.e., log_{10} or log_2.

Immediately preceding this Preface there is a list of acronyms, definitions and algebraic notation. I hope this helps. Unfortunately, though, there are far more quantities which require designation by a letter than there are letters, even when Latin is augmented by Greek. Consequently, there is inevitably some ambiguity which one hopes is alleviated by context.

I point out one abbreviation which is used extensively: "wrt" meaning "with respect to". Also, "iff" meaning the necessary and sufficient "if and only if".

Most importantly, in a book of this size and complexity, it is inevitable that there will be errors. If you spot any please let me know by email on RickAtMerlinHaven@hotmail.com.

1. Objectivity in Classical Physics

As this book is confined to classical physics, it is rather important to appreciate what is meant by "classical" in this context. Consequently, and rather unusually, we start by defining classical physics and emphasising the closely associated role of objectivity within classical physics. Objectivity provides science with much of its systematising power and predictive capability, and, crucially, creates credibility due to its repeatability by independent individuals. What is less appreciated is that objectivity also creates a self-imposed limitation. A metaphysical perspective is that these two features of objectivity, its power and its limitations, are two sides of the same coin and together constitute a metaphysical uncertainty principle. This issue is raised in this brief opening chapter only because it is so rarely mentioned. But, in common with other physics texts, the rest of the book will follow familiar objective lines.

Contents

1.1 What is Classical Physics?

Classical physics refers to physics prior to quantum theory. Classical physics includes everything normally classed as physics up to and including special and general relativity. The defining features of classical physics are,

- It takes place in a spacetime continuum,
- It is strictly objective,
- It is strictly deterministic,
- Matter and positive energy are localisable to points in spacetime with, in principle, arbitrary precision.

The first bullet point is deliberately vague regarding whether the spacetime in question need be four-dimensional in order to qualify as "classical". Kaluza's original five-dimensional theory, [1.1], was indeed classical. But, following Klein's contribution, [1.2], Kaluza-Klein theory became adopted into quantum theory/string theory, which are certainly not classical.

The stipulation of "positive energies" in the last bullet point is in recognition that gravitational energy density in general relativity is absolutely negative, and, as a result, is not localisable.

1.2 What is Objectivity?

A key feature which underpins physical science is the adoption of a strictly objective methodology, or as near as can be achieved. Part of this methodological approach is the elimination of subjective elements, in as far as possible. The metaphysical motivation for this is the (classical) conviction that there exists an external, physical world which is independent of observers. The aim of physics is to understand, quantify and provide predictive systems for the behaviour of this external reality. To do so requires any subjective elements introduced by observers to be eliminated in order to leave the objective core.

That observers introduce subjective elements into measurements is obvious and unavoidable. For example, the observed velocity of an object of study will depend upon the state of motion of the observer, i.e., velocity is necessarily relative. Similarly, it makes no sense to talk of "distance" unless one specifies from what (or from whom) the distance is to be measured. These instances are obvious, but it turns out that other quantities in physics also have a subjective, or relative, element though this would not always be apparent until the relevant physical theory has revealed it. For example, the time interval between two events turns out in relativity theory to have a subjective element, i.e., dependent on the observer. Even the simultaneity, or otherwise, of two spatially separated events may be something on which different observers will disagree and hence is a subjective matter.

The foundational paradigm of classical physics is that, when the subjective or relative elements have been excluded, what is left are "the elements of external reality", namely matter and energy and their various forms and behaviours. Even the distinction between matter and energy becomes blurred in relativity theory. Thus, in this perspective, matter and energy become the "objects" of external physical reality. The characteristic of "objects" is their objectivity, i.e., that they exist independently of any observer and have properties which are independent of any observer.

Hence, objectivity in physical science is both a methodology and a metaphysical presumption. The objective methodology provides physical science with its power. The metaphysical presumption of objectivity imposes a limitation on the reach of physical science.

Both these aspects of objectivity are crucial to the nature of the science which results. But note that this is a Faustian pact. By making the assumption from the start that an external, observer-independent reality exists and by restricting the sphere of discourse to what remains after the subjective is excluded, a limitation is imposed on the reach of such a science. In particular, having excluded the subjective one should not expect the resulting physics to be able to explain subjective experience, e.g., conscious awareness, qualia, etc. This is the logical fallacy involved in claiming that objective science (as described above) supports the contention that a materialistic metaphysics is of universal applicability, e.g., even in elucidating the nature of subjective awareness.

1.3 Materialism (or "Physicalism")

Neither the veracity of physics within its sphere of applicability, nor its utility, depend upon a metaphysical presumption that matter/energy is the only ontological primary. This is a logically separate assumption made by some people but does not form part of physics itself. There are profound problems with attempting to understand consciousness as an epiphenomenon of matter/energy. Happily that problem is excluded from physics by definition.

1.4 Objects

Having chosen to confine ourselves to the objective, it should not be surprising that our study then focusses on objects – those entities which remain after subjective elements are removed. But "objects" should not be misinterpreted to mean only ponderable matter or atoms and molecules. Since Faraday, the concept of the field has entered physics. Fields are very different from the lumpen stuff of everyday life. But fields are also objects in the sense that they are objective. They have a reality independent of any observer. This contrasts with the numbers used to describe them which will, in general, include subjective elements. This leads to the important clarification that…

1.5 Physical Things Are Not Numbers

"Richard Bradford" is my name, but it designates something (myself) which has an existence independent of this name. I would be the same person if I changed my name to John Smith.

Measurements are generally made in terms of numbers. An object may be described by numbers representing its position and orientation in space, as well as numbers representing its properties such as mass, charge, shape, volume, colour, etc. But this collection of numbers is not the object, but only a representation of the object. Moreover, in some cases, the numbers may be a particularly imperfect representation of the object because different sets of numbers may be obtained by different observers, i.e., the set of numbers is not objective.

The numbers which we use in physics to represent objects can include a subjective element dependent upon the observer or the observational process. But if we believe that the outside world is objectively real and independent of ourselves, then physics should be formulated in a manner which excludes any subjective element. How is this to be done?

If ever a theory was more inappropriately named, surely it is "relativity". The popular misconception that relativity theory consists of the view that "everything is relative" could hardly be more wrong. Einstein's philosophical perspective was that physics is about objectively real things which exist independently of ourselves. The theory of relativity is the quintessential theory of objectivity. What relativity theory tells us is how to account for the subjective elements of measurements made by a given observer in order to identify and extract the objective core. All observers must agree on the latter, objective aspects, and it is upon these objective aspects that physical theory must be built. We will see that vectors and tensor analysis are the key tools for formulating objectivity in mathematical terms, resulting in different observers who can agree on the objective physics despite measuring distinct "numbers".

For this reason, the next chapter serves to emphasise that simple vector analysis is a description of objective things, and that a vector is an object in its own right not merely a set of three numbers called coordinates.

1.6 Is Quantum Physics Objective?

Since this book is concerned only with classical physics, this is a question I feel entitled to evade. However, some of the key aspects of objectivist methodology are certainly carried over into quantum physics, both quantum mechanics and quantum field theory. The prime example is the use of tensor analysis (or group theory) as the central techniques to delineate the distinction between observer-dependent quantitative measurements and their objective core which is taken as an "object of reality".

On the other hand, the independent existence of the (ostensible) entities under study in quantum physics is more problematical. The infamous measurement problem and the associated unknowability of properties prior to measurement are aspects of this dissolving of independent existence. Also, in classical physics, "objectivity" subsumes local realism, a feature which is indubitably false in quantum physics due to the phenomenon of entanglement. Thankfully, we will have no need to address these slippery matters further here, being concerned with the strictly classical.

1.7 References

[1.1] Kaluza, T. (1921). *Zum Unitätsproblem in der Physik. Sitzungsber.* Preuss. Akad. Wiss. Berlin. (Math. Phys.): 966–972 (in German). English translation here: *[arXiv:1803.08616] On the Unification Problem in Physics*.

[1.2] Klein, O. (1926). *Quantentheorie und fünfdimensionale Relativitätstheorie*. Zeitschrift für Physik A 37(12): 895–906.

2. Euclidean Vectors in 3D Space

Having indulged in some philosophical orientation in chapter 1, here we get down to the business of how objects in spacetime are described mathematically, starting gently with ordinary (Euclidean) vectors in three-dimensional space. A host of fundamental theorems follow with little effort and without components or coordinates being necessary. This usefully emphasises that, with vectors, we are dealing with entities which are objects in their own right, not merely sets of numbers. The coordinate description does have important uses. Nevertheless, coordinates can sometimes obscure the key underlying fact that we are dealing with geometrical objects.

Contents

Starting, in this chapter, with vectors in three-dimensional Euclidean space, in later chapters we shall work up to tensors in 4-dimesional spacetime. In the case of vectors in three-dimensional Euclidean space we have the luxury of being able to rely on intuition as we are dealing with familiar geometry. Indeed, 3D Euclidean vectors serve us well for the development of the bulk of classical electromagnetic theory. This chapter also introduces some key notation that will be used throughout the book.

2.1 What are 3D Euclidean Vectors?

2.1.1 Vectors Without Position

Vectors are not merely sets of numbers, though numbers are required as part of their nature. A vector is a mathematical object in its own right. The proposal is that mathematical objects will be the appropriate language to use in the description of physical objects. A vector may be defined by having two properties,

- A magnitude (a real number);
- A direction in space (including an orientation along that direction).

3D Euclidean vectors will be denoted here by a bar over the symbol, such as $\bar{v}$. The magnitude of the vector will be denoted either by the same symbol without the bar, or, where emphasis or clarity is required, by $|\bar{v}| \equiv v$.

Vectors can be represented geometrically by an arrow. The length of the arrow is proportional to the vector's magnitude, whilst the direction and orientation of the arrow are direct representations of those of the vector.

Addition of two vectors is defined geometrically by,

Figure 2.1

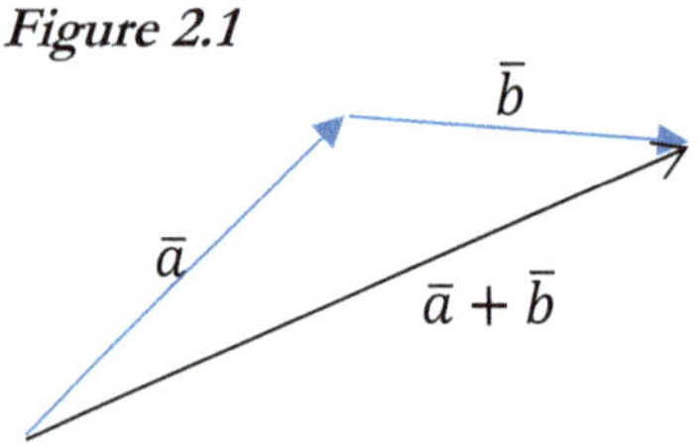

From this one can immediately derive the associativity of vector addition, i.e., that,

$$(\bar{a} + \bar{b}) + \bar{c} = \bar{a} + (\bar{b} + \bar{c}) \tag{2.1.1}$$

and hence that both can be written unambiguously as $\bar{a} + \bar{b} + \bar{c}$. If this isn't obvious then contemplate how the above two algebraic expressions are each interpreted in Figure 2.2 below, and hence that they have the same result.

Figure 2.2

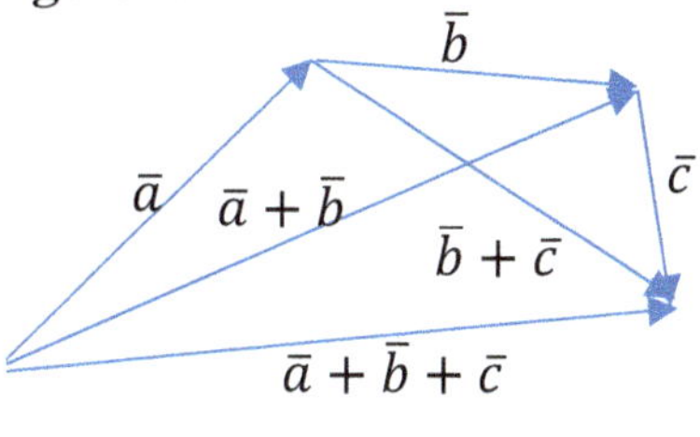

Recall that the definition of a vector does not constrain where it is placed, i.e., it is independent of position. Thus, the vectors $\bar{a}, \bar{b}$ may be duplicated and displaced to form a parallelogram, thus,

Figure 2.3

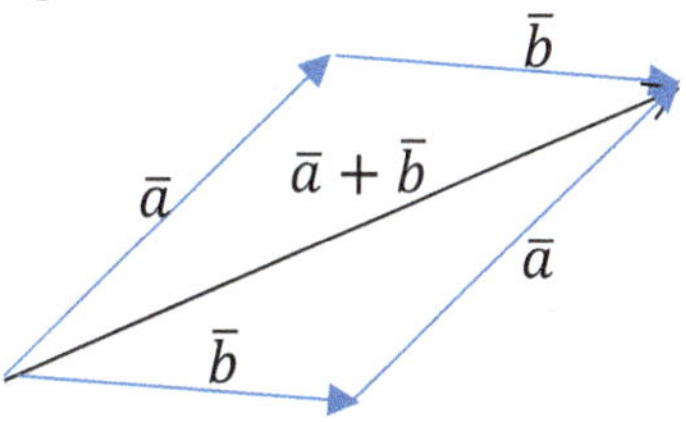

From this it follows that vector addition is commutative,

$$\bar{a} + \bar{b} = \bar{b} + \bar{a} \tag{2.1.2}$$

I emphasise that the concept of vectors as independent of position is essential to (2.1.2), and hence the whole of linear vector algebra that follows. For later purposes, note that the assumption that the underlying geometry is Euclidean is implicit in our assumed freedom to move vectors around without change, and hence the resulting proof that vector addition is commutative. If we were to study general relativity we would see that curved (hence non-Euclidean) spaces are such that (2.1.2) would not hold, precisely because we would then not be entitled to presume that "moving a vector without changing it" (called "parallel transport") had an unambiguous meaning. In fact, in non-Euclidean (e.g., Riemannian) geometry the difference between the two sides in (2.1.2) is proportional to the curvature of the underlying space (or spacetime). But for our purposes in this book we shall stick to Euclidean space or Minkowskian spacetime where (2.1.2) holds.

Multiplication of a vector, $\bar{a}$, by a real number, λ, to give $\lambda\bar{a}$ follows from (2.1.1), at least for integers and by extension for all real numbers, for example $\bar{a} + \bar{a} + \bar{a} = 3\bar{a}$. It follows that $\lambda\bar{a}$ is the vector with the same direction and orientation as $\bar{a}$ but whose magnitude is λa. The three types of distributive law can also be derived without difficulty from this geometrical definition of addition, i.e.,

$$(\lambda + \mu)\bar{a} = \lambda\bar{a} + \mu\bar{a} \tag{2.1.3a}$$

$$\lambda(\bar{a} + \bar{b}) = \lambda\bar{a} + \lambda\bar{b} \tag{2.1.3b}$$

$$\lambda(\mu\bar{a}) = \lambda\mu\bar{a} \tag{2.1.3c}$$

A unit vector is a vector with unit magnitude. The unit vector in the same direction and orientation as $\bar{a}$ will be denoted $\hat{a}$. Hence we can write $\bar{a} = a\hat{a}$ which thus displays the magnitude and direction of the vector separately.

2.1.2 Position Vectors

We may also use vectors as a means of specifying a point in 3D space. For this purpose we first choose an origin, O. The vector drawn from O to the point in question, P, then defines the position of the point. This "position vector" can be written $\overrightarrow{OP}$ although, where the origin is understood it may simply be denoted by $\bar{r}$, or some such symbol. We are free to use a different origin instead, say Q, but the same point will then be defined by a different position vector, namely by $\overrightarrow{QP} = \overrightarrow{QO} + \overrightarrow{OP}$ which, if we denote position vectors wrt Q as $\bar{s}$ would be written $\bar{s} = \overrightarrow{QO} + \bar{r}$.

2.2 The Scalar Product and Its Usefulness

Sometimes called the dot product or the inner product, the scalar product of two vectors $\bar{a} \cdot \bar{b}$, is defined as the real number formed as the product of their magnitudes and the cosine of the angle between them. Hence,

$$\bar{a} \cdot \bar{b} = ab\cos\theta \tag{2.2.1}$$

Figure 2.4

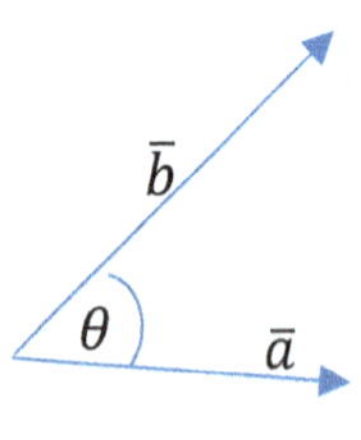

Commutativity of the scalar product follows immediately, $\bar{a} \cdot \bar{b} = \bar{b} \cdot \bar{a}$.

Thus taking the scalar product of a vector with a unit vector yields the component of the vector in the direction of the unit vector, $\bar{a} \cdot \hat{b} = a \cos \theta$.

The scalar product of a vector with itself is its square-magnitude $\bar{a} \cdot \bar{a} = a^2$.

The distributive property, $\bar{a} \cdot (\bar{b} + \bar{c}) = \bar{a} \cdot \bar{b} + \bar{a} \cdot \bar{c}$, also follows because the projection of the vector $\bar{b} + \bar{c}$ onto the direction of $\bar{a}$ is just the sum of the projections of $\bar{b}$ and $\bar{c}$ separately.

It follows from (2.2.1) that if two vectors are perpendicular (i.e., $\theta = \pi/2$) then their scalar product is zero. Also, if the scalar product is zero but both vectors have non-zero magnitude, then the vectors are perpendicular.

Some examples show the utility of the vector concept coupled with the scalar product in providing very simple proofs of familiar theorems.

2.2.1 The Diagonals of a Rhombus are Perpendicular.

A rhombus is a parallelogram with equal sides. The diagram below shows that the diagonals, as vectors, are $\bar{a} + \bar{b}$ and $\bar{a} - \bar{b}$. Hence their scalar product is $(\bar{a} + \bar{b}) \cdot (\bar{a} - \bar{b}) = a^2 - b^2 = 0$ because $a = b$. QED.

Figure 2.5

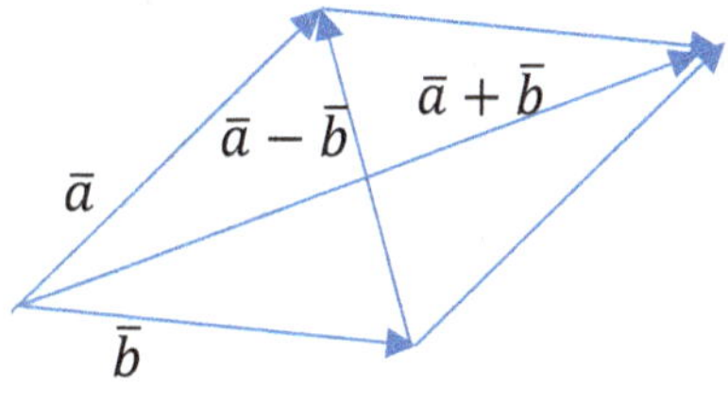

2.2.2 The Cosine Theorem

The cosine theorem for triangles, which always struck me as slightly odd as a small boy, becomes trivial in vectors. If the triangle is,

Figure 2.6

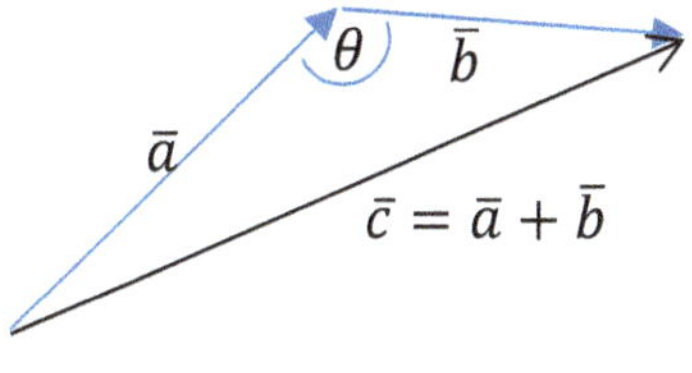

Then the squared length of the third side is just,

$$c^2 = (\bar{a} + \bar{b}) \cdot (\bar{a} + \bar{b}) = a^2 + b^2 + 2\bar{a} \cdot \bar{b} = a^2 + b^2 - 2ab\cos\theta$$

The minus sign comes from the fact that the angle between the vectors $\bar{a}$ and $\bar{b}$ is not the θ shown in the diagram but its supplementary angle, $\pi - \theta$.

Pythagoras's Theorem follows as a special case when $\theta = \frac{\pi}{2}$. This derivation therefore works as a proof of Pythagoras. Like all proofs of Pythagoras the assumption of the Euclidean nature of the geometry has been sneaked in somewhere. I have already indicated, above, one place where this has occurred, namely in the freedom to move vectors around freely and the resulting proof that vector addition is commutative.

2.2.3 The Circumcentre, O

The circumcentre, O, of a triangle is the centre of the circumscribed circle, i.e., the circle through the three points of the triangle. Proof that the circumcentre lies at the unique point of intersection of the three perpendicular bisectors of the triangle's sides is as follows.

Suppose we define the position of the three points of triangle, A, B, C, as the vectors $\bar{a}, \bar{b}, \bar{c}$ from O to A, B and C respectively. As $\bar{a}, \bar{b}, \bar{c}$ are all radii of the same circle they all have the same length. Denoting by $\overrightarrow{BA}$ the vector defined by the triangle's side from B to A then $\overrightarrow{BA} = \bar{a} - \bar{b}$. It follows from §2.2.1 that the vector $\bar{a} + \bar{b}$ is perpendicular to the side $\overrightarrow{BA}$. Moreover, the vector $\bar{a} + \bar{b}$ is that from O to the opposite vertex of the parallelogram formed by $\bar{a}, \bar{b}$. Hence moving exactly halfway along $\bar{a} + \bar{b}$ from O must result in the midpoint of side $\overrightarrow{BA}$. Hence, the circumcentre lies on the perpendicular bisector of side $\overrightarrow{BA}$. By symmetry this applies for all three sides of the triangle (which therefore all meet at the same point which must be O). QED.

2.2.4 The Orthocentre, H

The altitude of a triangle ABC through point C is the length of the line starting at C and ending on side AB (or an extension thereof) and perpendicular to side AB. The vector altitude is this line understood as a vector from AB, or its extension, to C.

Proof that the three vector altitudes of a triangle meet at a common point (defining the orthocentre, H) is as follows.

Consider the position of the three points of the triangle, A, B, C, as being defined by the vectors $\bar{a}, \bar{b}, \bar{c}$ from the circumcentre, O, to A, B and C respectively, as in §2.2.3. Now consider the point H defined by,

$$\overrightarrow{OH} = \bar{a} + \bar{b} + \bar{c}$$

The vector from C to H is thus $\overrightarrow{CH} = \bar{a} + \bar{b} + \bar{c} - \bar{c} = \bar{a} + \bar{b}$. This is perpendicular to side $\overrightarrow{BA} = \bar{a} - \bar{b}$, as in §2.2.3. Extending the vector $\overrightarrow{CH}$ back to side $\overrightarrow{BA}$ thus produces a vector which is perpendicular to $\overrightarrow{BA}$ and passes through C, i.e., it is the vector altitude through C. Because the definition of $\overrightarrow{OH} = \bar{a} + \bar{b} + \bar{c}$ is symmetrical under interchange of the vectors (or points) this result follows for all the points of the triangle. Hence, the three vector altitudes have H as their common point of intersection. QED.

2.2.5 The Centroid, G

The median of the triangle ABC through point C is the vector from C to the midpoint of side AB.

Proof that the three medians of a triangle meet at a common point (defining the centroid, G) and that G lies two-thirds of the way along each median from the vertex is as follows.

We again use vectors from the circumcentre, O, to define the triangle's points. The midpoint of side AB is $(\bar{a} + \bar{b})/2$ (see §2.2.3). Hence the vector from C to the midpoint of AB is $\frac{(\bar{a}+\bar{b})}{2} - \bar{c}$. The point two-thirds of the way along this vector thus lies at vector displacement $\frac{(\bar{a}+\bar{b})}{3} - \frac{2}{3}\bar{c}$ from point C and hence is at position vector wrt O of $\frac{(\bar{a}+\bar{b})}{3} - \frac{2}{3}\bar{c} + \bar{c} = \frac{(\bar{a}+\bar{b}+\bar{c})}{3}$. By symmetry the same result applies for the points two-thirds of the way along the other two medians, and hence the medians have a common point of intersection, the centroid, defined by $\overrightarrow{OG} = \frac{(\bar{a}+\bar{b}+\bar{c})}{3}$. QED.

2.2.6 The Circumcentre, Orthocentre and Centroid are Colinear

This follows immediately from §2.2.4 and §2.2.5.

2.2.7 The Incentre, I

The Incentre, I, is defined as the centre of the inscribed circle, i.e., the circle which is tangential to all three sides of the triangle. The incentre is the common point of intersection of the bisectors of the angles of the triangle.

This can be established as follows. Any point on the bisector at A has the same perpendicular distance from sides AB and AC, by symmetry. Similarly, any point on the bisector at B has the same perpendicular distance from sides BA and BC. And again, any point on the bisector at C has the same perpendicular distance from sides CA and CB. But the Incentre must have the same perpendicular distance from all three sides because those distances are all radii of the same circle, by definition. Hence, the Incentre is the common point of intersection of the three angular bisectors. QED.

An exercise for the reader is to prove that the position of the Incentre is,

$$\overrightarrow{OI} = \frac{a\bar{a} + b\bar{b} + c\bar{c}}{a + b + c}$$

Unlike the formulae for the orthocentre and the centroid, this expression is valid for an arbitrary origin.

2.3 The Cross Product and Its Usefulness

The second type of product of two vectors, $\bar{a} \times \bar{b}$, produces another vector rather than a scalar. To avoid being misleading, the cross product is not quite a vector, an enigmatic statement to be clarified later. Nevertheless, the cross product – so called because it is usually denoted $\bar{a} \times \bar{b}$ - is sometimes called the vector product. That this type of product appears to result in another vector is actually just an artefact of working in three dimensions. More correctly, as we will see later, this type of product really results in a tensor (or, even better, a 2-form), and even in 3D this is a better way of looking at it. Nevertheless, we stick with the traditional treatment for now.

The cross-product, $\bar{a} \times \bar{b}$, is defined as the vector which is perpendicular to the plane in which the vectors $\bar{a}$ and $\bar{b}$ lie and has magnitude equal to the product of their magnitudes times the sine of the angle between them. Thus, if $\hat{n}$ is the unit vector normal to the plane in which the vectors $\bar{a}$ and $\bar{b}$ lie then, referring to Figure 2.4, the cross product is,

$$\bar{a} \times \bar{b} = ab\hat{n} \sin\theta \qquad (2.3.1)$$

The alert reader will note that there is an ambiguity in which orientation we take the normal vector, $\hat{n}$. The convention usually adopted is the right-hand rule in which, if Figure 2.4 is used for illustration, $\hat{n}$ would be vertically upwards out of the plane of the page when the order of the cross-product is

$\bar{a} \times \bar{b}$. However, $\bar{b} \times \bar{a}$ would have the same magnitude but be oppositely oriented, so that we have,

$$\bar{a} \times \bar{b} = -\bar{b} \times \bar{a} \tag{2.3.1b}$$

For example, in the conventional right-handed coordinate system, we have $\hat{z} = \hat{x} \times \hat{y}$, whereas if we used $\hat{z} = \hat{y} \times \hat{x}$ this would produce a left-handed coordinate system. We assume the right-handed convention hereafter (with respect to which a left-handed coordinate system would produce apparently negative volumes).

Note that $\bar{a} \times \bar{a} = 0$ (2.3.1c)

2.3.1 The Vector Area (3D Only)

The magnitude of $\bar{a} \times \bar{b}$ is a times the length of the projection of vector $\bar{b}$ onto the plane normal to $\bar{a}$, and hence equals the area of the parallelogram defined by $\bar{a}, \bar{b}$. Alternatively, the area of the triangle defined by $\bar{a}, \bar{b}$ is half this. Hence, we say that their vector areas are respectively just $\bar{a} \times \bar{b}$ or $\frac{1}{2}\bar{a} \times \bar{b}$. Hence, by "vector area" we mean the area in question times a unit vector normal to the element of area (and respecting the right-hand rule). It is emphasised that the area of a 2D surface element can only be treated as a vector in three dimensions.

2.3.2 The Sine Theorem

There is a 'sister' theorem to the cosine theorem, namely the sine theorem, which follows from the identification of $\frac{1}{2}\bar{a} \times \bar{b}$ as the area of a triangle. Given the triangle defined by sides $\bar{a}, \bar{b}, \bar{c}$ and angles A, B, C thus,

Figure 2.7

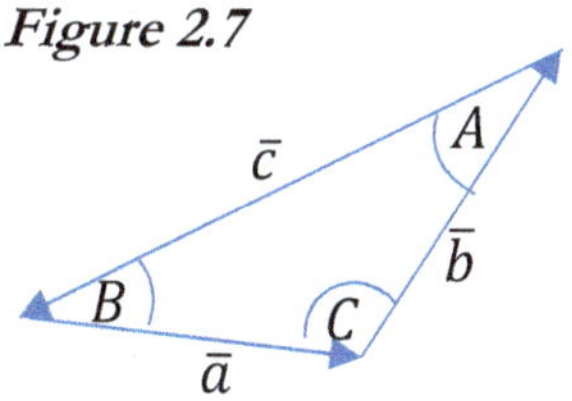

The area of the triangle can equally be expressed in three ways,

$$\frac{1}{2}\bar{a} \times \bar{b} = \frac{1}{2}\bar{b} \times \bar{c} = \frac{1}{2}\bar{c} \times \bar{a}$$

which gives, from (2.3.1),

$$ab\hat{n}\sin C = bc\hat{n}\sin A = ca\hat{n}\sin B$$

Dividing through by $abc\hat{n}$ gives the sine theorem,

$$\frac{sinA}{a} = \frac{sinB}{b} = \frac{sinC}{c} \qquad (2.3.2)$$

2.3.3 The Vector Area of a Closed Surface is Zero

We will be content to demonstrate this for the case of an arbitrary tetrahedron whose edges are defined by the vectors $\bar{a}, \bar{b}, \bar{c}, \bar{d}, \bar{e}, \bar{f}$ with the orientations shown below.

Figure 2.8

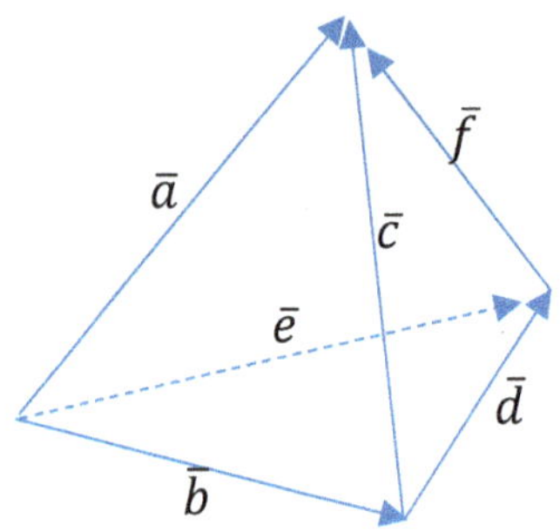

The sum of the four outward-pointing area elements, one for each face, is,

$$\frac{1}{2}\left[\bar{e} \times \bar{b} + \bar{d} \times \bar{c} + \bar{b} \times \bar{a} + \bar{a} \times \bar{e}\right] \qquad (2.3.3)$$

But the diagram shows that $\bar{c} = \bar{a} - \bar{b}, \bar{e} = \bar{d} + \bar{b}, \bar{f} = \bar{c} - \bar{d}$. Substitution of these in (2.3.3) and recalling that $\bar{a} \times \bar{b} = -\bar{b} \times \bar{a}$ shows that all terms cancel and so the total outward vector area is zero. As an exercise the reader could show that the same zero result is found for some other arbitrary closed surface comprising (say) six quadrilateral faces.

[We shall see later that this generalises to an arbitrary closed smooth curved surface for which $\oint d\bar{S} = 0$ where $d\bar{S}$ is the infinitesimal, outward-pointing vector area element, see (3.8.9). And ultimately, in the language of the exterior derivative, where each element of area is of the form $\mathbf{d}a \wedge \mathbf{d}b$ and $\mathbf{d}$ is the exterior derivative, that the total 'vector' area is zero follows from the Cartan-Stokes theorem and the fact that $\mathbf{d}^2 = 0$, matters we will eventually explain in the final chapter].

2.3.4 The Volume Element

It follows immediately that $(\bar{a} \times \bar{b}) \cdot \bar{c}$ is the volume of the parallelopiped defined by the three vectors $\bar{a}, \bar{b}, \bar{c}$. This is because $\bar{a} \times \bar{b}$ is the vector area of its base and its dot product with $\bar{c}$ means multiplying by the component of $\bar{c}$ projected normal to $\bar{a} \times \bar{b}$, hence yielding the volume. However, care must be taken because this is a ***signed*** volume. It will be positive if the projection of $\bar{c}$ along $\bar{a} \times \bar{b}$ is in the same sense as $\bar{a} \times \bar{b}$, but negative if it is in the opposite sense.

Note that if $\bar{a}, \bar{b}, \bar{c}$ are coplanar then $(\bar{a} \times \bar{b}) \cdot \bar{c}$ is zero, as it should be (i.e., the parallelopiped collapses to zero volume). Also note that, as this is a scalar volume we can permute the vectors so that,

$$(\bar{a} \times \bar{b}) \cdot \bar{c} = (\bar{b} \times \bar{c}) \cdot \bar{a} = (\bar{c} \times \bar{a}) \cdot \bar{b} \tag{2.3.4}$$

2.3.5 The Triple Product

Because $\bar{a} \times \bar{b}$ is a vector we can also form the vector product with another vector, $(\bar{a} \times \bar{b}) \times \bar{c}$. The triple product will be seen to arise in derivations later and a useful simplifying formula for it is,

$$(\bar{a} \times \bar{b}) \times \bar{c} = (\bar{a} \cdot \bar{c})\bar{b} - (\bar{b} \cdot \bar{c})\bar{a} \tag{2.3.5}$$

That this is correct can be seen most simply thus,

- $(\bar{a} \times \bar{b}) \times \bar{c}$ is perpendicular to $\bar{a} \times \bar{b}$ and hence lies in the plane defined by $\bar{a}, \bar{b}$ and can thus be expressed in the form $\alpha\bar{a} + \beta\bar{b}$;
- $(\bar{a} \times \bar{b}) \times \bar{c}$ is perpendicular to $\bar{c}$ and hence $(\alpha\bar{a} + \beta\bar{b}) \cdot \bar{c} = 0$ which means that the triple product can be expressed as $\mu[(\bar{a} \cdot \bar{c})\bar{b} - (\bar{b} \cdot \bar{c})\bar{a}]$ for some μ;
- But $(\bar{a} \times \bar{b}) \times \bar{c}$ is linear in each of a, b, c separately and hence μ is some numerical constant independent of $\bar{a}, \bar{b}, \bar{c}$. The special case $(\hat{x} \times \hat{y}) \times \hat{x} = \hat{y}$ then suffices to deduce $\mu = 1$. QED.

2.3.6 Solid Angle

Consider two non-colinear vectors $\bar{a}$ and $\bar{b}$ which are placed at a point P, thus defining a vector area $\bar{a} \times \bar{b}$ at P. Suppose the position vector of P from some origin is $\bar{r}$. The projection of the area perpendicular to $\bar{r}$ is thus $(\bar{a} \times \bar{b}) \cdot \hat{r}$. The solid angle subtended by the area at the origin is therefore,

$$\text{Solid angle} = \frac{(\bar{a} \times \bar{b}) \cdot \hat{r}}{r^2} \tag{2.3.6}$$

2.4 Basis Vectors and Coordinates

The fact that we have appealed to "3D", i.e., three dimensions, in the title of this chapter means that the concept of dimensionality has been assumed. For a more general vector space, any dimensionality may apply, but we restrict attention to 3D for now (and recall that only in 3D can the vector product be interpreted as a vector). In coordinate geometry, being 3D means, of course, that any point may be specified by three coordinates with respect to a set of three Cartesian axes, x, y, z. In the context of vectors, being 3D means that three vectors, $\hat{e}_1, \hat{e}_2, \hat{e}_3$, may be sufficient to express an arbitrary vector, i.e., any vector $\bar{r}$ might be written as $\bar{r} = r_1\hat{e}_1 + r_2\hat{e}_2 + r_3\hat{e}_3$ for some real numbers r_1, r_2, r_3.

But it is not the case that any three vectors $\hat{e}_1, \hat{e}_2, \hat{e}_3$ suffice for this. However any three vectors which are not coplanar (lying in the same plane) will suffice. Algebraically this is the requirement that there are no three numbers a, b, c such that $a\hat{e}_1 + b\hat{e}_2 + c\hat{e}_3 = 0$ other than $a = b = c = 0$. The three vectors $\hat{e}_1, \hat{e}_2, \hat{e}_3$ are then said to form a "basis" and the three numbers which define a given vector, $\bar{r}$, namely r_1, r_2, r_3, are the coordinates of the vector wrt this basis. To emphasise, the vector itself is $\bar{r} = r_1\hat{e}_1 + r_2\hat{e}_2 + r_3\hat{e}_3$, not merely the set of three coordinates, r_1, r_2, r_3.

In general the basis vectors need not be mutually orthogonal, as long as they are not coplanar. Nor is it necessary that the basis vectors be of unit length. A basis in which the vectors are all of unit length and they are all mutual orthogonal defines an "orthonormal basis". Such orthonormal basis vectors are simply unit vectors along Cartesian axes, although this set of axes may be oriented arbitrarily in 3D space.

An orthonormal basis thus has,

$$\hat{e}_i \cdot \hat{e}_j = \delta_{ij} \qquad (2.4.1)$$

where the Kronecker delta is defined such that $\delta_{ij} = 1$ when $i = j$ and otherwise it is zero.

The notation $\hat{x}, \hat{y}, \hat{z}$ will be assumed to refer to an orthonormal basis with the additional requirement that the axes are "right handed" (see §2.3) so that $\hat{z} = \hat{x} \times \hat{y}$.

Note that it follows from (2.2.1) and $\bar{r} = r_1\hat{e}_1 + r_2\hat{e}_2 + r_3\hat{e}_3$ that a unit vector has coordinates r_1, r_2, r_3 which are the cosines of the angles between the vector and the basis vectors of an orthonormal basis,

$$\bar{r} = r\hat{n} \quad \hat{n} = \cos\theta_1\,\hat{e}_1 + \cos\theta_2\,\hat{e}_2 + \cos\theta_3\,\hat{e}_3 \qquad (2.4.1b)$$

2.4.1 Basis Transformations

A vector may be written in the orthonormal basis $\hat{e}_1, \hat{e}_2, \hat{e}_3$ as,

$$\bar{r} = r_j\hat{e}_j \qquad (2.4.2)$$

where we have introduced the Einstein summation convention that repeated subscripts are summed, in the above case over 1, 2, 3. However, we can equally express the same vector in terms of a different orthonormal basis, $\hat{e}_1', \hat{e}_2', \hat{e}_3'$, in terms of which the coordinates will be r_1', r_2', r_3'. So that,

$$\bar{r} = r_j \hat{e}_j = r_k' \hat{e}_k' \tag{2.4.3}$$

Hence, we can express a vector as three coordinates in more than one way (in fact in an infinite number of ways) according to what basis is chosen. But it's the same vector, as (2.4.3) emphasises. To find the coordinates in the transformed basis, r_j', take the dot product of (2.4.3) with e_i' and use (2.4.1), giving,

$$r_i' = r_j \hat{e}_j \cdot \hat{e}_i' = R_{ij} r_j \tag{2.4.4}$$

where $R_{ij} = \hat{e}_j \cdot \hat{e}_i' = \hat{e}_i' \cdot \hat{e}_j$. But by symmetry we must therefore also have,

$$r_i = r_j' \hat{e}_j' \cdot \hat{e}_i = R_{ji} r_j' \tag{2.4.5}$$

(2.4.4) may be written in terms of the 3×3 matrix (R) if the coordinates are written as column matrices, thus,

$$\begin{pmatrix} r_1' \\ r_2' \\ r_3' \end{pmatrix} = (R) \begin{pmatrix} r_1 \\ r_2 \\ r_3 \end{pmatrix} \tag{2.4.4a}$$

Note that the LHS of (2.4.4a) is correctly termed a "column matrix", not a vector. It is not a vector, it is merely a set of three numbers. The vector is (2.4.4). I trust I have driven home that message enough now.

Similarly, the inverse transformation can, from (2.4.5), be written in terms of the transpose of (R), but also, from (2.4.4a), in terms of the inverse of (R), thus,

$$\begin{pmatrix} r_1 \\ r_2 \\ r_3 \end{pmatrix} = (R)^T \begin{pmatrix} r_1' \\ r_2' \\ r_3' \end{pmatrix} = (R)^{-1} \begin{pmatrix} r_1' \\ r_2' \\ r_3' \end{pmatrix} \tag{2.4.5a}$$

Since this must hold for an arbitrary vector it follows that the transformation matrix between two orthonormal bases must be orthogonal, i.e., such that its inverse equals its transpose,

$$(R)^T = (R)^{-1} \tag{2.4.6}$$

Now (2.4.3) must hold for any vector. So we may choose a vector equal to one of the $\hat{e}_j$, say $\hat{e}_n$, so that (2.4.3) gives $\hat{e}_n = r_k' \hat{e}_k'$ where (2.4.4) gives $r_k' = R_{kn}$ as $r_j = \delta_{jn}$ so that $\hat{e}_n = R_{kn} \hat{e}_k'$ consistent with the definition of

R_{kn}, and, by symmetry, $\hat{e}'_n = R_{nk}\,\hat{e}_k$. This means that the basis vectors and the coordinates transform the same way, i.e., in matrix notation,

$$\begin{pmatrix} \hat{e}'_1 \\ \hat{e}'_2 \\ \hat{e}'_3 \end{pmatrix} = (R)\begin{pmatrix} \hat{e}_1 \\ \hat{e}_2 \\ \hat{e}_3 \end{pmatrix} \quad \text{and} \quad \begin{pmatrix} \hat{e}_1 \\ \hat{e}_2 \\ \hat{e}_3 \end{pmatrix} = (R)^T \begin{pmatrix} \hat{e}'_1 \\ \hat{e}'_2 \\ \hat{e}'_3 \end{pmatrix} \tag{2.4.7}$$

Note that some sources will say that the basis vectors and the coordinates transform inversely, but this is because they consider the basis vectors arranged as a row matrix rather than as a column matrix. In component form we have $\hat{e}'_n = R_{nk}\,\hat{e}_k = \hat{e}_k(R^T)_{kn}$. In matrix notation,

$$(\hat{e}'_1 \quad \hat{e}'_2 \quad \hat{e}'_2) = (\hat{e}_1 \quad \hat{e}_2 \quad \hat{e}_3)(R)^T \tag{2.4.7b}$$

$$(\hat{e}_1 \quad \hat{e}_2 \quad \hat{e}_3) = (\hat{e}'_1 \quad \hat{e}'_2 \quad \hat{e}'_2)(R) \tag{2.4.7c}$$

This is a useful way of looking at the breakdown into components in matrix notation because then we have,

$$\bar{r} = r_j\hat{e}_j = (\hat{e}_1 \quad \hat{e}_2 \quad \hat{e}_3)\begin{pmatrix} r_1 \\ r_2 \\ r_3 \end{pmatrix} = r'_k\hat{e}'_k = (\hat{e}'_1 \quad \hat{e}'_2 \quad \hat{e}'_2)\begin{pmatrix} r'_1 \\ r'_2 \\ r'_3 \end{pmatrix} \tag{2.4.7d}$$

2.4.2 Rotations: Active and Passive

Any transformation from one orthonormal basis to another can be interpreted as either a rotation or an inversion, or a combination of the two. An inversion is a reversal of sign of an odd number of basis vectors (so, either one or all three in three dimensions). An inversion changes a right-handed system into a left-handed system, or vice versa. Provided we restrict attention to right-handed systems then orthonormal transformations are always rotations, and vice versa (see §2.4.3 for a proof).

In §2.4.1 the rotation considered was a "passive" rotation, meaning that the vector itself was not changed. Instead the basis (or coordinate system) in terms of which the vector may be expressed was rotated.

We may alternatively consider "active" rotations in which the vector is physically rotated. The word "physically" here appeals to the fact that, in developing physics, the vector will be a physical entity – such as the velocity of a particle or an electric field. Hence, an active rotation makes a real change in the physical world – in contrast to a passive rotation. In an active rotation the basis $\hat{e}_1, \hat{e}_2, \hat{e}_3$ is unchanged but the coordinates with respect to that basis change because we are dealing with a different vector. The rotated vector is thus,

$$\bar{r}' = r'_k \hat{e}_j \tag{2.4.8}$$

The new coordinates can be found by noting that they are the same as if we considered a passive rotation in the opposite sense. Suppose the transformation matrix, (R), in §2.4.1 represents a rotation by an angle α about an axis $\hat{n}$. Then in place of (2.4.4a) we have a similar expression but for a rotation by an angle $-\alpha$ about the axis $\hat{n}$. But this is just the inverse rotation, and hence $(R)^T$ by (2.4.6). Hence, for an active rotation we have,

$$\begin{pmatrix} r'_1 \\ r'_2 \\ r'_3 \end{pmatrix} = (R)^T \begin{pmatrix} r_1 \\ r_2 \\ r_3 \end{pmatrix} \tag{2.4.9}$$

2.4.3 The Dot Product is a Scalar

In §2.2 the dot product, or inner product, of two vectors was also called the "scalar product". The word "scalar" means that it is a quantity which does not change upon certain changes of the basis used to describe vectors – in this case transformations between different orthonormal bases, i.e., rotations and/or inversions. That the dot product is indeed scalar is clear from the definition, (2.2.1), as the lengths of the vectors and the angle between them are unchanged upon rotation (or inversion). However, this may also be seen algebraically as we have (recalling the Einstein summation convention),

$$\bar{a} \cdot \bar{b} = \left(a_j \hat{e}_j\right) \cdot \left(b_i \hat{e}_i\right) = a_j b_i \hat{e}_j \cdot \hat{e}_i = a_j b_i \delta_{ji} = a_i b_i \tag{2.4.10}$$

This establishes how the dot product is expressed in terms of the coordinates of the vectors. But the expression derived in the same way after rotation is,

$$a'_i b'_i = R_{ji} a_j R_{ki} b_k = a_j R_{ji} (R^T)_{ik} b_k = a_j \delta_{jk} b_k = a_j b_j \tag{2.4.11}$$

where we have used the orthogonality of (R), (2.4.6). Hence the dot product is invariant under rotations, i.e., a scalar.

In particular this shows that the dot product of a vector with itself is equal to its magnitude (or length) squared. This follows from (2.4.11) which gives $\bar{a} \cdot \bar{a} = a_j a_j$ which is the squared "length" of the vector by Pythagoras. In the case of vectors which are not defined geometrically (i.e., spanning between two points) but are other physical quantities, e.g., an electric field, this, $\bar{a} \cdot \bar{a} = a_j a_j$, functions as the definition of the vector's (squared) magnitude. It follows that the magnitude ("length") of any vector is invariant under orthonormal transformations. This allows the identification of orthonormal transformations (excluding inversions) with rotations.

A rotation may be defined as a transformation which preserves the magnitude of vectors, and also preserves the angle between any two vectors. The latter is equivalent to preserving the dot product between any two vectors (given their invariant magnitudes). We have seen in this section that orthonormal transformations have both these properties and hence are rotations. But the converse also follows because if $a_i \to R_{ij}a_j$ then the requirement that $a_i a_i = R_{ij}a_j R_{ik}a_k$ for any vector a_i requires the transformation matrix to be orthogonal, $(R)^T = (R)^{-1}$, (2.4.6), and then defining a basis transformation using (2.4.7) from an original basis which is orthonormal, $\hat{e}_i \cdot \hat{e}_j = \delta_{ij}$, (2.4.1), produces a new basis which is also orthonormal. QED.

2.4.4 The Vector Product Expressed in Coordinate Form

The coordinates of a vector product in the right-handed orthonormal basis $\hat{x}, \hat{y}, \hat{z}$ can be found with the aid of $\hat{z} = \hat{x} \times \hat{y}$. By cyclic symmetry we also have $\hat{x} = \hat{y} \times \hat{z}$ and $\hat{y} = \hat{z} \times \hat{x}$. Hence we find,

$$\bar{a} \times \bar{b} = (a_j \hat{e}_j) \times (b_i \hat{e}_i) = a_j b_i \hat{e}_j \times \hat{e}_i$$

Using also $\hat{x} \times \hat{y} = -\hat{y} \times \hat{x}$ this gives,

$$\bar{a} \times \bar{b} = (a_y b_z - a_z b_y)\hat{x} + (a_z b_x - a_x b_z)\hat{y} + (a_x b_y - a_y b_x)\hat{z} \quad (2.4.12)$$

It is emphasised that (2.4.12), and (2.4.10), are true only in orthonormal Cartesian bases in Euclidean space.

2.4.5 Is the Cross Product a Vector?

Strictly, no. But if you wish to think of it as such for the time being you'll not go far wrong.

From (2.4.12) the cross product has every appearance of being a vector as it takes the form $r_j \hat{e}_j$. If we rotate the coordinate system or perform an active rotation then the transformations are in complete accord with §2.4.1 and §2.4.2. So what is the problem?

The problem is that a true vector should have coordinates which change sign when the basis vectors are inverted from a right-handed to a left-handed system. Assuming that $\bar{a}$ and $\bar{b}$ are true vectors, and hence do change sign on inversion, it follows that $\bar{a} \times \bar{b}$ remains unchanged – and hence is not strictly a vector.

The cross product is sometimes called a pseudo-vector, or axial vector, because of this shortfall in respect of inversion. However, the more fundamental perspective is that the cross product is really an antisymmetric second rank tensor (or 2-form). Acceptance of this fact becomes unavoidable when dimensions greater than 3 are considered. Of these issues, more in chapters 27 and 30 which will deal with tensor analysis, the wedge product and the exterior derivative.

3. Vector Fields, Their Derivatives and Integral Theorems

In this chapter we consider fields, defined as entities which are specified at every point in the spacetime continuum (or a subspace thereof). The simplest field is a scalar field, which consists of a scalar number defined at every point in space and time. In chapter 2 we confined attention to discrete vectors. By associating a vector with every point in a spacetime continuum we get a vector field. By also imposing continuity of these vectors between neighbouring points we get a vector field upon which derivatives may be well defined. The resulting system of differentiable vector fields is one of the basic ingredients of classical field theory. The mathematical power of this system is considerable, being enhanced by a host of derivative identities and several key integral theorems which are discussed and proved in this chapter. In the final chapter we will see that these integral theorems are actually all special cases of a single overarching theorem.

Contents

3.1 What is a "Field"?

The word "field" is used in mathematical physics in a different sense to its usage in pure mathematics. In pure maths, a field is an algebraic structure (namely a set on which addition, subtraction, multiplication and division are defined with algebraic characteristics which mirror the properties of real numbers). That is not what is meant here. In physics, a field is a quantity which is defined at every point in a continuum, or some generalised form of "space", more correctly identified as a differentiable manifold. In due course this will mean spacetime (and even more generally, possibly a curved spacetime, though we will not address that possibility in this book). However, for the sake of this chapter, fields are defined in ordinary 3D Euclidean space. The generalisation to other differentiable manifolds will come later. In this chapter we shall use the word "space" to mean ordinary 3D Euclidean space.

3.2 Scalar Fields

A scalar is a number which is invariant under changes of basis. A scalar field consists of assigning a scalar to every point in space. In principle, the scalar could be a complex number. In quantum theory this is necessary, with spinless particles being represented by complex-valued scalar fields. However, we are not doing quantum physics in this book but classical physics so we shall usually consider only real-valued fields. (Occasionally it might be convenient to consider complex valued fields, simply as a mathematical shortcut, for example where waves are specified via a term $e^{-i(\omega t-\bar{k}\cdot\bar{r})}$, but the physical fields themselves will always ultimately be real).

3.3 Vector Fields

A vector field consists of assigning a vector to every point in space, and the dimensionality of the vector is the same as the dimensionality of the space in which it exists. This restriction is necessary so that the direction of the vector can be regarded as a direction within the space.

3.4 Differentiation of Fields

Differentiation of a field refers to differentiation wrt position in the space. The above definitions are so broad that differentiability cannot be assumed. That is, the scalar or vector might vary discontinuously from one point to an immediately adjacent point in the space. Except possibly at isolated points, or boundaries between regions, we shall assume sufficiently continuous and smooth behaviour that derivatives (and perhaps higher order derivatives) always exist.

3.5 The Gradient of a Scalar Field

The phrase "differentiation wrt position in the space" is ambiguous because it fails to specify the direction in space over which the difference is taken. Consider a point $\mathcal{P}$ in space and a neighbouring point $\mathcal{Q}$. The scalar field at any point, such as $\mathcal{P}$, may be denoted $f(\mathcal{P})$, a real number. If the distance between points $\mathcal{P}$ and $\mathcal{Q}$ is denoted $|\overrightarrow{\mathcal{PQ}}|$ then the derivative of the scalar field is,

$$LIM_{\mathcal{Q}\to\mathcal{P}} \frac{f(\mathcal{Q})-f(\mathcal{P})}{|\overrightarrow{\mathcal{PQ}}|} \tag{3.5.1}$$

For future reference, note that this definition requires that the distance between two points in space, $|\overrightarrow{\mathcal{PQ}}|$, has been defined. This is no problem in ordinary 3D Euclidean space, but if more general manifolds were considered (specifically curved spacetime) it would require a particular type of differentiable manifold, namely a Riemannian or metrical manifold. We will not address that in this book but it becomes crucial in general relativity.

The quantity (3.5.1) will vary according to the direction from which $\mathcal{Q}$ approaches $\mathcal{P}$. The gradient of a scalar field is defined using the direction which maximises (3.5.1), i.e., the direction of steepest increase. But, unlike (3.5.1) which is itself a scalar, the gradient of a scalar field is defined to be a vector, its direction being precisely the direction which maximises its magnitude. The gradient of a scalar field at a point $\mathcal{P}$, denoted $grad(f(\mathcal{P}))$, is therefore a vector field, and is defined as,

$$grad(f(\mathcal{P})) = MAX_{\hat{n}}\left(\left[LIM_{\delta s\to 0}\frac{f(\mathcal{P}+\hat{n}\delta s)-f(\mathcal{P})}{\delta s}\right]\hat{n}\right) \tag{3.5.2}$$

where δs is the infinitesimal length $|\overrightarrow{\mathcal{PQ}}|$ and $\hat{n}$ is its unit vector (direction). This may look rather frightening at first sight but it becomes clearer when expressed in a coordinate basis.

Consider, then, that the scalar field is written in terms of a function of the coordinates which define points in space. Since we are considering only ordinary 3D Euclidean space this means, wrt some given Cartesian axes, we can write $f(\mathcal{P}) = f(x,y,z)$. The derivative wrt s in direction $\hat{n}$ can therefore be found using the chain rule,

$$\frac{\partial}{\partial s} = \frac{\partial x}{\partial s}\frac{\partial}{\partial x} + \frac{\partial y}{\partial s}\frac{\partial}{\partial y} + \frac{\partial z}{\partial s}\frac{\partial}{\partial z} \tag{3.5.3}$$

But $\delta x = \delta s \cos\theta_x$, $\delta y = \delta s \cos\theta_y$, $\delta z = \delta s \cos\theta_z$ where $\theta_x, \theta_y, \theta_z$ are the angles between $\hat{n}$ and the basis vectors, and, moreover, from (2.4.1b) we have that $\hat{n} = \hat{x}\cos\theta_x + \hat{y}\cos\theta_y + \hat{z}\cos\theta_z$. Hence (3.5.3) becomes,

$$\frac{\partial}{\partial s} = n_x\frac{\partial}{\partial x} + n_y\frac{\partial}{\partial y} + n_z\frac{\partial}{\partial z} \tag{3.5.4}$$

And hence, using (2.4.10),

$$\frac{\partial}{\partial s} = \hat{n}\cdot\left(\hat{x}\frac{\partial}{\partial x} + \hat{y}\frac{\partial}{\partial y} + \hat{z}\frac{\partial}{\partial z}\right) \tag{3.5.5}$$

Defining the vector operator, "del",

$$\overline{\nabla} = \hat{x}\frac{\partial}{\partial x} + \hat{y}\frac{\partial}{\partial y} + \hat{z}\frac{\partial}{\partial z} \tag{3.5.6}$$

then (3.5.5) becomes simply,

$$\frac{\partial}{\partial s} = \hat{n}\cdot\overline{\nabla} \tag{3.5.7}$$

Be warned that all the above expressions may be subject to modification in spacetime, especially when curved, i.e., in the presence of gravity. Finally, using (3.5.7) the gradient, (3.5.2), becomes,

$$grad(f(\mathcal{P})), = MAX_{\hat{n}}\left[(\hat{n}\cdot\overline{\nabla}\mathrm{f})\hat{n}\right] \tag{3.5.8}$$

For an arbitrary vector it is not true that $(\hat{n}\cdot\bar{a})\hat{n}$ is the same as $\bar{a}$. But in (3.5.8) the direction $\hat{n}$ is being defined so as to maximise the magnitude of $(\hat{n}\cdot\bar{a})\hat{n}$, where $\bar{a} = \overline{\nabla}\mathrm{f}$, and this clearly occurs when $\bar{a}$ is in the direction $\hat{n}$ (because $\hat{n}\cdot\bar{a} = a\cos\theta$ which is maximum at $\theta = 0$). Hence (3.5.8) becomes simply,

$$grad(f(\mathcal{P})) = \overline{\nabla}\mathrm{f} = \hat{x}\frac{\partial f}{\partial x} + \hat{y}\frac{\partial f}{\partial y} + \hat{z}\frac{\partial f}{\partial z} \tag{3.5.9}$$

The difficult-looking (3.5.2) has thus been tamed to something quite simple. It is easy to be beguiled by this derivation and fail to see what has been achieved. We have proved that (3.5.9) is a vector in the direction of the steepest gradient ("rate of increase") of the scalar function, $f(x,y,z)$, something that is not immediately apparent (to me, anyway).

The utility of the gradient of a scalar field is considerable and occurs repeatedly in the physics of fields and continuous media. One example is the definition of force (a vector, $\bar{F}$) in terms of the associated potential energy (a scalar field, V) which is,

$$\bar{F} = -\bar{\nabla} V \tag{3.5.10}$$

Force is (minus) the gradient of potential energy, i.e., in the direction of steepest decrease (forces act to minimise potential energy).

Another example is the flux of heat in a solid (a vector field, $\bar{J}$) in terms of the temperature (a scalar field, T),

$$\bar{J} = -K\bar{\nabla}T \tag{3.5.11}$$

where K is the thermal conductivity. The minus sign is because heat flows from hot to cold. The same equation applies for the flux of current in a conducting solid in terms of the gradient of the electrical potential and the electrical conductivity.

3.6 The Divergence of a Vector Field

Just as the vector operator, del, $\bar{\nabla}$, (3.5.6), can be used to form a vector field from a scalar field, the same vector operator can be used to form a scalar field from a vector field, by taking the dot product. In 3D Euclidean space this defines the divergence of the vector field,

$$div(\bar{v}) = \bar{\nabla} \cdot \bar{v} = \frac{\partial v_x}{\partial x} + \frac{\partial v_y}{\partial y} + \frac{\partial v_z}{\partial z} \tag{3.6.1}$$

The intuitive meaning of the divergence will become clear when we consider integral theorems, below.

3.7 The Curl of a Vector Field

The consideration of $\bar{\nabla} \cdot \bar{v}$, the divergence, immediately suggests consideration also of the other form of product, the cross product, $\bar{\nabla} \times \bar{v}$, known as the curl of vector field $\bar{v}$. This creates another vector field out of the original vector field (though with the same limitation on the vectorial nature of $\bar{\nabla} \times \bar{v}$ as applies for other cross products, namely that it is defined in 3D only and is invariant under basis inversion rather than changing sign, see §2.4.5). From (2.4.12) it follows that, (3.7.1)

$$curl(\bar{v}) = \bar{\nabla} \times \bar{v} = \left(\partial_y v_z - \partial_z v_y\right)\hat{x} + (\partial_z v_x - \partial_x v_z)\hat{y} + \left(\partial_x v_y - \partial_y v_x\right)\hat{z}$$

where we have introduced the compact notation for the derivative operators wrt the spatial coordinates: $\partial_x = \frac{\partial}{\partial x}$; $\partial_y = \frac{\partial}{\partial y}$; $\partial_z = \frac{\partial}{\partial z}$, or, for any orthonormal basis with coordinates x_1, x_2, x_3, $\partial_i = \frac{\partial}{\partial x_i}$.

3.7.1 The Cartesian Coordinates of Vector Derivatives

Using the above compact notation, $\partial_i = \frac{\partial}{\partial x_i}$, and recalling the Einstein summation convention (which we will take as understood from here on), we have,

$$grad(f) = \hat{x}_i \partial_i f \quad div(\bar{v}) = \partial_i v_i \quad curl(\bar{v}) = \epsilon_{ijk} \hat{x}_i \partial_j v_k \tag{3.7.2}$$

where we have introduced the alternating tensor ϵ_{ijk} which is 1 if ijk is a positive permutation of 123, but -1 if it is a negative permutation, and zero if it is not a permutation of 123 at all, e.g., $\epsilon_{231} = 1$, $\epsilon_{321} = -1$, $\epsilon_{121} = 0$.

3.7.2 Solenoidal and Irrotational Vector Fields

Two identities follow immediately from (3.7.2): that $curl(grad(f)) \equiv 0$ and that $div(curl(\bar{v})) \equiv 0$. These follow from,

$$curl(grad(f)) = \epsilon_{ijk} \hat{x}_i \partial_j \partial_k f = 0 \tag{3.7.3}$$

and,
$$div(curl(\bar{v})) = \partial_i (\epsilon_{ijk} \partial_j v_k) = \epsilon_{ijk} \partial_i \partial_j v_k = 0 \tag{3.7.4}$$

In both cases the zero result follows from the antisymmetry of ϵ_{ijk} in any pair of its indices together with the commutation, hence symmetry, of the two derivatives.

A vector field whose $curl$ is identically zero is said to be irrotational, for reasons which will emerge below. From (3.7.3) it follows that any vector field which can be expressed as the $grad$ of some scalar field is necessarily irrotational. The converse is also the case: any irrotational vector field can be expressed as the $grad$ of some scalar field. This is proved in Appendix A4.

A vector field whose div is identically zero is said to be solenoidal, for reasons that will emerge when we discuss the magnetic fields of solenoids in §5.7.1. From (3.7.4) it follows that any vector field expressible as the $curl$ of some other vector field is necessarily solenoidal. The converse is also the case: any solenoidal vector field can be expressed as the $curl$ of some vector field. This is proved in Appendix A4.

Note that these identities hold only because we are assuming a simply connected underlying geometry, specifically 3D Euclidean space.

3.7.3 The Helmholtz Decomposition

The Helmholtz decomposition (or the Fundamental Theorem of Vector Calculus) states that, under certain conditions of smoothness in 3D, any vector field can be expressed as the sum of a solenoidal field and an irrotational field.

This is proved in Appendix A6. It follows that any sufficiently smooth vector field can be written,

$$\bar{v} = \bar{\nabla} f + \bar{\nabla} \times \bar{a} \tag{3.7.5}$$

for some scalar field, f, and some vector field, $\bar{a}$.

3.8 The Fundamental Vector Integral Theorems

3.8.1 Circulation and Stokes Theorem

Considering any curve in 3D Euclidean space, with end points $\mathcal{P}$ and $\mathcal{Q}$, it follows immediately from the definition of $grad$ that,

$$\int_{\mathcal{P}}^{\mathcal{Q}} \bar{\nabla} \mathrm{f} \cdot d\bar{s} = f(\mathcal{Q}) - f(\mathcal{P}) \tag{3.8.1}$$

where $d\bar{s}$ is the infinitesimal element of the curve as a vector, so that at each point along the curve $\bar{\nabla} f \cdot d\bar{s}$ is the component of $grad(f)$ along the curve, times ds, and hence equal to the change in f along that small element of the curve. So (3.8.1) follows. It also follows that, as long as the scalar field f is single valued, then if we integrate around any closed loop, so that $\mathcal{Q} = \mathcal{P}$, then,

$$\oint \bar{\nabla} f \cdot d\bar{s} = 0 \tag{3.8.2}$$

More generally, the "circulation" of a vector field, $\bar{v}$, around a given closed loop, C, is defined as $\oint \bar{v} \cdot d\bar{s}$. It follows that if the vector field can be expressed as a gradient, $\bar{v} = \bar{\nabla} f$, at all points along the closed loop in question, then it has zero circulation. If $\bar{v} = \bar{\nabla} f$ everywhere for some scalar field f, it follows that $\oint \bar{v} \cdot d\bar{s} = 0$ around any closed loop whatsoever. This is why a vector field with identically zero curl is said to be "irrotational" – it is because such a field can be expressed as the $grad$ of a scalar field (see §3.7.2 and Appendix A4) and therefore has zero circulation around any closed loop.

Now consider any two-dimensional surface, Γ, in general as curved as you like, whose boundary, denoted $\delta\Gamma$, is a closed loop. Stokes's theorem states that the surface integral of the curl of a vector field over any such surface equals the circulation of the vector field around its boundary, thus,

$$\int_{\Gamma} (\bar{\nabla} \times \bar{v}) \cdot d\bar{S} = \oint_{\delta\Gamma} \bar{v} \cdot d\bar{s} \tag{3.8.3}$$

where $d\bar{S}$ is an infinitesimal element of the vector surface area with a direction normal to the surface at that point (and such that it is right-handed wrt the local basis in the surface). Note that (3.8.3) again confirms that a vector field with identically zero curl is irrotational.

Closed loop integrals like (3.8.2) or the RHS of (3.8.3) are performed by going around the loop anticlockwise. If the area in (3.8.3) is not simply connected but instead has an internal boundary as well as an external boundary, then the internal boundary is traversed clockwise. That this must be the case is established by considering the doubly connected region to be the limit of a singly connected region. For example, in the limit that BA is equal and opposite to CD, below, then those parts of the line integral cancel and the result is an internal boundary traversed clockwise,

Figure 3.1

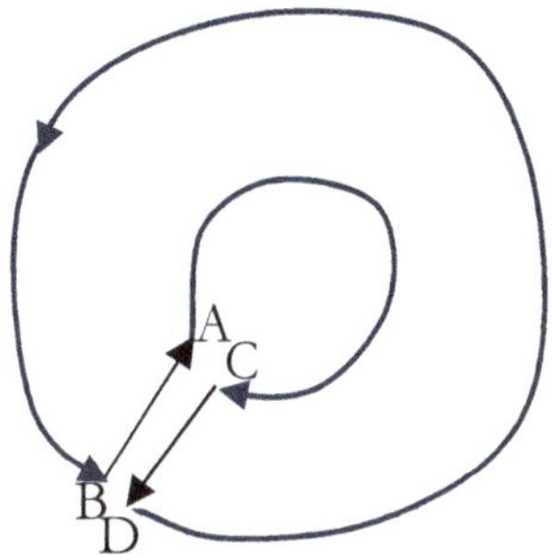

Stokes's theorem, (3.8.3), can be established, at least heuristically, by first proving Green's theorem (a proof is given in Appendix A3). Green's theorem is what Stokes's theorem reduces to if the surface is planar. If this planar surface lies in the x, y plane then Green's theorem states, for any differentiable functions v_x, v_y,

$$\int_{\mathrm{A}} \left(\frac{\partial v_y}{\partial x} - \frac{\partial v_x}{\partial y} \right) dxdy = \oint_{\delta \mathrm{A}} \left(v_x dx + v_y dy \right) \qquad (3.8.4)$$

where the area, A, in question lies fully in the x, y plane. We now consider the curved surface Γ in 3D to be divided into many tiny area elements, illustrated by Figure 3.2 below.

Figure 3.2

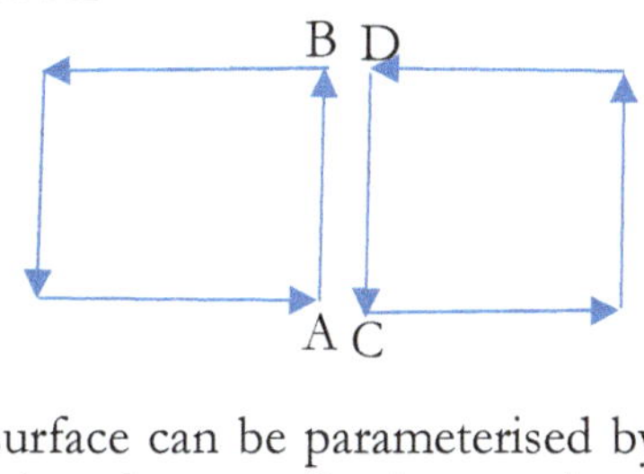

The surface can be parameterised by a pair of curvilinear coordinates η, ξ such that the contribution to the surface integral is given to sufficient accuracy by the local version of the RHS of (3.8.4), i.e., $\oint_{\delta \mathrm{A}} \left(v_\eta d\eta + v_\xi d\xi \right)$. The key to understanding Stokes's theorem is then to observe that the adjoining boundaries of two adjacent tiny area elements cancel in their contributions to this integral because they are traversed in the opposite direction. Thus, in the diagram below, as boundary AB is equal and opposite to DC and so their contributions to the line integral cancel.

Hence, the RHS of (3.8.3) equals the sum over the line integrals for each tiny area element, because the contributions from all the internal boundaries cancel leaving only the contribution from the external boundary. But each of the tiny line integrals equals the area integral for that tiny element given by the LHS of (3.8.4) – the sum of which is the LHS of (3.8.3). QED.

3.8.2 The Divergence Theorem (or Gauss's Theorem)

The divergence theorem states that the integral wrt volume of the divergence of a vector field over an arbitrary 3D region, V, equals the surface integral of the vector field over the boundary of the region, denoted δV, i.e.,

$$\int_V \bar{\nabla} \cdot \bar{v}\, d\tau = \oiint_{\delta V} \bar{v} \cdot d\bar{S} \tag{3.8.5}$$

where $d\tau$ is the infinitesimal element of 3D volume and $d\bar{S}$ is the outward pointing vector element of area. Again we may establish this theorem by considering the volume to be divided into many tiny cubes. For each cube it is readily checked that (3.8.5) holds. For example, the first term in $\bar{\nabla} \cdot \bar{v}$ is $\partial_x v_x$ which, when multiplied by $d\tau = dxdydz$ gives $\big(v_x(x + \Delta x) - v_x(x)\big)dydz$ which is the cube's contribution to the RHS of (3.8.5) arising from the faces normal to $\hat{x}$. The other two pairs of faces give the other two terms in $\bar{v} \cdot d\bar{A}$ and this establishes (3.8.5) for a small cube.

But because the outward pointing surface element is oppositely directed on adjoining faces of two adjacent small cubes, their combined contribution cancels from the RHS of (3.8.5). Hence, the sum of (3.8.5) over the whole volume ends up being equal to just the boundary contributions to the surface integral, as these are the only parts without a cancelling adjoining surface. The boundary surface integral is thus equal to the sum of all the cube's contributions to the volume integral, which is just the total volume integral. QED.

3.8.3 Other Vector Integral Theorems

A corollary of Stokes' theorem which can be derived using the identity (A1.10) in Appendix A1 by considering a vector field of the form $f\bar{a}$ where f is a scalar field and $\bar{a}$ is a constant vector, is,

$$\iint_\Gamma \bar{\nabla} f \times d\bar{S} = -\oint_{\delta\Gamma} f d\bar{s} \tag{3.8.6}$$

where we have also used (A1.2). Similarly, by considering in the divergence theorem a vector field of the form $\bar{v} \times \bar{a}$ where $\bar{a}$ is a constant vector we find, using (A1.11) and (A1.2),

$$\int_V (\bar{\nabla} \times \bar{v})d\tau = -\oiint_{\delta V} \bar{v} \times d\bar{S} \tag{3.8.7}$$

And by considering in the divergence theorem a vector field of the form $f\bar{a}$ where f is a scalar field and $\bar{a}$ is a constant vector we find, using (A1.9),

$$\int_V (\bar{\nabla} f)d\tau = \oiint_{\delta V} f d\bar{S} \tag{3.8.8}$$

When f is a constant this gives,

$$\oiint_{\delta V} d\bar{S} = 0 \tag{3.8.9}$$

which is the curved surface version of the polyhedron theorem which we alluded to in §2.3.3. These integral theorems have been collected together in Appendix A1 for ease of future reference.

3.8.4 Flux and the Equation of Continuity (or Conservation)

The equation of continuity embodies the conservation of a quantity at the local level, i.e., that conservation applies within every region of spacetime including tiny regions. The formulation requires the concept of "flux" which is the amount of the substance in question which flows through unit area in unit time. The "substance" may be anything which is conserved, such as electric charge, mass of a fluid, or the number of atoms of a given element.

The definition of flux involves area, which is a vector, hence flux is also a vector. If, at a given point, the density of the substance, i.e., its quantity per unit volume, is ρ, and its velocity is $\bar{v}$, then the flux, $\bar{J}$, is,

$$\bar{J} = \rho\bar{v} \tag{3.8.10}$$

This follows because, in unit time, the amount of the substance flowing through unit area whose normal is parallel to the velocity must be its density times the volume swept out by unit area in unit time which is just v. Hence, if we consider an element of vector area, $\delta\bar{S}$, oriented arbitrarily wrt the flux, then in time interval δt the amount of the substance passing through that area is $\bar{J} \cdot \delta\bar{S}\delta t$.

Both the density and the velocity in (3.8.10), and hence the flux also, may depend, in general, on both position, $\bar{r}$, and time, t. The net rate of outflow of the substance from some closed surface δV enclosing a volume V is $\oiint_{\delta V} \bar{J} \cdot d\bar{S}$. By the divergence theorem, (3.8.5), this equals $\int_V \bar{\nabla} \cdot \bar{J}\, d\tau$. But if the substance is conserved, at every spatial scale, then the net rate of outflow of substance from any closed surface must equal the rate at which the amount of the substance within the enclosed volume is decreasing. The latter is just $-\frac{\partial}{\partial t}\int_V \rho d\tau$ and so conservation requires that,

$$\frac{\partial}{\partial t}\int_V \rho d\tau + \int_V \overline{\nabla}\cdot\overline{J}\, d\tau = 0 \tag{3.8.11}$$

Moreover, this must hold for arbitrary volumes, V, and hence requires that,

$$\frac{\partial \rho}{\partial t} + \overline{\nabla}\cdot\overline{J} = 0 \tag{3.8.12}$$

This is the statement of conservation at the local level. If the substance is not conserved but may be subject to generation, then if the rate at which the substance is generated per unit volume is Q then (3.8.12) becomes,

$$\frac{\partial \rho}{\partial t} + \overline{\nabla}\cdot\overline{J} = Q \tag{3.8.13}$$

For example, (3.8.13) may be appropriate in the case that the "substance" is heat, which may be generated by chemical reaction or nuclear decay or electrical resistance to a current. Both (3.8.12) and (3.8.13) apply at every point in space and at all times.

3.9 Algebraic Identities Involving Grad, Div and Curl

An exercise for the reader is to prove the collection of identities which have been collected together for convenience in Appendix A1. These will be used repeatedly in derivations in subsequent chapters.

4. The Emergence of Electromagnetism and Maxwell's Equations

A brief history is presented of how electrical and magnetic phenomena were discovered and their interconnectedness finally appreciated. The basic experimental observations about electrical and magnetic phenomena are then used, together with Faraday's conception of "lines of force", to derive the Maxwell equations – the succinct and precise embodiment of electromagnetic theory and the crowning achievement of nineteenth century physics. In this chapter we consider only Maxwell's equations in free space. The effects of media in modifying electric and magnetic fields will be addressed in chapter 6. But one term will be omitted from these equations: the displacement current, which will be addressed in chapter 7. Following the development of vector calculus in chapter 3 we shall express the Maxwell equations in vector field notation. We shall see in later chapters that there are several other, more sophisticated, mathematical languages available for their expression and which have advantages in relativity theory.

Contents

4.1 A Brief History

The subject matter of this book is classical mathematical electromagnetism formulated in the continuum and in its mature form. We will not attempt a complete history of how this mature form of electromagnetic theory came about. But it must be acknowledged that, not only was the very existence of electrical and magnetic phenomena rather a surprise to historical man, but their nature as vector field entities (objects), requiring no substrate for their existence, was a revolutionary insight which emerged only late in the development of electromagnetic theory. As regards the lack of a substrate, the non-existence of a luminiferous aether would not become accepted until Einstein in the twentieth century.

The first indications of electrical phenomena were, of course, cases of static electricity. This is very familiar to us in the form of the sparks that can fly from our hands when wearing certain clothes or walking over certain carpet materials. It is familiar also in the trick of attaching a child's balloon to the wall by having first rubbed it on your jumper (sweater). Circa 600BC the ancient Greek philosopher Thales of Miletus wrote about his experiments rubbing various media, including amber, with animal fur. By this means, he noted, one could create sparks or attract light objects. The Greek for amber is, of course, elektron ($\eta\lambda\varepsilon\kappa\tau\rho o\nu$). Well before that, some ancient peoples had noted the shocks that certain fish could administer.

As for magnetism, the ability of lodestone (naturally occurring magnetite) to attract iron has been known for well over two thousand years. Before 200BC the Chinese had already used lodestone to make a magnetic compass. The West did not deploy this navigational aid until the 11th century.

A sequence of investigators wrote of their experiments with static electricity over the medieval and later periods, culminating in systematic experimentation by Gilbert in the 16th century and then by Boyle in the 17th century. The latter was able to show that the electrical effect did not depend upon the air as a medium. By the end of the 17th century a number of electrostatic generators had been constructed, giving experimenters access to very high voltage static charges (though they would not have used that terminology). The familiar Wimshurst machine and Holtz machine, and the Van de Graaff generator, were refinements of these early devices that came along only from the second half of the 19th century to the early 20th century.

The early 18th century saw experimenters classify different materials as "electrics" or "non-electrics" on the basis of their ability to hold static charge or not respectively. Thus, "electrics" were what we would now call insulators whereas "non-electrics" were conductors. Also early in the 18th century it was discovered that different materials when rubbed seemed to produce different types of electricity that would neutralise each other on contact. This gave rise to the "two fluids" theory of electricity, which would later mature into the idea of charges of opposite sign. A key device that also emerged in the mid-18th century was the Leyden jar, essentially a large capacitor (as we would now call it) capable of storing sufficient charge to deliver a severe electrical shock. The Leyden jar would become a standard piece of experimental equipment over the next two centuries.

For a long time there had been a suspicion that lightning was electrical in nature but the proof of it came only in 1752 when, following a suggestion by Benjamin Franklin, the Frenchman Thomas-François Dalibard drew electricity from a storm cloud using a 40-foot iron rod. It is not clear if Franklin himself ever conducted his famous kite experiment, though others did so. In the period 1752 to 1779 it became firmly established that lightning was essentially the same thing as static electricity. In the same period, experimenters began to claim that electricity and magnetism, though clearly distinct, were closely related.

It was appreciated by the mid-18th century that (static) electricity could be propagated long distances down wires and several attempts were made to measure the speed at which it travelled. The possibility of using electricity down wires as a means of long-distance communication arose straight away at this time, though these early attempts using static electricity were too crude to function well. It was not until well into the 19th century that the telegraph, based on current electricity, came into use.

The late 18th century saw the first quantitative theories emerging. Around 1784/5 Charles-Augustin de Coulomb formulated his famous law stating that the force of attraction between dissimilar charges, or of repulsion between like charges, varied as the inverse square of the distance and as the product of the magnitude of the charges.

A revolutionary step forward was made by Alessandro Volta towards the end of the 18th century that led to a break from the previous interest only in static electricity and the start of investigations into "voltaic electricity", or what we now call current electricity. The break-through was the realisation that different metals placed apart in an acidic liquid would develop opposite charges that would cause electricity to flow between them – not merely instantaneously, and hence discharging the terminals, as was the case for static electricity, but continuously. This quickly led to the idea of electrical potential (still popularly called "voltage", to schoolteachers' despair). Hence, the electrical cell was born and thence the battery: a pile of cells in series which Volta realised caused the total potential difference (voltage) to be the sum of the separate cells.

Large continuous currents now became available to the experimentalists, opening the way for rapid progress in the early 19th century. It was such batteries that allowed Nicholson and Carlisle to decompose water into hydrogen and oxygen by electrolysis in 1800. (That water was hydrogen and oxygen combined had been proved by Cavendish in 1781, Cavendish having also been the first to identify hydrogen earlier that century). That electric fish, static electricity and current ("voltaic") electricity were all the same thing continued to be doubted by some but their essential identity was ultimately established to Faraday's satisfaction by 1833.

The availability of current electricity led, in 1820 to Orsted being able to observe that a magnetised needle would be deflected when a wire carrying current was placed in its vicinity, thus suggesting that electrical currents created magnetic fields. In the same year, André-Marie Ampère published his three epochal laws regarding the forces between two currents,

(i) Two parallel portions of a circuit attract one another if the currents in them are flowing in the same direction, and repel one another if the currents flow in the opposite direction.
(ii) Two portions of circuits crossing one another obliquely attract one another if both the currents flow either towards or from the point of crossing, and repel one another if one flows to and the other from that point (as viewed along the acute angle of intersection).
(iii) When an element of a circuit exerts a force on another element of a circuit, that force always tends to urge the second one in a direction at right angles to its own direction.

This was a break-through in terms of quantifying the relationship between electricity and its mechanical effects. It led Maxwell to declare Ampère "the Newton of electricity". However, probably the most under-recognised contributor to the development of modern electromagnetic theory was George Green. He was the first person to create a mathematical theory of electricity and magnetism, and this formed the foundation for Maxwell's later definitive formulation. As a byproduct, Green's work also presented what we now know as Green's Theorem (see 3.8.4 and Appendix A3) and also what we now know as Green's functions, so central to quantum field theory and so helpful in solving differential equations (see chapter 15).

The towering figure in experimental work from 1831 was Michael Faraday. From previous workers it had been established that electricity and magnetism were related in that electrical currents induced magnetic fields and hence exerted forces on either permanent magnets or other currents (since they also acted as magnets). What Faraday added to this was the reverse relationship: electromagnetic induction, the phenomenon by which currents are induced in a conducting wire by a changing magnetic field in its vicinity. Induction was not immediately obvious because it does not occur if the magnetic field is steady and unvarying. Induced currents arise only when the magnetic field is changed. The magnetic field may be due to a permanent magnet – in which case a change in the field near a conductor is achieved simply by moving the magnet. Alternatively, the magnetic field may be due to another current in a circuit, in which case a change in the magnetic field can be created simply by varying that current (e.g., by switching it on and off).

Although Faraday was not mathematically sophisticated, his visualisation of electric and magnetic fields as consisting of lines of force was ahead of the mathematicians of his day and was to prove the lasting conception upon which the complete mathematical theory of electromagnetism would be formulated by James Clerk Maxwell. It was, in effect, Faraday who first thought in terms of "fields" which mediated between charges, currents and magnets, but it was Maxwell who took this idea and formulated it mathematically as we know it today.

4.2 Maxwell's Equations in Free Space

In this section we introduce Maxwell's equations in free space, i.e., in the absence of media. The intuitive appreciation of what these equations mean is obviously important in understanding electromagnetism. But the following discussion also illuminates the meaning of the vector derivatives and the associated integral theorems of chapter 3 in the context of electromagnetism.

Maxwell, in [4.1,4.2], expressed his equations mostly in coordinate form. Here we shall use vector notation, first used in electromagnetism by Heaviside. This is the first of four different mathematical descriptions of the electromagnetic field that we shall present in this book (the others are tensors, biquaternions and differential forms, as discussed in chapters 27-30).

4.2.1 The Electric Field and Its Divergence

Throughout this section we are assuming static charges.

By the late-18th century the concept of electric charge, and the idea that charge has two signs, had emerged from experimental investigations. This had permitted Coulomb to formulate his electrostatic law relating to the force on one charge, q_1, due to another, q_2, in a manner closely analogous to Newton's law of gravitation. Assuming stationary point charges, the vector force acting on the second charge due to the first is given by Coulomb's law,

Static conditions: $$\bar{F} = k\frac{q_1 q_2}{r_{12}^2}\hat{r}_{12} \qquad (4.2.1)$$

where r_{12} is the length of the vector from charge 1 to charge 2, so that $r_{12} = |\bar{r}_{12}|$ where $\bar{r}_{12} = \bar{r}_2 - \bar{r}_1$, and $\bar{r}_1, \bar{r}_2$ are the vector positions of the two charges wrt some arbitrary origin, and $\hat{r}_{12} = \frac{\bar{r}_{12}}{r}$ is the unit vector pointing towards charge 2 from charge 1. The constant k is positive so that (4.2.1) implies that like charges repel and opposite charges attract. The force on charge 1 due to charge 2 is obtained by swapping labels 1 and 2, and hence results in the same magnitude of force but of opposite sign, consistent with Newton's third law. If the space between the charges is empty then k is a universal constant. The effect of a medium between the charges will be discussed in chapter 6.

As written, (4.2.1) is a form of action-at-a-distance. But by the mid-19th century, largely thanks to Faraday's intuition, the means by which the electrostatic force is transmitted from one charge to the other had begun to be visualised in terms of lines of force connecting the charges. These lines of force were considered to radiate out from (or into) the charges. The inverse square force law could then be rationalised by considering that the force on a charge was proportional to the number of lines of force passing perpendicularly through unit area. If one adds to this the idea that the lines of force can only originate on positive charges, and only terminate on negative charges, then the number of lines of force passing through a sphere centred on a charge would be the same irrespective of the radius of the sphere. Hence, the area-density of the lines of force would reduce in inverse proportion to the area of the sphere, i.e., as the inverse square of the radius.

This gave rise to the concept of the "electric field strength" (or just "electric field"), $\bar{E}$, defined as the force acting on a unit charge placed in it. Equivalently, the force on a charge q_2 would be $\bar{F} = \bar{E}q_2$. As force is a vector, so is the electric field. From Coulomb's law, (4.2.1), it follows that the electric field at a point $\bar{r}_f$ due to a charge q_1 is,

$$\bar{E}(\bar{r}_f) = k\frac{q_1}{r_{1f}^2}\hat{r}_{1f} \tag{4.2.2}$$

where $\bar{r}_f$ is referred to as the "field point", $\bar{r}_{1f} = \bar{r}_f - \bar{r}_1$ and $\hat{r}_{1f} = \bar{r}_{1f}/r_{1f}$. Whilst the force, (4.2.1), is defined only at the charge on which it acts, the field, (4.2.2), is defined throughout space: it is a vector field in the sense discussed in chapter 3.

Formulation of the force law in terms of an electric field, $\bar{E}$, in the immediate vicinity of the charge experiencing the force has an important technical advantage over (4.2.1): unlike the latter it is no longer an "action at a distance" law. It remains true that the force on charge q_2 is $\bar{F} = \bar{E}q_2$ as long as $\bar{E}$ is the field at the location of q_2 at the time of interest. Under dynamic conditions, such as changing charges or moving charges, if Faraday's lines of force propagate at finite speed (which they do, as we shall see) then (4.2.1) remains valid only if q_1 and $\bar{r}_{12}$ are reinterpreted as "retarded" quantities, i.e., referenced to the magnitude and location of the charge producing the field at the time its field started to propagate from the charge to the point and time in question. These complications will be treated in detail in chapter 22.

In (4.2.2) we now have ready to hand an algebraic expression corresponding to Faraday's geometrical intuition that the lines of force only terminate on charges. This is simply that the divergence of the electric field is zero except at points where there is a charge. The key to understanding this is to consider any closed surface. In any region where the lines of force do not originate or terminate, all lines of force entering the closed surface must also egress from it. But, by definition of the electric field, the number of lines of force passing through an element of area whose normal vector is $d\bar{S}$ is just $\bar{E} \cdot d\bar{S}$, the sign of which determines the direction in which the lines of force pass through. It follows that the integral of $\bar{E} \cdot d\bar{S}$ over any closed surface which contains no charges must be zero. But by the divergence theorem, (3.8.5), we have,

$$\oiint_{\delta V} \bar{E} \cdot d\bar{S} = \int_V \bar{\nabla} \cdot \bar{E}\, d\tau = 0 \tag{4.2.3}$$

Moreover, this must be true for an arbitrary closed surface provided there are no charges within it, and this includes small surfaces which shrink nearly to a point. This can only be the case if the divergence of the electric field is zero, $\bar{\nabla} \cdot \bar{E} = 0$, everywhere within the region where there is no charge.

Conversely, where there is a positive charge there will be lines of force emanating from it, thus constituting an electric field proportional to the charge, (4.2.2). Any closed surface enclosing the positive charge, and assuming no other charges are within it, must have the same number of lines of force emerging from it as emerge from the charge. Hence $\oiint_{\delta V} \bar{E} \cdot d\bar{S}$ is proportional to the total charge within the surface. In fact, this works even if the charge is negative, in which case $\oiint_{\delta V} \bar{E} \cdot d\bar{S}$ is correspondingly negative. More generally, there may be a mixture of positive and negative charges within the surface, and $\oiint_{\delta V} \bar{E} \cdot d\bar{S}$ will then be the algebraic total, e.g., zero if the total negative charge equals the total positive charge.

The field due to a point charge, (4.2.2), and hence Coulomb's law, (4.2.1), can now be derived from $\oiint_{\delta V} \bar{E} \cdot d\bar{S} \propto q$ by considering the charge to be at the centre of a sphere of radius r. By symmetry, the field must have the same magnitude at all points on the sphere and be oriented radially. As the area of the sphere is $4\pi r^2$ it follows that $\bar{E} \propto \frac{q}{4\pi r^2}\hat{r}$ which is just (4.2.2). Coulomb's law then follows from the definition of the electric field in terms of the force induced on a second charge, $\bar{F} = \bar{E} q_2$.

In this book the units employed will be in the MKSA system. In this system the constant of proportionality in (4.2.2) is written as,

$$\bar{E} = \frac{q}{4\pi\varepsilon_0 r^2}\hat{r} \tag{4.2.4}$$

The rationale for including the 4π in the denominator of (4.2.4) is clear from the above discussion: it makes $\varepsilon_0 \oiint_{\delta V} \bar{E} \cdot d\bar{S} = q$ in MKSA units.

Note that (4.2.4) applies only in free space, not for fields within a medium. Here ε_0 is known as the "permittivity of free space", or simply "the electric constant". Its magnitude depends upon the units used to measure charge. Other systems of units absorb the constant term into the definition of charge so that (4.2.4) would become $\bar{E} = \frac{q}{r^2}\hat{r}$, but this will not be used here. Units and the magnitude of constants are discussed further in chapter 9.

Finally, as we have concluded that $\oiint_{\delta V} \bar{E} \cdot d\bar{S}$ takes the same value given the total charge, q, within an arbitrary closed surface, however this charge is distributed, it follows from the special case of the point charge at the centre of a sphere that its value will always be $\frac{q}{\varepsilon_0}$. But, by the divergence theorem, this must also equal $\int_V \bar{\nabla} \cdot \bar{E}\, d\tau$. This equivalence holds for arbitrary distributions of charge within an arbitrary region. Thus, it must be that $\bar{\nabla} \cdot \bar{E}$ is equal to $\frac{\rho}{\varepsilon_0}$ where ρ is the charge density (charge per unit volume), so that,

$$\int_V \bar{\nabla} \cdot \bar{E}\, d\tau = \int_V \frac{\rho}{\varepsilon_0} d\tau = \frac{q}{\varepsilon_0} \qquad (4.2.4b)$$

with complete generality. (We note in passing that we have implicitly assumed that no charge resides ***on*** the surface of the closed boundary. The case of surface charge density will be addressed in chapter 5).

Hence we arrive at the first of Maxwell's equations in free space,

$$\bar{\nabla} \cdot \bar{E} = \frac{\rho}{\varepsilon_0} \qquad (4.2.5)$$

Integrating (4.2.5) over an arbitrary volume and using the divergence theorem shows that the surface integral of the electric field over any closed surface equals the total charge contained within it (modulo the factor of $1/\varepsilon_0$, which is just a matter of the units adopted),

$$\oiint_{\delta V} \bar{E} \cdot d\bar{S} = \int_V \frac{\rho}{\varepsilon_0} d\tau = \frac{q}{\varepsilon_0} \qquad (4.2.5b)$$

Considering a continuous distribution of charge density, $\rho(\bar{r}_s)$, at the source point $\bar{r}_s$, and integrating the contributions from each element of charge, the total electric field at any field point, $\bar{r}_f$, is, with $\bar{r}_{sf} = \bar{r}_f - \bar{r}_s$,

$$\bar{E}(\bar{r}_f) = \frac{1}{4\pi\varepsilon_0}\left(\int \frac{\hat{r}_{sf}}{r_{sf}^2} \rho(\bar{r}_s)\, d\tau_s\right) \qquad (4.2.5c)$$

<u>Warning</u>: The reader is reminded that (4.2.5c) applies only for static distributions of charge. It may be tempting to imagine that the field at time t could be found simply by using the charge density at that time in the integral of (4.2.5c). But a moment's thought reveals that that would imply an instantaneous effect of a change in charge on a remote field point, i.e., effectively an infinite speed of propagation. As we will see, electromagnetic effects can propagate no faster than light speed, c, and this will necessitate a

change in (4.2.5c) under time dependent conditions, to be elucidated in chapter 22. However, (4.2.5c) remains adequate as long as any changes in the charge density are negligible over a timescale r_{sf}/c where r_{sf} is the largest distance of the source from the field point that contributes to the integral.

Note that these provisos apply only to integral expressions. The differential expression, (4.2.5), remains true even under time dependent conditions because both the field and the charge density refer to the same spatial point and the same time. Consequently, Maxwell's equations, the complete expression of which we are working towards, remain true under time dependent conditions because they are differential.

One of the most useful relationships which follows from (4.2.5) or (4.2.5b) is that for the electric field immediately adjacent to a surface bearing a surface charge per unit area of σ. Assuming there is zero field to one side, e.g., because the charge resides on the surface of a conductor, then the field is normal to the surface with magnitude,

$$E_n = \frac{\sigma}{\varepsilon_0} \tag{4.2.5d}$$

If there is no reason for the magnitude of the field to differ on the two sides, then the field will be equal and opposite on the two sides, both of magntidue half of (4.2.5d).

4.2.2 The Magnetic Flux Density and Its Divergence

As charge is to electricity, current is to magnetism – and current is moving charge. The truth of that simple sentence took many centuries to appreciate. Just as Coulomb's law was central to the emergence of the concept of the electric field, so Ampère's laws (§4.1) were central to the formulation of the magnetic field.

One difference which impacts the formulation of the magnetic force is that it depends, not just on the current, but the length of the element of current in question. Current is defined as the charge flowing past a point per unit time. Suppose a small amount of charge δq flows past a point in time δt, so that the current is $I = \delta q/\delta t$, and consider a length $\delta\bar{L}$ of that current. Note that this is a vector. The associated "current element" is thus $I\delta\bar{L}$. The magnetic force on a second current element, $I_2\delta\bar{L}_2$, due to the first, $I_1\delta\bar{L}_1$, is given by a law analogous to Coulomb's law and sometimes called the Biot-Savart law,

$$\bar{F} = \frac{\mu_0}{4\pi}\frac{I_1 I_2}{r_{12}^2}\delta\bar{L}_2 \times (\delta\bar{L}_1 \times \hat{r}_{12}) \tag{4.2.6}$$

where, as before, $\bar{r}_{12} = \bar{r}_2 - \bar{r}_1$ is the vector distance from current element 1 to current element 2, and $\hat{r}_{12} = \bar{r}_{12}/r_{12}$. The constant μ_0 is called the permeability of free space. As in §4.2.1 we may replace (4.2.6), which looks like action-at-a-distance, with Faraday's conception of a field as lines of force which mediate between the two current elements and which pervades the whole of space. Analogous to the electric field we define the "magnetic flux density", $\bar{B}$, at any location $\bar{r}_f$ due to current element $I_1\delta\bar{L}_1$ to be, in MKSA units,

$$\bar{B}(\bar{r}_f) = \frac{\mu_0}{4\pi}\frac{I_1}{r_{1f}^2}\left(\delta\bar{L}_1 \times \hat{r}_{1f}\right) \tag{4.2.7}$$

where $\bar{r}_{1f} = \bar{r}_f - \bar{r}_1$.

[Aside: There is a terminological inconsistency in referring to $\bar{B}$ as magnetic flux density. Unlike the general concept of flux (§3.8.4), which is the flow of a substance through unit area in unit time, the $\bar{B}$ field is just the number of magnetic field lines per unit area, without the rate element. Moreover adding the word "density" to "magnetic flux" is otiose and misleading as it suggests "per unit volume" rather than "per unit area" which is implicit in the word "flux" in any case. But the standard terminology is retained here because it is now universal].

We can now interpret the magnetic force on current element $I_2\delta\bar{L}_2$ as arising due to the magnetic flux density at its location, thus,

$$\bar{F} = I_2\delta\bar{L}_2 \times \bar{B}(\bar{r}_2) \tag{4.2.8}$$

The signs implied by (4.2.7) and (4.2.8) are important, and so the order of their cross products is important. Consistent with the definition of the vector cross product, the field, (4.2.7), is given by the right-hand rule, illustrated below in Figures 4.1-3. The reader should check that, as a result of the signs illustrated, two parallel current elements, in the same sense, experience a force of attraction towards each other. But, in contrast, if the current elements are parallel but in opposite senses then the force between them is repulsive. In both cases (4.2.6-8) are consistent with Ampère's laws: equal currents attract, opposite currents repel. For example, (4.2.7) results in the field direction illustrated by Figure 4.1.

Figure 4.1:

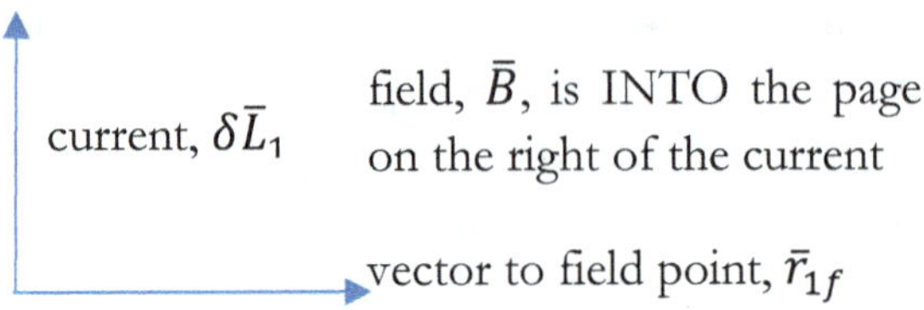

Hence, for a second parallel current placed to the right of the first, the force equation, (4.2.8), produces an attractive force when the currents are in the same sense. Alternatively, rotating one's perspective, we get Figure 4.2.

Figure 4.2:

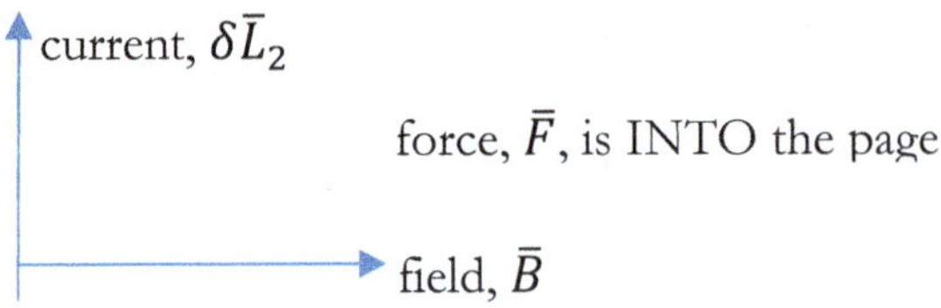

which, rotating one's perspective back again, gives Figure 4.3 which shows that the force on the second current is to the left in Figure 4.1, i.e., attractive because the second current was on the right of the first,

Figure 4.3:

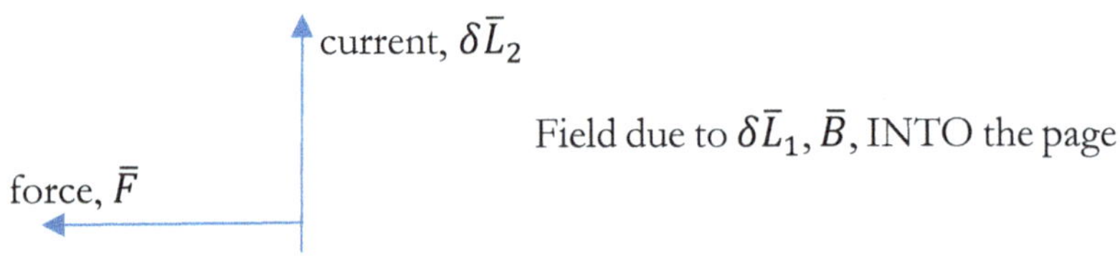

Conversely, the field to the left of the first current is OUT OF the page, and hence placing the second current to its left causes it to experience a force to the right, i.e., again attractive.

The reader should confirm that Ampère's other laws can also be deduced from (4.2.6-8). In particular, Ampère's third law is an immediate consequence of (4.2.8), the force on current element $I_2\delta\bar{L}_2$ is necessarily perpendicular to $\delta\bar{L}_2$.

Despite the comforting analogy with the force between charges, the reader should be wary of the Biot-Savart force law between small current elements considered as sources of magnetic field alone, i.e., with no electric field. This is because, at best, currents with overall neutrality exist as complete loops, i.e., as complete circuits. One symptom of taking too seriously the Biot-Savart law as regards small elements of circuits is that the force on current element 2 due to the magnetic field from current element 1 is not equal and opposite to the force on 1 due to 2, as the reader should confirm from (4.2.6). This is an apparent violation of Newton's third law. This problem vanishes, however, so long as we consider complete circuits. The total force on 2 due to 1 is found by integrating over both complete circuits, thus,

$$\bar{F} = \frac{\mu_0}{4\pi} I_1 I_2 \oint_{\Gamma_1} \oint_{\Gamma_2} \frac{\delta\bar{r}_f \times (\delta\bar{r}_s \times \hat{r}_{fs})}{r_{fs}^2} \tag{4.2.8b}$$

where Γ_1, Γ_2 are the two closed circuits. Here we have adopted the notation that $\bar{r}_s$ denotes the "source point", i.e., the location of the current element which gives rise to the magnetic field, whereas $\bar{r}_f$ refers to the "field point", i.e., the position of the second current element, and the vector $\bar{r}_{sf} = \bar{r}_f - \bar{r}_s$ runs from the source to the field point. Using identity (A1.3b) we have,

$$\delta\bar{r}_f \times (\delta\bar{r}_s \times \hat{r}_{fs}) = (\delta\bar{r}_f \cdot \hat{r}_{fs})\delta\bar{r}_s - (\delta\bar{r}_f \cdot \delta\bar{r}_s)\hat{r}_{fs} \tag{4.2.8c}$$

For any fixed $\bar{r}_s$ and $\delta\bar{r}_s$ on circuit 1 the first term in (4.2.8c) requires the integral over circuit 2 of $\oint_{\Gamma_2} \frac{d\bar{r}_f \cdot \hat{r}_{fs}}{r_{fs}^2}$. For any fixed $\bar{r}_s$ we can evaluate this integral by shifting the origin to that source point so that the required integral can be written,

$$\oint_{\Gamma_2} \frac{\hat{r} \cdot d\bar{r}}{r^2} = \int_A \left[\bar{\nabla} \times \left(\frac{\hat{r}}{r^2}\right)\right] \cdot d\bar{S} = 0 \tag{4.2.8d}$$

Here we have transformed the path integral over the closed circuit Γ_2 to an integral over some surface, A, bounded by Γ_2, using Stokes's theorem, (A1.18). But the integrand is then the $curl$ of a vector with only a radial coordinate which depends upon the radius alone, i.e., independent of the polar and azimuthal angles, and hence the expression for $curl$ in spherical polars (Appendix A3) shows that $\bar{\nabla} \times \left(\frac{\hat{r}}{r^2}\right)$ is identically zero. Finally, then, the total force on current circuit 2 due to the whole of current circuit 1 relates to the second term on the RHS of (4.2.8c) only, giving,

$$\bar{F} = -\frac{\mu_0}{4\pi} I_1 I_2 \oint_{\Gamma_1} \oint_{\Gamma_2} \frac{(\delta\bar{r}_f \cdot \delta\bar{r}_s)\hat{r}_{fs}}{r_{fs}^2} \tag{4.2.8e}$$

Swapping the circuits around, i.e., $s \leftrightarrow f$, leaves the integrand in (4.2.8e) unchanged apart from the change in sign due to $\hat{r}_{sf} = -\hat{r}_{fs}$. So the force on

1 due to 2 is equal and opposite to the force on 2 due to 1, so long as forces on, and due to, complete circuits are considered, hence restoring Newton's third law.

The moral is that one must be cautious about taking the Biot-Savart law too seriously as regards small current elements.

However, a current element may be envisaged, not as part of a circuit, but due to a moving charge in space. The current element, $I\delta\bar{L}$, can then be interpreted as the product of the charge, q, and its velocity, $\bar{v}$, if the length of the current element is identified with the distance the charge travels in the same time δt used to define the current, i.e.,

$$I\delta\bar{L} = \frac{q}{\delta t}\delta\bar{L} = q\frac{\delta\bar{L}}{\delta t} = q\bar{v} \tag{4.2.9}$$

The force law, (4.2.6), can then be reinterpreted as giving the magnetic force on a moving charge, q, due to a magnetic flux density at its location,

$$\bar{F} = q\bar{v} \times \bar{B} \tag{4.2.10}$$

Similarly, the magnetic field can be considered as arising from another moving charge in which case the magnetic flux density is given by

$$\bar{B}(\bar{r}_f) = \frac{\mu_0}{4\pi}\frac{q}{r_{sf}^2}(\bar{v} \times \hat{r}_{sf}) \tag{4.2.11}$$

The alert reader will spot that we have just reintroduced the same apparent violation of Newton's third law as above, namely that the force on moving charge 2 due to the magnetic field caused by moving charge 1 is,

$$\bar{F}_{21} = \frac{\mu_0}{4\pi}\frac{q_1 q_2}{r_{sf}^2}\left[\bar{v}_2 \times (\bar{v}_1 \times \hat{r}_{sf})\right] \tag{4.2.11b}$$

whereas the magnetic force on 1 due to 2 is,

$$\bar{F}_{12} = -\frac{\mu_0}{4\pi}\frac{q_1 q_2}{r_{sf}^2}\left[\bar{v}_1 \times (\bar{v}_2 \times \hat{r}_{sf})\right] \tag{4.2.11c}$$

and these are not equal and opposite. We can no longer appeal to complete circuits to resolve the problem. For now we draw comfort from the fact that, if we look at the forces from the centre-of-velocity frame of reference, which moves at velocity $\bar{v}_{cov} = \bar{v}_1 + \bar{v}_2$, so that the two charges then appear to move with equal but opposite velocities, then the forces (4.2.11b,c) are also equal and opposite, consistent with Newton's third law.

In the general case the reason why there is a force mismatch is related to the facts that, (i) there is also an electric field, (ii) the combination of the electric and magnetic fields means the e/m field possesses momentum, as we will see in §7.4,§7.5, as well as the charges themselves, and, (iii) it is the rate-of-change of the total momentum which equates with the total electromagnetic "force". Chapter 8 will show that the field momentum is in general non-zero for a pair of moving charges. It will also show that it is the total momentum of the particles plus the e/m field that is time independent, i.e., the rate-of-change of the momentum of the particles, which is their total net force, is equal and opposite to the rate-of-change of the field momentum. In Newtonian language, there is a net force on the pair of charged particles which is balanced by an equal and opposite 'force' acting on the field, whose momentum is therefore changing. This matter will be revisited again in chapter 28 in fully relativistic form. Relativity is the natural language in which to find the most cogent formulation of this issue because the problem arises dependent upon which inertial frame of reference is considered.

It is convenient now to introduce a continuous spatial distribution of moving charge with charge density $\rho(\bar{r}_s)$ at source point $\bar{r}_s$, and a velocity which might also vary with position, $\bar{v}(\bar{r}_s)$. The current density at $\bar{r}_s$ is,

$$\bar{J}(\bar{r}_s) = \rho(\bar{r}_s)\bar{v}(\bar{r}_s) \tag{4.2.12}$$

The current density is the flux of charge, i.e., the amount of charge flowing through unit area normal to the flow per unit time. The charge flowing through vector area $d\bar{S}$ per second is thus $\bar{J} \cdot d\bar{S}$, where this area element may be in any orientation. [Aside: The term "current density" is strictly misleading as its dimensions are current per unit area, not current per unit volume. However, we shall retain the use of "current density" because it is universally accepted. But the reader should bear in mind that what we mean by this term is strictly the charge flux, otherwise confusion will arise over units].

If we consider the magnetic flux density arising from the charge in a small volume $\delta\tau$ then $q\bar{v}$ in (4.2.11) can be replaced by $\bar{J}(\bar{r}_s)\delta\tau_s$ where the subscript on the element of volume, $\delta\tau_s$, emphasises that it relates to the source. Integrating over the whole distributed source gives the total magnetic flux density to be,

$$\bar{B}(\bar{r}_f) = \frac{\mu_0}{4\pi}\int \frac{d\tau_s}{r_{sf}^2}\left(\bar{J}(\bar{r}_s) \times \hat{r}_{sf}\right) \tag{4.2.13}$$

We now consider derivatives wrt the field point, so that $grad = \overline{\nabla}_f$. From (3.5.9) it is straightforward to show that,

$$\overline{\nabla}_f \left(\frac{1}{r_{sf}}\right) = -\frac{\hat{r}_{sf}}{r_{sf}^2} \tag{4.2.14}$$

So that (4.2.13) becomes,

$$\bar{B}(\bar{r}_f) = -\frac{\mu_0}{4\pi} \int \left(\bar{J}(\bar{r}_s) \times \overline{\nabla}_f \left(\frac{1}{r_{sf}}\right)\right) d\tau_s \tag{4.2.15}$$

Noting that as $\bar{J}(\bar{r}_s)$ is constant wrt the field point, vector identity (A1.10) means we can write

$$\bar{J}(\bar{r}_s) \times \overline{\nabla}_f \left(\frac{1}{r_{sf}}\right) = -\overline{\nabla}_f \times \left(\frac{1}{r_{sf}} \bar{J}(\bar{r}_s)\right) \tag{4.2.16}$$

Hence, the expression for the magnetic flux density due to some distributed current density becomes,

$$\bar{B}(\bar{r}_f) = \frac{\mu_0}{4\pi} \overline{\nabla}_f \times \left(\int \frac{1}{r_{sf}} \bar{J}(\bar{r}_s)\, d\tau_s\right) \tag{4.2.17}$$

Here we have used the independence of the field point from the source point to commute the order of differentiation and integration and so to take $\overline{\nabla}_f$ outside the integral.

We have established how a magnetic "field" (strictly, magnetic flux density) arises from moving charges (currents), but what can we say about its differential properties? (4.2.17) is a very useful expression in its own right, as it specifies the magnetic flux density due to any spatial distribution of current density. But because of vector identity (A1.7), namely that $div\ curl$ is identically zero, we get immediately,

$$\overline{\nabla} \cdot \bar{B} \equiv 0 \tag{4.2.18}$$

This is the second of Maxwell's equations. We could have anticipated that (4.2.18) must hold on the basis that there are no magnetic monopoles, i.e., no magnetic "charges" analogous to electric charge. Instead, the magnetic lines of force exist only in closed loops which neither originate nor terminate anywhere. This is the geometrical (or, rather, topological) meaning of (4.2.18).

Warning: (4.2.17) strictly applies only for steady currents, just as we saw for the electric field in terms of the charge density, (4.2.5c), otherwise this integral expression runs into the difficulty of implying an infinite speed of propagation. However, the Maxwell equation, (4.2.18), being a differential expression, is valid even when there is time dependence. Chapter 22 will provide the generalisation of (4.2.17) under time dependence.

4.2.3 The Curl of the B-Field when the E-Field is Unchanging

We have now derived two of Maxwell's equations, (4.2.5) and (4.2.18), and shown how these follow from the experimentally determined laws of Coulomb and Ampère. In this section we derive another of Maxwell's equations, but for the special case that any electric field is unvarying. This is an important stepping stone to the complete formulation.

Consider a source point lying on a current I, at vector position $\bar{r}_s$, and consider a field point at vector position $\bar{r}_f$ wrt the same origin. An element of the current along the circuit can be considered to be $\delta\bar{r}_s$. The magnetic flux density at $\bar{r}_f$ due to this element of current, is given by (4.2.7),

$$\bar{B}(\bar{r}_f) = \frac{\mu_0}{4\pi}\frac{I}{r_{sf}^2}(\delta\bar{r}_s \times \hat{r}_{sf})$$

We now consider the field point to be displaced by $\delta\bar{r}_f$ and form the dot product,

$$\bar{B}(\bar{r}_f)\cdot\delta\bar{r}_f = \frac{\mu_0}{4\pi}\frac{I}{r_{sf}^2}(\delta\bar{r}_s \times \hat{r}_{sf})\cdot\delta\bar{r}_f \tag{4.2.19}$$

From (A1.2) we can write $(\delta\bar{r}_s \times \hat{r}_{sf})\cdot\delta\bar{r}_f = (\delta\bar{r}_f \times \delta\bar{r}_s)\cdot\hat{r}_{sf}$. We now perform a trick by imagining that, instead of the field point moving by $\delta\bar{r}_f$ we imagine the current element moving by $-\delta\bar{r}_f$ as this must be equivalent. From (A1.1) we can interpret $\delta\bar{r}_f \times \delta\bar{r}_s$ as an area element at the source point, and from §A1.4, we know that $\frac{(\delta\bar{r}_f\times\delta\bar{r}_s)\cdot\hat{r}_{sf}}{r_{sf}^2}$ is the solid angle, $\delta\Omega$, subtended by this area element at the field point. Hence (4.2.19) becomes,

$$\bar{B}(\bar{r}_f)\cdot\delta\bar{r}_f = \frac{\mu_0}{4\pi}I\delta\Omega \tag{4.2.20}$$

Integrating (4.2.19) over the closed loop that is the current circuit (i.e., over $\delta\bar{r}_s$) and also integrating $\delta\bar{r}_f$ over a closed loop, C, is therefore equivalent to integrating the solid angle, $\delta\Omega$, in (4.2.20) over the whole 4π steradians. Hence, from (4.2.20) we get,

$$\oint_C \bar{B}(\bar{r}_f)\cdot d\bar{r}_f = \mu_0 I \tag{4.2.20b}$$

Here I is the total current that is linked by (i.e., passes through) the closed path C over which the line integral is performed. If the chosen integration loop encompasses no current (or no net current) then the RHS of (4.2.20) would be zero.

(Incidentally, (4.2.20b) illustrates why 4π was included in the denominator of (4.2.6), so that it cancels when integrated over solid angle).

A useful exercise for the reader is to consider a very long straight wire carrying a steady current and hence to show, by integrating (4.2.7) along the length of the wire, that the magnetic flux density at a distance r from the wire is,

Long straight wire: $$\bar{B} = \frac{\mu_0 I}{2\pi r}\hat{\phi} \tag{4.2.21}$$

where $\hat{\phi}$ denotes the unit vector which is circumferential wrt the wire as centre. It is clear that performing the line integral of (4.2.21) over a circle of radius r centred on the wire will give $\mu_0 I$, consistent with (4.2.20b). Indeed, (4.2.20b) is a far simpler means of deriving the key result, (4.2.21), than integrating (4.2.7).

An important corollary of (4.2.20b) is that if a closed loop encloses zero net current then the integral of the magnetic flux density along the loop is zero. If the path is a circle in a situation of axisymmetry so that $\bar{B}(\bar{r}_f) \cdot d\bar{r}_f$ is the same at all points along the circle, it follows that the circumferential component of $\bar{B}$ is zero at all points on the circle. If symmetry also dictates that there can be no component other than circumferential then it follows that $\bar{B}(\bar{r}_f) = 0$ everywhere on the circle. Finally, if there is no current linked anywhere within a cylindrical region and there is also cylindrical symmetry, it follows that $\bar{B} = 0$ everywhere within the cylinder. An important special case is a hollow cylinder bearing an axial current, of uniform current density. It follows that there is no magnetic field within it – only outside it.

Note that (4.2.20b) is the basis of a practical device used by electrical engineers to measure the current in a cable. This consists of a pair of semi-circular calipers which are used to encircle the cable and effectively measure the left-hand side of (4.2.20b) by detecting the magnetic field, thus finding the current in the cable without having to break into the circuit.

To find the curl of the B-field we simply use Stokes's Theorem, (A1.18), which allows (4.2.20b) to be rewritten as,

$$\int \left(\bar{\nabla} \times \bar{B}(\bar{r}_f)\right) \cdot d\bar{S} = \mu_0 I \tag{4.2.22}$$

Here the area integration is over any surface (which may be curved) and I is the total net current crossing the area. If we now consider the current to be distributed over space as a current density, $\bar{J}$, rather than necessarily confined to a one-dimensional circuit, then the total current within the boundary of the area of integration is just $\int \bar{J} \cdot d\bar{A}$. Since (4.2.22) is true for any arbitrary surface of integration it follows that,

Unchanging E-Field $$\bar{\nabla} \times \bar{B} = \mu_0 \bar{J} \tag{4.2.23}$$

This is not yet quite the third Maxwell equation, but it is what the third Maxwell equation reduces to in the case that any electric fields are unvarying. We will see in chapter 7 that a varying electric field can act as a source of magnetic field over and above that due to current, and accounting for this will be necessary in order to complete the third Maxwell equation.

That (4.2.23) cannot be true in general can be seen from the conservation of charge which, from (A1.5), requires that $\bar{\nabla} \cdot \bar{J} = -\frac{\partial \rho}{\partial t}$. This conflicts with (4.2.23) because $div\ curl$ is identically zero, (A1.7). Hence, (4.2.23) can only hold if all charge densities are constant, so that $\frac{\partial \rho}{\partial t} = 0$, in which case the electric field is everywhere constant also.

4.2.4 The Curl of the E-Field when the B-Field is Unchanging

We will see in the next section that a varying B-field can act as a source of electric field. But in the case that the only sources of an electric field are static charges, then (4.2.4) tells us that the electric field at $\bar{r}_f$ can be written,

$$\bar{E} = \sum_S \frac{q_s}{4\pi\varepsilon_0} \frac{\hat{r}_{sf}}{r_{sf}^2} \tag{4.2.24}$$

where the sum is over all charges, at various source points $\bar{r}_s$. But each of the terms $\frac{\hat{r}_{sf}}{r_{sf}^2}$, which varies over the different charges, can be written as in (4.2.14). Hence, (4.2.24) becomes,

$$\bar{E} = -\bar{\nabla}_f \left(\sum_S \frac{q_s}{4\pi\varepsilon_0} \frac{1}{r_{sf}} \right) \tag{4.2.24}$$

where the grad operator can be taken outside the sum because it operates only on the field variable, not the source variables. Hence, if the only source of the electric field are static charges, then the electric field is expressible as the $grad$ of a scalar and hence its $curl$ is identically zero, i.e., the electric field is an irrotational vector field (see §3.7.2).

Unchanging B-Field: $\bar{\nabla} \times \bar{E} \equiv 0$ (4.2.25)

This is not yet quite the fourth Maxwell equation but it is what the fourth Maxwell equation reduces to when any B-field is unvarying. We will see next that a varying B-field can act as a source of electric field over and above that due to charges, and this modifies (4.2.25). The physical phenomenon which leads to this extra term is…

4.3 Electromagnetic Induction

Whilst we shall leave the completion of the third Maxwell equation to chapter 7, the fourth Maxwell equation will be completed here by the incorporation of electromagnetic induction.

Electromagnetic induction was discovered by Michael Faraday in 1831, and independently the following year by Joseph Henry. Faraday noticed that when he moved a permanent magnet in the vicinity of an electric circuit, a surge in current was registered (e.g., by a galvanometer in the circuit). Experimentation showed that the moving magnet creates an "electro-motive force" (emf), i.e., a driving voltage gradient, in the circuit, and this causes the current. No emf was created if the magnet was stationary. Faraday showed the same effect arose if an electromagnet was used to create the magnetic field. In this case the magnetic field could be varied without motion being involved by varying the current in the electromagnet. The induced emf, or current, in the separate circuit was particularly marked if it shared the same iron core with the coils of the electromagnet.

The emf in a circuit is the cause of the current which flows continually around the circuit. The current is the result of electric fields, hence forces, acting on the charges. For the charges to move continually around a circuit, the electric field has to be oriented in a constant circulating sense. In other words, the integral of $\bar{E} \cdot d\bar{s}$ around the closed circuit must be non-zero and, it is this that is identified with the emf, i.e.,

$$\text{emf} = \oint_C \bar{E} \cdot d\bar{s} \tag{4.3.1}$$

Experimentation also showed that the induced emf depended on the rapidity of the change in the B-field and also upon its spatial extent within the region of the circuit. Both these effects were eventually formulated (by Heaviside following Faraday): the emf is proportional to $\frac{\partial \bar{B}}{\partial t} \cdot d\bar{S}$. As the B-field is considered to be a magnetic flux density, it follows that $\frac{\partial \bar{B}}{\partial t} \cdot d\bar{S}$ is the rate of change of magnetic flux through area $d\bar{S}$. The rate of change of magnetic flux through the interior of the circuit is thus $\int_S \frac{\partial \bar{B}}{\partial t} \cdot d\bar{S}$, where S denotes any curved surface whose boundary is the circuit, δS. The sign of the induced emf is such as to oppose the change in magnetic field. Taking this into account and then equating this with (4.3.1) gives us the Faraday-Heaviside law of electromagnetic induction,

$$\text{emf} = \oint_{\delta S} \bar{E} \cdot d\bar{s} = -\frac{\partial}{\partial t} \int_S \bar{B} \cdot d\bar{S} \tag{4.3.2}$$

Stokes's theorem, (A1.18), means that this can be written as,

$$\int_S (\bar{\nabla} \times \bar{E}) \cdot d\bar{S} = -\frac{\partial}{\partial t} \int_S \bar{B} \cdot d\bar{S} \tag{4.3.3}$$

Since this must hold for arbitrary circuits it follows that,

$$\bar{\nabla} \times \bar{E} = -\frac{\partial \bar{B}}{\partial t} \tag{4.3.4}$$

This is consistent with (4.2.25) when the magnetic field is unchanging. It is also consistent with (4.2.18), i.e., $\bar{\nabla} \cdot \bar{B} \equiv 0$, because $div\ curl$ is identically zero. (4.3.4) is the final form of the fourth Maxwell equation.

4.4 The Transformer

An important application of (4.3.4) is a transformer, in which two separate coils are wrapped onto the same iron core, one being called the primary coil and the other the secondary coil. It is important to appreciate that, to evaluate the flux or emf in a coil which has many turns, the integrals in (4.3.2,3) must be carried out over each turn of the coil. The value of $\int_A \bar{B} \cdot d\bar{A}$ taken over the N_p turns of a coil is called the "flux linkage" and will be N_p times the flux through one turn, Φ. Denoting the rate of change of the flux $\dot{\Phi}$, the emf in the first coil will be $V_p \propto N_p\dot{\Phi}$. The second coil, with N_s turns, shares the same core and hence has the same flux Φ through it, so that its flux linkage is $N_s\Phi$ so that the emf in the second coil is $V_s \propto N_s\dot{\Phi}$. From this the "transformer equation" follows,

$$\frac{V_s}{V_p} = \frac{N_s}{N_p} \tag{4.4.1}$$

Thus, a step-up transformer has proportionately more turns on the secondary (outlet) coil. The conservation of energy then implies that the ratio of the secondary to primary circuit currents is the reciprocal of (4.4.1), so that the product VI, which is the electrical power, is the same for both the primary and secondary coils. (This is the ideal position which ignores sources of energy loss, e.g., electrical resistance and magnetic hysteresis in the core).

We conclude this chapter having derived the first, second and fourth Maxwell equations, but having derived only a restricted form of the third Maxwell equation. That shortcoming will be remedied in chapter 7 with the crucial addition of the displacement current. But in the next chapter we first address electromagnetic potentials and the energy associated with static electromagnetic fields, as well as addressing some important types of sources of such fields, e.g., dipoles.

4.5 References

[4.1] Maxwell, James Clerk (1865). *A dynamical theory of the electromagnetic field*. Philosophical Transactions of the Royal Society of London. 155: 459–512.

[4.2] Maxwell, James Clerk (1873) *A Treatise on Electricity and Magnetism*. Nature. 7 (182): 478–480. 24 April 1873. Maxwell was revising the Treatise for a second edition when he died in 1879. The revision was completed by William Davidson Niven for publication in 1881. A third edition was prepared by J. J. Thomson for publication in 1892.

5. Static Fields: Electromagnetic Potentials, Dipoles, Capacitance, Inductance and Electromagnetic Energy

Before considering the effect of media on the electromagnetic fields and completing the Maxwell equations in the next two chapters, we turn our attention in this long chapter to electromagnetic potentials and the energy associated with static electromagnetic fields. Here we confine attention to static fields, i.e., unchanging in time. The opportunity is taken also to discuss the fields arising from dipole sources. This will expose some pathological issues when sources are idealised as point sources. Finally, this chapter introduces the concepts of capacitance and inductance and calculates these quantities for several simple geometries.

Contents

From chapter 4, the Maxwell equations for free space established so far are,

$$\bar{\nabla} \cdot \bar{E} = \frac{\rho}{\varepsilon_0} \qquad \bar{\nabla} \cdot \bar{B} = 0 \qquad \bar{\nabla} \times \bar{B} = \mu_0 \bar{J} \quad \bar{\nabla} \times \bar{E} = -\frac{\partial \bar{B}}{\partial t} \tag{5.1}$$

where the third of these equations is not yet the complete Maxwell equation as it relates only to the special case that the electric field is unvarying.

Consistent with this, throughout this chapter we shall assume fields which are unvarying in time, so the above four vector equations are valid, and, in this special case, $(\bar{\nabla} \times \bar{E}) = 0$.

5.1 The Electric Scalar Potential

The electric scalar potential applies only when the magnetic field is unvarying, hence the electric field becomes irrotational, i.e., $(\bar{\nabla} \times \bar{E}) = 0$. In this case, from Appendix A4, the electric field can be expressed as the $grad$ of some scalar field, Φ, known as the scalar potential.

$$\bar{E} = -\bar{\nabla}\Phi \tag{5.1.1}$$

The minus sign ensures that the electric field points towards reducing potential. This is analogous to the relationship between force and potential energy, V, where force points towards reducing potential energy, $\bar{F} = -\bar{\nabla}V$. Indeed, since the force on a charge, q, in an electric field is just the product of the charge and the field, by definition of the latter, it follows that we can interpret $q\Phi$ as the potential energy of the charge due to its presence in the electric field, i.e., $\bar{F} = q\bar{E} = -\bar{\nabla}(q\Phi)$. Hence the use of the word "potential" in electromagnetism derives from its relation to potential energy.

The other Maxwell equation involving the electric field, $\bar{\nabla} \cdot \bar{E} = \rho/\varepsilon_0$, together with (5.1.1) therefore gives,

$$\nabla^2\Phi = -\rho/\varepsilon_0 \tag{5.1.2}$$

which is known as Poisson's equation. Hence, in regions where there is no charge, the electric scalar potential obeys Laplace's equation,

$$\nabla^2\Phi = 0 \tag{5.1.3}$$

Problems in electrostatics are usually solved in terms of Φ, because, being a scalar, it is easier to work with than a vector. The electric field then follows from (5.1.1).

From the electric field due to a point charge, (4.2.4), we deduce the electric potential due to a point charge,

$$\Phi = \frac{q}{4\pi\varepsilon_0 r} \tag{5.1.4}$$

where r is the distance from the source to the field point at which Φ is evaluated. For a discrete set of point charges q_i located at vector positions $\bar{r}_i$, the potential at field point $\bar{r}_f$ is,

$$\Phi(\bar{r}_f) = \sum_i \frac{q_i}{4\pi\varepsilon_0 |\bar{r}_f - \bar{r}_i|} \tag{5.1.5}$$

For a continuous distribution of charge density $\rho(\bar{r})$ this becomes,

$$\Phi(\bar{r}_f) = \int \frac{\rho(\bar{r})}{4\pi\varepsilon_0 |\bar{r}_f - \bar{r}|} d\tau \tag{5.1.6}$$

Hence (5.1.6) is the integral inverse (i.e., the solution) of equ.(5.1.2). As we will see in chapter 15, $\frac{-1}{4\pi|\bar{r}_f - \bar{r}|}$ is an example of a Green's function.

Warning: The reader should beware that (5.1.6) is correct only when the charge density is static. Otherwise there would be an ambiguity in the time at which $\rho(\bar{r})$ in the integrand must be calculated, and the potential itself becomes time dependent. Recall the proviso, above, that the scalar potential correctly gives the electric field via (5.1.1) only when any magnetic field is unvarying. But, in general, time-varying charge density will produce a time-varying magnetic field. We will see in §7.2.1 that (5.1.1) is replaced by a more general expression for the electric field in terms of potentials where there is time dependence, and we will see in chapter 22 how (5.1.6) then changes to accommodate time-varying sources.

5.1.1 A Spherical Ball of Uniform Charge

As an exercise the reader should confirm that the solutions for the electric scalar potential and the electric field resulting from a uniform charge density ρ within a sphere of radius R centred on the origin (but no charge elsewhere) are,

$r \le R$: $\quad \Phi = \frac{\rho}{2}\left(R^2 - \frac{1}{3}r^2\right) \quad \bar{E} = \frac{1}{3}\rho\bar{r}$

$r \ge R$: $\quad \Phi = \frac{\rho R^3}{3r} \quad \bar{E} = \frac{\rho R^3}{3r^2}\hat{r}$

If you struggle with finding the potential at the origin note that it must equal the energy required to transfer a unit charge from infinity to the origin and hence must equal $\int_0^\infty E_r dr$ from which you will deduce $\Phi(0) = \rho R^2/2$ using the above expressions for the field. The field everywhere points outwards and correspondingly the gradient of the potential is everywhere negative, the potential being greatest at the origin.

5.2 The Electric Dipole Field

An electric dipole can be created by a negative point charge $-q$ separated from an equal and opposite positive point charge q by a small vector distance $\Delta\bar{s}$, which points from the negative charge towards the positive charge. The electric dipole moment is therefore a vector and is defined by,

$$\bar{p} = q\Delta\bar{s} \tag{5.2.1}$$

At a distance from the dipole which is very large compared with Δs, the point charge fields will tend to cancel. Nevertheless, there will be some net electric field because one charge will generally be closer to the field point than the other. From (5.1.5) the potential at $\bar{r}_f$ due to an electric dipole whose positive charge q is at $\bar{r}_1$, is,

$$\Phi(\bar{r}_f) = \frac{q}{4\pi\varepsilon_0}\left(\frac{1}{|\bar{r}_f - \bar{r}_1|} - \frac{1}{|\bar{r}_f - \bar{r}_1 + \Delta\bar{s}|}\right) \tag{5.2.2}$$

But, for small Δs, $\left(\frac{1}{|\bar{r}_f - \bar{r}_1|} - \frac{1}{|\bar{r}_f - \bar{r}_1 + \Delta\bar{s}|}\right) \approx -\Delta\bar{s}\cdot\bar{\nabla}\frac{1}{|\bar{r}_f - \bar{r}_1|} = -\Delta\bar{s}\cdot\bar{\nabla}\frac{1}{r}$ if we place the dipole at the origin of spherical polar coordinates, and the operator is acting on the field coordinate. Also, from (4.2.14) we have $\bar{\nabla}_f\left(\frac{1}{r}\right) = -\frac{\hat{r}}{r^2}$ so that the electric potential of a dipole is, assuming $r \gg \Delta s$,

$$\Phi(\bar{r}) = \frac{q}{4\pi\varepsilon_0}\Delta\bar{s}\cdot\frac{\hat{r}}{r^2} = \frac{\bar{p}\cdot\hat{r}}{4\pi\varepsilon_0 r^2} = \frac{p\cos\theta}{4\pi\varepsilon_0 r^2} = -\frac{1}{4\pi\varepsilon_0}\bar{p}\cdot\bar{\nabla}_f\left(\frac{1}{r}\right) \tag{5.2.3}$$

where the angle θ is that between $\bar{p}$ and $\bar{r}$ (and this will coincide with the polar angle if the dipole is aligned parallel to $\hat{z}$). The electric field of a dipole can be found in polar coordinates with the aid of the $grad$ operator in polars (see Appendix A2),

$$\bar{\nabla} = \hat{r}\partial_r + \frac{\hat{\theta}}{r}\partial_\theta + \hat{z}\partial_z \tag{5.2.4}$$

Hence, (5.1.1) and (5.2.3) give the field of a dipole oriented parallel to $\hat{z}$,

$$\bar{E}(\bar{r}) = \frac{p}{4\pi\varepsilon_0 r^3}\left(2\hat{r}\cos\theta + \hat{\theta}\sin\theta\right) = \frac{1}{4\pi\varepsilon_0 r^3}(3(\bar{p}\cdot\hat{r})\hat{r} - \bar{p}) \tag{5.2.5}$$

The first of the expressions in (5.2.5) is specific to the dipole lying along the polar $\hat{z}$ direction with θ the polar angle. The second expression is valid independent of the coordinates used and independent of the orientation of the dipole. It is worth noting that the latter important fact can be concluded immediately simply from the fact that the second expression is a vector, and hence independent of coordinate system. That it is a vector follows from the facts that both r and $\bar{p} \cdot \hat{r}$ are scalars.

5.3 Electrical Field Energy

Suppose we have a region within which there is some density of charge, ρ. As like charges repel, to bring such a region of charge together (from notional infinity) requires energy to overcome the repulsive force. Hence, there is a positive potential energy associated with such a region of charge. (We leave aside as a separate matter what might be holding the cluster of charge together against its Coulomb repulsion). We have already seen that the electric potential is just this potential energy per unit charge.

But in calculating the total energy of a finite collection of charge one must take into account that the electric potential increases as the charge increases. In fact, the linearity of the relationship between electrical potential (or field) and charge means $\Phi(q)\delta q$ integrates to $\frac{1}{2}\Phi q$, where the latter quantities are the final values when the whole collection of charge has been compiled. This expression holds at every point within the distribution of charge. The total potential energy is thus,

$$\text{Energy} = \frac{1}{2}\int \Phi(\bar{r})\rho(\bar{r})d\tau = \frac{\varepsilon_0}{2}\int \Phi(\bar{r})\bar{\nabla}\cdot\bar{E}\, d\tau \qquad (5.3.1)$$

where we have used the Maxwell equation $\bar{\nabla}\cdot\bar{E} = \rho/\varepsilon_0$. But the vector identity (A1.9) allows us to rewrite $\Phi\bar{\nabla}\cdot\bar{E} = \bar{\nabla}\cdot(\Phi\bar{E}) - \bar{E}\cdot\bar{\nabla}\Phi$. When the first term is substituted into the volume integral of (5.3.1), because it is a divergence, the divergence theorem allows us to rewrite it as the surface integral $\oiint \phi\bar{E}\cdot dS$. But if we are dealing with a finite region of charge then both the electric field and the electric potential will fall to zero at spatial infinity, their product being at most of order $1/r^3$, so this surface integral can be assumed zero if the volume integral is extended over the whole of space.

We are left with the volume integral over $-\bar{E}\cdot\bar{\nabla}\Phi = E^2$, by (5.1.1). That is, we have concluded that the potential energy of our finite region of charge is,

$$\text{Energy} = \frac{\varepsilon_0}{2}\int E^2 d\tau \tag{5.3.2}$$

Note the rather extraordinary thing that has happened in moving from (5.3.1) to (5.3.2). The former received non-zero contributions only from within the region where there is charge. Outside the region of charge, the integrands in (5.3.1) are zero. But (5.3.2) must be integrated both within the region of charge, and outside it, to infinity. This is necessary because we have appealed to the volume integration of $\overline{\nabla} \cdot (\Phi \overline{E})$ to be zero, but this only holds if integration is to infinity. In physical terms this means that, whilst potential energy is associated with charge, and hence localised to where the charge resides, the field energy is distributed throughout space because the field exists throughout space. That the two things are equal is actually quite remarkable (though absolutely necessary). Physically, the field energy is the more fundamental concept.

(5.3.1) was derived as the potential energy associated with the collection of charge. But Einstein's famous association of inertia with energy means that any such energy must have a physical manifestation. The equivalent quantity (5.3.2) tells us what this is. The potential energy is manifest as the energy implicit in the electric field itself. That is, for any electric field in free space,

$$\text{Energy density of an electric field} = u_E = \frac{\varepsilon_0}{2}E^2 \tag{5.3.3}$$

Note that this energy density, which occurs throughout space, is not in addition to the potential energy of the charge which creates it but is another way of interpreting the same quantity. The field is where the energy sometimes described as the charge's potential energy actually resides.

5.4 Examples of Electrostatic Energies

5.4.1 The Energy of a Sphere of Uniform Surface Charge

We first note from the divergence theorem in the form of (4.2.5b) that the electric field inside the spherical shell of charge must be zero. To spell this out, this is because symmetry means the field is the same at all points on a sphere placed within the sphere of charge and concentric with it. Hence, since the integral of the field over such a sphere equals the charge within it according to (4.2.5b), and this is zero, the field is zero everywhere within the shell of charge.

Hence we need consider only the integral (5.3.2) from the radius of the sphere, a, to infinity. Also from (4.2.5b), the electric field outside the sphere is identical to the field due to the same total charge placed at the centre of the sphere, and hence is given by (4.2.4). Squaring and integrating over volume gives,

$$\text{Energy} = \frac{1}{2} \cdot \frac{q^2}{4\pi\varepsilon_0 a} \tag{5.4.1}$$

5.4.2 The Energy of a Sphere of Uniform Charge Density

The contribution to the field energy from outside the sphere, of radius a, is just (5.4.1) again in terms of the total charge. However, the field inside the sphere is now non-zero because (4.2.5b) now tells us that the field at a radius $r < a$ is $E = \rho r/3\varepsilon_0$. Squaring and integrating over volume gives the energy due to the field within the sphere,

$$\text{Energy within sphere} = \frac{1}{10} \cdot \frac{q^2}{4\pi\varepsilon_0 a} \tag{5.4.2}$$

From which the total energy is obtained by adding (5.4.1), hence.

$$\text{Total energy} = \frac{3}{5} \cdot \frac{q^2}{4\pi\varepsilon_0 a} \tag{5.4.3}$$

5.4.3 The Divergent Energy of a Point Charge

A point charge can be regarded as the limit of either of the cases §5.4.1 or §5.4.2 as $a \to 0$, and hence the energy associated with a point charge is divergent.

5.4.4 The Ambiguous Self-Energy of a Dipole

The same problem is found for the self-energy of a dipole in the limit of zero size, because the dipole field diverges as $1/r^3$ at the origin, and hence the energy density diverges as $1/r^6$ and gives divergent contribution to the total energy. However, this is merely because an approximate solution has been used which is inapplicable near the dipole.

If we consider a finite dipole consisting of equal and opposite charges q at a distance s apart, their mutual potential energy just the energy of the charge $-q$ in the field of the charge q (or, equivalently, vice versa) which is $-\frac{q^2}{4\pi\varepsilon_0 s}$. This is negative as a consequence of the force between opposite charges being attractive. But this is odd because the expression (5.3.2) for the electrostatic energy in terms of the electric field is necessarily positive. What we have omitted is the self-energy of each of the contributing charges – but for a point charge §5.4.3 has noted that this is infinite (and positive for both because it involves q^2). Consequently, we arrive back at a (positive) divergent energy.

We could contrive a dipole with finite energy by considering the two charges to have a spatial distribution, rather than being point charges. For example, spheres of radii a with uniform surface charge would each have self-energy given by (5.4.1) so the total (absolute) energy of a dipole composed in this way would be, for $a \ll s$,

$$\frac{q^2}{4\pi\varepsilon_0}\left(\frac{1}{a}-\frac{1}{s}\right) \tag{5.4.4}$$

The alert reader will cry foul as we have assumed that the two sources of energy can simply be added. But this is surely invalid as it is the fields that add linearly, and the energy is related to the square of the field. Actually, it turns out that (5.4.4) can be justified via the integration of (5.3.2) using the correct total field, see [5.1], but only in the case that $a \ll s$. Nevertheless, (5.4.4) is rather arbitrary as the assumed spherical shells of charge could be spherical balls of charge, or indeed any other distribution, in which case a different total energy would result. (And in this context we note that if they were charged conducting spheres their surface charge density would not be uniform but concentrated on the regions facing the other charged sphere). In summary, the electrostatic energy of a dipole is indeterminate, depending upon its precise construction.

5.4.5 Capacitance

This section discusses capacitance in the absence of any dielectric media. By way of anticipating chapter 6, the effect of a dielectric within a capacitor can be taken into account by replacing ε_o wherever it occurs in the expressions below for capacitance with $\varepsilon_r\varepsilon_o$ where ε_r is the dielectric constant.

Capacitance refers to the ability to store charge, and hence electrostatic energy, by virtue of maintaining a potential difference over the item in question. In practice, the charge is invariably stored as equal and opposite charges associated with the parts of the item which differ in potential.

A capacitor (or condenser to use the older term) is an electronic device which is designed to have a certain tailored capacitance. Hence a capacitor is such that an applied potential difference across its two terminals causes it to become charged (or, more accurately, to separate equal and opposite charges within it). A capacitor can be modelled as a pair of parallel plates a distance D apart, with positive charge, Q, arising on one plate and an equal and opposite negative charge arising on the other. The capacitance, C, of the device is defined as the charge arising per unit potential difference. Hence, the charge due to a potential difference of V is,

$$Q = CV \tag{5.4.5a}$$

assuming linearity, which is a reasonable approximation in practice only for a certain range of working voltages, V.

The definition of potential, (5.1.1), tells us that, if the electric field between the two plates is assumed uniform, then it is $E = V/D$. But if we apply (4.2.5d) to this situation the field can also be written $E = \sigma/\varepsilon_0$ where σ is the charge per unit surface area of the plates, and hence $\sigma = Q/A$ if the total area of the plates is A. Equating the two expressions for the field gives $Q = \frac{\varepsilon_0 A}{D} V$ hence, for a capacitor with no medium between the plates,

$$C = \frac{\varepsilon_0 A}{D} \tag{5.4.5b}$$

The energy stored in a capacitor is derived simply from $Q = CV$. To move an additional δQ of charge to the plates, against the potential difference of V, requires energy $V\delta Q = \frac{Q}{C}\delta Q$ which integrates to,

$$\text{Energy} = \frac{1}{2}\frac{Q^2}{C} = \frac{1}{2}CV^2 \tag{5.4.6}$$

where the capacitance is given in terms of the geometry by (5.4.5b) for the case of a simple parallel plate capacitor.

The term V in the above discussion refers to the potential difference over the capacitor arising from the charge on it, but not including any external applied potential difference. If a capacitor, initially uncharged, has a potential difference, V_a, applied to it suddenly at time zero then current will flow initially in order to supply charge to the capacitor. The potential over the capacitor is $V = Q/C$ which rises from zero to V_a when the capacitor is fully charged and the current drops to zero. The net potential difference that drives the current, I, at any time is $V_a - V = V_a - \frac{Q}{C} = RI$ where R is the circuit resistance. Since $I = \frac{dQ}{dt}$ we thus have, for a constant applied 'voltage',

$$-\frac{I}{C} = R\frac{dI}{dt} \tag{5.4.6b}$$

This simple differential equation controls how the current varies over time after a potential difference is applied suddenly to an initially uncharged capacitor. The solution is,

$$I = I_0 e^{-t/RC} \tag{5.4.6c}$$

where I_0 is the maximum current, which flows "immediately" upon the potential being applied. (5.4.6c) shows how the current decays towards zero as the capacitor becomes fully charged, with a time constant of,

$$\tau = RC \tag{5.4.6d}$$

Integrating (5.4.6c) gives the charge on the capacitor at any time, and hence the potential over the capacitor at any time,

$$Q(t) = CV(t) = I_0 RC\left(1 - e^{-t/RC}\right) \quad (5.4.6e)$$

which shows the charge on the capacitor builds over time as $V(t) \to V_a$ and hence the initial current is $I_0 = V_a/R$ and the full charge is $Q_0 = CV_a$.

(5.4.6b) also applies if a charged capacitor is allowed to discharge through a resistance R, i.e., shorted out at time zero with no applied potential. The solution is then,

$$Q = CV = Q_0 e^{-t/RC} \quad (5.4.6f)$$

$$I = -I_0 e^{-t/RC} \quad (5.4.6g)$$

which shows how the capacitor discharges and how the current decreases towards zero from its maximum, $I_0 = \frac{V_0}{R}$, which arises "immediately" and where $V_0 = \frac{Q_0}{C}$ is the initial potential over the capacitor when fully charged. The minus sign in (5.4.6g) indicates the discharging current is in the opposite direction to the charging current.

See Figure 5.4 for a graph of the current on charging and discharging a capacitor (contrasted with that of an inductor). See also the discussion which follows that Figure regarding what "immediately" means is this context.

Although capacitors can be a purposefully fabricated component for use in an electrical circuit, the phenomenon of capacitance is of general applicability. Nor is the capacitance of a device necessarily always something that is welcome but can merely be something that has to be taken into account when anticipating the electrical behaviour of the device. Capacitance is usually associated with a device which consists of two parts between which the electric field arises due to opposite charges on the two parts. However, this is not always the case and it is possible to speak of the capacitance of a single part. In such cases the role of the other part is played by the surroundings, e.g., the "capacitance to earth". But even an item remote from anything else has a capacitance, as the capacitance is defined as the ratio of the charge on it to its potential.

Some examples of the calculation of capacitance for two-part devices follow.

5.4.5.1 Concentric Cylinders

Suppose the inner and outer cylinders have radii a and b respectively. For the same reason as argued in §5.4.1, any charge on the outer cylinder produces no field within it. Hence, we need consider only the charge, q, per unit length of the inner cylinder when calculating the field between the cylinders. Now consider the field on a notional cylinder of radius r between the cylinders. By (4.2.5b) the field must be proportional to q divided by the area per unit length, $2\pi r$, specifically a radial field $E_r = \frac{q}{2\pi\varepsilon_0 r}$. But the radial field is (minus) the radial gradient of the potential, so $\frac{\partial\Phi}{\partial r} = -\frac{q}{2\pi\varepsilon_0 r}$ and integrating gives the potential difference between the cylinders to be,

$$\Phi(a) - \Phi(b) = V = \frac{q}{2\pi\varepsilon_0} log\left(\frac{b}{a}\right) \tag{5.4.7}$$

So the capacitance per unit length, $C = q/V$, is.

$$C = \frac{2\pi\varepsilon_0}{log\left(\frac{b}{a}\right)} \tag{5.4.8}$$

This covers the important practical example of coaxial cables or coaxial transmission lines.

5.4.5.2 Concentric Spheres

Suppose the inner and outer spheres have radii a and b respectively. For the same reason as argued in §5.4.1, any charge on the outer sphere produces no field within it. Hence, we need consider only the charge, Q, on the inner sphere when calculating the field between the spheres. Now consider the field on a notional sphere of radius r between the spheres, whose area is thus $4\pi r^2$. The field is thus radial with magnitude $E_r = \frac{Q}{4\pi\varepsilon_0 r^2}$. But the radial field is (minus) the radial gradient of the potential, so $\frac{\partial\Phi}{\partial r} = -\frac{q}{4\pi\varepsilon_0 r^2}$ and integrating gives the potential difference between the cylinders to be,

$$\Phi(a) - \Phi(b) = V = \frac{Q}{4\pi\varepsilon_0}\left(\frac{1}{a} - \frac{1}{b}\right) \tag{5.4.9}$$

So the capacitance, $C = Q/V$, is,

$$C = \frac{4\pi\varepsilon_0 ab}{b-a} \tag{5.4.10}$$

5.4.5.3 Straight Parallel Wires

This is another realistic example of practical importance, e.g., applicable to parallel transmission lines or twin cables. We assume the radius of the wires, a, is very small compared with the distance between their axes, D. Suppose the charge on one wire is λ per unit length, and equal and opposite on the other. The field due to the first wire on a cylinder of radius r is $E_r = \frac{\lambda}{2\pi\varepsilon_0 r}$. But there is also a field arising from the other wire, of opposite charge, which will be $E_r = -\frac{\lambda}{2\pi\varepsilon_0(D-r)}$. Integrating $\frac{\partial\Phi}{\partial r} = \frac{\lambda}{2\pi\varepsilon_0(D-r)} - \frac{\lambda}{2\pi\varepsilon_0 r}$ gives the potential at the first wire to be $\frac{\lambda}{2\pi\varepsilon_0} log\left(\frac{D-a}{a}\right)$. The potential is equal and opposite at the other wire so that the potential difference is $V = \frac{\lambda}{\pi\varepsilon_0} log\left(\frac{D-a}{a}\right)$. The capacitance per unit length, $C = \lambda/V$, is thus,

$$C = \frac{\pi\varepsilon_0}{log\left(\frac{D-a}{a}\right)} \approx \frac{\pi\varepsilon_0}{log\left(\frac{D}{a}\right)} \tag{5.4.11}$$

5.4.6 The Energy, Force and Moment acting on an Electric Dipole in an External Electric Field

We have seen in §5.1 that the potential energy of a charge q in an electric potential Φ is just $q\Phi$. Hence, for a dipole consisting of a positive charge q at position $\bar{r} + \Delta\bar{r}$ and a charge $-q$ at position $\bar{r}$ the total energy is $q\Phi(\bar{r} + \Delta\bar{r}) - q\Phi(\bar{r})$. But for small $\Delta\bar{r}$ this can be written $q\bar{\nabla}\Phi \cdot \Delta\bar{r} = -q\bar{\mathrm{E}} \cdot \Delta\bar{r}$ by (5.1.1). But $q\Delta\bar{r}$ is just the dipole moment, $\bar{p}$, hence the energy of a permanent dipole due to being placed in an external electric field is,

Permanent dipole: $$U = -\bar{p} \cdot \bar{E} \tag{5.4.12}$$

Note that (5.4.12) does not include any self-energy of the dipole (see §5.4.4). Note also that if the dipole is aligned with the field the potential energy is negative, but positive if it is oppositely oriented.

The force acting on the dipole due to the electric field is $\bar{F} = -\bar{\nabla}U$. Using identity (A1.12), and when there is either no magnetic field or only a static magnetic field, so that $\bar{\nabla} \times \bar{E} = 0$, we get,

$$\bar{F} = (\bar{p} \cdot \bar{\nabla})\bar{E} \tag{5.4.13}$$

There is thus no force on a dipole in a uniform field, because then the force on its positive and negative charges would cancel. But if the field is spatially varying, the force on the positive charge will in general differ from the force on the negative charge, hence producing the net force (5.4.13).

In general there will also be a moment (or torque) acting on the dipole. This is clear because (5.4.12) indicates that minimum energy is obtained if the dipole is aligned with the field. If θ is the angle between the dipole and the field then (5.4.12) gives $U = -pE\cos\theta$, so that the torque is $\frac{\partial U}{\partial\theta} = pE\sin\theta$. Moreover, the moment acts about the axis which is perpendicular to both the dipole and the field so that, in general, we have, for the vector moment or torque,

$$\bar{T} = \bar{p} \times \bar{E} \qquad (5.4.14)$$

noting that the sign of (5.4.14) is such that the dipole tends to align with the field (as opposed from anti-aligning), resulting in minimising the energy, (5.4.12). The torque is greatest when the dipole is perpendicular to the field (when the energy is zero). Although the torque is zero when the dipole is anti-aligned with the field, this is an unstable equilibrium with positive energy and the slightest perturbation will make the dipole tend to align with the field.

These relationships, from (5.4.12), assume a permanent dipole. But if the dipole were to be induced in proportion to the applied field, $\bar{p} = a\bar{E}$, for some constant, a, then integrating an increment of energy $\delta U = -\delta\bar{p} \cdot \bar{E}$ leads to,

Induced dipole: $$U = -\frac{1}{2}\bar{p} \cdot \bar{E} = -\frac{a}{2}E^2 \qquad (5.4.15)$$

The factor of $\frac{1}{2}$ then works through into the expressions for force which becomes,

$$\bar{F} = \frac{a}{2}(\bar{E} \cdot \bar{\nabla})\bar{E} \qquad (5.4.15b)$$

But there is no torque acting on such a dipole because, by assumption, it is always aligned with the field.

Note that throughout $\bar{E}$ is the applied ("external") field, not including the field due to the dipole.

5.4.7 The Force, Moment and Energy between Two Electric Dipoles

In this section we assume a pair of permanent electric dipoles (as opposed to induced dipoles) interacting through their electric fields. Static conditions are assumed so that the net force and moment acting on the pair of dipoles considered together are zero. Static conditions with forces and moments acting between the dipoles presupposes that other sources of force and moment are being applied to the dipoles to keep them stationary. To simplify the algebra we confine attention to dipoles, $\bar{p}_1$ and $\bar{p}_2$, which are coplanar. We take this to be the x, z plane with dipole $\bar{p}_1$ in the direction $\hat{z}$. We shall derive expressions for,

- The force, $\bar{F}_2$, acting on dipole $\bar{p}_2$ due to the field from $\bar{p}_1$;
- The force $\bar{F}_1$, acting on dipole $\bar{p}_1$ due to the field from $\bar{p}_2$ which is necessarily $\bar{F}_1 = -\bar{F}_2$;
- The moment, $\bar{M}_2$, acting on dipole $\bar{p}_2$ due to the field from $\bar{p}_1$;
- The moment $\bar{M}_1$, acting on dipole $\bar{p}_1$ due to the field from $\bar{p}_2$ can then be found by imposing equilibrium, giving $M_1 + M_2 + |\bar{r} \times \bar{F}_2| = 0$ where $\bar{r}$ is the vector displacement between the dipoles. The sign convention here is that the moments, and the couple $\bar{r} \times \bar{F}_2$, will be positive if clockwise (i.e., the positive $\hat{\phi}$ direction in spherical polars);
- The mutual potential energy of interaction between the dipoles. Note that this is the energy associated with the pair of dipoles, rather than with either dipole individually. It excludes the (ill-defined) self-energy of the individual dipoles.

Figure 5.1: Evaluating Forces and Moments between Two Dipoles

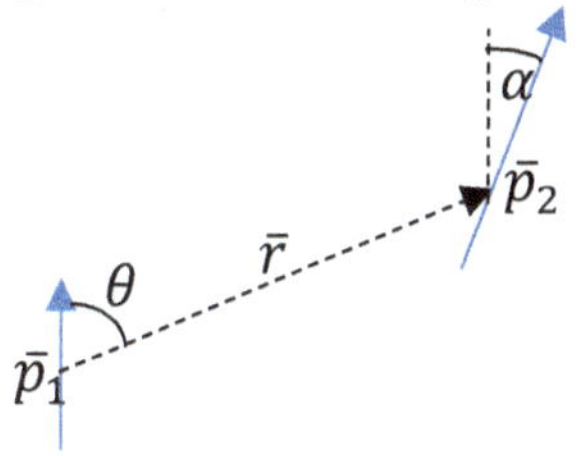

In polars on the x, z plane the angle between $\bar{p}_1$ and $\bar{r}$ is θ, whilst the angle between $\bar{p}_1$ and $\bar{p}_2$ is α, as illustrated by Figure 5.1. Hence we see that,

$$\bar{p}_2 = p_2\big(\hat{r}\cos(\theta - \alpha) - \hat{\theta}\sin(\theta - \alpha)\big) \tag{5.4.16}$$

From (5.4.13) the force on $\bar{p}_2$ due to the field from $\bar{p}_1$ is $(\bar{p}_2 \cdot \bar{\nabla})\bar{E}_1$ and $\bar{E}_1$ is given by (5.2.5). From $\bar{\nabla}$ in polars from Appendix A2 we get,

$$\bar{p}_2 \cdot \bar{\nabla} = p_2\left[\cos(\theta - \alpha)\,\partial_r - \frac{\sin(\theta-\alpha)}{r}\partial_\theta\right] \tag{5.4.17}$$

Applying this to $\bar{E}_1 = \frac{p_1}{4\pi\varepsilon_0 r^3}\big(2\hat{r}\cos\theta + \hat{\theta}\sin\theta\big)$ requires the operation of the derivatives on the unit vectors. It is left as an exercise for the reader to show that,

$$\partial_r\hat{r} = \partial_r\hat{\theta} = 0 \quad \text{and} \quad \partial_\theta\hat{r} = \hat{\theta},\ \partial_\theta\hat{\theta} = -\hat{r} \tag{5.4.18}$$

After a little algebra we arrive at, (5.4.19)

$$\bar{F}_2 = -3F_0\{\hat{r}[\cos\theta\cos(\theta-\alpha) + \cos(2\theta-\alpha)] + \hat{\theta}\sin(2\theta-\alpha)\}$$

where, $$F_0 = \frac{p_1 p_2}{4\pi\varepsilon_0 r^4} \qquad (5.4.20)$$

For four special cases the force is given in Table 5.1. The moment acting on $\bar{p}_2$ is, following (5.4.14), $\bar{M}_2 = \bar{p}_2 \times \bar{E}_1$, hence, from (5.4.16) and (5.2.5) and using $\hat{r} \times \hat{\theta} = -\hat{\theta} \times \hat{r} = \hat{\phi}$,

$$\bar{M}_2 = p_2\big(\hat{r}\cos(\theta-\alpha) - \hat{\theta}\sin(\theta-\alpha)\big) \times \frac{p_1}{4\pi\varepsilon_0 r^3}\big(2\hat{r}\cos\theta + \hat{\theta}\sin\theta\big)$$

or, $$\bar{M}_2 = \frac{p_1 p_2}{4\pi\varepsilon_0 r^3}\hat{\phi}\{\sin\theta\cos(\theta-\alpha) + 2\cos\theta\sin(\theta-\alpha)\}$$

or, $$\bar{M}_2 = F_0 r\hat{\phi}\left\{\frac{3}{2}\sin(2\theta-\alpha) - \frac{1}{2}\sin\alpha\right\} \qquad (5.4.21a)$$

From $\bar{M}_1 + \bar{M}_2 + \bar{r} \times \bar{F}_2 = 0$ we deduce that,

$$\bar{M}_1 = F_0 r\hat{\phi}\left\{\frac{3}{2}\sin(2\theta-\alpha) + \frac{1}{2}\sin\alpha\right\} \qquad (5.4.21b)$$

Finally, the mutual potential energy, from (5.4.12), is $U = -\bar{p}_2 \cdot \bar{E}_1$, hence,

$$U = -p_2\big(\hat{r}\cos(\theta-\alpha) - \hat{\theta}\sin(\theta-\alpha)\big) \cdot \frac{p_1}{4\pi\varepsilon_0 r^3}\big(2\hat{r}\cos\theta + \hat{\theta}\sin\theta\big)$$

or, $$U = -\frac{p_1 p_2}{4\pi\varepsilon_0 r^3}\{2\cos\theta\cos(\theta-\alpha) - \sin\theta\sin(\theta-\alpha)\}$$

or, $$U = -F_0 r\left\{\frac{3}{2}\cos(2\theta-\alpha) + \frac{1}{2}\cos\alpha\right\} \qquad (5.4.22)$$

The moments and mutual energies for the four special cases are also included in Table 5.1.

Table 5.1: Forces, Moments and Energy for Two Dipoles

θ	α	$\bar{F}_2 = -\bar{F}_1$ #	M_2 *	M_1 *	U
0	0	$-6F_0\hat{r}$	0	0	$-2F_0 r$
$\pi/2$	0	$3F_0\hat{r}$	0	0	$F_0 r$
$\pi/2$	$\pi/2$	$-3F_0\hat{\theta}$	$F_0 r$	$2F_0 r$	0
0	$\pi/2$	$3F_0\hat{\theta}$	$-2F_0 r$	$-F_0 r$	0

**Positive means clockwise (vector moment in direction $\hat{r} \times \hat{\theta} = \hat{\phi}$)*

#Positive $\bar{F}_2$ means repulsive, negative $\bar{F}_2$ means attractive (and the reverse for $\bar{F}_1$)

5.4.8 The Force, Moment and Energy between a Charge and a Dipole

The same approach as above may be used to find the force and moment on dipole $\bar{p}_2$ in the field of a point charge, $\bar{E}_1 = \frac{q\hat{r}}{4\pi\varepsilon_0 r^2}$. We find, taking the dipole to be aligned with the z axis ($\alpha = 0$),

$$\bar{F}_2 = -\frac{qp_2}{4\pi\varepsilon_0 r^3}\left(2\hat{r}\cos\theta + \hat{\theta}\sin\theta\right) \tag{5.4.23}$$

This is in fact immediately obvious because it is necessarily equal and opposite to the force acting on the charge from the dipole field. The moment on the dipole is,

$$\bar{M}_2 = \frac{qp_2}{4\pi\varepsilon_0 r^2}\hat{\phi}\sin\theta \tag{5.4.24}$$

We would expect the moment acting on the charge to be zero as a point entity cannot sustain a moment. This is in fact the case as the moment of the force $\bar{F}_2$ about the charge is equal and opposite to $\bar{M}_2$. The mutual potential energy is,

$$U = -\frac{qp_2}{4\pi\varepsilon_0 r^2}\cos\theta \tag{5.4.25}$$

Hence, as would be expected, the potential energy is negative if the dipole is pointing away from the charge ($\theta = 0$) but positive if it points towards the charge ($\theta = \pi$).

5.5 The Magnetic Potentials and Magnetic Dipoles

The magnetic potential of most general applicability is a vector potential, $\bar{A}$, discussed below in §5.5.3. But first we consider a scalar magnetic potential, which, whilst of limited applicability, is helpful in some circumstances.

5.5.1 The Magnetic Scalar Potential

A magnetic scalar potential, ϕ_m, can be defined by analogy with (5.1.1), i.e.,

$$\bar{B} = -\mu_0\bar{\nabla}\phi_m \tag{5.5.1}$$

The alert reader will note that the expression for $\bar{B}$ as a gradient means that $\bar{\nabla}\times\bar{B} = 0$ which conflicts with the Maxwell equations, (5.1), specifically $\bar{\nabla}\times\bar{B} = \mu_0\bar{J}$ and hence a scalar potential can be used only where there is no current, $\bar{J} = 0$, and when any electric field is static. With this proviso, inserting (5.5.1) into $\bar{\nabla}\cdot\bar{B} \equiv 0$, from (5.1), gives,

$$\nabla^2\phi_m \equiv 0 \tag{5.5.2}$$

So the magnetic scalar potential obeys Laplace's equation, analogous to the electric scalar potential in the absence of charges, (5.1.3). This can be very useful because it means that some problems in magnetism can be solved immediately by direct analogy with the electrical case if the solution to that is known. An example is the magnetic dipole field…

5.5.2 The Magnetic Dipole (via the Scalar Potential)

From (4.2.20) the magnetic flux density due to a current circuit is,

$$\bar{B}(\bar{r}_f) \cdot \delta\bar{r}_f = \frac{\mu_0}{4\pi} I\delta\Omega \qquad (4.2.20)$$

where the solid angle is here assumed to be that associated with the whole circuit of current, to ensure that $\bar{B}$ is the total magnetic flux density. Following the discussion in §4.2.3 this solid angle is illustrated by Figure 5.2 as the annular region covered by the blue arrows

In Figure 5.2 we can consider three areas, each of which subtends some solid angle at the field point $\bar{r}_f$,

(a) the annual region whose solid angle at the field point is $\delta\Omega$;
(b) the area within the first copy of the current circuit, whose solid angle at the field point is Ω_0;
(c) the area within the second (displaced) copy of the current circuit, whose solid angle at the field point is Ω.

Figure 5.2: Illustration of the Areas Defining the Solid Angle, δΩ

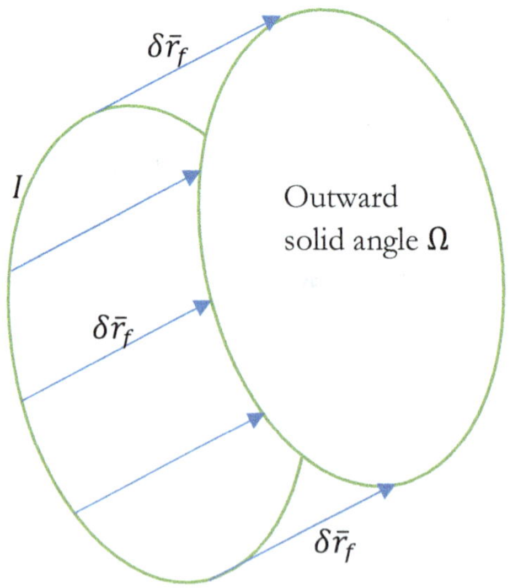

The green loop is the current circuit, the blue arrows indicate a notional displacement of the field point (reinterpreted as a virtual displacement of the current loop)

But for a field point lying outside the region illustrated, the total vector of solid angle of the closed surface consisting of (a)+(b)+(c) is zero. Hence, we conclude that $\delta\Omega = \Omega_0 - \Omega$ (where Ω and $\delta\Omega$ are outward pointing and Ω_0 is inward pointing). But $\overline{\nabla}\phi_m \cdot \delta\bar{r}_f = \Delta\phi_m$ is the change in the magnetic scalar potential when the field point is moved by $d\bar{r}_f$ so using (5.5.1) in (4.2.20) gives,

$$\Delta\phi_m = \frac{1}{4\pi} I(\Omega - \Omega_0) \tag{5.5.3}$$

Hence we can adopt $\phi_m = \frac{1}{4\pi} I\Omega$ noting that we are free to choose an arbitrary datum for ϕ_m as (5.5.1) means that only its gradient matters physically.

The magnetic dipole moment associated with a current loop is defined as,

$$\bar{m} = I\bar{S} \tag{5.5.3b}$$

where $\bar{S}$ is the area of the (small) current loop, so that $\Omega = \frac{\bar{S}\cdot\hat{r}}{r^2}$ where $\bar{r} = \bar{r}_f - \bar{r}_m$ and $\bar{r}_m$ is the position of the current loop (taken to be sufficient small to have a unique location, i.e., point-like).

The magnetic scalar potential of a magnetic dipole is thus, in direct analogy with (5.2.3),

$$\phi_m(\bar{r}) = \frac{\bar{m}\cdot\hat{r}}{4\pi r^2} = \frac{m\cos\theta}{4\pi r^2} = -\frac{1}{4\pi}\bar{m}\cdot\overline{\nabla}\left(\frac{1}{r}\right) \tag{5.5.4}$$

where the angle θ is that between $\bar{m}$ and $\bar{r}$ (and this will coincide with the polar angle if the dipole is aligned parallel to $\hat{z}$). The external magnetic flux density of a magnetic dipole is found exactly as for the electric dipole, (5.2.5), i.e.,

$$\bar{B}(\bar{r}) = \frac{\mu_0 m}{4\pi r^3}\left(2\hat{r}\cos\theta + \hat{\theta}\sin\theta\right) = \frac{\mu_0}{4\pi r^3}(3(\bar{m}\cdot\hat{r})\hat{r} - \bar{m}) \tag{5.5.5}$$

The scalar potential has thus provided a rapid route to this non-trivial result. Next we look at the magnetic vector potential.

5.5.3 The Magnetic Vector Potential

The vector potential, $\bar{A}$, is defined such that,

$$\bar{B} = \overline{\nabla} \times \bar{A} \tag{5.5.6}$$

Note that, unlike the magnetic scalar potential, this is consistent with Maxwell's equations, (5.1), and, in fact can replace the equation $\overline{\nabla}\cdot\bar{B} = 0$ which is a consequence of (5.5.6).

The magnetic flux density is related to a current density $\bar{J}$ which causes it by (4.2.17) which is a $curl$ and so we can write immediately,

$$\bar{A}(\bar{r}_f) = \frac{\mu_0}{4\pi}\left(\int \frac{1}{r_{sf}} \bar{J}(\bar{r}_s)\, d\tau_s\right) \tag{5.5.7}$$

where $\bar{r}_{sf} = \bar{r}_f - \bar{r}_s$ runs from the source to the field point. A special case of this is a single small current element $I\delta\bar{L}$ for which we get the vector potential,

$$\bar{A}(\bar{r}_f) = \frac{\mu_0}{4\pi}\frac{I\delta\bar{L}}{r_{sf}} \tag{5.5.7b}$$

This is just the magnetic field of a small current element, consistent with the Biot-Savart law, in vector potential form, i.e., taking the $curl$ of (5.5.7b) reproduces (4.2.7) noting that $\bar{\nabla} \times \frac{\delta\bar{L}}{r} = \frac{\delta\bar{L} \times \hat{r}}{r^2}$.

In the same way that (5.1.6) for the electric scalar potential is the solution to the Poisson equation, (5.1.2), we have the equivalent Poisson equation for the magnetic vector potential,

$$\nabla^2 \bar{A} = -\mu_0 \bar{J} \tag{5.5.8}$$

where both functions and the operator relate to the field point. The relationship between (5.5.8) and its solution (5.5.7) may be expressed in terms of the Green's function of the Laplacian operator which we noted in §5.1, namely $G(\bar{r}_f, \bar{r}_s) = -1/4\pi|\bar{r}_f - \bar{r}_s|$. The solution (5.5.7) can then be written $\bar{A}(\bar{r}_f) = -\mu_0\left(\int G(\bar{r}_f, \bar{r}_s)\bar{J}(\bar{r}_s)\, d\tau_s\right)$.

In Cartesian coordinates, $\nabla^2 = \partial_x^2 + \partial_y^2 + \partial_z^2$. The alert reader will have noted that because $\bar{\nabla} \times \bar{B} = \mu_0 \bar{J}$ is a Maxwell equation when the electric field is unchanging, from (5.5.6) it follows that,

$$\bar{\nabla} \times (\bar{\nabla} \times \bar{A}) = \mu_0 \bar{J} \tag{5.5.9}$$

but the identity (A1.17) tells us that $\bar{\nabla} \times (\bar{\nabla} \times \bar{A}) = \bar{\nabla}(\bar{\nabla} \cdot \bar{A}) - \nabla^2 \bar{A}$ so there is an apparent mismatch between (5.5.8) and (5.5.9) as the former does not include the term $\bar{\nabla}(\bar{\nabla} \cdot \bar{A})$. This is because the vector potential is determined only up to the gradient of some scalar function. That is, because $curl\ grad$ is identically zero, it follows that any gradient may be added to the vector potential without affecting the magnetic field, i.e., the replacement $\bar{A} = \bar{A}' + \bar{\nabla} f$ leaves $\bar{B}$ unchanged. As $\bar{\nabla} \cdot \bar{A} = \bar{\nabla} \cdot \bar{A}' + \nabla^2 f$ we can find a function f such that,

$$\bar{\nabla} \cdot \bar{A} = 0 \tag{5.5.10}$$

Such an f exists because it is the solution of the Poisson equation $\nabla^2 f = -\bar{\nabla}\cdot\bar{A}'$ which is $\int \frac{\bar{\nabla}\cdot\bar{A}'(\bar{r}_s)}{4\pi|\bar{r}_f-\bar{r}_s|}d\tau_s$. Hence we can assume the "gauge condition", (5.5.10), which is known as the Coulomb gauge. Note, however, that it will be convenient to adopt a different gauge condition when time variation of the fields is considered (see chapter 10), in which case neither (5.5.10) nor (5.5.8) will continue to hold.

We note in passing that if the current is confined to a surface then (5.5.7) would be,

$$\bar{A}(\bar{r}_f) = \frac{\mu_0}{4\pi}\left(\int \frac{1}{r_{sf}}\bar{\varsigma}(\bar{r}_s)\, dS_s\right) \tag{5.5.10b}$$

where $\bar{\varsigma}$ is the current flowing across unit length lying within the surface and perpendicular to the current flow (so the current defines the direction of the vector, $\bar{\varsigma}$).

<u>Warning</u>: Just as for the electric field in terms of charge density, (5.1.6), the expressions (5.5.7) and (5.5.10b) are not valid under time dependent conditions, the generalisation of them being addressed in chapter 22. Any results derived from them therefore implicitly assume that any changes in the current are negligible over a timescale given by the maximum distance between the current and the field point divided by the speed of light.

5.5.4 The Magnetic Dipole (Vector Potential Method)

In the case that the current source is confined to a circuit (i.e., one-dimensional and closed), (5.5.7) becomes

$$\bar{A}(\bar{r}_f) = \frac{\mu_0}{4\pi}\oint I\frac{d\bar{r}_s}{r_{sf}} = -\frac{\mu_0}{4\pi}\iint_\Gamma \bar{\nabla}\left(\frac{1}{r_{sf}}\right)\times Id\bar{S} \tag{5.5.11}$$

where we have used vector integral identity (A1.20) to transform the line integral into a surface integral. The surface in question, Γ, is any surface taking the circuit as its boundary. The gradient in (5.5.11) is wrt the source point but equals the gradient wrt the field point if we change the sign. Now if the current loop is very small compared with the distance to the field point then r_{sf} is roughly constant over the integral and we can write,

$$\bar{A}(\bar{r}) = \frac{\mu_0}{4\pi}\bar{\nabla}\left(\frac{1}{r}\right)\times\iint_\Gamma Id\bar{S} \tag{5.5.12}$$

where we have shifted the origin to the position of the current loop. But $\bar{m} = \iint_\Gamma Id\bar{S}$ is just the magnetic dipole moment of the current loop, so we have,

$$\bar{A}(\bar{r}) = \frac{\mu_0}{4\pi}\bar{\nabla}\left(\frac{1}{r}\right) \times \bar{m} = \frac{\mu_0}{4\pi}\frac{\bar{m}\times\hat{r}}{r^2} = \bar{\nabla} \times \left(\frac{\mu_0 \bar{m}}{4\pi r}\right) \qquad (5.5.13)$$

where the last expression results from identity (A1.10) and the fact that $\bar{\nabla} \times \bar{m} = 0$. Recall that the gradient in (5.5.12,13) is wrt the field variable. An exercise for the reader is to check that (5.5.13) results in same magnetic flux density as derived using the scalar potential, namely (5.5.5).

5.6 Magnetic Energy

5.6.1 The Energy of a Magnetic Dipole in an External Magnetic Field: The Force and Torque on the Dipole

In this section we consider a fixed magnetic dipole, either a permanent magnet or a small loop of current in which the current is contrived to be kept constant independent of any external fields. Here we derive the energy associated with placing such a fixed dipole in a pre-existing (or "external") magnetic field. This does not include the self-energy of the dipole which will be considered later.

From (4.2.8) we know that the force on a current-carrying circuit, C, due to placing it in an external magnetic field is,

$$\bar{F} = I \oint_C d\bar{s} \times \bar{B} \qquad (5.6.1)$$

where $d\bar{s}$ is an element of length of the circuit, C. The external work required to displace the whole circuit by $\delta\bar{r}$ is thus,

$$\delta W = -\bar{F} \cdot d\bar{r} = -I \oint_C (d\bar{s} \times \bar{B}) \cdot d\bar{r} \qquad (5.6.2)$$

But by (A1.2), $(d\bar{s} \times \bar{B}) \cdot d\bar{r} = (d\bar{r} \times d\bar{s}) \cdot \bar{B}$ and $d\bar{r} \times d\bar{s}$ is an element of the area swept out by the circuit as it is displaced, $d\bar{S}$ (i.e., the annular region covered by the blue arrows in Figure 5.2). Hence $\oint_C (d\bar{s} \times \bar{B}) \cdot d\bar{r} = \oint_C \bar{B} \cdot d\bar{S}$. But because $\bar{\nabla} \cdot \bar{B} \equiv 0$, by the divergence theorem the quantity $\bar{B} \cdot d\bar{S}$ integrates to zero over the whole of the surface shown in Figure 5.2, i.e., including integrals over the surfaces whose boundary is the circuit (the two green circuits in Figure 5.2). But the integral of $\bar{B} \cdot d\bar{S}$ over one of the surfaces bounded by the circuit is the magnetic flux linked by the circuit, φ. Hence, it follows that the work done to move the circuit (i.e., the magnetic dipole) is $\delta W = -I\delta\varphi$ where $\delta\varphi$ is the change in the flux linkage due to the movement of the circuit by $\delta\bar{r}$. The energy required to move the circuit in a region with no external field is zero, so the energy of a current loop due to being placed in an external magnetic field can be taken to be,

$$U = -I\varphi \tag{5.6.3}$$

where φ is the magnetic flux linked by the circuit. The sign in (5.6.3) is important. It indicates that the circuit will tend to move, or rotate, so as to increase the flux linked. The force and torque on the dipole is thus,

$$\bar{F} = I\bar{\nabla}\varphi \qquad \text{and} \qquad T = I\frac{\partial\varphi}{\partial\theta} \tag{5.6.4}$$

Here, the angle θ is wrt the axis about which the torque acts. Hence, the torque equation provides a vectorial torque, despite appearances.

If the current loop is sufficiently small that the magnetic field can be taken as constant across it, then $\varphi \approx \bar{B} \cdot \Delta\bar{S}$, and the dipole moment is $\bar{m} = I\Delta\bar{S}$ so the energy of a fixed dipole in an external magnetic field is,

$$U = -\bar{B} \cdot \bar{m} \tag{5.6.5}$$

which is directly analogous to (5.4.12). This can now be taken to apply for a permanent magnet as well as a current loop. The force on such a dipole due to a magnetic field is thus,

$$\bar{F} = -\bar{\nabla}U = \bar{\nabla}(\bar{B} \cdot \bar{m}) = (\bar{m} \cdot \bar{\nabla})\bar{B} \tag{5.6.6}$$

which is analogous (5.4.13). Note that the last expression follows from the constancy of $\bar{m}$ and the fact that $\bar{\nabla} \times \bar{B} = 0$ outside the dipole, (A1.14). It follows that a dipole experiences no force in a field which is uniform in its vicinity. A force only arises due to the gradient of the field.

We note that, by analogy with the force on an electric dipole in an electric field, the force on a magnetic dipole in a magnetic field is the same as would result if a magnetic dipole were considered as a pair of equal and opposite magnetic monopoles a small distance apart.

The torque will be non-zero in general even for a constant field. As (5.6.5) can be written $U = -Bm\cos\theta$, it follows that the torque about the axis perpendicular to both $\bar{B}$ and $\bar{m}$, specifically parallel to $\bar{B} \times \bar{m}$, is $T = -Bm\sin\theta$, hence.

$$\bar{T} = \bar{m} \times \bar{B} \tag{5.6.6b}$$

directly analogous to (5.4.14). Hence, the torque acts to tend to decrease θ and so to bring the dipole into alignment with the magnetic field, thus minimising the potential energy, from (5.6.5). The torque is greatest when the dipole is perpendicular to the field.

5.6.2 The Energy of a Magnetic Field

(5.3.3) is the energy density associated with an electric field in free space. Here we derive a corresponding expression for a magnetic field. To do so we consider a current loop whose current is slowly increased. Suppose the current at some time is fI where I is the final current to be established and $0 \leq f \leq 1$. When the final current is established the magnetic flux density is $\bar{B}$ and the dipole moment is $\bar{m}$. At some intermediate time when the field is $f\bar{B}$ we increase the dipole moment by $\bar{m}\delta f$. The work required to do so is $f\bar{B} \cdot \bar{m}\delta f$. Integrating over $0 \leq f \leq 1$ gives the total work done to establish the final dipole to be $\frac{1}{2}\bar{B} \cdot \bar{m}$.

We now consider the magnetic dipole to be composed of some circulating distribution of current density, $\bar{J}$. A small section across this current flow has area $\Delta\sigma$ whilst the line vector element in the direction of the flow is $\delta\bar{s}$. A small element of area normal to the local plane of the circulating current, hence providing a contribution to the dipole moment, is $\delta\bar{S}$ (Figure 5.3)

The current flowing through the small area $\Delta\sigma$ is $J\Delta\sigma$ and the normal area vector is $\delta\bar{S}$ so the contribution of the latter to the dipole moment is $J\Delta\sigma\delta\bar{S}$. The work done to establish this is thus $\frac{1}{2}\bar{B} \cdot J\Delta\sigma\delta\bar{S}$. Hence, in terms of the vector potential the work done is,

$$W = \frac{J\Delta\sigma}{2}\int(\bar{\nabla} \times \bar{A}) \cdot \delta\bar{S} = \frac{J\Delta\sigma}{2}\oint \bar{A} \cdot d\bar{s} \qquad (5.6.7)$$

Figure 5.3: The Composition of the Distributed Dipole

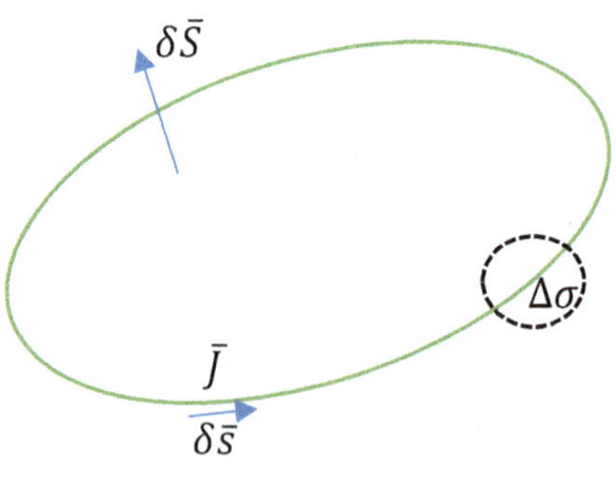

Here Stokes's theorem has been used to covert the area integral to a line integral. But note that $\Delta\sigma\, d\bar{s} = \hat{t}d\tau$ where $d\tau$ is a volume element (because $\Delta\sigma$ is perpendicular to $d\bar{s}$) and $\hat{t}$ is a unit vector parallel to the flow, i.e., locally parallel to both $d\bar{s}$ and $\bar{J}$. Hence, if we consider a volumetric distribution of current density, then (5.6.7) becomes,

$$W = \frac{1}{2}\int \bar{A} \cdot \bar{J} d\tau \tag{5.6.8}$$

We can now substitute the Maxwell equation $\bar{\nabla} \times \bar{B} = \mu_0 \bar{J}$, (5.1), (assuming no electric field) to get,

$$W = \frac{1}{2\mu_0}\int \bar{A} \cdot (\bar{\nabla} \times \bar{B}) d\tau \tag{5.6.9}$$

The integral in (5.6.9) may be extended to the whole of space, though the integrand is zero where there is no current. Vector identity (A1.11) allows us to rewrite the integrand as,

$$\bar{A} \cdot (\bar{\nabla} \times \bar{B}) = \bar{B} \cdot (\bar{\nabla} \times \bar{A}) - \bar{\nabla} \cdot (\bar{A} \times \bar{B})$$

But because the field associated with a localised dipole will itself be localised, and hence zero at infinity, the volume integral of the divergence will be zero when extended to the whole of space (because the divergence theorem means it equals the surface integral of $\bar{A} \times \bar{B}$ at infinity and the product of A and B falls off at least as fast as $1/r^3$). Hence,

$$W = \frac{1}{2\mu_0}\int \bar{B} \cdot (\bar{\nabla} \times \bar{A}) d\tau = \frac{1}{2\mu_0}\int B^2 \, d\tau \tag{5.6.10}$$

This applies for any localised distribution of current, and so we conclude that,

$$\text{Energy density of a magnetic field } u_M = \frac{1}{2\mu_0} B^2 \tag{5.6.11}$$

which is directly analogous to (5.3.3).

5.6.3 The Force, Moment and Energy between Two Magnetic Dipoles

The same expressions for the force, moment and energy between two magnetic dipoles, interacting through their magnetic fields, apply as found for electric dipoles in §5.4.7 provided we redefine,

$$F_0 = \frac{\mu_0 m_1 m_2}{4\pi r^4} \tag{5.6.12}$$

The expressions for electric field and the force, moment and energy associated with electric dipoles, i.e., (5.2.5), (5.4.19), (5.4.21a,b), (5.4.22), can therefore also be used for the magnetic dipole equivalents with the replacements $p \to m$ and $1/\varepsilon_0 \to \mu_0$.

5.6.4 Inductance

In this section we assume the absence of any magnetic media. By way of anticipating chapter 6, the effect of a ferromagnetic or paramagnetic material within an inductor can be taken into account by replacing μ_o wherever it occurs in the expressions below for inductance with $\mu_r\mu_o$ where μ_r is the relative permeability of the medium.

Just as capacitance refers to the storage of electrostatic energy due to a potential difference across the item in question, so inductance refers to the storage of magnetostatic energy due to the flow of current. We know that a steady current causes a steady magnetic field, e.g., see (4.2.17), and we know that a magnetic field is associated with an energy density, (5.6.11). Inductance quantifies the magnetostatic energy stored for a given current.

The Faraday-Heaviside law of induction, §4.3, tells us that the emf induced in a circuit equals the rate of change of the magnetic flux linked to it, and acts in a sense so as to oppose that change and so is written with a minus sign, $V_{induced} = -\frac{d\varphi}{dt}$. But the linked flux is proportional to the magnetic field which in turn is proportional to the current giving rise to it. Consequently, the induced emf, $V_{induced}$, is proportional to the rate of change of that current, but in opposition to it. Inductance, L, is the associated constant of proportionality,

$$V_{induced} = -\frac{d\varphi}{dt} = -L\frac{dI}{dt} \tag{5.6.13}$$

Consider a conducting item over which a potential different, V_a, is applied at time zero and consider the finite, if perhaps very short, time over which the current builds to its final steady value, $I_a = V_a/R$, where R is its resistance. Whilst the current is increasing, (5.6.13) gives the induced "back emf" which dies away as the current becomes steady, so the total potential across the item at any time is,

$$V = V_a + V_{induced} = V_a - \frac{d\varphi}{dt} = V_a - L\frac{dI}{dt} \tag{5.6.14}$$

Integrating (5.6.13) also shows that the inductance can be defined equivalently as the flux linked per unit current,

$$\varphi = LI \tag{5.6.14b}$$

Because V_a is the known (applied) potential difference, whilst the time-varying V might not be known without calculation or use of a special measuring device, (5.6.14) is often written,

$$V_a = V + L\frac{dI}{dt} = RI + L\frac{dI}{dt} \tag{5.6.15}$$

where I is the instantaneous, time-varying current. (5.6.15) can be thought of as the applied potential difference consisting of the sum of the resistance part, proportional to the instantaneous current, and the inductance part proportional to the rate-of-change of current (and in the same sense as that rate of change). The latter, inductance part, can thus be written,

$$V_{a,induced} = L\frac{dI}{dt} \tag{5.6.16}$$

We have laboured this point to explain why texts generally write the induced 'voltage' as in (5.6.16), i.e., without the minus sign that appears in (5.6.13). It is a matter of whether one is referring to the emf, which is a back-emf, (5.6.13), or the applied 'voltage', (5.6.16). If the resistance is small and the inductance large, $V_a \approx L\frac{dI}{dt}$ is approximately what a measuring device will register, albeit reducing over time as the current tends to a steady value. Texts often elide this point.

In passing we note that (5.6.15) is a differential equation in the current, the solution of which tells us how the current builds from zero if we assume that the applied potential difference rises instantaneously from zero to a constant value, V_a, at time zero. The solution is readily confirmed to be,

$$I = I_a\left(1 - e^{-Rt/L}\right) \tag{5.6.17}$$

where t is the time. Time will be measured in seconds if resistance and inductance are measured in MKSA units, see chapter 9. (5.6.17) shows how the current builds from zero to its steady value, I_a. Similarly, if the applied voltage is suddenly removed from steady conditions at time zero, the decay of the current over a finite time is given by,

$$I = I_a e^{-Rt/L} \tag{5.6.18}$$

The characteristic time, τ, over which the current changes by a fraction $1/e$ of its steady value is,

$$\tau = \frac{L}{R} \tag{5.6.19}$$

These results, (5.6.17-19), apply only when capacitance can be neglected. It is worth noting that the time variation of current for the case of an inductor (with negligible capacitance) is opposite to that of a capacitor (with negligible inductance). When a potential is applied suddenly to an inductor, the current increases from zero to a steady non-zero value; when a potential is applied suddenly to a capacitor, the current attains its maximum value essentially "immediately" and then decreases towards zero. The contrast between an inductor and a capacitor is illustrated by Figure 5.4.

Figure 5.4: Current versus Time, an Inductor and a Capacitor Contrasted for a Suddenly Applied Potential Difference (Charging) followed by Suddenly Removing the Potential at Time 4τ (Discharging by Shorting-Out). *Illustrated for* $\tau = 1$ *in both cases (i.e.,* $\frac{L}{R} = \frac{R}{C} = 1$*).*

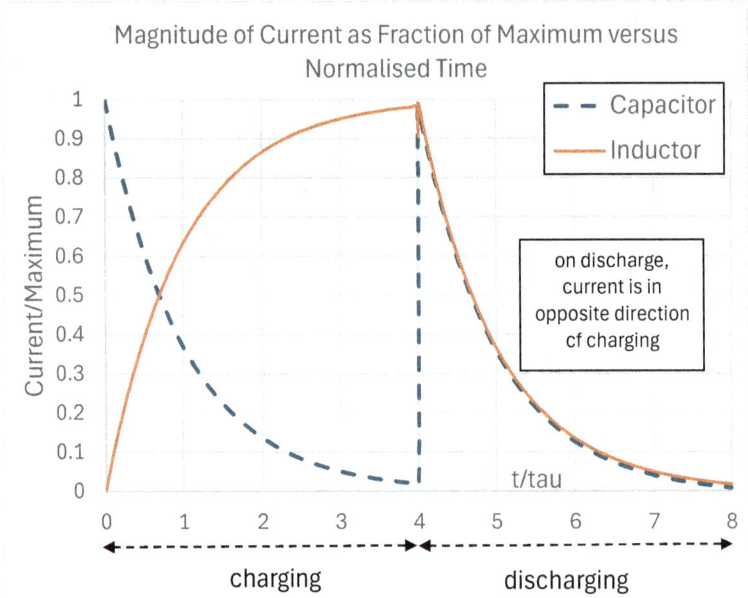

[Aside: Here and elsewhere when a current is assumed to arise "immediately" we know that this cannot be literally true because current consists of the motion of electrons, and electrons have finite inertia. However, typical electrical forces on electrons are huge compared with their mass, i.e., their acceleration is potentially prodigious but is terminated almost "immediately" by attaining a terminal drift velocity due to resistance (i.e., atomic scale interactions). Consequently, when electrons are subject to a sudden potential gradient they attain their terminal drift velocity typically in the order of a femtosecond. The whole issue of the "speed of electricity" is more subtle than many sources suggest, it being necessary to distinguish between electron velocity, electron drift speed and the rate of propagation of electrical potential or field. The latter will have to await the careful analyses of chapters 19 and 20, but the simpler matter of electron velocities can be addressed in elementary fashion, see Appendix A.8].

It is a familiar experience that, on unplugging a mains powered device, the power light on the device often does not go out instantly but takes a few seconds – perhaps many seconds – to fade gradually. This shows directly the effect of a finite time constant in the order of seconds when discharging the inductance of the device – the inductance usually arising from the coils of an external or internal transformer. In principle, devices with very large capacitance might show a similar slow-discharge effect, but in practice it is generally the inductance of transformer coils that dominate the effect.

A distinction is usually drawn between self-inductance and mutual inductance. The former applies when the induced emf refers to the same circuit in which the current is changing, and hence the associated linked flux arises from that circuit itself (self-linkage). Mutual inductance refers to an emf induced in circuit A, due to a changing flux linked to circuit A, but in which that flux is due to the current in a separate circuit B. The following section derives the inductance for a few simple geometries. These examples incidentally provide the solution for the magnetic flux density arising from simple circuit geometries.

5.7. Magnetic Field and Inductance of Simple Geometries

5.7.1 The Solenoid

A solenoid is a current-carrying wire formed into coils wrapped around the surface of a cylinder. An ideal solenoid is densely wound, so that the current is circumferential to a very good approximation, and has a uniform number of windings per unit axial length. The magnetic flux density inside a solenoid is very simple to calculate if the solenoid is very long compared to its diameter. We apply (4.2.20b), $\oint_C \bar{B}(\bar{r}_f) \cdot d\bar{r}_f = \mu_0 I$, to a rectangular integration path illustrated in Figure 5.5.

Figure 5.5

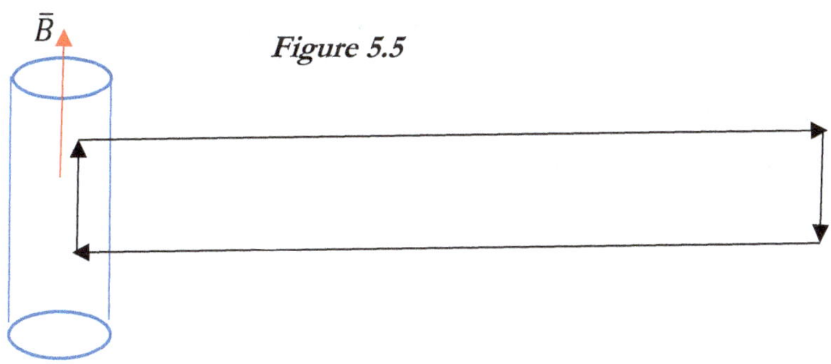

It consists of an axial part of length L within the solenoid, two radial parts which extend to infinity, and finally closed at infinity by an axially oriented part. At the latter part the field is zero and hence it does not contribute to the integral. Moreover, the radial and circumferential components of the field are zero by symmetry, so the radial parts of the integration path do not contribute either. As the field is purely axial, the integral is simply Bl. The total current linked within the closed integration path is $NI = nlI$ where N is the number of turns of wire in length l, and $n = N/l$ is the number of turns per unit length. Hence we get simply,

$$B = \mu_0 nI \qquad (5.7.1)$$

where it is understood that the field is axially oriented. Note that the integration path was not assumed to lie on the axis but could be at any radial position within the solenoid. Hence, the magnetic field is uniform throughout the interior of the solenoid, independent of radial or axial position. This applies, however, only so long as the axial position is not too near the ends of the solenoid. Near the ends there will be end effects.

The self-inductance of a solenoid per unit length can be found using (5.6.14b), noting that the flux linked per turn is $BA = \mu_0 nIA$ where A is the internal cross-sectional area of the solenoid. The flux linked per unit length is thus $\varphi = \mu_0 n^2 IA = LI$ hence,

Inductance per unit length: $$L = \mu_0 n^2 A \qquad (5.7.2)$$

The magnetic field outside a long solenoid is small – theoretically zero for an infinite solenoid. However, because $\overline{\nabla} \cdot \bar{B} = 0$ the lines of magnetic flux exist only in closed loops. The lines of flux must therefore return from the top end of the solenoid to the bottom end in a large external loop. But the longer the solenoid, the more spread-out these return loops are, so that the flux density may be arbitrarily small. That there must be a return loop for the flux should never be forgotten, though. That the flux exists in closed loops is why $\overline{\nabla} \cdot \bar{B} = 0$ is referred to as solenoidal.

5.7.2 Two Parallel Wires (Twin Cable)

Here we envisage a pair of parallel wires which form part of the same circuit and hence both are carrying the same current, I, but in opposite directions. Any linked flux is therefore that occurring in the space between the wires.

We know from (4.2.21) that the magnetic flux density due to a current flowing in one long straight wire is $\bar{B} = \frac{\mu_0 I}{2\pi r}\hat{\phi}$. If the radius of both wires is R and the distance between their centres is D, the total flux linked per unit length arising from the field produced by one wire is,

$$\varphi = \int_R^{D-R} \frac{\mu_0 I}{2\pi r} dr = \frac{\mu_0 I}{2\pi} log\left(\frac{D-R}{R}\right) \tag{5.7.3}$$

But both wires contribute an equal amount of flux (and in the same sense between the wires because their currents flow in opposite directions). Hence the inductance per unit length is,

$$L = \frac{\mu_0}{\pi} log\left(\frac{D-R}{R}\right) \tag{5.7.4}$$

Note that this is divergent as the wire becomes infinitely thin (but so is its resistance, and not merely logarithmically). The reason is that the field is formally divergent as $r \to 0$ if such a wire continued to carry the same, non-zero, current – but this is clearly unphysical as the current density would be divergent.

5.7.3 Concentric Cylinders / Coaxial Cable

This is essentially the same as §5.7.2 with $R \to R_1$ now interpreted as the outer radius of the inner cylinder and $D - R \to R_2$ is now interpreted as the inner radius of the outer cylinder. However, there is a qualitative difference from the twin cable case because only the inner cylinder produces a magnetic field in the gap between the cylinders (because we have seen in §4.2.3 that the outer cylinder produces no field within it). Hence the inductance per unit length of a coaxial cable or pair of concentric cylinders is,

$$L = \frac{\mu_0}{2\pi} log\left(\frac{R_2}{R_1}\right) \tag{5.7.5}$$

Note that this estimate neglects the flux within the material of the wires themselves. This will be accurate when the wires are thin. An exact solution can be devised for the case that the inner cylinder has inner and outer radii R_{1i}, R_{1o} and the outer cylinder has inner and outer radii R_{2i}, R_{2o} together with the assumption that the current density is uniform in both wires and that the total current in the wires is equal and opposite, i.e., they form a closed circuit. It is left as an exercise for the reader to show that the resulting inductance per unit length is,

$$L = \frac{\mu_0}{2\pi}\left\{\frac{1}{R_{1o}^2 - R_{1i}^2}\left[R_{1o}^2 log\frac{R_{2i}}{R_{1o}} - R_{1i}^2 log\frac{R_{2i}}{R_{1i}}\right] + \frac{R_{2o}^2}{R_{2o}^2 - R_{2i}^2} log\frac{R_{2o}}{R_{2i}}\right\} \tag{5.7.6}$$

In practice, however, (5.7.5) is very accurate, and even for thick cylinders overestimates the inductance by only a few percent, see Figure 5.6.

Figure 5.6: *The inductance per unit length of a pair of coaxial cylinders of equal finite thickness. The inductance is normalised by equ.(5.7.5) and the thickness is normalised by* R_{1o}

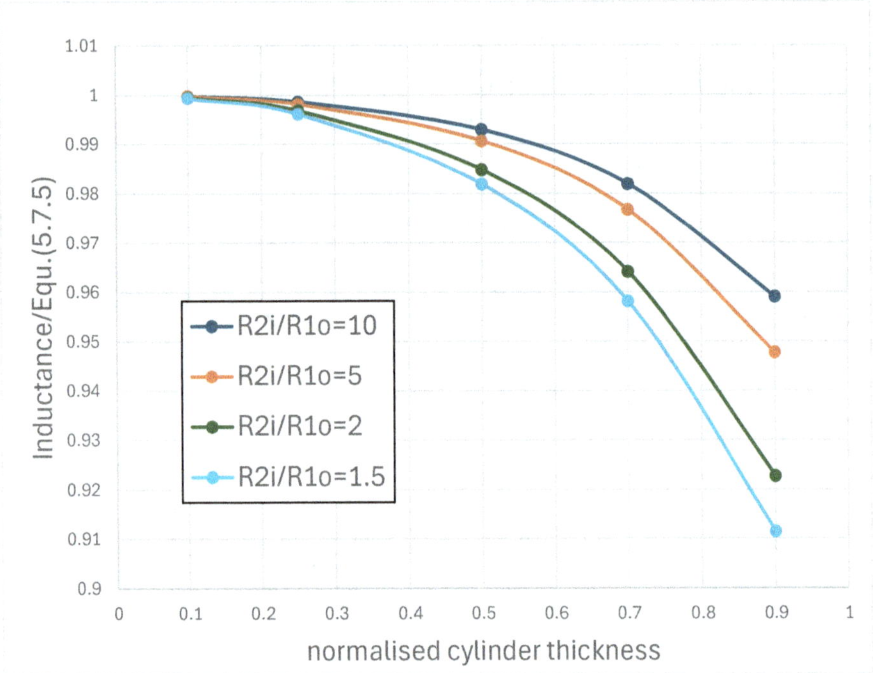

5.7.4 A Single Straight Cylinder or Wire

Many sources give a result for the inductance of a single wire, or cylinder, carrying a current. But the present author regards this as a dangerously misleading undertaking. It is important to include the return path for the current, i.e., the complete circuit, in order for the inductance to be well defined. For example, in (5.7.6) we cannot obtain a sensible inductance for the inner cylinder alone by pushing the outer cylinder off to infinity, $R_{2i}, R_{2o} \to \infty$. The result would be divergent (because the flux is divergent if $B \propto 1/r$ is integrated to infinity). It is crucial to truncate the flux at a finite distance, which requires the return path for the current to be included.

5.7.5 A Circular Planar Coil

The coil of radius R is assumed to have N turns. The magnetic flux density at the centre of the coil can be found by integrating the Biot-Savart expression, (4.2.7), all elements of length along the coil being the same, namely $r\delta\theta$ where r is the coil radius. Moreover, symmetry means that all elementary contributions to the magnetic field at the centre are oriented axially and so their magnitudes are simply additive. Accounting for the N turns, the flux density at the centre is thus $\frac{\mu_0 NI}{4\pi}\int_0^{2\pi}\frac{1}{R^2}Rd\theta = \frac{\mu_0 NI}{2R}$. Unfortunately this does not hold away from the centre, as the lack of axisymmetry undermines that simple derivation. However, if we are satisfied with an approximate calculation we can proceed by assuming that the flux density does not differ from uniformity too much, in which case the self-linked flux, again accounting for the N turns, is about $\varphi = \frac{\pi}{2}\mu_0 N^2 IR$. An approximate expression for the self-inductance is thus,

Approximate: $$L = \frac{\pi}{2}\mu_0 N^2 R \qquad (5.7.7)$$

The exact solution for the field off the axis of the coil is available and is expressed in terms of complete elliptic integrals. The solution is given in [5.2]. The magnetic flux density diverges as the radial position approaches the coil, i.e., $r \to R$, as shown in Figure 5.7. A fitted approximation for the B field has been included in Figure 5.7 for comparison. It is given by,

Fit: $$B(r) = \frac{\mu_0 NI}{2R}\left(1 + \tan\left[\frac{\pi}{2}\left(\frac{r}{R}\right)^{2.4}\right]\right) \qquad (5.7.8)$$

We shall return to the exact solution for the planar coil in chapter 20.

Figure 5.7: *Magnetic flux density on the plane of a coil as a function of the radial position, normalised by the field at the centre,* $\frac{\mu_0 NI}{2R}$

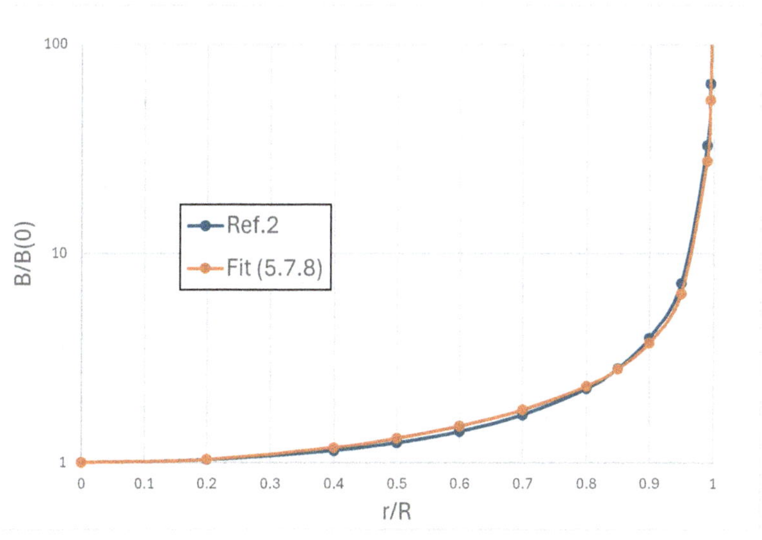

5.7.6 Extension Reels

There is a common household safety issue associated with inductance which is worth highlighting here. Mains extension cables, wound on a convenient drum, are midway between a planar coil and an ideal solenoid – actually a short solenoid. (5.7.2) and (5.7.7) would give the same result if the axial length of the solenoid was the same as the diameter, but typically the drums are shorter than that. Hence we can use (5.7.7) as a rough guide to the inductance of the fully wound cable. A typical domestic extension reel will have 25 metres of cable wound on a drum so that the average coil diameter is about 18cm, and hence has 44 turns. (5.7.7) gives the self-inductance to be 3.4×10^{-4} Henry (H), where we have used $\mu_0 = 4\pi \times 10^{-7}$ H/m. Units and the values of the universal constants will be discussed in detail in chapter 9. For now merely note that the Henry is the SI/MKSA unit for inductance. The significance of this inductance will be made clear in chapter 18 when we address AC circuits.

For now note the safety warning which such extension reels invariably include (in the UK at least): Unwound 13A (3.12 kW), wound 5A (1.2 kW). As I know to my cost, this is a warning you should not ignore. Attaching a 3 kW load to such a cable in the nearly fully wound condition will lead to the cable melting, writing-off your extension lead and potentially causing a serious fire. These reels often claim to have a thermal cut-out. Both safety considerations and personal experience suggest you should not rely on such cut-outs.

5.7.7 One Solenoid Inside Another (Mutual Inductance)

All the preceding examples have been of self-inductance. We now consider a couple of examples of mutual inductance between two separate circuits. In this first example we envisage a smaller solenoid inside a larger solenoid. The former is assumed to be both of smaller radius and of shorter length so that the flux density within the smaller solenoid due to the larger is uniform. We are interested in the mutual inductance, ignoring the self-inductance of either solenoid. If the number of turns per unit length of the larger solenoid is n_2 and the number of turns per unit length of the smaller solenoid is n_1, and the length and cross-sectional area of the smaller solenoid are L_1 and A_1, the mutual inductance follows from (5.7.1),

$$L = \mu_0 n_1 n_2 A_1 L_1 \tag{5.7.9}$$

5.7.8 Two Planar Coils on the Same Axis

Here we consider the mutual inductance of a small circular coil, of radius r_1 and N_1 turns, on the same axis as a larger circular coil, of radius $r_2 > r_1$ and N_2 turns. The coils are initially assumed coplanar. If the small coil is sufficiently small that the flux density due to the larger coil is essentially uniform across it, then the flux density is $\frac{\mu_0 N_2 I_2}{2r_2}$ and so the flux linked to the small coil is $\varphi = \frac{\mu_0 N_1 N_2 I_2}{2r_2}\pi r_1^2$ so the mutual inductance is,

$$L = \frac{\mu_0 \pi N_1 N_2}{2} \cdot \frac{r_1^2}{r_2} \tag{5.7.10}$$

We can generalise this result to the case that the small coil is at an axial distance D from the centre of the larger coil, but still on-axis and parallel. We first need to find the flux density due to the larger coil at the axial distance D. This can be done from the Biot-Savart field, (4.2.7), all elements of length along the coil being the same, namely $r_2\delta\theta$. But the resulting element of $\bar{B}$-field is now at an angle α to the axis where $\cos\alpha = \frac{r_2}{\sqrt{r_2^2+D^2}}$. The component of the field normal to the axis will integrate to zero, by symmetry. The total axial field is, after accounting for the N_2 turns and noting that $\cos\alpha$ is constant wrt the integration variable, as is the source-field distance,

$$\frac{\mu_0 N_2 I_2}{4\pi}\int_0^{2\pi}\frac{\cos\alpha}{r_2^2+D^2} r_2 d\theta = \frac{\mu_0 N_2 I_2}{2} \cdot \frac{r_2^2}{\left(r_2^2+D^2\right)^{3/2}} \tag{5.7.11}$$

Multiplying (5.7.11) by $\pi N_1 r_1^2$ gives the flux linked to the small coil, and hence the mutual inductance is,

$$L = \frac{\mu_0 \pi N_1 N_2}{2} \cdot \frac{r_1^2 r_2^2}{\left(r_2^2+D^2\right)^{3/2}} \tag{5.7.12}$$

If the plane of the small coil is at an arbitrary angle β to the plane of the larger coil, but still on the same axis, then the mutual inductance is obtained simply by multiplying (5.7.12) by $\cos\beta$.

5.7.9 The Magnetic Energy of Inductors

So far we have glossed over an issue that may have occurred to the reader as a problem with mutual inductance. Mutual inductance may be defined as the flux linked to the first circuit by the flux density generated by the second circuit. Or the mutual inductance might alternatively be defined as the flux linked to the second circuit by the flux density generated by the first circuit.

Unless these always produce the same result, then the mutual inductance would be ill-defined. That they do necessarily always produce the same result is proved in this section as a by-product of the quantification of the magnetic field energy stored by inductance.

Consider firstly the energy stored by self-inductance. The power delivered to the circuit in question is $V_a I$ where V_a is the potential difference applied and I is the current. But (5.6.15) gives $V_a = RI + L\frac{dI}{dt}$ so the supplied power is,

$$\text{Power} = RI^2 + LI\frac{dI}{dt} \tag{5.7.13}$$

The first term in (5.7.13) is the power dissipated as heat due to resistance. Integrating the second term over time yields the energy stored as a magnetic field due to the circuit's self-inductance,

$$\text{Stored energy due to self-inductance} = \frac{1}{2}LI^2 \tag{5.7.14}$$

The reason why this energy can be considered as stored, rather than dissipated, is that the same analysis applies also on discharging the inductor (as per Figure 5.4) yielding the same energy given back. This is, of course, an idealisation based on neglecting resistance losses which will, in reality, always occur.

Consider now the mutual inductance between two circuits. Suppose the potential induced in circuit 1 due to changing the current in circuit 2 is given by $V_1 = L_{12}\frac{dI_2}{dt}$. The power in circuit 1, ignoring the resistive part, is $V_1 I_1 = L_{12} I_1 \frac{dI_2}{dt}$. Time integration of this gives,

$$\text{Stored energy due to mutual inductance} = L_{12} I_1 I_2 \tag{5.7.15}$$

The calculation of the same stored energy could also be carried out by starting with the potential induced in circuit 2 due to changing the current in circuit 1 as given by $V_2 = L_{21}\frac{dI_1}{dt}$, yielding,

$$\text{Stored energy due to mutual inductance} = L_{21} I_1 I_2 \tag{5.7.16}$$

But (5.6.34) and (5.6.35) must be equal so we conclude that $L_{12} = L_{21}$.

The total stored energy of the two circuits, due to both their self-inductances and their mutual inductance, is,

$$\text{Stored energy due to inductance} = \frac{1}{2}L_1 I_1^2 + \frac{1}{2}L_2 I_2^2 + L_{12} I_1 I_2 \tag{5.7.17}$$

This can be extended to any number of circuits which have magnetic linking,

Stored energy due to inductance $= \frac{1}{2}\sum_i L_i I_i^2 + \sum_i \sum_{j>i} L_{ij} I_i I_j$ (5.7.18)

5.8 References

[5.1] Bradford, R.A.W. (2013). *Classical Physics Needs Renormalisation Too! http://rickbradford.co.uk/ClassicalPhysicsNeedsRenormalisationToo.pdf*, 1 April 2013.

[5.2] Montgomery, D.B., and Terrell, J. (1961). *Some Useful Information for the Design of Air-Core Solenoids. Armed Service Technical Information Agency*. AFOSR-1525, November 1961. A convenient presentation and code with numerically calculated graphs is here.

6. Electromagnetic Fields in Media

So far we have addressed only electromagnetic fields in a vacuum (which is often a good enough approximation for e/m fields in atmospheric air). But electromagnetic fields can be modified greatly when they pass through a medium. This gives rise to a variant of the electric field strength ($\bar{E}$) *called the "electric displacement"* ($\bar{D}$)*, and a variant of the magnetic flux density* ($\bar{B}$) *called the "magnetic field strength"* ($\bar{H}$)*. These new concepts are defined here in terms of the corresponding properties of the medium: the electric susceptibility, the relative permittivity (also called the dielectric constant), the magnetic susceptibility and the relative permeability. Finally, the boundary conditions applicable to the four fields* $\bar{B}, \bar{D}, \bar{E}, \bar{H}$ *are derived when the boundary lies between media of differing dielectric constant or magnetic susceptibility, including the influence of possible surface charges or surface current densities.*

Contents

Throughout this chapter the media in question are assumed to be homogeneous, isotropic and a continuum, thus avoiding complications associated with molecular or atomic structure of real materials which will necessarily depart from homogeneity and isotropy at small size scales. In particular, the assumption of isotropy allows the material properties (electric and magnetic susceptibility and the relative permittivity and permeability) to be introduced as scalars rather than as second rank tensors, which would more generally be the case. We shall also assume linearity between the applied field and its effect on the medium, which is another way of saying that the above material properties are assumed to be constants independent of field strength. These conditions define LIH (Linear, Isotropic and Homogeneous) media. These assumptions are made in the service of pedagogic simplicity rather than empirical accuracy and can easily be relaxed once the basic concepts are understood.

6.1 Electrical Fields in Dielectric Media

A medium is said to be dielectric if each element of its volume tends to polarise under the action of an applied electric field. It is to be expected that such polarisation will occur if the molecules of the material are either themselves polarised or contain charges that can displace to cause polarisation. In the latter case the application of an electric field will tend to move positive charges in its direction, and negative charges in the opposite direction, hence resulting in polarisation. In the former case the polarised molecules, effectively small dipoles, will initially be randomly oriented and hence result in no net dipole moment. We have seen in chapter 5 that these molecular dipoles will tend to align with an applied field, thus resulting in a net dipole moment. In either case what results is a certain dipole moment per unit volume, $\bar{P}$, called the "polarisation", so that the dipole moment associated with volume element $\delta\tau$ is $\delta\bar{p} = \bar{P}\delta\tau$.

The polarisation induced by an applied electric field, $\bar{E}$, is assumed to be linearly related to the field via a material constant, the electric susceptibility, χ_e, thus,

$$\bar{P} = \chi_e \varepsilon_0 \bar{E} \tag{6.1.1}$$

Recalling that the dimensions of dipole moment are charge times length, and recalling (4.2.4) that applies in the MKSA system of units, the reader should check that the units of polarisation are the same as $\varepsilon_0 E$ and hence that the electric susceptibility, χ_e, is a dimensionless number.

It is important to emphasise that the field in (6.1.1) does not include any field created by the polarisation itself. This is often denoted by the term "external field". It is conceptually simplest to assume that whatever process is used to create the external field can be contrived to maintain it constant whilst the polarisation takes place. However, we must acknowledge that the polarisation will also cause an electric field to arise, which we denote $\bar{E}_P$. We know what electric potential arises from a dipole, from (5.2.3), and integrating over volume we get the total potential due to the polarised medium to be,

$$\Phi(\bar{r}_f) = \frac{1}{4\pi\varepsilon_0}\int \bar{P}\cdot\bar{\nabla}_s\left(\frac{1}{r}\right)d\tau \tag{6.1.2}$$

Here the sign has changed from that of (5.2.3) because we have changed the gradient operator to act on the source point, over which we are integrating. From identity (A1.9) we have $\bar{P}\cdot\bar{\nabla}_S\left(\frac{1}{r}\right) = \bar{\nabla}_S\cdot\left(\frac{\bar{P}}{r}\right) - \frac{1}{r}\bar{\nabla}_S\cdot\bar{P}$. Considering the polarisation at any location $\bar{r}_s$ the term r then means $r = \left|\bar{r}_f - \bar{r}_s\right|$

The integral over volume of the first term can be transformed using the divergence theorem into a surface integral, thus we get,

$$\Phi(\bar{r}_f) = \frac{1}{4\pi\varepsilon_0}\left\{-\int_V \frac{1}{|\bar{r}_f-\bar{r}_s|}\bar{\nabla}_S\cdot\bar{P}d\tau + \oint_{\delta V}\frac{\bar{P}\cdot d\bar{S}}{|\bar{r}_f-\bar{r}_s|}\right\} \tag{6.1.3}$$

where $d\tau$ and $d\bar{S}$ refer to integrations wrt the source point $\bar{r}_s$. Now (5.1.5) shows how the electric potential arises from a distribution of charges, and (5.1.6) shows what this becomes if the charge exists as a volume distribution. If there were also a distribution of charge over a surface, with surface charge density per unit area σ, then (5.1.6) would generalise to,

$$\Phi(\bar{r}_f) = \int\frac{\rho(\bar{r})}{4\pi\varepsilon_0|\bar{r}_f-\bar{r}_s|}d\tau + \int\frac{\sigma(\bar{r})}{4\pi\varepsilon_0|\bar{r}_f-\bar{r}_s|}dS \tag{6.1.4}$$

Comparison with (6.1.3) shows that the effect of the polarisation is identical to that of a charge density of $\rho_P = -\bar{\nabla}\cdot\bar{P}$ plus a surface charge density of $\sigma_p = \bar{P}\cdot\hat{n}$ where $\hat{n}$ is the outward normal to the surface of the medium. Thus, if we replace the actual charge density, ρ, which causes the external field, with $\rho + \rho_P$, and any surface charge density with $\sigma + \sigma_p$, then treating these modified charge densities as if they were real will result in the correct potential and hence the correct total electric field. Hence, (4.2.5) becomes, for the total field,

$$\overline{\nabla} \cdot \overline{E} = \frac{\rho + \rho_P}{\varepsilon_0} = \frac{\rho - \overline{\nabla} \cdot \overline{P}}{\varepsilon_0} \tag{6.1.5}$$

or,

$$\overline{\nabla} \cdot (\varepsilon_0 \overline{E} + \overline{P}) = \rho \tag{6.1.6}$$

This motivates the introduction of the new field quantity, the "electric displacement", defined as,

$$\overline{D} = \varepsilon_0 \overline{E} + \overline{P} \tag{6.1.7}$$

So that

$$\overline{\nabla} \cdot \overline{D} = \rho \tag{6.1.8}$$

is always true, even in the presence of a dielectric medium, unlike (4.2.5) which applies only *in vacuo*. Using (6.1.1) we have,

$$\overline{D} = (1 + \chi_e)\varepsilon_0 \overline{E} = \varepsilon_r \varepsilon_0 \overline{E} \tag{6.1.9}$$

Here $\varepsilon_r = 1 + \chi_e$ defines the "relative permittivity", which is a dimensionless number. The relative permittivity is also called the dielectric constant. The combination $\varepsilon = \varepsilon_r \varepsilon_0$ is called the permittivity of the material.

Note that (4.2.5d) for the normal electric field due to a surface charge density will now become a difference of field on the two sides of $\Delta E_n = \frac{\sigma + \sigma_p}{\varepsilon_0}$, so the normal E-field will be discontinuous over the boundary of a dielectric in an applied field even if there is no real surface charge.

The relative permittivity of plastics, rubber, glass and other solid electrical insulators is typically in the range 2 to 10. For organic liquids, e.g., alcohols, it is typically 30 to 40, whilst in ionic liquids, e.g., water and aqueous solutions, 50 to 150 is typical. There are exceptional substances with far higher permittivity, from thousands to hundreds of thousands.

6.2 Electrostatic Energy with Dielectric Material

Imagine a small increase in the charge distribution, $\delta\rho$, resulting in a small increase in the electric displacement, $\delta\overline{D}$. (6.1.8) gives $\overline{\nabla} \cdot \delta\overline{D} = \delta\rho$. But electrostatic energy is the product of charge and potential, so that the increase in the total electrostatic energy over all space is,

$$U_E = \int_V \Phi \delta\rho d\tau = \int_V \Phi \overline{\nabla} \cdot \delta\overline{D} d\tau \tag{6.2.1}$$

Recalling from (A1.9) that $\overline{\nabla} \cdot (\Phi \delta\overline{D}) = \Phi \overline{\nabla} \cdot \delta\overline{D} + \delta\overline{D} \cdot \overline{\nabla}\Phi$ and also from (5.1.1) that $\overline{E} = -\overline{\nabla}\Phi$ then,

$$U_E = \int_V (\overline{\nabla} \cdot (\Phi \delta\overline{D}) + \overline{E} \cdot \delta\overline{D}) d\tau = \int_V \overline{E} \cdot \delta\overline{D} d\tau + \oint_{\delta V} \Phi \delta\overline{D} \cdot d\overline{S} \tag{6.2.2}$$

Provided the field is confined to a finite region the surface integral will be zero. Then integrating over the increasing build-up of charge, since $\bar{E}$ and $\bar{D}$ are proportional for linear dielectrics, we get,

$$U_E = \frac{1}{2}\int_V \bar{E} \cdot \bar{D} d\tau \qquad (6.2.3)$$

In other words, the electrostatic energy density which generalises (5.3.3) when dielectric material is present, is,

Energy density of an electric field = $u_E = \frac{1}{2}\bar{E} \cdot \bar{D} = \frac{\varepsilon_r \varepsilon_0}{2} E^2$ (6.2.4)

This may also be written $u_E = \frac{\varepsilon_0}{2} E^2 + \frac{1}{2}\bar{E} \cdot \bar{P}$ (6.2.5)

This can easily be misinterpreted to mean that the potential energy per unit volume of the dielectric due to its dipole moment is $\frac{1}{2}\bar{E} \cdot \bar{P}$. But we know from (5.4.15) that there is a minus sign in the correct expression (and this is essential in causing the dipole to tend to align with the field). The resolution of this conundrum is that $\bar{E}$ in this section includes the field due to the dipole, as indicated by (6.1.5), whereas (5.4.15) does not. If the two are consistent then (6.2.4), or (6.2.5), minus its value when there is no dielectric, should agree with $-\frac{1}{2}\bar{E}_0 \cdot \bar{P}$, including the minus sign, where $\bar{E}_0$ is the field without any dielectric. Hence, we need to find,

$$U_E = \frac{1}{2}\int_V (\bar{E} \cdot \bar{D} - \bar{E}_0 \cdot \bar{D}_0) d\tau$$

$$= \frac{1}{2}\int_V (\bar{E} \cdot \bar{D}_0 - \bar{E}_0 \cdot \bar{D}) d\tau + \frac{1}{2}\int_V (\bar{E} + \bar{E}_0) \cdot (\bar{D} - \bar{D}_0) d\tau \qquad (6.2.6)$$

The utility of this transformation is that the second integral is zero because (6.1.8) shows that $\bar{D}$ depends only on the (real) charge density, but not on the presence of the dielectric, i.e., $\bar{D} = \bar{D}_0$. Hence, we get,

$$U_E = \frac{1}{2}\int_V (\bar{E} - \bar{E}_0) \cdot \bar{D}_0 d\tau \qquad (6.2.7)$$

But $\bar{D} = \bar{D}_0 = \varepsilon_0 \bar{E}_0 = \varepsilon_0 \bar{E} + \bar{P}$ so that $\varepsilon_0(\bar{E} - \bar{E}_0) = -\bar{P}$ and (6.2.7) becomes,

$$U_E = -\frac{1}{2}\int_V \bar{E}_0 \cdot \bar{P} d\tau \qquad (6.2.8)$$

QED, confirming (5.4.15), especially as regards the sign.

6.3 Magnetic Fields in Magnetic Media

Magnetic media are characterised by a magnetic dipole density, $\bar{M}$, called its "magnetisation", so that the dipole moment of a volume element $\delta\tau$ is $\delta\bar{m} = \bar{M}\delta\tau$. This magnetisation may be permanent or induced only when an external field is applied. The formulation of the effect of magnetic materials on magnetic fields can be developed in analogy with the effect of dielectrics on electric fields. However, we do not start with the definition of magnetic susceptibility but only introduce that quantity later, for reasons which will become clear shortly.

From (5.5.13) the magnetic vector potential arising due to the magnetisation of a medium of volume V is,

$$\bar{A}(\bar{r}) = \frac{\mu_0}{4\pi}\int_V \bar{M} \times \bar{\nabla}\left(\frac{1}{r}\right) d\tau \tag{6.3.1}$$

where we have changed the sign of (5.5.13) because the gradient is taken to operate on the source variable. We now use identity (A1.10) which gives us,

$$\bar{\nabla} \times \left(\frac{\bar{M}}{r}\right) = \frac{1}{r}\bar{\nabla} \times \bar{M} - \bar{M} \times \bar{\nabla}\frac{1}{r}$$

and hence,

$$\bar{A}(\bar{r}) = \frac{\mu_0}{4\pi}\int_V \left(\frac{1}{r}\bar{\nabla} \times \bar{M} - \bar{\nabla} \times \left(\frac{\bar{M}}{r}\right)\right) d\tau \tag{6.3.2}$$

Using identity (A1.21) the second term may be transformed into a surface integral, giving,

$$\bar{A}(\bar{r}) = \frac{\mu_0}{4\pi}\left\{\int_V \frac{1}{r}\bar{\nabla} \times \bar{M} d\tau + \oint_{\delta V} \frac{\bar{M}}{r} \times d\bar{S}\right\} \tag{6.3.3}$$

From (5.5.7) and (5.5.10b) we see that the vector potential produced by the magnetised medium is equivalent to a current density, $\bar{J}_M$, plus a surface current density, $\bar{\varsigma}_M$, where,

$$\bar{J}_M = \bar{\nabla} \times \bar{M} \quad \text{and} \quad \bar{\varsigma}_M = \bar{M} \times \hat{n} \tag{6.3.4}$$

where $\hat{n}$ is the outward normal of the surface of the medium. It follows that the magnetic flux density, $\bar{B}_M$, arising from the magnetisation, which is the $curl$ of (6.3.3), equals that which would arise from real current densities given by (6.3.4). The total magnetic flux density can thus be written $\bar{B} = \bar{B}_0 + \bar{B}_M$ where $\bar{B}_0$ is the "external" flux density which would exist prior to the introduction of the medium (and is assumed maintained constant). Hence, from the Maxwell equation $\bar{\nabla} \times \bar{B} = \mu_0 \bar{J}_{total}$, (5.1), which applies when any electric fields are static, we have,

$$\bar{\nabla} \times \bar{B} = \mu_0(\bar{J} + \bar{J}_M) = \mu_0(\bar{J} + \bar{\nabla} \times \bar{M}) \tag{6.3.5}$$

Hence,
$$\bar{\nabla} \times \left(\frac{\bar{B}}{\mu_0} - \bar{M}\right) = \bar{J} \tag{6.3.6}$$

We therefore define the "magnetic field strength", $\bar{H}$, as,

$$\bar{\mathrm{H}} = \frac{\bar{B}}{\mu_0} - \bar{M} \tag{6.3.7}$$

So that,
$$\bar{\nabla} \times \bar{H} = \bar{J} \tag{6.3.8}$$

Just as (6.1.8), $\bar{\nabla} \cdot \bar{D} = \rho$, always applies even in the presence of dielectric material (whereas $\bar{\nabla} \cdot \bar{E} = \rho/\varepsilon_0$ applies only in the absence of such material) so (6.3.8) always applies even in the presence of magnetic material (whereas $\bar{\nabla} \times \bar{B} = \mu_0 \bar{J}$ applies only in the absence of such material). To be clear, ρ is the real charge density (not including ρ_P) and $\bar{J}$ is the real current density (not including $\bar{J}_M$).

Now we see why we had to delay the introduction of the magnetic susceptibility, χ_m. Just as the electrical susceptibility was defined in terms of the electric field strength, $\bar{E}$, (6.1.1), so the magnetic susceptibility is defined in terms of the magnetic field strength, $\bar{H}$, and so we had to delay until the latter was defined. So now we define magnetic susceptibility via,

$$\bar{M} = \chi_m \bar{H} \tag{6.3.9}$$

From (6.3.7)
$$\bar{B} = (1 + \chi_m)\mu_0 \bar{H} = \mu_r \mu_0 \bar{H} \tag{6.3.10}$$

where $\mu_r = 1 + \chi_m$ is the "relative permeability" of the magnetic material. The reader should check, from for example (4.2.7), that the units of $\frac{\bar{B}}{\mu_0}$ and $\bar{\mathrm{H}}$ and $\bar{M}$ are all current/length, and hence that χ_m and μ_r are dimensionless numbers.

6.4 Magnetic Field Energy with Magnetic Materials

Outside any magnetic material, hence where $\bar{M} = 0$, the field energy density due to an "external" flux density is given by $\frac{1}{2\mu_0}B^2 = \frac{1}{2}\bar{B} \cdot \bar{H}$, see (5.6.11). This expression also applies within magnetic material if we ignore the field created by the magnetisation, i.e., ignoring the energy to create the magnetisation the energy density within the medium would be $\frac{1}{2}\bar{B}_0 \cdot \bar{H}$, noting that $\bar{H}$ is unchanged by the medium. But we know from §5.6.2 that the energy required to establish an induced dipole by an "external" flux density, $\bar{B}_0$, is $\frac{1}{2}\bar{B}_0 \cdot \bar{m}$ and this becomes an energy requirement per unit volume of $\frac{1}{2}\bar{B}_0 \cdot \bar{M}$ for induced magnetisation. From (6.3.9) this can be written $\frac{1}{2}\chi_m \bar{B}_0 \cdot \bar{H}$. The total energy density within the magnetic material is therefore $\frac{1}{2}\bar{B}_0 \cdot \bar{H} + \frac{1}{2}\chi_m \bar{B}_0 \cdot \bar{H} = \frac{1}{2}(1+\chi_m)\mu_0 \bar{H} \cdot \bar{H} = \frac{1}{2}\bar{B} \cdot \bar{H}$. Hence we conclude that the expression for the energy density of a magnetic field,

$$U_M = \frac{1}{2}\bar{B} \cdot \bar{H} \tag{6.4.1}$$

applies both within and outside of magnetic materials. This is analogous to (6.2.4) for the energy density of the electric field including the effect of any dielectric.

6.5 Types of Magnetic Material

It is beyond the scope of this work to elucidate the atomic level origins of the different types of magnetic media. However, it is worth the reader being aware that such media fall into five classes, as follows.

Ferromagnetic media are characterised by having large relative permeability (large magnetic susceptibility). So-called "hard" ferromagnetic materials are capable of forming strong permanent magnets, in contrast to "soft" ferromagnetic materials. Initially named after iron, of course, there are also other ferromagnetic elements (not to mention compounds and alloys), including nickel and cobalt. Some typical magnetic susceptibilities at room temperature are: pure iron up to 200,000, typical iron 6,000, mu-metal up to 60,000, annealed ferritic stainless steel 1,000 to 1,800 and nickel: 100 to 600. In contrast, austenitic stainless steel has a magnetic susceptibility of about 0.05 and hence is paramagnetic rather than ferromagnetic. (In practice, notionally austenitic stainless steels are likely to contain some grains which are ferritic, e.g., delta-ferrite or magnetite, and so the effective magnetic susceptibility will be larger than this).

Contrary to popular belief, pure neodymium is not used for the powerful permanent magnets in speakers and wind turbine generators, as it has a relative permeability of only about 1.05 (i.e., paramagnetic). Rather it is the compound with iron and boron, $Nd_2Fe_{14}B$, which is very strongly ferromagnetic for reasons of the crystalline structure of this compound, not the intrinsic properties of pure neodymium.

Ferrimagnetism is distinguished from ferromagnetism in having rather smaller magnetic susceptibilities, but well above the paramagnetic level. In terms of structure, ferromagnetic materials have atomic or molecular level spins aligned whereas ferrimagnetic materials have spins of alternating sign, but one sign is stronger than the other. Ferrimagnetic media can display permanent magnetisation, in fact spontaneous magnetisation due to their atomic structure. The most familiar example is magnetite (Fe_3O_4). This is the compound which gives lodestone its magnetisation and hence was important in the original discovery of magnetism.

Antiferromagnetic media do not retain permanent magnetisation, though they may behave in a similar fashion to ferrimagnets when an external magnetic field is applied. An example is chromium. At very low temperatures, neodymium is also antiferromagnetic.

Ferromagnetic and ferrimagnetic behaviour persists only below a certain material-dependent critical temperature called the Curie temperature. Above the Curie temperature the prevailing thermal energies are sufficient to breakdown the long-range order that lies behind these magnetic behaviours. The theory of the Curie temperature would take us beyond electromagnetism into statistical thermodynamics and is treated in standard works on that subject.

Paramagnetic materials are only very weakly attracted to magnets, forming internal induced magnetic fields in the same sense as the applied field, but only feebly. As an indication of how weakly compared to iron, which has a magnetic susceptibility of up to 200,000, here are some susceptibilities for paramagnetic materials: platinum 0.00026, aluminium 0.000022, sodium 0.0000072.

Diamagnetic materials are repelled by magnets, the induced magnetism being in opposition to the applied field. Their magnetic susceptibilities are therefore negative, but invariably very small in magnitude. Examples are silver -0.000026, diamond -0.000021, graphite -0.000016, copper -0.000009.

6.6 Capacitance and Inductance Revisited

Capacitance and inductance were discussed in detail in §5.4.5, §5.6.4 and §5.7 where it was assumed that no dielectric or magnetic media were present. Here we merely remind the reader that all the formulae derived therein will also apply in the presence of media if the permittivity or permeability of the vacuum, ε_0 or μ_0, are replaced by the permittivity or permeability of the media in question, i.e., $\varepsilon_r\varepsilon_0$ and $\mu_r\mu_0$.

6.7 Boundary Conditions (Static Fields)

In the case of static fields in the possible presence of dielectric and/or magnetic media, the Maxwell equations of (5.1) become, using (6.1.8) and (6.3.8),

$$\bar{\nabla} \cdot \bar{D} = \rho \qquad \bar{\nabla} \cdot \bar{B} \equiv 0 \qquad \bar{\nabla} \times \bar{H} = \bar{J} \qquad \bar{\nabla} \times \bar{E} = 0 \tag{6.7.1}$$

So, with media present we are dealing with four types of electromagnetic field, $\bar{B}, \bar{D}, \bar{E}, \bar{H}$ rather than just two as in the case of fields *in vacuo.*

It is important to appreciate that the equations (6.7.1) impose boundary conditions on the fields. Specifically, the divergence equations impose boundary conditions on the component of the field normal to any chosen boundary. Consider $\bar{\nabla} \cdot \bar{D} = \rho$. By the divergence theorem we have,

$$\int_V \bar{\nabla} \cdot \bar{D}\, d\tau = \int_V \rho d\tau = Q = \oint_{\delta V} \bar{D} \cdot d\bar{S} \tag{6.7.2}$$

where Q is the total charge within the integration volume V. Choosing this region to be a vanishingly short cylinder whose ends are on either side of the boundary of interest, and choosing the end caps to also be small so that the field $\bar{D}$ can be taken as constant over them, the surface integral becomes simply $\Delta D_n dS$, where ΔD_n is the difference between the normal component of $\bar{D}$ on one side (A) and that on the other (B), $\Delta D_n = D_{An} - D_{Bn}$. But, if there is no surface charge on the boundary, as the volume of the integration region shrinks to zero the contained charge Q becomes zero. In this case we would conclude that $\Delta D_n = 0$. However, if there is a surface charge density (per unit area) of σ on the boundary, then $\Delta D_n dS = Q = \sigma dS$ and hence $\Delta D_n = \sigma$. Note that σ is the real surface charge, not including any polarisation effect. Hence we conclude that,

- The component of $\bar{D}$ normal to a boundary, D_n, is discontinuous by the surface charge density, σ;
- Hence, D_n is continuous if there is no surface charge on the boundary;
- Equivalently, $\varepsilon_r E_n$ is discontinuous by σ/ε_0;
- Hence $\varepsilon_r E_n$ is continuous if there is no surface charge on the boundary;
- Hence E_n will be discontinuous even if there is no surface charge on the boundary if the materials on the two sides of the boundary have differing relative permittivity (dielectric constant), ε_r;
- In contrast, B_n is always continuous (because $\bar{\nabla} \cdot \bar{B} \equiv 0$).

In a similar manner, the $curl$ equations impose boundary conditions via Stokes's Theorem. Consider a rectangular integration path δA enclosing area A,

$$\int_A \bar{\nabla} \times \bar{H} \cdot d\bar{S} = \int_A \bar{J} \cdot d\bar{S} = \oint_{\delta A} \bar{H} \cdot d\bar{s} \tag{6.7.3}$$

For a long rectangular path oriented parallel to the boundary of interest the line integral becomes, for a suitably small path, just $\Delta H_{t1} ds$ where ΔH_{t1} is the difference over the boundary of the component of $\bar{H}$ parallel to the rectangular path and hence tangential to the boundary, $\Delta H_{t1} = H_{At1} - H_{Bt1}$. The subscript 1 here acknowledges that there are two orthogonal directions which are both tangential to the surface, which we are tacitly labelling by 1 and 2. If there is no surface current on the boundary, the integral $\int_A \bar{J} \cdot d\bar{S}$ tends to zero as the rectangle is shrunk to zero area, hence $\Delta H_{t1} = 0$ in this case. But if there is a surface current, $\bar{\varsigma}$, then the integral $\int_A \bar{J} \cdot d\bar{S} = \zeta_{t2} ds$ as the 'thickness' of the rectangular path shrinks onto the boundary. In this case we get $\Delta H_{t1} ds = \zeta_{t2} ds$, i.e., $\Delta H_{t1} = \zeta_{t2}$. Note that ζ_{t2} is the component of the surface current which is perpendicular to the chosen rectangular integration path. Hence it is tangential to the boundary but perpendicular to the length of the rectangle, i.e., orthogonal to the direction to which ΔH_{t1} relates. Hence,

- H_{t1} is discontinuous by the surface current ζ_{t2};
- Hence, H_{t1} is continuous if there is no surface current;
- Equivalently, B_{t1}/μ_r is discontinuous by the surface current $\mu_0 \zeta_{t2}$;
- Hence, B_{t1}/μ_r is continuous if there is no surface current;
- Hence, B_{t1} will be discontinuous even if there is no surface current if the materials on the two sides of the boundary have differing relative permeability, μ_r;
- In contrast, E_{t1} is always continuous under static conditions (because $\bar{\nabla} \times \bar{E} = 0$).

All the above also hold if tangential direction 1 is replaced by tangential direction 2 (and ζ_{t2} is replaced by $-\zeta_{t1}$).

The above boundary conditions will be crucial in determining specific solutions of Maxwell's equations.

7. The Displacement Current, the Complete Maxwell Equations and the Momentum Flux of the Electromagnetic Field

Here Maxwell's equations are finally completed. To do so requires the introduction of one further phenomenon: the displacement current as a source term for the magnetic field. The displacement current must be introduced into the Maxwell equations to make them consistent with the conservation of charge. This chapter also derives the Poynting vector for the energy flux or momentum flux of the electromagnetic field.

Contents

7.1 The Need for the Displacement Current

The Maxwell's equations which we have developed so far, in the presence of media and with time variation, are given by (6.7.1) with the fourth equation replaced by that of (5.1) to account for time variation, hence,

$$\bar{\nabla}\cdot\bar{D} = \rho \qquad \bar{\nabla}\cdot\bar{B} = 0 \qquad \bar{\nabla}\times\bar{H} = \bar{J} \qquad \bar{\nabla}\times\bar{E} = -\frac{\partial\bar{B}}{\partial t} \tag{7.1.1}$$

As noted in chapter 4, the third equation is not yet complete, but we complete it in this chapter. That there is a problem with the third equation in (7.1.1) as it stands is immediately apparent from the equation of continuity (or conservation) of charge, (3.8.12), which we can write as,

$$\bar{\nabla}\cdot\bar{J} = -\frac{\partial\rho}{\partial t} \tag{7.1.2}$$

But the third equation in (7.1.1) as it stands gives, $\bar{\nabla}\cdot\bar{J} = \bar{\nabla}\cdot(\bar{\nabla}\times\bar{H}) \equiv 0$, which is in flat contradiction with (7.1.2) unless the charge density is constant. If the charge density is varying then it appears that we must add a term to the third equation in (7.1.1), say $\bar{\nabla}\times\bar{H} = \bar{J} + \bar{C}$, and then require that $\bar{\nabla}\cdot\bar{C} = \frac{\partial\rho}{\partial t}$ so that the modified equation becomes consistent with the identity $\bar{\nabla}\cdot(\bar{\nabla}\times\bar{H}) \equiv 0$.As we know that $\bar{\nabla}\cdot\bar{D} = \rho$ the obvious choice is $\bar{C} = \frac{\partial\bar{D}}{\partial t}$ thus giving a modified equation $\bar{\nabla}\times\bar{H} = \bar{J} + \frac{\partial\bar{D}}{\partial t}$.

7.2 The Complete Maxwell's Equations

Hence, finally, the complete Maxwell equations are,

$$\begin{array}{|ll|} \hline \bar{\nabla}\cdot\bar{D} = \rho & \bar{\nabla}\cdot\bar{B} = 0 \\ \bar{\nabla}\times\bar{H} = \bar{J} + \frac{\partial\bar{D}}{\partial t} & \bar{\nabla}\times\bar{E} = -\frac{\partial\bar{B}}{\partial t} \\ \hline \end{array} \tag{7.2.1}$$

That this continues to be a good theory, when the displacement current term $\frac{\partial\bar{D}}{\partial t}$ is added to the third equation, is to be judged by its implications.

7.2.1 Electromagnetic Potentials Revisited

The magnetic vector potential, $\bar{A}$, was introduced in (5.5.6), being defined by $\bar{B} = \bar{\nabla}\times\bar{A}$. Since the final Maxwell equations still include $\bar{\nabla}\cdot\bar{B} = 0$, this definition of the magnetic vector potential remains valid. In contrast, the electric scalar potential, Φ, was introduced in (5.1.1), defined by $\bar{E} = -\bar{\nabla}\Phi$. This was appropriate only under conditions of static fields so that the fourth

Maxwell equation would then be $\bar{\nabla} \times \bar{E} = 0$. But the final Maxwell equation is $\bar{\nabla} \times \bar{E} = -\frac{\partial \bar{B}}{\partial t}$ which is inconsistent with $\bar{E} = -\bar{\nabla}\Phi$ as this implies $\bar{\nabla} \times \bar{E} = 0$. Instead, writing $\bar{\nabla} \times \bar{E} = -\frac{\partial \bar{B}}{\partial t} = -\bar{\nabla} \times \frac{\partial \bar{A}}{\partial t}$ motivates expressing the electric field as,

$$\bar{E} = -\frac{\partial \bar{A}}{\partial t} - \bar{\nabla}\Phi \tag{7.2.2}$$

which is now consistent with the Maxwell equations, (7.2.1).

7.3 Solving Maxwell's Equations

For the set of equations (7.2.1) to be soluble, given some distribution of source charges and currents, the constitutive relations which define the $\bar{D}$ and $\bar{H}$ fields in terms of the $\bar{E}$ and $\bar{B}$ fields must be specified. The simplest form of these assume LIH media and are as given by (6.1.9) and (6.3.10),

$$\bar{D} = \varepsilon_r \varepsilon_0 \bar{E} \qquad\qquad \bar{H} = \bar{B}/\mu_r \mu_0 \tag{7.3.1}$$

Providing the relative permittivity, ε_r, and relative permeability, μ_r, can be treated as constants, (7.2.1) together with (7.3.1) potentially provide a soluble system of linear equations, given the source terms. (The cautious "potentially" here reflects that the existence of a solution, let alone a unique solution, may depend upon the boundary conditions and sources specified).

7.3.1 Real Materials

(7.3.1) is all one needs in the case of fields in a vacuum, when $\varepsilon_r = \mu_r = 1$. However, it is rather a pedagogic fiction to regard ε_r and μ_r as constants in real media. Leaving aside the matter of isotropy and homogeneity, these material properties will generally depend upon the magnitude of the fields and also upon the frequency of the field variations. Under these conditions the system of equations (7.2.1) and (7.3.1) become non-linear and are likely to be soluble only by incremental numerical procedures. This is analogous to the stress analysis of solids as regards the transition from the ideal of linear elasticity to the realistic behaviour of deformable metals in which incremental plasticity imposes non-linear constitutive relations.

An even more fundamental difficulty arises in the case of ferromagnetism. If an initially unmagnetized sample of ferromagnetic material, below its Curie temperature, is subject to an increasing external magnetic field which is then removed, the sample will have become permanently magnetised. The material has behaved in an irreversible fashion. There is therefore no fixed relationship between the $\bar{H}$ and $\bar{B}$ fields but only an incremental relationship which is not reversible. The situation is analogous to the absence of a fixed relationship between stress and strain in solids once plasticity enters the picture: a plastically strained material will retain its strain as a permanent set even when the stress is removed.

7.3.2 Wave Equations for the Scalar and Vector Potentials

In the case of static fields, the electric scalar potential obeys a Poisson equation in terms of the source charge density, (5.1.2), which becomes a Laplace equation, (5.1.3), in regions where there is no charge. Similarly, the magnetic vector potential obeys a vector Poisson equation, (5.5.8), in terms of the source current density, and becomes a vector Laplace equation in regions which are free of current. But in the dynamic case, when time derivatives cannot be ignored, these foundational equations for the static potentials must be generalised to include time derivatives.

The definition of the vector potential, $\bar{B} = \bar{\nabla} \times \bar{A}$, means that Maxwell equation $\bar{\nabla} \cdot \bar{B} = 0$ is automatically fulfilled. Similarly, the modified definition of the scalar potential, $\bar{E} = -\frac{\partial \bar{A}}{\partial t} - \bar{\nabla}\Phi$, means that Maxwell equation $\bar{\nabla} \times \bar{E} = -\frac{\partial \bar{B}}{\partial t}$ is automatically fulfilled. Substitution of $\bar{E} = -\frac{\partial \bar{A}}{\partial t} - \bar{\nabla}\Phi$ into Maxwell equation $\bar{\nabla} \cdot \bar{D} = \rho$ together with (7.3.1) gives,

$$\nabla^2\Phi + \bar{\nabla} \cdot \partial_t \bar{A} = -\frac{\rho}{\varepsilon_r \varepsilon_0} \tag{7.3.2}$$

Similarly, substitution of $\bar{B} = \bar{\nabla} \times \bar{A}$ into the only remaining Maxwell equation, $\bar{\nabla} \times \bar{H} = \bar{J} + \frac{\partial \bar{D}}{\partial t}$, together with (7.3.1) and the assumption that any media are both homogeneous and have time-independent permittivity gives,

$$\left(\frac{1}{v^2}\partial_t^2 - \nabla^2\right)\bar{A} = \mu_0\mu_r\bar{J} - \bar{\nabla}\left(\frac{1}{v^2}\partial_t\Phi + \bar{\nabla} \cdot \bar{A}\right) \tag{7.3.3}$$

where,

$$v = \frac{c}{\sqrt{\mu_r \varepsilon_r}} = \frac{1}{\sqrt{\mu_r \varepsilon_r \mu_0 \varepsilon_0}} \tag{7.3.4}$$

At this point we note (as we did previously in chapter 5) that there is freedom to choose different potentials without changing the fields. Adding a $Grad$ term to the vector potential, i.e., $\bar{A} \to \bar{A}' = \bar{A} + \bar{\nabla} f$, leaves $\bar{B}$ unchanged and will also leave $\bar{E}$ unchanged if we also change the scalar potential $\Phi \to \Phi' = \Phi - \partial_t f$. Hence this so-called "gauge transformation", defined by an arbitrary scalar field f, leaves the fields invariant. Moreover, we can find a particular gauge transformation, i.e., a particular f, such that in the transformed gauge we have,

$$\frac{1}{v^2}\partial_t\Phi' + \bar{\nabla} \cdot \bar{A}' = 0 \tag{7.3.5}$$

That such a function f exists follows from the fact that,

$$\frac{1}{v^2}\partial_t\Phi' + \bar{\nabla}\cdot\bar{A}' = \frac{1}{v^2}\partial_t\Phi + \bar{\nabla}\cdot\bar{A} - \left(\frac{1}{v^2}\partial_t^2 - \nabla^2\right)f \tag{7.3.6}$$

and hence that (7.3.5) is equivalent to,

$$\left(\frac{1}{v^2}\partial_t^2 - \nabla^2\right)f = \frac{1}{v^2}\partial_t\Phi + \bar{\nabla}\cdot\bar{A} \tag{7.3.7}$$

(7.3.7) is a wave equation for f with a source term given by the potentials prior to the gauge transformation (i.e., in an arbitrary gauge). But a wave equation with a given source term has a solution, i.e., an f obeying (7.3.7) exists, because a solution can be written in terms of a Green's function for the wave operator, the d'Alembertian operator $\boxdot = \frac{1}{v^2}\partial_t^2 - \nabla^2$. The existence of a solution depends also upon other boundary conditions, as will the issue of uniqueness, but in very general unrestrictive cases a solution does exist (see chapter 15 for further details on Green's functions). It follows that we can freely impose the condition, (7.3.5), known as the Lorentz gauge, in most circumstances. In this case we see that (7.3.2) and (7.3.3) become,

$$\boxdot\,\Phi = \left(\frac{1}{v^2}\partial_t^2 - \nabla^2\right)\Phi = \frac{\rho}{\varepsilon_r\varepsilon_0} \tag{7.3.8a}$$

$$\boxdot\,\bar{A} = \left(\frac{1}{v^2}\partial_t^2 - \nabla^2\right)\bar{A} = \mu_0\mu_r\bar{J} \tag{7.3.8b}$$

(We write $\boxdot$ rather than ∂^2 here because the wave velocity v in a medium will differ from c and $\partial^2 = \frac{1}{c^2}\partial_t^2 - \nabla^2$).

These are the (inhomogeneous) wave equations with source terms and become the standard (homogeneous) wave equations where there are no sources. The physical implications of (7.3.8a,b) will be addressed in chapters 10-12 and 22.

7.4 The Poynting Vector for Electromagnetic Energy Flux

We already have the expression for the energy density of an electromagnetic field in linear media, namely the sum of (6.2.4) and (6.4.1),

$$u_{EM} = \frac{1}{2}\bar{E}\cdot\bar{D} + \frac{1}{2}\bar{B}\cdot\bar{H} \tag{7.4.1}$$

Or, more generally, the increment in the energy density of an electromagnetic field when the magnetic flux density increases by $\delta\bar{B}$ and the electric displacement increases by $\delta\bar{D}$ is,

$$\delta u_{EM} = \bar{E}\cdot\delta\bar{D} + \bar{H}\cdot\delta\bar{B} \tag{7.4.2}$$

Or, the rate of increase of the energy density of an electromagnetic field is,

$$\frac{\delta u_{EM}}{\delta t} = \bar{E} \cdot \frac{\delta \bar{D}}{\delta t} + \bar{H} \cdot \frac{\delta \bar{B}}{\delta t} \tag{7.4.3}$$

Now we have the Maxwell equations in fully time-dependent form, (7.2.1), we can find the flux of electromagnetic energy when the fields are time varying. This can be deduced by noting firstly that $\bar{J} \cdot \bar{E}$ is the rate at which the field does work on the charges comprising the current, per unit volume. This is derived by considering a region of cross-section A containing charge density ρ and letting this move a distance δx perpendicular to the cross-section and parallel to $\bar{E}$ in time δt. The charge $\rho A \delta x$ thus has a force $E\rho A \delta x$ acting on it and so the energy of the charge increases by $E\rho A \delta x^2$ (in time δt). The rate of the field doing work on the charge is thus,

$\frac{E\rho A \delta x^2}{\delta t} = E\rho v A \delta x = EJA\delta x$. So the rate of the field doing work on the charge per unit volume is $EJ = \bar{J} \cdot \bar{E}$ because we assumed $\bar{J}$ was parallel to $\bar{E}$. (The component of $\bar{J}$ perpendicular to $\bar{E}$ has no work done upon it because the corresponding displacement is perpendicular to the force on the charge).

But if $\bar{J} \cdot \bar{E}$ is the rate at which the field does work on the charge, per unit volume, then the rate at which the charge does work on the field is $-\bar{J} \cdot \bar{E}$. Integrating this over an arbitrary volume, V, must therefore equal the rate at which the electromagnetic energy within V increases plus the rate at which electromagnetic energy flows out through the boundary of V. Calling the electromagnetic energy flux $\bar{N}$ we thus have,

$$-\int_V \bar{J} \cdot \bar{E}\, d\tau = \int_V \left(\bar{E} \cdot \frac{\delta \bar{D}}{\delta t} + \bar{H} \cdot \frac{\delta \bar{B}}{\delta t}\right) d\tau + \oint_{\delta V} \bar{N} \cdot d\bar{S} \tag{7.4.4}$$

But the third Maxwell equation, (7.2.1), is $\bar{J} = \bar{\nabla} \times \bar{H} - \frac{\partial \bar{D}}{\partial t}$ and so $\bar{E} \cdot \bar{J} = \bar{E} \cdot (\bar{\nabla} \times \bar{H}) - \bar{E} \cdot \frac{\partial \bar{D}}{\partial t}$. Now identity (A1.2) means we can rewrite,

$\bar{E} \cdot (\bar{\nabla} \times \bar{H}) = \bar{H} \cdot (\bar{\nabla} \times \bar{E}) - \bar{\nabla} \cdot (\bar{E} \times \bar{H})$ so that the LHS of (7.4.4) becomes,

$$-\int_V \bar{J} \cdot \bar{E}\, d\tau = \int_V \left(-\bar{H} \cdot (\bar{\nabla} \times \bar{E}) + \bar{\nabla} \cdot (\bar{E} \times \bar{H}) + \bar{E} \cdot \frac{\partial \bar{D}}{\partial t}\right) d\tau$$

Using the fourth Maxwell equation, $\bar{\nabla} \times \bar{E} = -\frac{\partial \bar{B}}{\partial t}$, this becomes,

$$-\int_V \bar{J}\cdot\bar{E}\,d\tau = \int_V \left(\bar{H}\cdot\frac{\partial\bar{B}}{\partial t} + \bar{\nabla}\cdot(\bar{E}\times\bar{H}) + \bar{E}\cdot\frac{\partial\bar{D}}{\partial t}\right)d\tau$$

Using the divergence theorem this becomes,

$$-\int_V \bar{J}\cdot\bar{E}\,d\tau = \int_V \left(\bar{H}\cdot\frac{\partial\bar{B}}{\partial t} + \bar{E}\cdot\frac{\partial\bar{D}}{\partial t}\right)d\tau + \oint_{\delta V}(\bar{E}\times\bar{H})\cdot d\bar{S} \qquad (7.4.5)$$

Comparing with (7.4.4) we conclude that we can interpret $\bar{E}\times\bar{H}$ as the required flux of electromagnetic energy,

$$\bar{N} = \bar{E}\times\bar{H} \qquad (7.4.6)$$

known as the Poynting vector. The reader is reminded that "flux" refers to the amount of a substance, in this case energy, flowing through unit area in unit time, the vector of flux being in the same direction as the flow. The reader should show that the dimensions (units) of $\bar{N}$ are $\mathrm{Wm^{-2}}$, as required. The energy flux is therefore also power per square metre.

The alert reader will note that we are not really entitled to conclude that (7.4.6) is the unique flux of electromagnetic energy at every point in space, only that its integral over any closed surface (i.e., the last term in (7.4.5)) must equal the net electromagnetic energy flowing out of the surface per unit time. Thus, any vector field whose divergence is identically zero could be added to (7.4.6).

One such vector is simply a constant multiple of $\bar{B}$. But if we replaced (7.4.6) with $\bar{N} = \bar{E}\times\bar{H} + a\bar{B}$ this would imply, for $\bar{E} = 0$, that a static magnetic field alone would have a flux of energy. There is no experimental reason to suspect this. Another option, in regions where there is no charge, would be a constant multiple of $\bar{D}$ because in that case we would have $\bar{\nabla}\cdot\bar{D} = 0$. But this suffers from the same problem: when $\bar{H} = 0$ it would lead to an energy flux for a static electric field alone, for which there is no experimental evidence.

Hence we are led to take (7.4.6) seriously as the correct expression for electromagnetic energy flux, the cautionary note sounded by many authors notwithstanding.

7.5 Electromagnetic Momentum Flux (Radiation Pressure)

Anticipating Einstein's famous 1905 equivalence of energy and mass: $E = mc^2$, the energy flux of the electromagnetic field must be equivalent to a mass flux if we divide (7.4.6) by c^2. As long as we are considering fields in the vacuum, not within media, then the propagation speed of this mass-energy is just c. (Again we are anticipating the result that light can be identified with an electromagnetic wave, which we will address in detail in chapters 10 and 11). Hence the mass flux can be converted to a momentum flux by multiplying by c. We conclude that the momentum flux associated with electromagnetic fields in vacuo must be simply,

$$\bar{N}_{mf} = \frac{1}{c}\bar{\mathrm{E}} \times \bar{H} \tag{7.5.1}$$

Momentum flux is the momentum through unit area in unit time. Because the rate of change of momentum is force, it follows that, if the momentum represented by $\bar{N}_{mf}$ were brought to a halt by being fully absorbed by a perfect black body, then $\bar{N}_{mf}$ is also the radiation pressure exerted on that surface, hence the notation. However, if the incident momentum flux is reflected, then the recoil force will increase the total radiation pressure on the surface. For example, for perfect reflection in which the field amplitudes are undiminished in the reflected beam, the total radiation pressure will be twice that given by (7.5.1). Both these idealised examples assume normal incidence, i.e., that $\bar{N}_{mf}$ is parallel to the surface normal. For oblique incidence, see chapter 11.

The remarkable thing about the expression (7.5.1) is that it associates momentum with purely static fields, as long as both the electric and the magnetic fields are present. Can this be right? In chapter 8 we present a gedanken experiment due to Feynman that suggests that, yes, it is not only right but essential in order to preserve the conservation of angular momentum.

Note that the field momentum per unit volume times its velocity (c) must equal $\bar{N}_{mf}$, the momentum flux, so the field momentum per unit volume is therefore $\bar{E} \times \bar{B}/c^2\mu_r\mu_0$. Using a result to be derived in chapter 10, namely that $c^2 = 1/\varepsilon_0\mu_0$, the field momentum per unit volume *in vacuo* is,

in vacuo: $$\bar{N}_{momentum\_density} = \varepsilon_0\bar{E} \times \bar{B} \tag{7.5.2}$$

For the momentum flux in a medium see chapter 11.

8. Feynman's Disc Paradox and the Conservation of Momentum with Moving Charges

We closed the last chapter with a note of uncertainty regarding whether the Poynting vector does indeed correctly represent a real physical momentum flux of the electromagnetic field. Doubt arises because of the curious fact that purely static electric and magnetic fields apparently give rise to momentum. In this chapter the reality of this curious fact is seen to be essential if the conservation of angular momentum is to continue to hold. This is illustrated by a gedanken experiment due to Feynman. The reality of the field momentum is further illustrated by the difficulty encountered in chapter 4 where we saw that, in an arbitrary inertial frame of reference, a pair of moving charges experiences a net magnetic force, i.e., the forces on the two particles are not equal and opposite. Here we resolve that paradox by showing that the e/m field momentum in that situation is generally non-zero, and it is the total momentum of the particles plus the field that is conserved. Hence, there is a net force on the pair of particles because the equal and opposite balancing force acts on the field, changing the field momentum.

Contents

8.1 Feynman's Disc Paradox

In his famous physics lectures, [8.1], Richard Feynman introduced an illustrative conundrum based on a gedanken experiment. The situation is depicted in Figure 8.1.

Figure 8.1. The experimental setup envisaged by Feynman, [8.1], Fig.17-5. The disc is initially stationary with current I flowing in the coil.

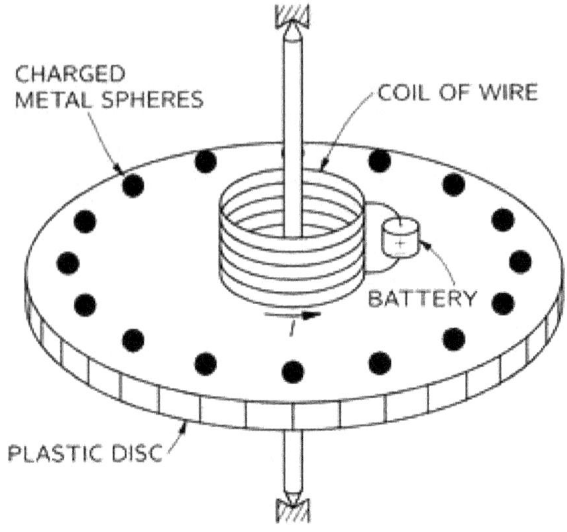

Fig. 17–5. Will the disc rotate if the current I is stopped?

Initially the disc is stationary. It has a large number of equal positive charges fixed and equally spaced around its periphery, and so these charges are also stationary initially. There is a coil, or solenoid, whose centre-line lies on the axle through the disc. The disc is free to turn about bearings which support the axle above and below. Initially there is a steady current in the coil, thus producing a steady magnetic field – approximating a magnetic dipole field. There is no force acting on the charges initially other than their mutual repulsion, which, by symmetry, is in the radial direction. The charges are stuck to the disc, so this mutual repulsive force is equilibrated by a reaction at the bond (and a small mechanical stress in the disc).

At time zero the source of electrical power driving the current is disconnected. The current, and hence the magnetic field, drop quicky to zero. A changing magnetic field produces an induced electric field through the Maxwell equation $\overline{\nabla} \times \tilde{E}_I = -\frac{\partial \bar{B}}{\partial t}$ where $\tilde{E}_I$ is the induced electric field (i.e.,

Faraday's law of induction). Axisymmetry (at least in the limit of a circumferentially uniform distribution of charges) and the fact that the initial magnetic field was vertical at the charges, means that only the vertical component of $\overline{\nabla} \times \tilde{E}_I$ is non-zero whilst the magnetic field is falling, and has magnitude $\frac{1}{r}\partial_r\left(r\tilde{E}_{I\theta}\right)$ (This is obtained from the expression for the z component of $curl$ in cylindrical polar coordinates when there is no θ dependence, see Appendix A2). As this equals $-\frac{\partial \bar{B}}{\partial t}$, which is non-zero for a short time when the current is cut-off, there must be a tangential component, $\tilde{E}_{I\theta}$, to the induced electric field, and hence a tangential force acting on the charges which is the same size, and in the same circulating sense, all around, and hence produces a torque which will cause the disc to rotate (assuming zero friction).

The apparent paradox is that the situation is initially static, and hence appears to have zero angular momentum, and yet after turning off the current to the coil we gain angular momentum, thus apparently violating the conservation of angular momentum.

This puzzle can be resolved by accepting the result of the last chapter, namely that a static electric field and a static magnetic field really do possess momentum, if they are not parallel, and that this is in accord with a field momentum flux given by the Poynting vector $\overline{N}_{mf} = \bar{E} \times \bar{B}/c\mu_0$. Here, $\bar{E}$ is the electric field due to the charges, not to be confused with the induced electric field, $\tilde{E}_I$. On the disc $\bar{E}$ is radial, whereas the magnetic field is vertical, and so the field momentum is tangential. But tangential momentum around a circle means angular momentum. Hence, the system does possess angular momentum initially even while it is mechanically stationary.

Note that the field momentum per unit volume times its velocity (c) must equal $\overline{N}_{mf}$, the momentum flux, so the field momentum per unit volume is therefore $\bar{E} \times \bar{B}/c^2\mu_0$. Using a result to be derived in chapter 10, namely $c^2 = 1/\varepsilon_0\mu_0$, the field momentum per unit volume is $\varepsilon_0\bar{E} \times \bar{B}$.

What remains is to confirm that the field angular momentum before time zero equals the mechanical angular momentum after time zero. The proof is as follows (noting that Feynman did not provide a proof in his lecture notes).

The induced field obeys $\overline{\nabla} \times \tilde{E}_I = -\frac{\partial \bar{B}}{\partial t}$ but the static field of the charges obeys $\varepsilon_0\overline{\nabla} \cdot \bar{E} = \rho$. Conversely, because the induced field corresponds to no charge and the static field corresponds to no magnetic field, we must also have $\overline{\nabla}\cdot\tilde{E}_I = 0$ and $\overline{\nabla}\times\overline{E} = 0$.

The force acting on the charges, per unit volume, is $\rho\tilde{E}_I$ and so the torque is $T = \int \bar{r}\times\rho\tilde{E}_I\, dV$. We can replace the charge density from $\varepsilon_0\overline{\nabla}\cdot\overline{E} = \rho$ and since torque is the rate of change of angular momentum the induced mechanical angular momentum is,

$$\bar{L}_m = \int \varepsilon_0(\overline{\nabla}\cdot\overline{E})\bar{r}\times\tilde{E}_I\, dVdt \tag{8.1}$$

The initial field angular momentum is $\bar{L}_f = \int \bar{r}\times(\varepsilon_0\overline{E}\times\overline{B})dV$. The magnetic field can be re-expressed as a time-integral over the induced electric field from $\overline{\nabla}\times\tilde{E}_I = -\frac{\partial\overline{B}}{\partial t}$ which gives us,

$$\bar{L}_f = \int \bar{r}\times\left(\varepsilon_0\overline{E}\times\left(\overline{\nabla}\times\tilde{E}_I\right)\right) dV\, dt \tag{8.2}$$

(Note that the minus sign disappears because the time integral equals the final minus the initial conditions). The challenge is to prove that (8.1) and (8.2) are equal. This is rather harder work than it has any right to be. Here I have adapted an approach of Lombardi, [8.2].

To conclude that (8.1) and (8.2) are equal requires the additional conditions $\overline{\nabla}\times\overline{E} = 0$ and $\overline{\nabla}\cdot\tilde{E}_I = 0$. These can be incorporated into (8.1) and (8.2) by symmetrising both, so,

$$\bar{L}_m = \varepsilon_0 \int [(\overline{\nabla}\cdot\overline{E})\bar{r}\times\tilde{E}_I + (\overline{\nabla}\cdot\tilde{E}_I)\bar{r}\times\overline{E}]\, dVdt \tag{8.3}$$

$$\bar{L}_f = \varepsilon_0 \int \bar{r}\times\left[\overline{E}\times\left(\overline{\nabla}\times\tilde{E}_I\right) + \tilde{E}_I\times(\overline{\nabla}\times\overline{E})\right]dV\, dt \tag{8.4}$$

The i component of the integrand of (8.3) is $\epsilon_{ijk}x_j\left(E_{J,J}\tilde{E}_{Ik} + \tilde{E}_{IJ,J}E_k\right)$ whereas that of (8.4) is $\epsilon_{ijk}x_j\left[E_J\left(\tilde{E}_{IJ,k} - \tilde{E}_{Ik,J}\right) + \tilde{E}_{IJ}\left(E_{J,k} - E_{k,J}\right)\right]$, where, as usual, repeated indices are summed (but noting that the subscript I is not an index but denotes the induced component of the electric field). In deriving the latter expression you may find it helpful to note the following tensor identity,

$$\epsilon_{klm}\epsilon_{mnp} \equiv \delta_{kn}\delta_{lp} - \delta_{kp}\delta_{ln} \tag{8.5}$$

We have also introduced the comma notation which compactly denotes derivatives wrt the coordinates, e.g., $v_{i,j} \equiv \partial_j v_i \equiv \frac{\partial v_i}{\partial x_j}$. Subtracting the two integrands gives the integral we need to prove is zero. Terms simplify as,

$$\epsilon_{ijk}x_j\left[-\left(E_J\tilde{E}_{IJ}\right)_{,k} + \left(E_J\tilde{E}_{Ik}\right)_{,J} + \left(E_k\tilde{E}_{IJ}\right)_{,J}\right] \tag{8.6}$$

Both angular momenta, (8.3) and (8.4), are in the z direction (axial), so we need only consider $i = z$. The integrand, (8.6), is thus, (8.7)

$$x\left[-\left(E_J\tilde{E}_{IJ}\right)_{,y} + \left(E_J\tilde{E}_{Iy}\right)_{,J} + \left(E_y\tilde{E}_{IJ}\right)_{,J}\right] - y\left[-\left(E_J\tilde{E}_{IJ}\right)_{,x} + \left(E_J\tilde{E}_{Ix}\right)_{,J} + \left(E_x\tilde{E}_{IJ}\right)_{,J}\right]$$

That the volume integral of (8.7) is zero follows in two steps. The first step notes that integrals of the form, $\int x\frac{\partial f}{\partial y}dV$ are zero,

$$\int x\frac{\partial f}{\partial y}dV = 0 \tag{8.8}$$

if the integral is over all space and the function f falls off at infinity at least as fast as $1/r^3$. This applies here because both $\bar{E}$ and $\tilde{E}_I$ fall off as $1/r^2$ and so their product falls of as $1/r^4$. To demonstrate (8.8), carry out the y integral at fixed x, z giving $x\big(f(x,\infty,z) - f(x,-\infty,z)\big)$. Providing $f(x,\pm\infty,z)$ reduce at least as fast as $1/x^3$ then the x integral goes as $1/x$ or smaller, and so also gives zero over the infinite domain. Clearly we also have $\int y\frac{\partial f}{\partial x}dV = 0$, $\int x\frac{\partial f}{\partial z}dV = 0$ and $\int y\frac{\partial f}{\partial z}dV = 0$. Hence, on integrating (8.7) we are left only with the terms,

$$x\left[\left(E_x\tilde{E}_{Iy}\right)_{,x} + \left(E_y\tilde{E}_{Ix}\right)_{,x}\right] - y\left[\left(E_y\tilde{E}_{Ix}\right)_{,y} + \left(E_x\tilde{E}_{Iy}\right)_{,y}\right] \tag{8.9}$$

which can be rearranged as,

$$\left(x\partial_x - y\partial_y\right)\left(E_x\tilde{E}_{Iy} + E_y\tilde{E}_{Ix}\right) \tag{8.10}$$

Integration by parts confirms for any function which falls off sufficiently fast, as the above does, then,

$$\int y\partial_y f\, dV = \int x\partial_x f\, dV \tag{8.11}$$

over infinite limits. This follows because under this assumption, integration by parts gives,

$$\left(\int y\partial_y f\, dy\right)dxdz = -\int fdydxdz$$

which therefore equals $\int x\partial_x f\, dV$ which is just the same. <u>QED</u>.

Hence the final mechanical angular momentum does indeed equal the initial field angular momentum. This confirms that the Poynting vector is just the field momentum that is required to ensure that the conservation of angular momentum continues to hold. If we did not acknowledge that the initial, static, fields possessed momentum then there would be no way in which the conservation of angular momentum could be saved.

8.1.1 A Fallacy Resolved

When I first heard of this explanation of Feynman's Disc paradox I had a problem with it. I reasoned that the coil could be a long solenoid, in which case the magnetic field would be zero (or as small as we might like) outside it. But if the solenoid, being comprised of conducting wires, effectively shielded its inside from the external charges, there would be no static E-field inside it. Consequently, I argued, the Poynting vector $\bar{E} \times \bar{H}$ would be zero everywhere and the initial field angular momentum would be zero. On the other hand, the induced emf at the charges depends only upon the rate of change of total flux linkage within the circle of the charges (and hence including within the solenoid) – and so there would still be an induced torque causing the disc to turn even though the B-field outside the solenoid was zero. So we would get non-zero mechanical angular momentum from an initial angular momentum which was zero – oh dear!

The fallacy of my reasoning, I now realise, is that although the Poynting vector could be as small as we want both inside the solenoid and also outside it up to the circle of charges, what we are interested in is $\bar{L}_f = \int \bar{r} \times (\varepsilon_0 \bar{E} \times \bar{B}) dV$ where the integral is over the whole of space, i.e., to infinity. Now lines of B-flux are closed loops, so the vertical upward flux within the solenoid must be matched by an equal vertical downwards flux outside the solenoid somewhere – though that may mean a very small flux density (i.e., B) but extending over very large distances. In other words, the initial field angular momentum may be very "dilute" but also extending to distances a long way from the disc and the charges. One needs to have more respect for infinite integrals than my qualitative argument portrayed. One lives and learns! Physically, the lesson is that the field momentum may reside a long way away from the experimental apparatus – rather surprisingly.

8.2 The Field Momentum of Two Moving Charges

In chapter 4 it was noted that, in an arbitrary frame of reference, the magnetic forces acting on a pair of moving charges are not equal and opposite as one might have expected from Newton's third law. It was hinted that the resolution of this problem lay in the momentum possessed by the e/m field itself, but without proof. Here we show firstly that the total e/m field momentum in such a situation is generally non-zero. Secondly, we resolve the paradox by showing that it is the total momentum of the particles plus the e/m field that is time independent, i.e., the rate-of-change of the momentum of the particles, which is their total net force, is equal and opposite to the rate-of-change of the field momentum. In Newtonian language, there is a net force on the pair of charged particles which is balanced by an equal and opposite 'force' acting on the field, whose momentum is therefore changing.

8.2.1 The Field Momentum is Generally Non-Zero

Consider charges q_1, q_2 moving with velocities $\bar{v}_1, \bar{v}_2$ and instantaneous positions $\bar{r}_1, \bar{r}_2$ so that the instantaneous position of 2 wrt 1 is vector is $\bar{r}_{12} = \bar{r}_2 - \bar{r}_1$. In this section we show that the total field momentum is, in general, non-zero. We work in the non-relativistic limit such that the e/m fields can be considered to propagate with infinite speed.

Writing $\bar{s}_1 = \bar{r} - \bar{r}_1$, $\bar{s}_2 = \bar{r} - \bar{r}_2$ the electric field at position $\bar{r}$ due to the two charges is,

$$\bar{E} = \frac{q_1\bar{s}_1}{4\pi\varepsilon_0 s_1^3} + \frac{q_2\bar{s}_2}{4\pi\varepsilon_0 s_2^3} \tag{8.2.1}$$

and the magnetic flux density is,

$$\bar{B} = \frac{\mu_0 q_1\bar{v}_1\times\bar{s}_1}{4\pi s_1^3} + \frac{\mu_0 q_2\bar{v}_2\times\bar{s}_2}{4\pi s_2^3} \tag{8.2.2}$$

The Poynting vector for the momentum density is thus, from (7.5.2),

$$\bar{N}_{md} = \varepsilon_0\left(\frac{q_1\bar{s}_1}{4\pi\varepsilon_0 s_1^3} + \frac{q_2\bar{s}_2}{4\pi\varepsilon_0 s_2^3}\right)\times\left(\frac{\mu_0 q_1\bar{v}_1\times\bar{s}_1}{4\pi s_1^3} + \frac{\mu_0 q_2\bar{v}_2\times\bar{s}_2}{4\pi s_2^3}\right) \tag{8.2.3}$$

What we are interested in is the total momentum obtained by integrating (8.2.3) over the whole volume of space,

$$\bar{P} = \frac{\mu_0}{(4\pi)^2}\int\left(\frac{q_1\bar{s}_1}{s_1^3} + \frac{q_2\bar{s}_2}{s_2^3}\right)\times\left(\frac{q_1\bar{v}_1\times\bar{s}_1}{s_1^3} + \frac{q_2\bar{v}_2\times\bar{s}_2}{s_2^3}\right)d^3r \tag{8.2.4}$$

We start by considering what (8.2.4) gives in the centre-of-velocity (cov) frame of reference in which $\bar{v}_1 = -\bar{v}_2 = \bar{v}$,

$$\bar{P}_{cov} = \frac{\mu_0}{(4\pi)^2}\int\left(\frac{q_1\bar{s}_1}{s_1^3} + \frac{q_2\bar{s}_2}{s_2^3}\right)\times\left(\frac{q_1\bar{v}\times\bar{s}_1}{s_1^3} - \frac{q_2\bar{v}\times\bar{s}_2}{s_2^3}\right)d^3r \tag{8.2.5}$$

The integrand is the sum of the following four terms,

$$\mathbb{I}_1 = \frac{q_1^2\bar{s}_1\times(\bar{v}\times\bar{s}_1)}{s_1^6} \qquad \mathbb{I}_2 = -\frac{q_2^2\bar{s}_2\times(\bar{v}\times\bar{s}_2)}{s_2^6} \tag{8.2.6a}$$

$$\mathbb{I}_3 = \frac{q_1q_2\bar{s}_2\times(\bar{v}\times\bar{s}_1)}{s_1^3s_2^3} \qquad \mathbb{I}_4 = -\frac{q_1q_2\bar{s}_1\times(\bar{v}\times\bar{s}_2)}{s_1^3s_2^3} \tag{8.2.6b}$$

Using identity (A1.3b) we have,

$$\bar{s}_2\times(\bar{v}\times\bar{s}_1) = (\bar{s}_1\cdot\bar{s}_2)\bar{v} - (\bar{s}_2\cdot\bar{v})\bar{s}_1 \tag{8.2.7a}$$

$$\bar{s}_1\times(\bar{v}\times\bar{s}_2) = (\bar{s}_1\cdot\bar{s}_2)\bar{v} - (\bar{s}_1\cdot\bar{v})\bar{s}_2 \tag{8.2.7a}$$

The first terms on the RHS of (8.2.7a,b) will cancel between $\mathbb{I}_3$ and $\mathbb{I}_4$ so that the contribution of these two terms to the total momentum becomes,

$$\bar{P}_{3+4,cov} = \frac{\mu_0q_1q_2}{(4\pi)^2}\int\frac{(\bar{s}_1\cdot\bar{v})\bar{s}_2-(\bar{s}_2\cdot\bar{v})\bar{s}_1}{s_1^3s_2^3}d^3r \tag{8.2.8}$$

To evaluate the integral we can shift the origin to the position $\bar{r}_c = -(\bar{r}_1+\bar{r}_2)/2$ so that $\bar{s}_{1,2}$ become $\bar{r}\pm\frac{1}{2}\bar{r}_{12}$ so that,

$$(\bar{s}_1\cdot\bar{v})\bar{s}_2 - (\bar{s}_2\cdot\bar{v})\bar{s}_1 = (\bar{r}_{12}\cdot\bar{v})\bar{r} - (\bar{r}\cdot\bar{v})\bar{r}_{12} \tag{8.2.9}$$

So,

$$\bar{P}_{3+4,cov} = \frac{\mu_0q_1q_2}{(4\pi)^2}\int\frac{(\bar{r}_{12}\cdot\bar{v})\bar{r}-(\bar{r}\cdot\bar{v})\bar{r}_{12}}{s_1^3s_2^3}d^3r \tag{8.2.10a}$$

$$\text{where, } s_1^3s_2^3 = \left|\bar{r}+\frac{1}{2}\bar{r}_{12}\right|^3\left|\bar{r}-\frac{1}{2}\bar{r}_{12}\right|^3 \tag{8.2.10b}$$

Noting that $\bar{v}$ and $\bar{r}_{12}$ are constant over the integral we see that the integrand in (8.2.10a) is odd, i.e., it changes sign when $\bar{r}\to-\bar{r}$, consequently it integrates to zero. The total momentum in the cov frame is therefore given by the first two terms, (8.2.6a), alone. As $\bar{s}_1\times(\bar{v}\times\bar{s}_1) = s_1^2\bar{v} - (\bar{s}_1\cdot\bar{v})\bar{s}_1$ the first term gives a contribution to the field momentum of,

$$\bar{P}_{1,cov} = \frac{\mu_0q_1^2}{(4\pi)^2}\int\frac{\bar{v}-(\hat{s}_1\cdot\bar{v})\hat{s}_1}{s_1^4}d^3r \tag{8.2.11}$$

This can be evaluated by shifting the origin to the position of charge 1, i.e., $\bar{r}_1$, so that $\bar{s}_1\to\bar{r}$ and the numerator becomes $\bar{v}-v_r\hat{r} = v_\theta\hat{\theta}+v_\phi\hat{\phi}$ where we have expressed the cov velocity $\bar{v}$ in terms of its polar coordinates. Without loss of generality we can take the direction of $\bar{v}$ as the polar axis so that $v_\theta = -v\sin\theta$ and $v_\phi = 0$. (8.2.11) becomes,

$$\bar{P}_{1,cov} = \frac{\mu_0q_1^2}{(4\pi)^2}\int\frac{-v\sin\theta}{r^4}\hat{\theta}d^3r \tag{8.2.11}$$

Using $\hat{\theta} = -\hat{z}\sin\theta + \cos\theta(\hat{x}\cos\phi + \hat{y}\sin\phi)$ the $\hat{x}$ and $\hat{y}$ components integrate to zero leaving,

$$\bar{P}_{1,cov} = \frac{\mu_0 q_1^2 v}{(4\pi)^2} 2\pi\hat{z} \int \frac{dr}{r^2} \int_0^{\pi} \sin^3\theta \,.\, d\theta = \frac{\mu_0 q_1^2 \bar{v}}{6\pi} \int \frac{dr}{r^2} \tag{8.2.12}$$

The radial integral is divergent for a point source, not surprisingly as the energy associated with the field of a point source is also divergent. But if we assume a charge of finite radius, a, it becomes finite,

$$\bar{P}_{1,cov} = -\frac{\mu_0 q_1^2 \bar{v}}{6\pi a} \tag{8.2.13}$$

The integral of $\mathbb{I}_2$ is evaluated in the same way so that, finally, the total field momentum in the cov frame is,

$$\bar{P}_{cov} = \frac{\mu_0 (q_2^2 - q_1^2)\bar{v}}{6\pi a} \tag{8.2.14}$$

Hence, the field momentum is non-zero even in the cov frame when the charges are different, only being zero when the charges are equal or equal and opposite. Comparing with the non-relativistic momentum of a mass at velocity $\bar{v}$ we see that $\frac{\mu_0(q_2^2-q_1^2)}{6\pi a}$ appears to be the inertia of the e/m field. If we take, say, $q_2 = 2e$, $q_1 = e$ and a to be of typical atomic size (one Angstrom, 10^{-10}m) then the field 'mass' is only 0.000056 times the electron mass, so only a small correction. But if we take a to be of typical nuclear size (one fermi, 10^{-15}m) then the field mass is 5.6 times the electron mass but still only 0.003 of the proton mass.

Nevertheless, the field momentum is non-zero and we might expect odd things to happen at the smallest size scale (equivalent to high energies) but further consideration of such matters would take us into quantum field theory.

Even when the charges are equal, in an arbitrary frame of reference we expect non-zero field momentum. This is most easily demonstrated numerically, For example, with charges $q_1 = 5e, q_2 = 8e$, with a spacing in the x-direction of 0.7 fermi and $v_1 = 0.0005c, v_2 = 0.0008c$, the velocity of both charges being in the z-direction, integration was carried out over a grid of cubes considering distances from both charges, in any direction, being at least three times their separation. The cubes containing the charges were omitted from the integral as the integrand is divergent there. The size of the cube acts as a rough surrogate for the effective size of the charges. So as the number of integration steps (i.e., cubes) increases, the size of the cubes decreases and the expectation is that the resulting field momentum will scale as 1/cube size.

Figure 8.2 confirms that that is the case to very good accuracy, and gives the resulting field momentum in kgm/s, which is of the order 10^{-20} for the range of charge "sizes" considered. This

Figure 8.2: Numerically integrated field momentum (kgm/s) versus reciprocal charge size (1/fm) for the example case described in the text

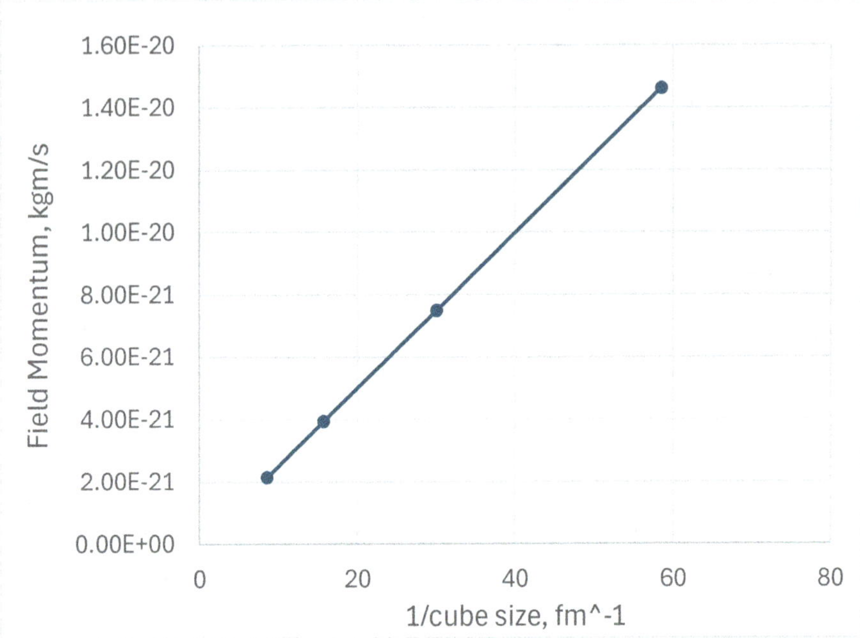

8.2.2 Total Momentum of Field plus Particles is Conserved

We now consider the general case. The strategy is to express the time derivative of the combined momentum of the particles and the field as the volume integral of a divergence which can then be transformed to a surface integral at infinity which vanishes. To do this we need to express the mechanical momentum in terms of field quantities. This is accomplished by noting that the force on a given particle $\bar{F} = q(\bar{E} + \bar{v} \times \bar{B})$, from (4.2.10) and the definition of the electric field, can be expressed as a force per unit volume acting on a distribution of charge density, $\rho(\bar{E} + \bar{v} \times \bar{B})$. Any number of particles can be swept up in the distribution, ρ, so their total rate-of-change of momentum is therefore,

$$\frac{d\bar{P}_{particles}}{dt} = \int \rho(\bar{E} + \bar{v} \times \bar{B})d\tau \tag{8.2.15}$$

But $\rho\bar{v} = \bar{J}$ is the current density, and we can therefore replace the distribution and its velocity by using the Maxwell equations $\rho = \varepsilon_0 \bar{\nabla} \cdot \bar{E}$ and $\bar{J} = \frac{1}{\mu_0} \bar{\nabla} \times \bar{B} - \varepsilon_0 \frac{\partial \bar{E}}{\partial t}$ giving,

$$\frac{d\bar{P}_{particles}}{dt} = \int \left(\varepsilon_0 (\bar{\nabla} \cdot \bar{E}) \bar{E} + \left(\frac{1}{\mu_0} \bar{\nabla} \times \bar{B} - \varepsilon_0 \frac{\partial \bar{E}}{\partial t} \right) \times \bar{B} \right) d\tau \qquad (8.2.16)$$

We can replace $\frac{\partial \bar{E}}{\partial t} \times \bar{B} = \frac{\partial}{\partial t} (\bar{E} \times \bar{B}) - \bar{E} \times \frac{\partial \bar{B}}{\partial t}$ and noting $\varepsilon_0 \int \bar{E} \times \bar{B}\, d\tau$ is the field momentum gives, (8.2.17)

$$\frac{d\bar{P}_{particles}}{dt} + \frac{d\bar{P}_{field}}{dt} = \int \left(\varepsilon_0 (\bar{\nabla} \cdot \bar{E}) \bar{E} + \frac{1}{\mu_0} (\bar{\nabla} \times \bar{B}) \times \bar{B} + \varepsilon_0 \bar{E} \times \frac{\partial \bar{B}}{\partial t} \right) d\tau$$

We can now substitute from the Maxwell equation $\frac{\partial \bar{B}}{\partial t} = -\bar{\nabla} \times \bar{E}$ and also add into the integrand $\frac{1}{\mu_0} (\bar{\nabla} \cdot \bar{B}) \bar{B}$ since this is identically zero by the final Maxwell equation, giving,

$$\frac{d\bar{P}_{particles}}{dt} + \frac{d\bar{P}_{field}}{dt} = \qquad (8.2.18)$$

$$\int \left(\varepsilon_0 (\bar{\nabla} \cdot \bar{E}) \bar{E} + \frac{1}{\mu_0} (\bar{\nabla} \times \bar{B}) \times \bar{B} - \varepsilon_0 \bar{E} \times (\bar{\nabla} \times \bar{E}) + \frac{1}{\mu_0} (\bar{\nabla} \cdot \bar{B}) \bar{B} \right) d\tau$$

Making use of identity (8.5), the ith component of the above integrand is found to be,

$$\frac{1}{\mu_0} \left\{ \frac{1}{c^2} \left(E_{j,j} E_i - E_j E_{j,i} + E_j E_{i,j} \right) + B_{j,j} B_i - B_j B_{j,i} + B_j B_{i,j} \right\} \qquad (8.2.19)$$

This is readily seen to be identical to $\partial_j M_{ij}$ where the Maxwell tensor M_{ij} is defined as,

$$M_{ij} = \tilde{E}_i \tilde{E}_j + B_i B_j - \tfrac{1}{2} (\tilde{E}^2 + B^2) \delta_{ij} \qquad \text{where } \tilde{E} = \bar{E}/c \qquad (8.2.20)$$

(Aside, as a preview of chapter 28: This 3x3 matrix is tensorial wrt rotations and is the spatial part of the 4x4 matrix which we will see in chapter 28 is tensorial wrt Lorentz transformations and known as the stress-energy-momentum tensor)

The dimensions (units) of M_{ij} are seen to be energy density, or equivalently force per unit area or pressure or stress. So, finally we have shown that,

$$\left(\frac{d\bar{P}_{particles}}{dt} + \frac{d\bar{P}_{field}}{dt} \right)_i = \int \partial_j M_{ij} d\tau \qquad (8.2.21)$$

The volume integral of the divergence is converted to a surface integral at infinity using Gauss's theorem. As the square of the fields will reduce at infinity faster than $1/r^2$ this surface integral is zero and we conclude,

$$\frac{d\bar{P}_{particles}}{dt} + \frac{d\bar{P}_{field}}{dt} = 0 \tag{8.2.22}$$

Hence the total momentum of particles and field, $\bar{P}_{particles} + \bar{P}_{field}$, is conserved.

The reason why there is a net force on the particles as a group, i.e., the reason why $\frac{d\bar{P}_{particles}}{dt} \neq 0$, is because the particles are not the whole dynamical system. However, there is no net force on the combined particles-plus-field.

Alternatively, if we choose to truncate the integration at a finite distance, then the surface integral $\oiint M_{ij} dS_j = \int \partial_j M_{ij} d\tau$ is the net force acting over the surface in question and (8.2.21) then tells us that this equals the net force acting on the combined particles and fields within that surface.

8.3 References

[8.1] Feynman, R. (1971). The Feynman Lectures on Physics, Volume II, chapter 17, §17-4. The link is to the free online version.

[8.2] Gabriel Lombardi. (1983). Feynman's disk paradox. American Journal of Physics 51(3):213-214, March 1983. http://doi.org/10.1119/1.13272 (Available at https://www.researchgate.net/publication/243489921_Feynman's_disk_paradox)

9. Units, Constants and Illustrative Magnitudes

So far we have not stated the actual magnitudes of the universal constants, nor have we given any indication of the order of magnitude of the many quantities discussed in the preceding chapters. We remedy those shortcomings here. The important issue of units is also addressed, using throughout the MKSA (or SI) system. Unfortunately, the student will find many different units in use in practice, e.g., in texts, including some in which some universal constants do not appear at all, for example because they are absorbed into the units used for charge. In this book we have opted to avoid confusing the reader with a multiplicity of options for units, using MKSA throughout. Many examples are given of the numerical magnitudes of forces, energies, field strengths, etc., arising in simple archetypal electromagnetic situations so that the reader can gain some intuitive appreciation of typical orders of magnitude.

Contents

9.1 Summary of MKSA Basis Units

Table 9.1: MKSA Basis Units

Dimension	Unit	Symbol
Length	metre	m
Mass	kilogram	kg
Time	second	s
Current	ampère	A = C/s
Temperature	kelvin	K

The ampère is invariably contracted to "amp". Personally, I think of the basis quantity as being the coulomb (C), a quantity of electrical charge, which is formally defined as an amp times one second in the MKSA system.

For use below we introduce the notation $[X]$ to mean "the units of X". For any quantity, $[X]$ can always be reduced to a product of powers of m, kg, s, A (and possibly, K). In what follows I prefer to use kg, m, s, C.

The derived MKSA units of electromagnetic quantities are listed in Table 9.2, together with their expressions in the kg, m, s, C basis.

Note that units are written as normal (upright) text, not as italics. In contrast, algebraic quantities are always italics and will generally mean something different. For example, C means "coulomb", whereas the meaning of C will vary with context and might mean, for example, "capacitance", the unit of which is F, the farad, and a farad = C/V = $\mathrm{kg^{-1}m^{-2}s^2C^2}$.

Note that the unit of moment (or torque) is identical to the joule but is not usually expressed as such, but rather as Nm = $\mathrm{kgm^2s^{-2}}$, because moment and energy are distinct physical concepts. (Geometrically, the distinction is that moment relates to a length perpendicular to the force whereas energy relates to a length parallel to the force).

The unit of angular momentum is $\mathrm{kgm^2s^{-1}}$ which is the same as Js and hence the same as the units for Planck's constant, h, or $\hbar = h/2\pi$. In nuclear physics or particle physics, referring to a particle as having "spin ½" or "spin 1" means that its intrinsic angular momentum is that multiple of $\hbar$.

The units of energy density, pressure and stress are all the same, $\mathrm{kgm^{-1}s^{-2}}$.

The derivations of the units for the electromagnetic quantities are laid out in sections §9.1.1 to §9.1.14 which follow Table 9.2.

Table 9.2: The MKSA Derived Units. *The last column gives the definition in terms of the basis units, taken as* kg, m, s, C *The convention is that the name of the unit takes a small initial letter, even when it is derived from a proper name (to distinguish the unit from the person). The symbols for derived units, however, are capitals. Symbols for units are not italicised.*

Item	Unit	Symbol	In Basis
Charge	coulomb	C	C
Force	newton	N	$\mathrm{kgms^{-2}}$
Energy	joule = Nm	J	$\mathrm{kgm^2s^{-2}}$
Power	watt = J/s	W	$\mathrm{kgm^2s^{-3}}$
Angular momentum	-	-	$\mathrm{kgm^2s^{-1}}$
Moment (torque)	-	-	$\mathrm{kgm^2s^{-2}}$
Electrical potential	volt = J/C	V	$\mathrm{kgm^2s^{-2}C^{-1}}$
Capacitance	farad = C/V	F	$\mathrm{kg^{-1}m^{-2}s^2C^2}$
Permittivity	farads/metre	F/m	$\mathrm{kg^{-1}m^{-3}s^2C^2}$
Electric field strength, *E*	V/m = N/C	-	$\mathrm{kgms^{-2}C^{-1}}$
Electric dipole moment	-	-	Cm
Polarisation	-	-	$\mathrm{Cm^{-2}}$
Electric Displacement, *D*	-	-	$\mathrm{Cm^{-2}}$
Resistance	ohm	Ω	$\mathrm{kgm^2s^{-1}C^{-2}}$
Resistivity	-	Ωm	$\mathrm{kgm^3s^{-1}C^{-2}}$
Conductance	siemen	S	$\mathrm{kg^{-1}m^{-2}sC^2}$
Conductivity	-	S/m	$\mathrm{kg^{-1}m^{-3}sC^2}$
Inductance	henry	H = Ωs	$\mathrm{kgm^2C^{-2}}$
Permeability	henry/m	H/m	$\mathrm{kgmC^{-2}}$
Magnetic flux density, *B*	tesla	T = Wb/m²	$\mathrm{kgs^{-1}C^{-1}}$
Magnetic flux	weber	Wb=Vs	$\mathrm{kgm^2s^{-1}C^{-1}}$
Magnetic dipole moment	-	J/T	$\mathrm{m^2s^{-1}C}$
Magnetisation	-	A/m	$\mathrm{m^{-1}s^{-1}C}$
Magnetic field strength, *H*	-	A/m	$\mathrm{m^{-1}s^{-1}C}$

9.1.1 Potential (volt)

§5.1 shows that we can interpret the product of charge and potential as the potential energy of the charge due to its presence in an electric field. Hence the volt is defined as a joule per coulomb.

$$\text{volt} = \frac{\text{joule}}{\text{coulomb}} = \mathrm{kgm^2s^{-2}C^{-1}} \tag{9.1.1}$$

9.1.2 Capacitance (farad)

§5.4.5 defines capacitance as the amount of charge stored per unit potential difference over a capacitor, $Q = CV$. Hence,

$$[\text{capacitance}] = \frac{\text{coulomb}}{\text{volt}} = \text{farad} = \text{kg}^{-1}\text{m}^{-2}\text{s}^2\text{C}^2 \qquad (9.1.2)$$

9.1.3 Permittivity

The permittivity of the vacuum is defined via Coulomb's Law, (4.2.1-4), from which we see that its units are,

$$[\text{permittivity}] = [\varepsilon_0] = \left[\frac{q_1 q_2}{F r_{12}^2}\right] = \text{kg}^{-1}\text{m}^{-3}\text{s}^2\text{C}^2 = \frac{\text{farad}}{\text{metre}} \qquad (9.1.3)$$

9.1.4 Electric Field Strength (E-field)

§5.1 gives the electric field as the gradient of the potential, hence,

$$[\text{electric field strength}] = \frac{\text{volt}}{\text{metre}} = \text{kgms}^{-2}\text{C}^{-1} = \text{N/C} \qquad (9.1.4)$$

Hence, this is consistent with the force on an electrical charge in an electrical field being $F = qE$ (§4.2.1).

9.1.5 Electric Dipole Moment

Electric dipole moment is defined in §5.2 as the product of the charge on a pair of equal and opposite charges times their distance apart. Hence, Cm.

9.1.6 Polarisation

The (electric) polarisation of a dielectric material is defined in §6.1 as the dipole moment per unit volume induced by an applied electric field. Hence, the units of polarisation are dipole moment per unit volume = Cm^{-2}.

9.1.7 Electric Displacement (D-field)

The (electric) displacement, discussed in §6.1, is an alternative description of the electric field as it is modified by a dielectric medium. Defined by (6.1.7) it has the same units as polarisation, which is also the same as the product of permittivity times electric field strength.

The reader should confirm that the product of the E-field and the D-field has units of energy density.

9.1.8 Resistance, Resistivity, Conductance, Conductivity

For completeness we anticipate the basics of electric circuit theory to be discussed in more detail in chapter 18. The resistance of a circuit element is the potential needed to drive unit current through it. So,

$$[\text{resistance}] = \frac{\text{volt}}{\text{amp}} = \text{ohm } (\Omega) = \text{kgm}^2\text{s}^{-1}\text{C}^{-2} \tag{9.1.5}$$

Note that the power dissipated as heat in a resistor is RI^2 and so ohms times amps^2 = watts, as the reader should confirm.

The resistance of a circuit component depends upon the material and also upon the dimensions of the item composed of that material. In contrast, the "resistivity" is a material property and is common to all components made of that material. The resistance increases in proportion to the length of the component (in the direction of the current flow) but decreases in proportion to the cross-sectional area. Hence,

$$\text{resistance} = \text{resistivity} \times \frac{\text{length}}{\text{area}} \tag{9.1.6}$$

From which we see that,

$$[\text{resistivity}] = [\text{resistance} \times \text{length}] = \Omega\text{m} = \text{kgm}^3\text{s}^{-1}\text{C}^{-2}$$

Conductivity is defined as the reciprocal of resistivity, whilst conductance is the reciprocal of resistance.

9.1.9 Inductance

The concept of inductance was introduced in §5.6.4. In a circuit element it is the constant of proportionality which relates the induced potential difference (emf) over a component to the rate of change of current through it. Hence,

$$[\text{inductance}] = \frac{\text{volt}}{\text{amp}.s^{-1}} \equiv \text{henry} = \text{kgm}^2\text{C}^{-2} = \Omega\text{s} \tag{9.1.7}$$

which defines the henry.

9.1.10 Permeability

The units of the permeability of the vacuum, μ_0, follows from the Biot-Savart Law, (4.2.6), which gives,

$$[\mu_0] = \left[\frac{\text{Force}}{\text{Current}^2}\right] = \text{NA}^{-2} = \text{kgmC}^{-2} = \text{H/m} \tag{9.1.8}$$

9.1.11 Magnetic Flux Density (B-field)

Comparing (4.2.7) with the Biot-Savart Law, (4.2.6), the units of the magnetic flux density are seen to be those of force divided by a current element ($I\delta L$). Hence,

$$[\text{magnetic flux density}] = \text{kg}s^{-1}\text{C}^{-1} \equiv \text{Wb.}\,\text{m}^{-2} \equiv \text{tesla} \quad (9.1.9)$$

which defines the tesla. The weber, the unit of magnetic flux, is thus,

$$\text{weber, Wb} = \text{kgm}^2\text{s}^{-1}\text{C}^{-1} \qquad \text{T} = \text{Wb/m}^2 \quad (9.1.10)$$

9.1.12 Magnetic Dipole Moment

§5.5.2 defines the magnetic dipole moment of a current loop of area $\bar{S}$ to be $\bar{m} = I\bar{S}$. Hence the unit of magnetic dipole moment is $\text{Am}^2 = \text{m}^2\text{s}^{-1}\text{C}$ which equals J/T.

9.1.13 Magnetisation

§6.3 defines magnetisation as the magnetic dipole moment per unit volume induced by an applied magnetic field. Hence its unit is $\text{m}^{-1}\text{s}^{-1}\text{C}$.

9.1.14 Magnetic Field Strength (H-field)

The H-field is defined by (6.3.7) and thus has the same units as magnetisation, $\text{m}^{-1}\text{s}^{-1}\text{C}$. The reader should confirm that this is also the unit of the magnetic flux density divide by permeability.

The reader should confirm that the product of the B-field and the H-field has units of energy density.

9.2 The Universal Constants

Table 9.3 gives the values of the universal constants most relevant in this book, with their MKSA units.

In the MKSA system, the amp is defined before the other electromagnetic units. The amp is defined as the current required to flow in both of a pair of infinitely long parallel wires 1 metre apart to cause a force between them of 2×10^{-7} Newtons per metre of their length.

We know from (4.2.8) and (4.2.21) that the force per unit length between two parallel wires would be $\frac{\mu_0 I_1 I_2}{2\pi r}$ which becomes $\frac{\mu_0}{2\pi}$ if the currents are both 1 amp and the distance of the wires apart is 1 metre. Hence the definition of the amp requires that $\frac{\mu_0}{2\pi} = 2 \times 10^{-7}$ and hence that $\mu_0 = 4\pi \times 10^{-7}$. That the units are $\mathrm{H/m}$ follows from §(9.1.10).

The coulomb is defined as the charge flowing past a point in one second if the current is one amp.

In Coulomb's Law, (4.2.1), the units of force, length and charge are therefore defined, leaving the magnitude of the vacuum permittivity, ε_0, to be determined by experiment. In principle this could be done by measuring the force between two charges at various distances apart and then using (4.2.1) to find ε_0. That the units of permittivity are $\mathrm{F/m}$ follows from (9.1.3).

The Bohr magneton is the magnetic dipole moment that would arise, in classical (i.e., pre-quantum) physics, if a point particle equal in mass and charge to the electron was executing a circular orbit at constant speed and with an angular momentum equal to Planck's reduced constant, $\hbar$. Note that the orbital angular momentum of an electron in an atom in an $L = 1$ state has an angular momentum of $\hbar$, so that the Bohr magneton is the expected magnetic dipole due to the orbital motion but based on classical physics. The Bohr magneton is given by $\mu_B = \frac{e\hbar}{2m_e}$. This may be derived as follows.

If the speed is v the angular momentum is $m_e vr$ where r is the radius of the orbit. Its periodic time is $2\pi r/v$ and hence the current is $I = \frac{ev}{2\pi r}$ so the magnetic dipole is $I\pi r^2 = \frac{erv}{2}$. Hence the magnetic dipole moment divided by the angular momentum is $\frac{erv}{2m_e vr} = \frac{e}{2m_e}$. So, for an orbital angular momentum equal to $\hbar$ the corresponding magnetic dipole moment according to classical physics for a circular orbit would be $\mu_B = \frac{e\hbar}{2m_e}$, i.e., the Bohr magneton, the value of which is given in Table 9.5.

The ratio of the magnetic dipole moment to the angular momentum is called the gyromagnetic ratio and, on the basis of the above argument, is about $\frac{e}{2m_e} = 8.794 \times 10^{10}\ \mathrm{C/kg}$ for the orbital motion of a classical particle with the same charge and mass as the electron.

The electron and proton also have magnetic moments due to their spin (a spinning charge constituting a current loop if it has non-zero size). The neutron also has a magnetic moment, despite overall neutrality, suggesting that it has unequal radial distributions of positive and negative charges. The magnetic dipole of the neutron is in the opposite sense to its spin, suggesting that the negative charge predominates at larger radii.

The measured magnetic moment of the electron given in Table 9.3 has been normalised by the Bohr magneton. This is called the "g-ratio" of the electron and would have been expected to be close to 1 based on the above argument. That it is instead close to 2 was first explained by Dirac on the basis of his quantum equation (see [9.1]). Note that because the spin angular momentum of an electron is $\hbar/2$ having a g-ratio close to 2 means that its spin magnetic moment is roughly one Bohr magneton.

In Table 9.3 the magnetic dipole moments of the proton and neutron are given after normalisation by the nuclear magneton (defining their g-ratios). The latter, given in Table 9.5, is defined in terms of the proton mass, $\mu_N = \frac{e\hbar}{2m_p}$. The g-ratios of the proton and neutron differ from the naively expected value of unity even more than that of the electron.

Note that the signs of the g-ratios in Table 9.3 reflect whether the magnetic dipole is aligned with, or opposite to, the spin.

Table 9.3: The Universal Constants

Constant	Value
Permittivity of the vacuum, ε_0	$8.854187816 \times 10^{-12}$ F/m
Permeability of the vacuum, μ_0	$4\pi \times 10^{-7}$ H/m
Charge of the electron or proton, e	$1.602176634 \times 10^{-19}$ C
Velocity of light, c	2.99792458×10^{8} m/s
Planck's constant, h	$6.62607015 \times 10^{-34}$ Js
Planck's constant / $2\pi, \hbar$	$1.054571818 \times 10^{-34}$ Js
Gravitational constant, G	6.6743×10^{-11} kg^{-1}m^{3}s^{-2}
g_e, Electron magnetic moment / μ_B*	-2.002319304361
g_p, Proton magnetic moment / μ_N*	5.5856946893
g_n, Neutron magnetic moment / μ_N*	-3.82608552
Electron mass, m_e	$9.1093837015 \times 10^{-31}$ kg
Proton mass, m_p	$1.672621898 \times 10^{-27}$ kg
Neutron mass, m_n	$1.674927471 \times 10^{-27}$ kg

**These dimensionless constants are also known as g-ratios or g-factors. The Bohr magneton and the nuclear magneton, in terms of which they are defined, as given in Table 9.5.*

Table 9.4 lists some constants which may be regarded as merely conventions. Thus, Boltzmann's constant, k_B, essentially defines the kelvin temperature scale in terms of the thermodynamic temperature, τ, whose dimensions are those of energy, so that temperature in kelvin is τ/k_B.

It is now usual to define a "mole" as the quantity of a substance containing Avogadro's number of molecules. But originally the mole was defined as the quantity of a pure element or compound whose mass was its molecular weight in grams. For example, a mole of CO_2 would be 44 grams because, using the atomic masses of 12 and 16 for carbon and oxygen respectively, $12 + 2 \times 16 = 44$. Avogadro's number is then defined as the number of molecules in a mole of any substance.

It is useful to know that a mole of any substance contains the same number of molecules, namely Avogadro's number. However, for this to equate to (say) 44 grams of CO_2 the carbon dioxide in question would need to be isotopically pure, consisting entirely of carbon-12 and oxygen-16. Traces of other isotopes will occur for most elements (e.g., carbon-13, carbon-14, oxygen-17 and oxygen-18) so that 44 grams of naturally occurring CO_2 will not contain exactly Avogadro's number of molecules but slightly fewer.

Table 9.4: Conventional Constants

Constant	Value
Boltzmann's constant, k_B	1.38065×10^{-23} J/K
Avogadro's number, N_A	6.0221×10^{23}

Table 9.5: Derived Constants

Constant	Definition	Value
Bohr magneton, μ_B	$e\hbar/2m_e$	$9.274009994 \times 10^{-24}$ J/T
Nuclear magneton, μ_N	$e\hbar/2m_p$	$5.0507837393 \times 10^{-27}$ J/T

9.3 Illustrative Magnitudes

Unless otherwise stated, in the illustrations which follow, the fields, charges, currents, etc., are assumed to lie within the vacuum (for which ambient air is generally a good enough approximation).

9.3.1 Force between Two Point Charges

The electrostatic force between two point charges of one coulomb at a distance of one metre is $\frac{q_1 q_2}{4\pi\varepsilon_0 r^2} \to \frac{1}{4\pi\varepsilon_0} = 9 \times 10^9$ N, which is 916,000 tonnes force. You can take from this either that the electrostatic force is extremely strong or that one coulomb is a very large charge in the context of static electricity. This is despite one amp being a very ordinary current, and equal to one coulomb per second.

9.3.2 Force on a Point Charge in an Electric Field

The force is given by $F = qE$ hence, in an electric field of $1\ \mathrm{V/m}$ a point charge of $1\ \mathrm{C}$ experiences an electrostatic force of $1\ \mathrm{N}$. Comparison with 9.3.1 indicates that a one coulomb charge must create a very large electric field.

9.3.3 Electric Field at One Metre from a Point Charge

At a distance of one metre from a point charge of one coulomb the electric field is, from (4.2.4), $E = \frac{q}{4\pi\varepsilon_0 r^2} \to \frac{1}{4\pi\varepsilon_0} = 9 \times 10^9\ \mathrm{V/m}$, hence reproducing the force on a unit charge given in 9.3.1. This is an enormous field strength, some 3,000 or more times larger than the breakdown strength in dry atmospheric air, i.e., the field required to produce arcing in air.

9.3.4 Electric Field due to Uniform Surface Charge Density

Sufficiently close to a conductor with a surface charge density of $1\ \mathrm{C/m^2}$ the electric field, which is normal to the surface, is given by (4.2.5d) and is thus $\frac{1}{\varepsilon_0} \approx 1.1 \times 10^{11}\ \mathrm{V/m}$. This example, and §9.3.1 and §9.3.3, illustrate that, as regards as the generation of electrostatic forces and fields, one coulomb is a huge charge. This contrasts with current electricity in which one coulomb per second, i.e., one amp, is a very "ordinary" magnitude of current.

9.3.5 Magnetic Force between Two Current Elements

The force between a pair of parallel current elements of length 1 metre, both bearing currents of 1 amp, and 1 metre apart is $F = \frac{\mu_0}{2\pi} = 2 \times 10^{-7}\ \mathrm{N}$. Note how tiny this is – some 17 orders of magnitude smaller than the electrostatic force of §9.3.1.

9.3.6 Force on Current Element in Unit Magnetic Flux Density

The force on a current of 1 amp in a uniform magnetic field of flux density 1 tesla is given by (4.2.8) to be $F = BI$ per unit length, and hence is $1\ \mathrm{N/m}$. Comparison with §9.3.5 implies that 1 tesla is a strong magnetic field.

9.3.7 Magnetic Flux Density due to a Current in a Straight Wire

The magnetic flux density at a distance of 1 metre from a current of 1 amp flowing in a straight wire is given by (4.2.21) to be $B = \frac{\mu_0}{2\pi} = 2 \times 10^{-7}\ \mathrm{T}$. This is tiny, two orders of magnitude smaller than the Earth's magnetic field at the Earth's surface. However the field at 1cm would be comparable.

9.3.8 The Earth's Magnetic Field

The naturally occurring magnetic field at the Earth's surface is strongest near the poles where it is about 6×10^{-5} T. At the equator it is roughly half that. As would be expected for a roughly dipole field, the direction of the field approaches the vertical at the poles but is horizontal at the equator. The field is vertically upwards (i.e., radially outwards) at the south pole and vertically downwards (radially inwards) at the north pole. In other words, the geographical north pole is a south magnetic pole and vice-versa. (To avoid misleading the reader, the magnetic poles do not align precisely, only rather crudely, with the geographic poles. Moreover, the poles flip magnetic sign randomly with typical intervals of a few hundred thousand years).

9.3.9 The Magnetic Field of an NMI Machine

Typical hospital NMI (nuclear magnetic imaging) scanners use a magnetic B-field (flux density) of between 1.5 and 3 tesla, which is a strong field. Research machines are currently up to 7 tesla.

9.3.10 What is the Magnetic Dipole Moment of an Electron?

The electron has spin ½, which means its angular momentum is $\frac{\hbar}{2} = 5.2729 \times 10^{-35}$ Js (or $\text{kgm}^2\text{s}^{-1}$). The magnetic dipole moment is obtained by multiplying the Bohr magneton by the electron's g-ratio/2, and hence is about $9.2848 \times 10^{-24}\, J/T$, very close to one Bohr magneton.

9.3.11 What is the Magnetic Dipole Moment of a Proton?

The proton has spin ½, which means its angular momentum is $\frac{\hbar}{2} = 5.2729 \times 10^{-35}$ Js (or $\text{kgm}^2\text{s}^{-1}$), the same as the electron despite its far larger mass. The magnetic dipole moment is obtained by multiplying the nuclear magneton by the proton's g-ratio/2, and hence is about $1.4106 \times 10^{-26}\, J/T$. The dipole moment of the electron is therefore 658 times larger than that of the proton, mostly as a consequence of the proton's greater mass despite the same angular momentum.

9.3.12 What is the Magnetic Field due to an Electron's Dipole?

The classical radius of the orbit of an electron in the ground state of the hydrogen atom (the Bohr radius) is 5.2917×10^{-11} m. At this distance, what is the magnetic field due to the electron's dipole moment? At the peak of the angular distribution, namely $\theta = 0$, (5.5.5) gives the flux density to be $B = \frac{2\mu_0 m}{4\pi r^3} = 12.53$ T. So, at such a close range the electron's dipole produces a magnetic field rather stronger than that of an NMI scanner.

9.3.13 What is the Magnetic Field due to a Proton's Dipole?

Using the dipole moment from §9.3.11, and at a distance equal to the Bohr radius, $B = \frac{2\mu_0 m}{4\pi r^3} = 0.019039$ T. Though far smaller than the field due to an electron, this is still 317 times stronger than the Earth's surface magnetic field.

9.3.14 The Magnetic Potential Energy of an Electron near a Proton

Ignoring the electrostatic energy due to the Coulomb interaction, the magnetic energy of an electron due to the magnetic field of a proton at a distance of r (or equivalently the energy of the proton in the magnetic field of the electron) is $\frac{2\mu_0 m_p m_e}{4\pi r^3}$, see (5.5.5) and (5.6.5). At a distance of one Bohr radius this is 1.767×10^{-25} J (1.105×10^{-6}eV). This is the potential energy if the particles lie at the peak of the dipole field (at $\theta = 0$) and their spins are aligned so that their dipole moments are anti-aligned, which arrangement yields the maximum positive potential energy. The potential energy is minus this if their spins are anti-aligned so that their dipole moments are aligned.

9.3.15 The Electric Potential Energy of an Electron near a Proton

The Coulomb electrostatic potential energy of a pair of charges separated by one Bohr radius is, using (5.1.4), $\frac{q_e q_p}{4\pi\varepsilon_0 r} = -4.360 \times 10^{-18}$ J (27.2eV). This is necessarily always negative as the electron and proton have opposite signs. It completely swamps the magnetic energy associated with the electron and proton dipole moments, being $\sim 2.5 \times 10^7$ times larger.

9.3.16 What Field can a Permanent Magnet Produce?

A basic fridge magnet would probably have a flux density of 0.001 to 0.01 tesla. But a strong neodymium magnet can produce a flux density of up to about 1.4 T. (Note that so-called neodymium magnets are not pure neodymium but a compound with iron and boron, $Nd_2Fe_{14}B$, whose tetragonal crystalline structure is key to its powerful magnetizability).

9.3.17 The Magnetic Field of a Coil

§5.7.5 has given the magnetic flux density at the centre of a circular coil of radius r, namely $B = \frac{\mu_0 NI}{2r}$ where N is the number of turns of wire in the coil. To produce a flux density of 1 tesla within a 1 metre diameter coil of 2000 turns would therefore need a current of nearly 400 amps. Unless superconducting, such a current would likely imply large power consumption unless fairly large copper windings were used – which implies both great weight and great cost in terms of copper given 2000 large windings being required. This is why NMI machines are heavy, expensive and require special electrical supplies.

9.3.18 What Field can a Superconducting Electromagnet Produce?

The superconducting magnets at CERN's LHC (Large Hadron Collider) deliver 8.3 T from a current of over 11,000 A at a temperature below 2 K. The central superconducting solenoid in the ITER fusion reactor delivers 13.5 T from a current of 46,000 A. An electromagnet producing 32 T is planned using a high temperature superconductor.

9.3.19 The Magnetic Field within an Atomic Nucleus

The peak field within an atomic nucleus due to the nucleus's own dipole moment is given roughly by $B = \frac{2\mu_0 m}{4\pi r^3}$ where r is now of order of the size of the nucleus. This is about 5 orders of magnitude smaller than atomic size, and because the field goes as $1/r^3$ the field could exceed that of §9.3.13 by up to 15 orders of magnitude. Realistically, flux densities are likely to be of order 10^{12} T, very roughly.

9.3.20 Example Electric Field Energies

The energy density in an electric field is given by (5.3.3). In one cubic metre the energy of a 1 V/m field would be $\frac{\varepsilon_0}{2} \approx 4.43 \times 10^{-12}$ J.

At an electric field strength of 3×10^6 V/m air will cease to be an insulator, undergoing "dielectric breakdown". At and above such field strengths, sparks will fly through the air as the field pulls neutral atoms apart and creates ions and free electrons, which can then conduct current. In a cubic metre, a field of 3×10^6 V/m will have an energy of 40 J. (NB: air containing moisture will breakdown at lower field strengths). It is remarkable that sparking occurs when the energy density is so modest.

9.3.21 Example Magnetic Field Energies

The energy density in a magnetic field is given by (5.6.11). In one cubic metre the energy of a 1 T field would be $\frac{1}{2\mu_0} \approx 4 \times 10^5$ J. The reason why a 1 T magnetic field has a very large energy but a 1 V/m electric field has a very tiny energy is simply because 1 V/m is a tiny electric field but 1 T is a very large magnetic field. It is salutary to consider the prodigious energy density you are placing your head into when having a brain scan in an NMI device.

9.4 Sticking a Balloon to a Wall

Readers will be familiar with sticking a child's balloon to a wall with static electricity. The balloon is rubbed on a suitable fabric, e.g., your jumper (sweater), to build sufficient charge. It will then cling to an ordinary interior wall, often remaining there for many hours. A homework problem I used to set students was to estimate the charge and the potential of the balloon. Before setting them loose on the calculation I would ask them first what they would guess to be the rough order of magnitude of the potential: a microvolt? A millivolt? One volt? Ten Volts? A hundred volts? A thousand volts? Ten thousand volts? A hundred thousand volts? The last few options would draw laughter...

I don't make any great claims for the accuracy of the following rough estimate – you may be able to do better. But it's good enough. The basis of the calculation is that,

- The balloon is attracted to the wall because its charge induces an opposite charge in the wall, and,
- A vertical force balancing the balloon's weight arises from friction, and I assume a coefficient of friction of unity so that this friction force equals the total horizontal force of attraction.
- I further assume that the surface charge density induced on the wall has the same magnitude (but opposite sign) to the surface charge density on the balloon, σ.
- Finally, I assume the force of attraction can be estimated to be the same as that which would apply for two parallel planes, each of area πr^2 where r is the radius of the balloon, with surface charge density σ.

The electric field due to the surface charge density σ is roughly $\frac{\sigma}{\varepsilon_0}$ (this being exact if the idealised planes of charge were of infinite extent). The attractive force generated by the idealised total charge of $\pi r^2 \sigma$ would therefore be $\pi r^2 \sigma^2 / \varepsilon_0$. Taking, as stated, the limiting friction to be equal to this, and because that must exceed the weight of the balloon, w, we have $\frac{\pi r^2 \sigma^2}{\varepsilon_0} \geq w$.

I shall take the balloon's radius to be 10 cm = 0.1 m. Sources I have seen put the mass of a deflated balloon at between 0.5 g and 3 g. For sake of argument I shall assume 1 g, noting that underestimating the weight will underestimate the charge and potential of the balloon. (Note that the gauge pressure of the air inside the balloon is likely to be about 3,000 Pa, and so the absolute pressure is only about 3% above atmospheric pressure, which is 0.1 MPa. Hence the mass of the air inside the balloon minus the mass of displaced air only adds an effective mass of about 0.15 gram after allowing for buoyancy forces, and this I shall neglect for conservatism). Hence I take $w \approx 0.01$ N.

Hence $\sigma = \sqrt{\frac{\varepsilon_0 w}{\pi r^2}} \approx 1.7 \times 10^{-6}\ C/m^2$ and the total charge on the balloon is then $q = 2.1 \times 10^{-7}$ C. This small charge will not surprise. However, the corresponding potential of the balloon is given by (5.1.4), being the same as that of a point charge q at a distance r, hence $\phi = \frac{q}{4\pi\varepsilon_0 r} = 19{,}000$ V.

So, the initial laughter at the idea that the balloon might be at, say, 10,000 volts was inappropriate – actually the balloon's potential is probably greater than that.

A moment's reflection shows that one should have guessed that static electricity typically produces very high voltages. It is a common experience that, after walking over certain floor coverings, one's own body can be charged – and that this can result in sparks when touching other objects such as the handrail on a staircase. But the breakdown field strength of dry air is 3×10^6 V/m or 3,000 V/mm. Consequently, even if the sparks fly only when one's hand is within a few millimetres of the handrail, one is still looking at a potential of the order of 10,000 volts.

The reason why we find this surprising is that our intuition is trained on circuit electricity, in which sources of 10,000 volts would need to be treated with the greatest respect as the danger they pose is considerable. The difference is that such a source is generally capable of delivering this high voltage sustainably. In contrast, a static source discharges and neutralises itself extremely quickly. The difference then relates to the energy that may be delivered. A sustained 10,000 volts can push a sustained current into your body, thus resulting potentially in substantial energy deposition and causing a lot of damage. But what energy is available from our balloon charged to 19,000 volts? The energy is just $q\phi = 4 \times 10^{-3}$ J, so not much. Nevertheless, static electricity of this kind can be dangerous. For example, a spark from a static electric discharge could initiate a major fire if it occurs in an environment with easily inflammable materials, especially gases. And static discharges can be fatally damaging to delicate silicon chip integrated circuits.

9.5 Electromagnetic Material Properties and Components

9.5.1. Relative Permittivity

The relative permittivity (also known as the dielectric constant) of some materials is typically as follows, bearing in mind that these will vary with frequency, temperature, field strength and the particular structure of the sample of material.

- Metals: 1
- Air: 1.0006
- Paper: 1.4 to 3.5
- Paraffin wax: 2 to 2.5
- Plastics: typically between 2 and 4
- Glass: typically 4 to 10
- Diamond: 5 to 10
- Germanium: 16
- Ethanol: 24
- Water (pure): 78

And for some compounds with exceptionally high relative permittivity,

- Lead lanthanum titanate: 1200 – 1700
- Lead zirconate titanate (PZT): 1000 - 2000
- Lead magnesium niobate (PMN): 3000 - 8000

9.5.2 Relative Permeability

The relative permeability (the magnetic susceptibility plus 1) of some materials is typically as follows, bearing in mind that these will vary with frequency, temperature, field strength and the particular structure of the sample of material.

- Bismuth:0.99983 (diamagnetic), magnetic susceptibility -1.7×10^{-4}
- Copper: 0.999991 (diamagnetic), magnetic susceptibility -9×10^{-6}
- Superconductors: 0 (i.e., a magnetic susceptibility of -1, the magnetic field being repelled, known as the Meissner effect)
- Platinum:1.00026 (paramagnetic), magnetic susceptibility 2.6×10^{-4}
- Austenitic stainless steel: < 1.05 ("non-magnetic stainless steel")
- Neodymium magnet material: 1.05
- Mild steel (carbon steel): 100
- Nickel: 100 – 600
- Ferritic stainless steel: 1000 – 1800 ("magnetic stainless steel")
- Pure iron (99.8%): 5,000*
- Mu-metal: 20,000 – 60,000
- Permalloy: 100,000

*Some sources, e.g., [9.2], list 99.95% pure iron, annealed in hydrogen, as having a relative permeability of up to 200,000.

9.5.3 Electrical Conductivity

Some conductivities at 20°C are given below in $\mathrm{S/m}$, but these will vary with temperature. They may also be different in different orientations for crystalline samples.

- Silver: 6.30×10^7
- Copper: 5.96×10^7
- Gold: 4.11×10^7
- Iron: 1.03×10^7
- Platinum: 0.943×10^7
- Carbon steel: 0.70×10^7
- Titanium: 0.24×10^7
- Austenitic stainless steel: 0.14×10^7
- Graphite: 3×10^5 (in-plane), 3×10^2 (perpendicular)
- Sea water: 4.8
- Tap water: $< 5 \times 10^{-2}$ (deionised, 4×10^{-5})
- Wood: $< 10^{-14}$ (dry), $10^{-4} - 10^{-3}$ (damp)
- Glass: $< 10^{-11}$
- Diamond: 10^{-13}
- Air: $10^{-9} - 10^{-15}$ (depending on humidity)
- PTFE: $< 10^{-23}$

9.6 References

[9.1] Bjorken, James D,., and Drell, Sidney D. (1964). *Relativistic Quantum Mechanics*. McGaw-Hill.

[9.2] Nave, C.R. (2024). *Magnetic Properties of Ferromagnetic Materials*. Georgia State University. Retrieved 8/8/24.

10. The Great Unification: Electromagnetic Waves and the Nature of Light

Here we arrive at the nineteenth century's greatest scientific triumph in physics, and the example par excellence of unification: the identification of light as a particular case of an electromagnetic wave. This was necessarily presaged by the realisation that the combined electric and magnetic fields would display wave properties – a purely theoretical realisation that results from the complete Maxwell equations. The theory already developed shows how such electromagnetic waves, both visible light and other wavelengths, will propagate energy and momentum through space. Whether said space could be said to be empty or full of some "luminiferous aether" was not decided on the basis of Maxwell's formulation alone. In the mid-19th century it was generally assumed that some sort of luminiferous aether was required as the medium which carried the e/m waves, in much the same manner as air carries sound. Only in the late 19th and early 20th century was the idea of a luminiferous aether abandoned, not least due to the Michelson-Morley experiment's negative outcome and finally by relativity theory and the realisation that fields were physical entities in their own right, not excitations of some medium.

Contents

10.1 Electromagnetic Waves in the Vacuum

In chapter 7 we arrived at the final version of Maxwell's equations,

$$\bar{\nabla}\cdot\bar{D}=\rho \qquad \bar{\nabla}\cdot\bar{B}=0 \qquad \bar{\nabla}\times\bar{H}=\bar{J}+\frac{\partial\bar{D}}{\partial t} \qquad \bar{\nabla}\times\bar{E}=-\frac{\partial\bar{B}}{\partial t} \qquad (10.1.1)$$

In the vacuum we have also that,

$$\bar{D}=\varepsilon_0\bar{E} \qquad \bar{B}=\mu_0\bar{H} \qquad (10.1.2)$$

If there are no charges or currents present then we also have $\rho=0$ and $\bar{J}=0$ so that equs.(10.1.1) become,

$$\bar{\nabla}\cdot\bar{E}=0 \qquad \bar{\nabla}\cdot\bar{B}=0 \qquad \frac{1}{\mu_0}\bar{\nabla}\times\bar{B}=\varepsilon_0\frac{\partial\bar{E}}{\partial t} \qquad \bar{\nabla}\times\bar{E}=-\frac{\partial\bar{B}}{\partial t} \qquad (10.1.3)$$

In this order these will be referred to as equations (1), (2), (3) and (4).

Note firstly that equs.(1) and (3) are consistent. As $DivCurl$ is identically zero, equ.(3) implies (assuming commutativity of the operators) $\frac{\partial\bar{\nabla}\cdot\bar{E}}{\partial t}=0$, and this is ensured by equ.(1). Similarly, equs.(2) and (4) are consistent for the same reason.

Taking the time derivative of equ.(4) then substituting from equ.(3) and using identity (A1.17) gives,

$$\frac{\partial^2\bar{B}}{\partial t^2}=-\bar{\nabla}\times\frac{\partial\bar{E}}{\partial t}=-\bar{\nabla}\times\frac{1}{\mu_0\varepsilon_0}\bar{\nabla}\times\bar{B}=-\frac{1}{\mu_0\varepsilon_0}\left[\bar{\nabla}(\bar{\nabla}\cdot\bar{B})-\nabla^2\bar{B}\right] \qquad (10.1.4)$$

Using also equ.(2) then gives,

$$\nabla^2\bar{B}=\mu_0\varepsilon_0\frac{\partial^2\bar{B}}{\partial t^2} \qquad (10.1.5)$$

This is the equation which the magnetic field must obey in a region containing no media, no charges and no currents. The reader will recognise it as the wave equation in 3D. If the reader is not familiar with the wave equation, that the solutions of (10.1.5) have a wave nature can be seen from its general solution. This is developed by seeking a plane wave solution of the form,

$$\bar{B}=\bar{B}_0 exp\{i(\omega t-\bar{k}\cdot\bar{r})\} \qquad (10.1.6)$$

(If the reader is unsettled by using a complex-valued field, just assume that we mean the real part – or the imaginary part). For this function, $\nabla^2\bar{B}=-k^2\bar{B}$ and $\frac{\partial^2\bar{B}}{\partial t^2}=-\omega^2\bar{B}$, so that (10.1.5) is obeyed so long as,

$$k^2=\mu_0\varepsilon_0\omega^2 \qquad (10.1.7)$$

But the speed of such a wave is ω/k. If this is not immediately obvious, the wave speed is the speed of a point of constant phase, say zero, so that we have $\omega t - kr = 0$ and so $\frac{r}{t} = speed = \frac{\omega}{k}$. Hence we conclude that solution (10.1.6) is a wave of the magnetic field propagating with speed,

$$c = \frac{1}{\sqrt{\mu_0 \varepsilon_0}} \tag{10.1.8}$$

The most general solution for the magnetic field in the vacuum with no sources is thus an arbitrary superposition of plane waves of the form (10.1.6) with any $\bar{k}$ and some amplitude $\bar{B}_0$ chosen for each $\bar{k}$, but understanding that the frequency is given by (10.1.7). Algebraically,

$$\bar{B} = \int_{-\infty}^{\infty} \bar{\beta}(\bar{k}) exp\{i(\omega t - \bar{k} \cdot \bar{r})\} d^3k \tag{10.1.9}$$

where $\bar{\beta}(\bar{k})$ is some vector function. It is not arbitrary but must obey certain constraints which are derived below.

What about the electric field? Taking the time derivative of equ.(3) then substituting from equ.(4) and again using identity (A1.17) gives,

$$\mu_0 \varepsilon_0 \frac{\partial^2 \bar{E}}{\partial t^2} = \bar{\nabla} \times \frac{\partial \bar{B}}{\partial t} = -\bar{\nabla} \times (\bar{\nabla} \times \bar{E}) = [-\bar{\nabla}(\bar{\nabla} \cdot \bar{E}) + \nabla^2 \bar{E}] \tag{10.1.10}$$

Using also equ.(1) then gives,

$$\nabla^2 \bar{E} = \mu_0 \varepsilon_0 \frac{\partial^2 \bar{E}}{\partial t^2} = \frac{1}{c^2} \frac{\partial^2 \bar{E}}{\partial t^2} \tag{10.1.11}$$

So the electric field in the vacuum obeys the same wave equation as the magnetic field, propagating with the same speed given by (10.1.8). Assume a plane wave solution for the electric field, of the form,

$$\bar{E} = \bar{E}_0 exp\{i(\omega' t - \bar{k}' \cdot \bar{r})\} \tag{10.1.12}$$

Substitution in (10.1.11) again requires $k'^2 = \mu_0 \varepsilon_0 \omega'^2$. However equs.(3) and (4) can only hold if $\bar{k}' = \bar{k}$ and $\omega' = \omega$ so we rewrite,

$$\bar{E} = \bar{E}_0 exp\{i(\omega t - \bar{k} \cdot \bar{r})\} \tag{10.1.12b}$$

Equations (3) and (4) also impose restrictions on the amplitudes, $\bar{B}_0$ and $\bar{E}_0$, giving respectively,

$$-\bar{\mathrm{k}} \times \bar{B}_0 = \mu_0 \varepsilon_0 \omega \bar{E}_0 \quad \text{and} \quad \bar{\mathrm{k}} \times \bar{E}_0 = \omega \bar{B}_0 \tag{10.1.13}$$

These equations tell us that $\bar{\mathrm{k}}, \bar{E}, \bar{B}$ are all mutually orthogonal. On the other hand, $\bar{\mathrm{k}}$ is the direction of propagation of the plane wave whose magnetic and electric components are given by (10.1.6) an (10.1.12b). So we conclude that the electromagnetic field in the source-free vacuum takes the form of a superposition of plane waves whose magnetic and electric field components are perpendicular to each other and also transverse to the direction of propagation. The relative orientation of the fields in a plane polarised plane wave are illustrated in Figure 10.1, in which the wave is propagating in the direction of the red arrow.

In terms of the electromagnetic potentials this plane wave solution can be expressed more compactly as,

$$\Phi = 0 \quad \text{and} \quad \bar{A} = i\bar{A}_0 exp\{i(\omega t - \bar{k}\cdot\bar{r})\} \tag{10.1.14}$$

where we must choose $\bar{A}_0$ to be perpendicular to $\bar{k}$. All the above equations are then obeyed with $\bar{E}_0 = \omega\bar{A}_0$ and $\bar{B}_0 = \bar{k}\times\bar{A}_0$.

Figure 10.1: Plane Polarised Electromagnetic Plane Wave

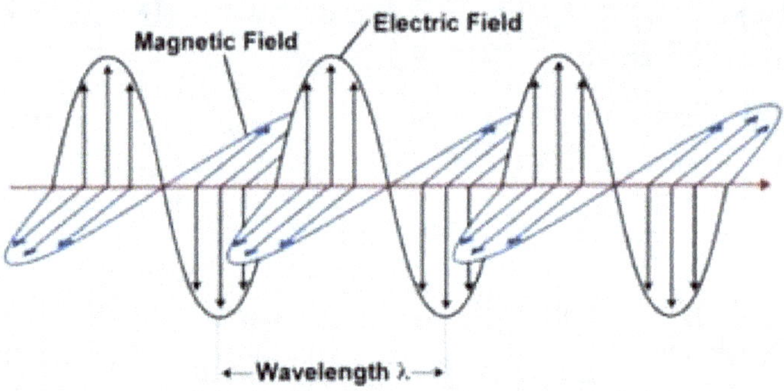

Note that (10.1.13) means that $\bar{k}, \bar{E}, \bar{B}$, in that order, form a right-handed system. Hence, in Figure 10.1, if $\bar{k}$ is in the positive $\hat{x}$ direction and $\bar{E}$ is in the positive $\hat{z}$ direction, then $\bar{B}$ is in the direction $-\hat{y}$, as shown. The amplitudes of the magnetic and electric components are not independent but related by equs.(10.1.13). So the equations of our example plane polarised plane wave are,

$$\bar{E} = E_0\hat{z}\cos(\omega t - kx) \quad \text{and} \quad \bar{B} = -\frac{E_0}{c}\hat{y}\cos(\omega t - kx) \tag{10.1.14}$$

Thus, at any given point in space, the electric field oscillates sinusoidally between $-E_0$ and $+E_0$ with a periodic time of $2\pi/\omega$. Similarly, at any given time, the electric field varies with position between $-E_0$ and $+E_0$ with a wavelength of $2\pi/k$.

More generally, if the magnetic field is given by the superposition (10.1.9) then the electric field is given by,

$$\bar{E} = -c^2\int_{-\infty}^{\infty}\frac{\bar{k}}{\omega}\times\bar{\beta}(\bar{k})exp\{i(\omega t - \bar{k}\cdot\bar{r})\}d^3k \tag{10.1.15}$$

where $\bar{\beta}(\bar{k})$ is arbitrary except the requirement that,

$$\bar{k} \cdot \bar{\beta}(\bar{k}) = 0 \tag{10.1.16}$$

which is required to ensure that $\bar{\nabla} \cdot \bar{B} = 0$. The equation $\bar{\nabla} \cdot \bar{E} = 0$ is obeyed by virtue of the form of (10.1.15).

10.1.1 Circular Polarisation

In a plane polarised plane wave, the electric field varies in time and space but its direction is always and everywhere the same. The same is true of the magnetic field. But it is important to recall that the general solution of the vacuum field equations is a superposition of such waves – which will then usually cease to be a plane polarised plane wave. An alternative polarisation for a plane wave is circular polarisation. An example of a circularly polarised plane wave is,

$$\bar{E} = E_0[\hat{z}\cos(\omega t - kx) + \hat{y}\sin(\omega t - kx)] \tag{10.1.17a}$$

$$\bar{B} = \frac{E_0}{c}[-\hat{y}\cos(\omega t - kx) + \hat{z}\sin(\omega t - kx)] \tag{10.1.17b}$$

As can readily be seen, this is obtained as the superposition of a pair of plane polarised plane waves, with electric vectors which are perpendicular and 90° out of phase (as are the magnetic vectors). Note that the field vectors in (10.1.17a,b) are orthogonal to the direction of propagation, which is $\hat{x}$. Also, the electric and magnetic field vectors are everywhere and always perpendicular, i.e., $\bar{E} \cdot \bar{B} = 0$.

Equs.(10.1.17a,b) are circularly polarised in the clockwise direction as seen as the wave approaches the observer. To obtain anticlockwise circular polarisation, change the + signs in both of (10.1.17a,b) to $-$ signs.

10.1.2 Plane Waves versus Non-Plane Waves

Though circularly polarised, the wave in §10.1.1 is still a plane wave. A plane wave has an unvarying set of parallel planes such that the phase, and hence the magnitude of the fields, are uniform across each plane at a given time, and the wave propagation direction is normal to these planes. The characteristic of plane waves is thus that they propagate everywhere and always in the same Cartesian direction, and hence the fields are functions of $\omega t \pm \bar{k} \cdot \bar{r}$ with a fixed $\bar{k}$ only. But this need not be the case. An example is a spherical wave in which the wavefronts (the surfaces of constant phase at a given time) are concentric spheres and the propagation direction is radial (inwards or outwards). One example is a source located at the origin which produces a spherical wave which propagates radially outwards. If idealised as a point source, the potentials and fields will be divergent at the origin. The derivation of the spherical wave solutions in the electromagnetic case is presented in chapter 24.

10.2 Identification of Light as an Electromagnetic Wave

Waves of the type derived in §10.1 arise as an automatic consequence of the Maxwell equations. That light is an example of such an electromagnetic wave follows by evaluating the wave speed, (10.1.8), given the permittivity and permeability of free space. Using the values for the latter given in chapter 9, the speed is found to be 2.99792458×10^8 m/s which is precisely the speed of light.

Of course, the precision of this agreement is because, in modern derivations of the universal constants, this agreement is constrained by this assumed identification. But recall that both the permittivity and the permeability of the vacuum can, in principle, be found from the corresponding force laws (essentially the Coulomb and Biot-Savart laws) in experiments involving static electric fields only and steady magnetic fields only respectively. Such experimental determinations do not involve any waves, nor even time dependence, and certainly not any light. Consequently, it is quite stunning that the electric and magnetic constants obtained in this manner do predict the speed of light.

It is worth pausing to be suitably reverential at this point. The great theme of physics is unification. To understand as much as possible from as few principles as possible is our perpetual goal. Surely Maxwell's achievement in identifying the electromagnetic nature of light must be the unification *par excellence.* Yesterday electromagnetism and light were two entirely separate phenomena, yet today they are the same thing. Personally, I count this as the most impressive unification in physics.

In Maxwell's day, the precision with which the permittivity and the permeability of the vacuum (or equivalent quantities) were known was limited. Nevertheless, the available measurements permitted (10.1.8), or its equivalent in the notation of the time, to predict the speed of light to within about 3%. As Maxwell himself put it, "*we can scarcely avoid the conclusion that light consists in the transverse undulations of the same medium which is the cause of electric and magnetic phenomena*". It is also worth emphasizing that this was crucially a triumph of theory. The electromagnetic nature of light was not an experimental discovery. It was discovered because the complete description of the electromagnetic field through Maxwell's equations permitted a wave-like solution via the simple manipulation given above.

Without doubt it is Maxwell who deserves the chief accolade in this triumph of understanding. But, as usual, the complete history reveals much nuance and many precursors of partial understanding.

In 1855 Weber and Kohlrausch had already determined experimentally, and published, that $\frac{1}{\sqrt{\mu_0 \varepsilon_0}}$ was 3.104×10^8 m/s, [10.1], hence within 3.5% of the now known speed of light. (I have taken the liberty of describing as $\frac{1}{\sqrt{\mu_0 \varepsilon_0}}$ the equivalent in the terminology of the day). However, whilst Weber and Kohlrausch derived this from electromagnetic considerations they did not link it to light. Nevertheless, by 1857 they had noted that electricity propagates along a wire of negligible resistance at the speed of light (which Fizeau had measured in 1848 to be 3.133×10^8 m/s. In 1850 Foucault had also measured the speed of light, and, in a later experiment in 1862 obtained the result 2.98×10^8 m/s, within 0.6% of the now accepted value [10.2]).

So Weber and Kohlrausch, and also Kirchhoff who carried out similar analyses, were not keen to identify light as being an electromagnetic phenomenon. In contrast, Maxwell's contemporary, Ludvig Lorenz (not to be confused with Hendrik Lorentz) did explicitly propose the electromagnetic nature of light. So did Riemann, of all people, and published a paper to that effect (a paper he later withdrew as it contained a mathematical error: he had falsely assumed the commutation of two integrals). Riemann noted that he believed Gauss was of the same opinion and had his own theory which, as far as I am aware, was never published. However, it was not until Maxwell that the full formulation of the theory and its conceptualisation in terms of fields of force successfully elucidated the physical nature of light fully. This was built firmly upon Faraday's visualisation of the fields as "lines of force", an innovation that Maxwell alone at the time took seriously and which gave us the concept of vector fields. For the complete fascinating history see [10.2,3]. It is not for nothing that Einstein acknowledged his chief debt being to Maxwell.

10.3 The Energy and Momentum of Electromagnetic Waves

In this section we continue to consider electromagnetic waves *in vacuo* only. We have seen in (7.4.6) that the flux of electromagnetic energy is given by the Poynting vector, $\bar{N} = \bar{E} \times \bar{H}$. Hence, for our example plane wave of (10.1.14) the flux of energy is,

$$\bar{N} = -E_0\hat{z}\cos(\omega t - kx) \times \frac{E_0}{\mu_0 c}\hat{y}\cos(\omega t - kx) = \hat{x}\frac{E_0^2}{\mu_0 c}\cos^2(\omega t - kx)$$

Averaging over either time or space this becomes,

$$\langle\bar{N}\rangle = \hat{x}\frac{E_0^2}{2\mu_0 c} = \frac{1}{2}c\varepsilon_0 E_0^2\hat{x} \tag{10.3.1}$$

We recognise this as the peak electrical energy density $\frac{1}{2}\varepsilon_0 E_0^2$, as given by (5.3.3), propagating at speed c in the $\hat{x}$ direction. Why the peak energy density, not the spatial average energy density? Well, actually it is also the spatial average energy density propagating at speed c in the $\hat{x}$ direction, providing the magnetic energy density, given by (5.6.11), $\frac{1}{2\mu_0}B_0^2$, is also taken into account. Because (10.1.13) tells us that $E_0 = cB_0$, it follows that the electrical and magnetic energy densities are the same, $\frac{1}{2}\varepsilon_0 E_0^2 = \frac{1}{2\mu_0}B_0^2$, and hence the total is $\varepsilon_0 E_0^2$ and when this is averaged we arrive back at $\frac{1}{2}\varepsilon_0 E_0^2$.

The average momentum flux is just (10.3.1) divided by c and hence $\frac{1}{2}\varepsilon_0 E_0^2\hat{x}$. Note that the units of momentum flux are the same as energy density.

10.4 Electromagnetic Waves Generated by Sources

We saw in §7.2.1 that the scalar and vector potentials for use with the full, time-dependent Maxwell equations, (10.1.1), are given by,

$$\bar{E} = -\frac{\partial\bar{A}}{\partial t} - \bar{\nabla}\phi \quad \text{and} \quad \bar{B} = \bar{\nabla}\times\bar{A} \tag{10.4.1}$$

These make Maxwell equations (2) and (4) identities. There remain equations (1) and (3), which are the equations with source terms, charge density, ρ, and its flux, the current density, $\bar{J}$. Substituting from (10.4.1) these become,

$$-\varepsilon_0\bar{\nabla}\cdot(\partial_t\bar{A} + \bar{\nabla}\phi) = \rho \quad \text{and} \quad \frac{1}{\mu_0}\bar{\nabla}\times(\bar{\nabla}\times\bar{A}) = \bar{J} - \varepsilon_0\partial_t(\partial_t\bar{A} + \bar{\nabla}\phi)$$

Using (A1.17) these become,

$$\partial_t\bar{\nabla}\cdot\bar{A} + \nabla^2\phi = -\frac{\rho}{\varepsilon_0} \tag{10.4.2}$$

and

$$\bar{\nabla}(\bar{\nabla}\cdot\bar{A}) - \nabla^2\bar{A} = \mu_0\bar{J} - \frac{1}{c^2}\partial_t(\partial_t\bar{A} + \bar{\nabla}\phi) \tag{10.4.3}$$

Rearranging (10.4.3) gives,

$$\nabla^2 \bar{A} - \frac{1}{c^2}\partial_t^2 \bar{A} = -\mu_0 \bar{J} + \bar{\nabla}\left[\bar{\nabla}\cdot\bar{A} + \frac{1}{c^2}\partial_t \phi\right] \tag{10.4.4}$$

whilst (10.4.2) gives,

$$\nabla^2 \phi - \frac{1}{c^2}\partial_t^2 \phi = -\frac{\rho}{\varepsilon_0} - \partial_t\left[\bar{\nabla}\cdot\bar{A} + \frac{1}{c^2}\partial_t \phi\right] \tag{10.4.5}$$

10.4.1 Gauge Freedom

Equs.(10.4.4,5) may be simplified because we are entitled to assume that,

$$\bar{\nabla}\cdot\bar{A} + \frac{1}{c^2}\partial_t \phi = 0 \tag{10.4.6}$$

This is known as the "Lorenz gauge"[1]. That we are able to impose this condition without loss of physical content follows from gauge invariance, namely that the transformations, for some scalar field f,

$$\bar{A} \to \bar{A} + \bar{\nabla} f \quad \text{and} \quad \phi \to \phi - \partial_t f \tag{10.4.7}$$

leave the $\bar{E}$ and $\bar{B}$ fields unchanged. Moreover, the expression $\bar{\nabla}\cdot\bar{A} + \frac{1}{c^2}\partial_t \phi$ transforms as follows,

$$\bar{\nabla}\cdot\bar{A} + \frac{1}{c^2}\partial_t \phi \to \bar{\nabla}\cdot\bar{A} + \frac{1}{c^2}\partial_t \phi + \left(\nabla^2 f - \frac{1}{c^2}\partial_t^2 f\right) \tag{10.4.8}$$

Hence, if we choose the gauge function, f, to be the solution to the equation $\left(\nabla^2 f - \frac{1}{c^2}\partial_t^2 f\right) = -g$ where g is the scalar function $\bar{\nabla}\cdot\bar{A} + \frac{1}{c^2}\partial_t \phi$ then after transformation we have (10.4.6). That such an f exists follows from the explicit solution for it in terms of an integral of g times the Green's function for the d'Alembertian operator $\nabla^2 - \frac{1}{c^2}\partial_t^2$ (see chapter 15). (Strictly we have ridden rough-shod over the boundary conditions required to make that argument rigorous).

Thus, adopting the Lorenz gauge we find from (10.4.4,5) that the equations determining the scalar and vector potentials in terms of the sources are,

$$\frac{1}{c^2}\partial_t^2 \bar{A} - \nabla^2 \bar{A} \equiv \partial^2 \bar{A} = \mu_0 \bar{J} \tag{10.4.9a}$$

---

[1] *The Lorenz gauge is named after the Danish physicist Ludvig Lorenz and should not to be confused with the Dutch physicist Hendrik Lorentz, born some 24 years later, who gave us the Lorentz transformation. That both were noted for their work in electromagnetic theory adds to the potential confusion. It was Lorenz, not Lorentz, who was among the first to recognise the electromagnetic nature of light. This was already established by the time Lorentz was adult.*

$$\frac{1}{c^2}\partial_t^2\phi - \nabla^2\phi = \partial^2\phi = \frac{\rho}{\varepsilon_0} \qquad (10.4.9b)$$

where the d'Alembertian operator is defined as $\partial^2 \equiv \frac{1}{c^2}\partial_t^2 - \nabla^2$. In the case that there are no sources terms, $\rho = \bar{J} = 0$, these become the wave equation, so in this case $\bar{E}, \bar{B}, \bar{A}$ and ϕ all obey the wave equation. However, (10.4.9a,b) add crucially to our understanding because they tell us how the potentials (and hence the fields) may be generated by source terms. Without this we would have no way of calculating how such waves arise. Chapter 22 will take this up in greater detail to derive the radiation fields from oscillating charges and dipoles.

10.5 An Intimation of Relativity

We close this chapter by returning to an observation made at its start. The following are the Maxwell equations in vacuo without sources,

$$\bar{\nabla}\cdot\bar{E} = 0 \qquad \bar{\nabla}\cdot\bar{B} = 0 \qquad \bar{\nabla}\times\bar{B} = \frac{1}{c^2}\frac{\partial\bar{E}}{\partial t} \qquad \bar{\nabla}\times\bar{E} = -\frac{\partial\bar{B}}{\partial t}$$

The observation is that, as $DivCurl$ is identically zero, equ.(3) implies (assuming commutativity of the operators) $\frac{\partial\bar{\nabla}\cdot\bar{E}}{\partial t} = 0$, and this is ensured by equ.(1) and hence equs.(1) and (3) are consistent. But note how close this is to being able to derive equ.(1) from equ.(3). The latter gives $\frac{\partial\bar{\nabla}\cdot\bar{E}}{\partial t} = 0$ which means that we can write $\bar{\nabla}\cdot\bar{E} = f(\bar{r})$ where $f(\bar{r})$ is a function of the spatial coordinates alone, i.e., independent of time.

From this we can, if we are already versed in relativity theory, anticipate that the requirement for relativistic covariance will oblige us to regard $f(\bar{r})$ as a constant (which can then be taken as zero). The reason is that the Lorentz transformation – the essence of special relativity – mixes space and time coordinates. But this would mean that a function seen as $f(\bar{r})$ would be seen as $f(t', \bar{r}')$ by an observer in relative motion wrt the first. But relativistic covariance requires the formal expression of the laws of physics to be the same for all inertial observers, so that the absence of a time dependence in one frame conflicts with the time dependence in another frame. The only resolution is that $f(\bar{r})$ is actually a constant.

This simple argument, based on the principle of relativistic covariance and the specifics of the Lorentz transformation, allows equ.(1) to be derived from equ.(3). In exactly the same manner, equ.(2) can be derived from equ.(4). Hence we may anticipate that special relativity will reduce the number of independent Maxwell equations from 4 to 2, i.e., just equs.(3) and (4).

Note that this remains true when sources are added because the modified equ.(3), namely $\overline{\nabla} \times \overline{H} = \overline{J} + \frac{\partial \overline{D}}{\partial t}$, allows the modified equ.(1), $\overline{\nabla} \cdot \overline{D} = \rho$, to be derived given the continuity, or conservation of charge, equation, (3.8.12), $\frac{\partial \rho}{\partial t} + \overline{\nabla} \cdot \overline{J} = 0$. We shall develop these matters in greater detail when we turn to relativity proper in chapters 26 – 30.

10.6 References

[1] Assis, A.K.T. (2003). *On the First Electromagnetic Measurement of the Velocity of Light by Wilhelm Weber and Rudolf Kohlrausch*. In: Volta and the History of Electricity, edited by F.Bevilacqua and E.A.Giannetto, Università degli Studi di Pavia and Editore Ulrico Hoepli, Milano, 2003, pp. 267-286.

[2] Keithley, J.F. (1998). *The Story of Electrical and Magnetic Measurements: From 500 BC to the 1940s*. John Wiley & Sons, 18 December 1998.

[3] Rosenfeld, L. (1956). *The velocity of light and the evolution of electrodynamics*. Nuovo Cimento Vol.4 (Suppl 5), 1630–1669. Also obtainable here.

11. Electromagnetic Waves in Media: Reflection, Refraction and Absorption

The form of electromagnetic waves in an LIH medium are obtained and used to derive the laws of reflection and refraction. The boundary conditions to which the fields are subject are seen to be fully compatible with (and indeed to require) the well-known laws of geometrical optics. This should be viewed as a completion of the theoretical case for identifying light with electromagnetic waves. (Had electromagnetic theory implied different behaviours on reflection or refraction, the identification of e/m waves with light would have failed, a key issue that is often neglected). Unlike geometrical optics, the understanding of the structure of light provided by electromagnetic theory allows the amplitudes and phases of the reflected and refracted waves to be determined. A distinction is drawn between reflection from dielectrics and reflection from good conductors (metals). The reason for the high reflectivity of metals at and below optical frequencies is elucidated.

Contents

Throughout this chapter the media in question are assumed to be linear, isotropic and homogeneous (LIH). The assumption of linearity means that the fields in such media can be derived from the assumed proportionality of $\bar{D}$ and $\bar{E}$, and also that of $\bar{B}$ and $\bar{H}$, together with the constancy of the corresponding coefficients of proportionality, the electric permittivity and the magnetic permeability. This is convenient in an elementary exposition but hardly indicative of real materials except within a narrow range of field strengths.

The assumption of isotropy means that the permittivity and the permeability can be taken to be scalars. For anisotropic media, e.g., single crystals, they would become second rank tensors, but we shall not address such cases in this book. Isotropy is often a reasonable approximation for either amorphous materials, or materials composed of many randomly oriented crystalline grains, provided average behaviour over a large enough spatial scale is adequate in the context. Isotropy is always false at sufficiently small scales.

The assumption of homogeneity applies on a large enough scale that the microscopic details of the crystalline structure, or even the molecular or atomic structure, of real materials need not concern us. The reader should be aware, though, that the assumption of LIH behaviour is adopted only for pedagogic convenience. The fascinating, and sometimes commercially useful, behaviour of real materials frequently lies in their departure from the LIH ideal.

11.1 The Modified Wave Equation in LIH Media

The derivation of the wave equation in §10.1 proceeds in essentially the same way when the fields arise in an LIH medium. The Maxwell equations, in the absence of charges or currents, are again as given by (10.1.1), but, as explained in chapter 6, equations (10.1.2) are replaced by,

$$\bar{D} = \varepsilon_r \varepsilon_0 \bar{E} \qquad \bar{B} = \mu_r \mu_0 \bar{H} \tag{11.1.1}$$

It follows immediately that the wave equations (10.1.5) and (10.1.11) become,

$$\nabla^2 \bar{B} = \mu_r \mu_0 \varepsilon_r \varepsilon_0 \frac{\partial^2 \bar{B}}{\partial t^2} = \frac{\mu_r \varepsilon_r}{c^2} \frac{\partial^2 \bar{B}}{\partial t^2} = \frac{1}{v^2} \frac{\partial^2 \bar{B}}{\partial t^2} \tag{11.1.2a}$$

and,

$$\nabla^2 \bar{E} = \frac{1}{v^2} \frac{\partial^2 \bar{E}}{\partial t^2} \tag{11.1.2b}$$

where, $$v = \frac{c}{\sqrt{\mu_r \varepsilon_r}} \tag{11.1.2c}$$

is the velocity of the waves in the medium. The refractive index, n, of the medium is defined as,

$$n = \sqrt{\mu_r \varepsilon_r} \qquad \text{so that} \quad v = \frac{c}{n} \tag{11.1.3}$$

Materials in which electromagnetic waves can propagate invariably have a relative permeability extremely close to unity. Consequently, the refractive index is generally written $n = \sqrt{\varepsilon_r}$. Materials have permittivities greater than 1, hence the refractive index is also greater than 1 which means $v < c$. As we will see in chapters 26-27, $v \leq c$ is a requirement of the relativity principle which thus provides a theoretical proof that materials must always be such that $\mu_r \varepsilon_r \geq 1$.

Plane wave solutions are thus of exactly the same form as derived in §10.1, namely,

$$\bar{B} = \bar{B}_0 exp\{-i(\omega t - \bar{k} \cdot \bar{r})\} \qquad \bar{E} = \bar{E}_0 exp\{-i(\omega t - \bar{k} \cdot \bar{r})\} \tag{11.1.4}$$

where $\bar{B}$ and $\bar{E}$ are orthogonal and $\bar{k}$ is in the direction $\bar{E} \times \bar{B}$ which is the direction of motion of the wave: the direction in which energy and momentum are carried by it. The sole difference is that the amplitudes of the fields are now related by $vB_0 = E_0$ rather than $cB_0 = E_0$ and the frequency and wavevector are related by $vk = \omega$ rather than $ck = \omega$. This latter condition means that the wavevector, and the wavelength, will change when an electromagnetic wave passes from the vacuum into a medium, e.g., when light enters into glass or other transparent material from the air or vacuum.

11.2 The Laws of Reflection and Refraction

11.2.1 Snell's Law: The Usual Argument

It is worth reprising the usual geometrical argument for deriving Snell's Law of refraction, familiar from school physics and apparently reliant only upon knowing that the speed of the wave is related to the refractive index by $v = c/n$. Considering Figure 11.1, a beam of light is incident on a plane boundary between two different media at an angle of θ_i to its normal (the angle of incidence). The beam is of finite width, so one side of it reaches the boundary (at A) before the other side (at B). The ray which reaches B has to travel an additional distance of $D = w \sin \theta_i$ where w is the width of the footprint of the beam on the boundary, as shown. Hence, the ray on the other side reaches A at a time $w \sin \theta_i / v_1$ earlier than the other ray reaches B, where

$v_1 = c/n_1$ is the wave velocity in the first (upper) medium. But as the wave in the second (lower) medium travels at a different speed, $v_2 = c/n_2$, it has travelled a distance $D' = (v_2/v_1)w \sin\theta_i$ when the other ray has just reached B. But we also have, from simple geometry, $D' = w \sin\theta_t$ where θ_t is the angle of refraction (to the boundary normal, as shown). Hence we must have,

$$D' = w \sin\theta_t = (v_2/v_1)w \sin\theta_i \quad \text{hence} \quad \frac{\sin\theta_i}{\sin\theta_t} = \frac{v_1}{v_2} = \frac{n_2}{n_1} \tag{11.2.1}$$

This is Snell's Law. Hence, if the beam travels from a region of lower to a region of higher refractive index ($n_2 > n_1$) the refracted beam moves closer to the boundary normal than it was ($\theta_r < \theta_i$).

Figure 11.1: Illustrating Snell's Law (for $n_2 > n_1$)

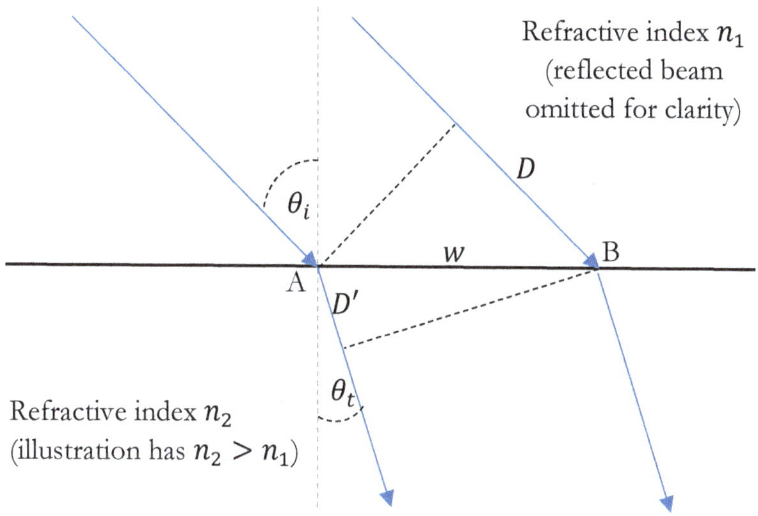

However, this argument can be criticised on the grounds that it relies on Huygens Principle, or something equivalent, i.e., how the wave propagates from point A. But even if it is taken as a valid derivation of Snell's Law, that is not in what we are principally interested. We wish to confirm that the electromagnetic understanding of light is fully consistent with the known behaviour of light, and specifically that the boundary conditions to which the electromagnetic fields are subject are all satisfied whilst also providing full consistency with geometrical optics. This is tackled next.

11.2.2 Field Boundary Conditions Cause Reflection and Refraction

The geometry of the incident, reflected and transmitted (refracted) beams with respect to a plane boundary between two different media is illustrated in Figure 11.2. The unit vectors $\hat{n}_i$, $\hat{n}_r$, $\hat{n}_t$ are in the directions of those beams, so that the associated wave vectors are $\bar{k}_i = k_i\hat{n}_i$, $\bar{k}_r = k_r\hat{n}_r$ and $\bar{k}_t = k_t\hat{n}_t$. The unit vectors $\hat{m}_i, \hat{m}_r, \hat{m}_t$ are perpendicular to the unit vectors $\hat{n}_i, \hat{n}_r, \hat{n}_t$ respectively, as shown. Their full definition will emerge shortly.

Although Figure 11.2 appears to show the incident, reflected and refracted beams to be coplanar, this will not be assumed but is the first thing to be proved. This coplanarity is known as the First Law of Reflection and Refraction.

Figure 11.2: Deriving Snell's Law from Electromagnetic Theory

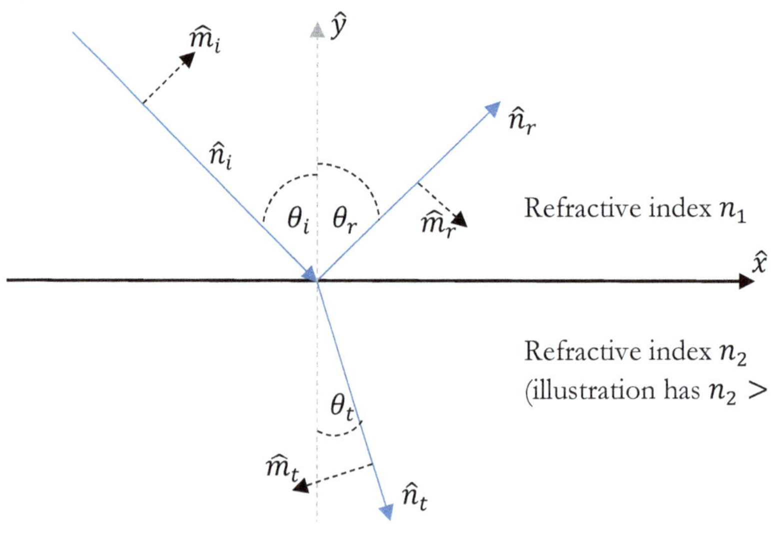

A heuristic argument for the coplanarity of the three beams (and the boundary normal) is as follows. The momentum flows are in the directions given by the unit vectors $\hat{n}_i, \hat{n}_r, \hat{n}_t$. The conservation of momentum requires that the momentum flow delivered by the incident beam must balance with that carried away by the combination of the reflected and refracted beams. Hence, representing the magnitudes of the momentum flows by $\alpha_i, \alpha_r, \alpha_t$ we must have $\alpha_i\hat{n}_i = \alpha_r\hat{n}_r + \alpha_t\hat{n}_t$, i.e., that the beams are coplanar.

This alone does not establish that the boundary normal also lies within that same plane. But symmetry would appear to require it, on the grounds that if the plane containing the three beams was skewed to one side of the boundary normal there would seem to be nothing to determine whether that should be to the left or to the right. This argument, however, is spurious because the handedness of the fields and their momentum flow, i.e., $\bar{E}, \bar{B}$ and $\bar{E} \times \bar{B}$, does provide a preferred rotational direction.

11.2.2.1 Establishing the First Law

More importantly for our present purposes, the preceding argument does not confirm that the field boundary conditions are consistent with the laws of reflection and refraction, or any other laws from geometrical optics. To do so we must start with the fields themselves, which in obvious notation are,

$$\bar{E}_i = \bar{E}_{i0} exp\{-i(\omega_i t - \bar{k}_i \cdot \bar{r})\} \tag{11.2.1a}$$

$$\bar{E}_r = \bar{E}_{r0} exp\{-i(\omega_r t - \bar{k}_r \cdot \bar{r})\} \tag{11.2.1b}$$

$$\bar{E}_t = \bar{E}_{t0} exp\{-i(\omega_t t - \bar{k}_t \cdot \bar{r})\} \tag{11.2.1c}$$

with similar expressions for the B-fields and where the amplitudes are connected by,

$$v_1 B_{i0} = E_{i0} \quad v_1 B_{r0} = E_{r0} \quad v_2 B_{t0} = E_{t0} \tag{11.2.2}$$

Assuming there is no charge on the boundary, nor any surface currents, then the boundary conditions, from §6.7, are simply that all four of the following quantities are continuous over the boundary,

- The tangential components of $\bar{E}$ and $\bar{H}$, and,
- The normal components of $\bar{D}$ and $\bar{B}$.

Note that the assumed absence of surface charges and currents essentially means that we are considering non-conducting boundaries (e.g., dielectrics like glass, etc.). The issue of reflection at a metallic surface is different and is addressed in §11.8.

Consider firstly the tangential component of $\bar{E}$. The boundary is defined as lying on the (x, z) plane. Hence the x component of $\bar{E}$ must be continuous over the boundary, and this must be true across the whole boundary and for all times. From (11.2.1a-c) we therefore require, (11.2.3)

$$E_{i0x} exp\{-i(\omega_i t - \bar{k}_i \cdot \bar{r})\} + E_{r0x} exp\{-i(\omega_r t - \bar{k}_r \cdot \bar{r})\} = E_{t0x} exp\{-i(\omega_t t - \bar{k}_t \cdot \bar{r})\}$$

This must hold on $y = 0$ for all t, x, z. This immediately requires that all the frequencies be the same,

$$\omega_i = \omega_r = \omega_t = \omega \tag{11.2.4}$$

which also means that the wavelength of the transmitted wave will differ from that of the incident and reflected waves because,

$$\omega = v_2 k_t = v_1 k_i = v_1 k_r \tag{11.2.5}$$

which also means that $k_r = k_i$. As the wavevector is related to the wavelength by $k = 2\pi/\lambda$, if the second medium has the larger refractive index, hence slower wave speed, then k_t will be larger than $k_i = k_r$ and the wavelength smaller.

But (11.2.3) also requires, across the whole (x, z) boundary plane,

$$\bar{k}_i \cdot \bar{r} = \bar{k}_r \cdot \bar{r} = \bar{k}_t \cdot \bar{r} \tag{11.2.6}$$

We are free to define the x axis such that the incident beam lies in the (x, y) plane, so that $\bar{k}_i \cdot \bar{r} = k_{ix}x + k_{iy}y$ and from (11.2.6) it then follows that there can be no z dependence in $\bar{k}_r \cdot \bar{r}$ or $\bar{k}_t \cdot \bar{r}$ either. But this means $\bar{k}_r$ and $\bar{k}_t$ must also lie in the (x, y) plane. This establishes the First Law of Reflection and Refraction, and hence that Figure 11.2 is correct in depicting the three beams to be coplanar and coplanar with the boundary normal, $\hat{y}$.

11.2.2.2 Establishing the Other Laws of Reflection and Refraction

The dot products in (11.2.6) can be written in terms of the angles thus,

$$\bar{k}_i \cdot \bar{r} = k_i(x \sin\theta_i - y\cos\theta_i) \tag{11.2.7a}$$

$$\bar{k}_r \cdot \bar{r} = k_i(x \sin\theta_r + y\cos\theta_r) \tag{11.2.7b}$$

$$\bar{k}_t \cdot \bar{r} = k_t(x \sin\theta_t - y\cos\theta_t) \tag{11.2.7c}$$

On the boundary plane, $y = 0$, (11.2.6) must be an identity in x and hence these give,

$$k_i \sin\theta_i = k_i \sin\theta_r = k_t \sin\theta_t \tag{11.2.8}$$

As $k_r = k_i$ the first of these gives us the First Law of Reflection, that the angles of incidence and reflection are equal,

$$\theta_r = \theta_i \tag{11.2.9}$$

Also, (11.2.8) together with (11.2.5) give,

$$\frac{\sin\theta_i}{\sin\theta_t} = \frac{k_t}{k_i} = \frac{v_1}{v_2} = \frac{n_2}{n_1} \tag{11.2.10}$$

which is Snell's Law, or the Second Law of Refraction, (11.2.1), derived again via the required field boundary conditions.

So far, so good, but we still need to establish that all the field boundary conditions are consistent with each other and with known optics. The byproducts of this will be expressions for the amplitudes of the reflected and transmitted waves, and confirmation therefrom that energy is conserved. With these latter results the electromagnetic field theory of light exceeds the capability of simple geometrical optics.

11.2.2.3 Consistency of the Field Boundary Conditions

Linearity of the field equations means that an arbitrary incident polarisation can be obtained by a superposition of the following two cases,

- Case A: The incident electric field is normal to the plane of incidence (so that the incident magnetic field lies in the plane of incidence). Polarisation is, by convention, described by the orientation of the electric field, so case A is called "normal polarisation";
- Case B: The incident magnetic field is normal to the plane of incidence (so that the incident electric field lies in the plane of incidence). This case is called "parallel polarisation".

It therefore suffices to consider these two cases separately.

From this point on we drop the subscript $_0$ on the field amplitudes, it being understood that by "fields" we refer to their vector amplitudes, i.e., without the exponential (wave phase) factor unless needed.

Case A: Incident Electric Field Normal to the Plane of Incidence

This is often called "s" polarisation. By assumption we have $\bar{E}_i = E_i\hat{z}$. However, we cannot assume that the reflected and/or the refracted waves also have their electric field oriented parallel to $\hat{z}$, rather this must be proved from the boundary conditions.

Continuity of E (tangential)

As the electric field is always normal to the beam, and using the unit vector definitions in Figure 11.2, we can put,

$$\bar{E}_r = E_r\hat{z} + E_{rm}\widehat{m}_r \text{ and } \bar{E}_t = E_t\hat{z} + E_{tm}\widehat{m}_t \tag{11.2.11}$$

The tangential projections involve the z components and the x components. Continuity of the z components gives,

$$E_i + E_r = E_t \tag{11.2.12}$$

The unit vectors can be written,

$$\hat{m}_r = \hat{x}\cos\theta_r - \hat{y}\sin\theta_r \tag{11.2.13a}$$

$$\hat{m}_t = -\hat{x}\cos\theta_t - \hat{y}\sin\theta_t \tag{11.2.13b}$$

Continuity of the x components therefore gives,

$$E_{rm}\cos\theta_r = -E_{tm}\cos\theta_t \tag{11.2.14}$$

Continuity of D (normal)

This means continuity of ε_r times the normal component of E. There is no normal (i.e., $\hat{y}$) component of the incident electric field, so we get,

$$\varepsilon_{r1}E_{rm}\sin\theta_r = \varepsilon_{r2}E_{tm}\sin\theta_t \tag{11.2.15}$$

But (11.2.14) and (11.2.15) are in conflict because the former requires E_{rm} to be of the opposite sign to E_{tm}, whereas the latter requires that they are of the same sign. This can be resolved only if,

$$E_{rm} = E_{tm} = 0 \tag{11.2.16}$$

So (11.2.11) now tells us that, if the electric field of the incident wave is perpendicular to the plane of incidence, then so are the electric fields of the reflected and the refracted waves. No inconsistency arises from the two boundary conditions considered so far if this is so. However, we must also establish the consistency of the remaining boundary conditions.

Continuity of H (tangential)

This additional constraint will be seen to lead to a specific requirement for the amplitudes of the reflected and transmitted (refracted) beams.

We now appeal to the fact that materials in which electromagnetic waves propagate will almost always have μ_r very close to unity, though it is sufficient that $\mu_{r1} = \mu_{r2}$. The continuity of the tangential component of $\bar{H}$ then implies the continuity of the tangential component of $\bar{B}$. From Figure 11.2, because the electric fields of all the beams have been taken to be oriented in the positive z direction we have,

$$\bar{B}_i = -B_i\hat{m}_i \quad \bar{B}_r = B_r\hat{m}_r \quad \bar{B}_t = B_t\hat{m}_t \tag{11.2.17}$$

(The signs result from requiring that $\bar{E} \times \bar{B}$ be in the positive direction of travel of the corresponding beam). Using (11.2.13a,b) and $\hat{m}_i = \hat{x}\cos\theta_i + \hat{y}\sin\theta_i$ and noting that it is the x component that is tangential to the boundary, we require,

$$-B_i\cos\theta_i + B_r\cos\theta_r = -B_t\cos\theta_t \qquad (11.2.18)$$

But we can substitute for the magnetic field amplitudes from (11.2.2), giving,

$$E_i\cos\theta_i - E_r\cos\theta_r = (E_i - E_r)\cos\theta_i = \frac{v_1}{v_2}E_t\cos\theta_t \qquad (11.2.19)$$

But we can substitute for E_t from (11.2.12) giving,

$$(E_i - E_r)\cos\theta_i = \frac{v_1}{v_2}(E_i + E_r)\cos\theta_t \qquad (11.2.20)$$

Rearranging and using (11.2.10) gives the reflection coefficient, defined as the ratio of the reflected wave amplitude and the incident wave amplitude, to be,

$$r_A = \frac{E_r}{E_i} = -\frac{\sin(\theta_i-\theta_t)}{\sin(\theta_i+\theta_t)} \qquad (11.2.21)$$

From (11.2.12) the transmission coefficient, t_A, defined as the ratio of the refracted wave amplitude and the incident wave amplitude, must be,

$$t_A = \frac{E_t}{E_i} = 1 + r_A = \frac{2\sin\theta_t\cos\theta_i}{\sin(\theta_i+\theta_t)} \qquad (11.2.22)$$

Thus, we see that the field boundary conditions resulting from the electromagnetic nature of light allows the amplitudes of the reflected and refracted waves to be determined as a function of the angles of incidence and refraction.

The subscript $_A$ on r_A and t_A refers to these expressions applying only in Case A, i.e., when the electric field vectors are perpendicular to the plane of incidence. There is one further boundary condition which must be respected.

Continuity of B(normal)

The normal component is the y component and hence (11.2.17) and (11.2.13a,b) give,

$$-B_i\sin\theta_i - B_r\sin\theta_r = -B_t\sin\theta_t \qquad (11.2.23)$$

(11.2.2) then gives,

$$(E_i + E_r)\sin\theta_i = \frac{v_1}{v_2}E_t\sin\theta_t \qquad (11.2.24)$$

Using Snell's Law, (11.2.10), simplifies this to $E_i + E_r = E_t$, which is the same as (11.2.12). Hence this final boundary condition is consistent with the rest and provides no further independent constraint or requirement.

Case B: Incident Magnetic Field Normal to the Plane of Incidence

This is often called "p" polarisation. By assumption we have $\bar{B}_i = E_i\hat{z}$. However, we cannot assume that the reflected and/or the refracted waves also have their magnetic field oriented parallel to $\hat{z}$, rather this must be proved from the boundary conditions. We shall again ensure that all four boundary conditions are fulfilled, i.e., the continuity of *E*(tangential), *H*(tangential), *B*(normal) and *D*(normal).

Because $\bar{B}_i = B_i\hat{z}$ we can put $\bar{E}_i = E_i(\hat{x}\cos\theta_i + \hat{y}\sin\theta_i)$. As the fields are always perpendicular to the wavevector we can write,

$$\bar{B}_r = B_r\hat{z} + B_{rm}(\hat{x}\cos\theta_i - \hat{y}\sin\theta_i) \tag{11.2.25a}$$

$$\bar{B}_t = B_t\hat{z} - B_{tm}(\hat{x}\cos\theta_t + \hat{y}\sin\theta_t) \tag{11.2.25b}$$

Continuity of *H*(tangential)

As before we assume that materials in which electromagnetic waves propagate will almost always have μ_r very close to unity, though it is sufficient that $\mu_{r1} = \mu_{r2}$. So we can take *B*(tangential) to be continuous also. The tangential components x and z therefore give, from (11.2.25a,b),

$$B_{rm}\cos\theta_i = -B_{tm}\cos\theta_t \tag{11.2.26}$$

and,

$$B_i + B_r = B_t \tag{11.2.27}$$

Continuity of *B*(normal)

The y components from (11.2.25a,b) give,

$$-B_{rm}\sin\theta_i = -B_{tm}\sin\theta_t \tag{11.2.28}$$

Equs.(11.2.26) and (11.2.28) are inconsistent unless we take,

$$B_{rm} = B_{tm} = 0 \tag{11.2.29}$$

which proves that the magnetic fields of the reflected and transmitted beams are also normal to the plane of incidence. Consequently we can now write the electric fields as,

$$\bar{E}_r = -E_r(\hat{x}\cos\theta_i - \hat{y}\sin\theta_i) \tag{11.2.30a}$$

$$\bar{E}_t = E_t(\hat{x}\cos\theta_t + \hat{y}\sin\theta_t) \tag{11.2.30b}$$

The signs in these expressions are obtained by ensuring that $\bar{E} \times \bar{B}$ is in the direction of the wave propagation.

Continuity of E(tangent)

Continuity of the x components of the electric fields gives,

$$E_i\cos\theta_i - E_r\cos\theta_i = E_t\cos\theta_t \tag{11.2.31}$$

Now (11.2.27) can be re-expressed in terms of the electric fields recalling (11.2.2). Hence,

$$E_i + E_r = \frac{v_1}{v_2}E_t = \frac{\sin\theta_i}{\sin\theta_t}E_t \tag{11.2.32}$$

Substituting for E_t in (11.2.31) from (11.2.32) gives,

$$E_i - E_r = \frac{\cos\theta_t \sin\theta_t}{\cos\theta_i \sin\theta_i}(E_i + E_r) \tag{11.2.33}$$

Rearranging gives the reflection coefficient for Case B,

$$r_B = \frac{E_r}{E_i} = \frac{\sin\theta_i\cos\theta_i - \sin\theta_t\cos\theta_t}{\sin\theta_i\cos\theta_i + \sin\theta_t\cos\theta_t} = \frac{\tan(\theta_i - \theta_t)}{\tan(\theta_i + \theta_t)} \tag{11.2.34}$$

From this we obtain the transmission coefficient using (11.2.32), thus,

$$t_B = \frac{E_t}{E_i} = \frac{2\sin\theta_t\cos\theta_i}{\sin\theta_i\cos\theta_i + \sin\theta_t\cos\theta_t} \tag{11.2.35}$$

The expressions (11.2.21), (11.2.22), (11.2.34) and (11.2.35) for the reflection and transmission coefficients are known as the Fresnel relations.

The reflection coefficients, r_A, (11.2.21) and r_B, (11.2.34), are plotted against angle of incidence in Figure 11.4 ($n_1 < n_2$) and Figure 11.5 ($n_1 > n_2$).

Special cases are as follows,

Grazing Incidence ($\theta_i \to \pi/2, n_1 < n_2$)

For $n_1 < n_2$, (11.2.21) and (11.2.34) shows that in the limit $\theta_i \to \pi/2$ the reflection coefficients are $r_A \to -1, r_B \to -1$, indicating that all the incident energy is reflected. Consistently, (11.2.22) and (11.2.35) show that $t_A \to 0$ and $t_B \to 0$.

Critical Angle ($n_1 > n_2$)

For $n_1 > n_2$ we will see in §11.5.1 that there is no transmitted beam for any incident angle greater than the critical angle, $sin^{-1}(n_2/n_1)$. Hence we consider what happens at this angle of incidence, when $\theta_t = \pi/2$. (11.2.21) and (11.2.34) then show that $r_A = 1,\ r_B = 1$. (The transmitted wave amplitudes do not tend to zero but the conservation of energy is not violated because the transmitted beam cross-section becomes zero at the critical angle, see Figure 11.3, i.e., there is no transmitted propagating wave. Note, however, there is an evanescent field within medium 2 to which the non-zero transmission coefficients relate, see §11.6 for details).

Normal Incidence ($\theta_i \to 0$)

In this case, for both $n_1 < n_2$ and $n_1 > n_2$ we find $r_B = -r_A = \frac{n_2 - n_1}{n_2 + n_1}$ and $t_B = t_A = \frac{2n_1}{n_2 + n_1}$. The reader can check that these reflection and transmission coefficients respect the conservation of energy, as specified by (11.4.2b).

11.3 The Phases of the Reflected and Transmitted Waves

This section applies for $n_1 < n_2$ and $n_1 > n_2$ but, in the latter case, only for angles of incidence less than the critical angle, $\theta_i < \sin^{-1}(n_2/n_1)$. For angles greater than the critical angle the phase changes will be addressed in §11.6. In the cases considered here the reflected and transmitted waves are either in-phase or in exact anti-phase

All the waves appear to be in phase because they all have the same exponent on the boundary. However, they may be in exact anti-phase (i.e., a phase difference of 180°) if their reflection or transmission coefficient is negative. Thus, for Case A (normal polarisation) the transmitted wave is always in phase with the incident wave, from (11.2.22), whereas the reflected wave will be in anti-phase, (11.2.21), if the second medium has the higher refractive index, so that $\theta_i > \theta_t$. On the other hand, if the incident beam is in the higher refractive index medium, both the reflected and the transmitted waves will be in phase with it.

For Case B (parallel polarisation) the transmitted wave is again always in phase with the incident wave, from (11.2.35). For $n_1 < n_2$ the reflected wave is in-phase below the Brewster angle, $\theta_i < \theta_B$, (see §11.5.2) but in anti-phase above it. For $n_1 > n_2$ the reverse is the case.

For a general state of polarisation these results imply that the reflected wave will be neither in phase nor in exact anti-phase. In this general case the reflected wave will be at some intermediate phase angle between 0 and 180°.

The derivation in this general case is usually carried out using complex valued field amplitudes as a convenient way of handing the phase differences. The interested reader should consult one of the classic texts for details, e.g., [11.1-3], especially [11.3].

11.4 Conservation of Energy in Reflection and Transmission

So far we have assumed wave propagation without any absorption, and so the energy in the incident wave must be conserved when the wave splits into a reflected and a transmitted wave. One cannot check this by falsely assuming that energy conservation means that $N_i = N_r + N_t$, where these terms are the magnitudes of the respective Poynting vectors. This is because the Poynting vector gives the energy per second per unit area, not the total energy. Instead one imagines an incident beam of finite width and considers the 'footprint' that it makes on the boundary – a footprint which all three beams share. One then works on the basis of the energy entering and exiting this 'footprint', which depends upon the projection of the area of the footprint which is normal to the wavevector. For the three beams this means multiplying the magnitudes of the Poynting vectors by $\cos\theta_i$, $\cos\theta_r$, $\cos\theta_t$ respectively, and recalling that $\theta_r = \theta_i$.

The magnitudes of the Poynting vectors are given by $N_i = \frac{E_i^2}{\mu_0 v_1}$, etc. Hence the conservation of energy requires,

$$\frac{E_i^2}{\mu_0 v_1}\cos\theta_i = \frac{E_r^2}{\mu_0 v_1}\cos\theta_i + \frac{E_t^2}{\mu_0 v_2}\cos\theta_t \qquad (11.4.1)$$

Dividing through by the LHS this becomes, for Case A,

$$r_A^2 + \frac{v_1\cos\theta_t}{v_2\cos\theta_i}t_A^2 = 1 \qquad (11.4.2a)$$

i.e.,

$$r_A^2 + \frac{\sin\theta_i\cos\theta_t}{\sin\theta_t\cos\theta_i}t_A^2 = 1 \qquad (11.4.2b)$$

Substitution of the reflection and transmission coefficients from (11.2.21) and (11.2.22) shows that (11.4.2b) is an identity. Similarly, substitution of (11.2.34) and (11.2.35) shows that (11.4.2b) is also an identity in Case B. This confirms that energy is conserved.

11.5 The Critical and Brewster Angles

11.5.1 The Critical Angle (Total Internal Reflection)

In the case that the light passes from a material of smaller to a material of greater refractive index, $n_2 > n_1$ then Snell's Law, (11.2.10), always has a solution for θ_t which is necessarily less than θ_i. In other words, some light always penetrates into the material of higher refractive index and is transmitted.

But in the reverse case, when the light is incident on a boundary with a lower refractive index material (e.g., light in glass meeting an air boundary) the situation is different. Because we now have $n_2 < n_1$, when light is transmitted into the second material Snell's Law tells us that $\theta_t > \theta_i$. But this clearly becomes impossible when $\theta_i = 90^o$. In fact the physical limit must be reached when $\theta_t = 90^o$, which cannot be any larger. Snell's Law tells us that the angle of incidence at which this happens is,

$$\theta_i = \theta_c = \sin^{-1}\left(\frac{n_2}{n_1}\right) \tag{11.5.1}$$

This is the "critical angle". Light incident with $\theta_i > \theta_c$ can have no transmitted beam and so undergoes "total internal reflection" in which there is only a reflected beam.

Are the reflection coefficients of (11.2.21) and (11.2.34) consistent with the conclusion that light is totally internally reflected when $\theta_i \geq \theta_c$? These expressions as written become inapplicable for $\theta_i > \theta_c$ because θ_t is no longer defined. However, at the critical angle, $\theta_i = \theta_c$ we have $\theta_t = 90^o$ and we see that (11.2.21) and (11.2.34) give $r_A = -1$ and $r_B = 1$, consistent with all the energy being reflected.

The reader may be bewildered because the transmission coefficients, given by (11.2.22) an (11.2.35), are not zero at $\theta_t = 90^o$. But recall that the energy carried by a beam depends also on its cross-section, not just the wave amplitude. But as $\theta_t \to 90^o$ the width of the transmitted beam becomes zero (see Figure 11.3) and so the energy transmitted becomes zero despite t_A and t_B being non-zero.

Figure 11.3: The Transmitted Beam Width becomes Zero at the Critical Angle of Incidence

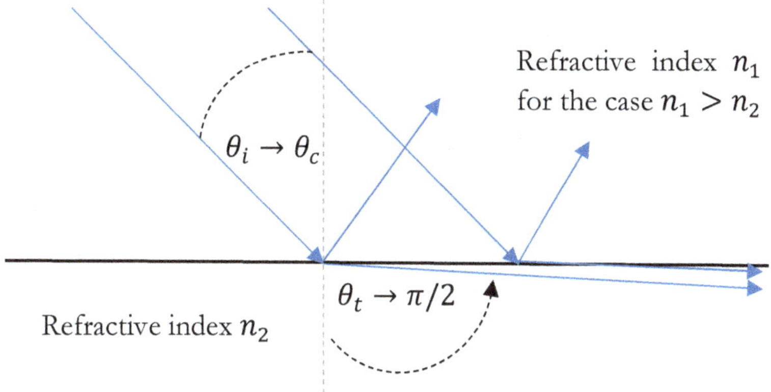

11.5.2 Polarisation on Reflection (Brewster Angle)

We see from (11.2.34) that r_B will be zero if $\theta_i + \theta_t = \pi/2$. In contrast, (11.2.21) for r_A has no zero at any angle of incidence. Hence, there is an angle of incidence, $\theta_i = \theta_B$, called Brewster's angle, at which a parallel polarised incident beam produces no reflected beam, but there is no such angle for normally polarised incident light. Consequently, if the incident beam is in a general state of polarisation, reflection at the Brewster angle provides a means of filtering out the parallel polarised component and providing a reflected beam which is purely normally polarised.

$$\theta_i = \theta_B = \tan^{-1}\left(\frac{n_2}{n_1}\right) \tag{11.5.2}$$

At the Brewster angle the transmitted beam is not fully polarised but will contain a larger proportion of parallel polarisation than the incident beam. However, if the incident beam is parallel polarised, at the Brewster angle all the incident energy emerges in the transmitted beam.

Unlike the critical angle, (11.5.1), which applies only for $n_2 < n_1$, Brewster's angle, (11.5.2), and the associated polarisation phenomenon applies whether $n_2 < n_1$ or $n_2 > n_1$.

11.6 Beyond the Critical Angle

In the preceding sections we have glossed over exactly what happens for $n_2 < n_1$ at angles of incidence greater than the critical angles, $\theta_i > \theta_c$. Expressions for the transmitted beam like (11.2.11) and (11.2.25b) no longer mean what they appear to mean because the transmitted beam angle, θ_t, does not exist. The reader will naturally leap to the conclusion that there is no transmitted beam and hence that the satisfaction of the boundary conditions carried out in §11.2 must be revisited by omitting the transmitted beam entirely. Confusion will ensue because there is no solution on this basis.

The statement that there is no transmitted beam at angles $\theta_i > \theta_c$ is only correct if the word "beam" is understood to mean a propagating wave which conveys energy. However, this does not mean that, just within the surface of the second medium, the electric and magnetic fields are zero. In fact, they cannot be if the boundary conditions are to be satisfied. What has happened is that a propagating wave has turned into a decaying wave because the component of $\bar{k}_t \cdot \bar{r}$ in the z direction, namely $k_t z \cos\theta_t$, has become an imaginary quantity. To see this note that,

$$\cos\theta_t = \sqrt{1 - sin^2\theta_t} = \sqrt{1 - \left(\frac{n_1}{n_2}\right)^2 sin^2\theta_i} = ib \tag{11.6.1}$$

where, $b^2 = \left(\frac{n_1}{n_2}\right)^2 sin^2\theta_i - 1$. So $\cos\theta_t$ is purely imaginary when $\theta_i > \theta_c$ and $\sin\theta_i$ is real but greater than $\frac{n_2}{n_1}$, i.e., θ_t is not a physically possible angle. Hence the "transmitted" wave factor $\exp\{i(\omega t - \bar{k}_t \cdot \bar{r})\}$ will now include a factor $\exp\{-k_t yb\}$ which causes the "wave" to decay exponentially into the depth of the second medium (but is nevertheless non-zero at the boundary). This is known as an evanescent wave. By the ruse of replacing $\cos\theta_t$ with ib we avoid having to work laboriously through the derivations of §11.2 again because they remain formally correct.

Hence, for normal polarisation (case A) we have, (11.6.2)

$$r_A = -\frac{\sin(\theta_i-\theta_t)}{\sin(\theta_i+\theta_t)} = -\frac{\sin\theta_i\cos\theta_t-\sin\theta_t\cos\theta_i}{\sin\theta_i\cos\theta_t+\sin\theta_t\cos\theta_i} = -\frac{\cos\theta_t-\left(\frac{n_1}{n_2}\right)\cos\theta_i}{\cos\theta_t+\left(\frac{n_1}{n_2}\right)\cos\theta_i} = \frac{a-ib}{a+ib}$$

where $a = \left(\frac{n_1}{n_2}\right)\cos\theta_i$. Hence we see that, as expected, r_A has magnitude (modulus) of 1, because all the energy is reflected in total internal reflection, but it is now a complex quantity,

$$r_A = e^{i\alpha} \quad \text{where} \quad \cos\alpha = \frac{a^2-b^2}{a^2+b^2}, \quad \sin\alpha = -\frac{2ab}{a^2+b^2} \tag{11.6.3}$$

This angle, α, is the phase of the reflected beam wrt the incident beam. At the critical angle, $\theta_i = \theta_c$, it is zero, whereas as $\theta_i \to \pi/2$, $\alpha \to -\pi$. At intermediate angles of incidence the phase of the reflected wave is plotted in Figure 11.6, below.

For parallel polarisation (case B) we have,

$$r_B = -\frac{\tan(\theta_i-\theta_t)}{\tan(\theta_i+\theta_t)} = -\left(\frac{a-ib}{a+ib}\right)\left(\frac{\cos\theta_i\cos\theta_t-\sin\theta_i\sin\theta_t}{\cos\theta_i\cos\theta_t+\sin\theta_i\sin\theta_t}\right) =$$

$$-\left(\frac{a-ib}{a+ib}\right)\left(\frac{\cos\theta_i\cos\theta_t-\frac{n_1}{n_2}\sin^2\theta_i}{\cos\theta_i\cos\theta_t+\frac{n_1}{n_2}\sin^2\theta_i}\right) = -\left(\frac{a-ib}{a+ib}\right)\left(\frac{a\cos\theta_t-\left(\frac{n_1}{n_2}\right)^2\sin^2\theta_i}{a\cos\theta_t+\left(\frac{n_1}{n_2}\right)^2\sin^2\theta_i}\right)$$

$$= -\left(\frac{a-ib}{a+ib}\right)\left(\frac{iab-(b^2+1)}{iab+(b^2+1)}\right) = \left(\frac{a-ib}{a+ib}\right)\left(\frac{c-ib}{c+ib}\right) = e^{i\alpha}e^{i\beta} \qquad (11.6.4)$$

where $c = \frac{b^2+1}{a}$. Hence, the phase change for parallel polarisation is,

$$\alpha_B = \alpha + \beta \qquad (11.6.5)$$

where α is given by (11.6.3) and,

$$\cos\beta = \frac{c^2-b^2}{c^2+b^2}, \quad \sin\beta = -\frac{2cb}{c^2+b^2} \qquad (11.6.6)$$

A little algebra shows that (11.6.3) and (11.6.5) are equivalent to the expressions given in [11.3], section 1.5, equ.(60), which I include for completeness,

$$\tan\frac{\alpha_A}{2} = -\frac{\sqrt{\sin^2\theta_i-\left(\frac{n_2}{n_1}\right)^2}}{\cos\theta_i} \qquad (11.6.7a)$$

$$\tan\frac{\alpha_B}{2} = -\frac{\sqrt{\sin^2\theta_i-\left(\frac{n_2}{n_1}\right)^2}}{\left(\frac{n_2}{n_1}\right)^2\cos\theta_i} \qquad (11.6.7b)$$

But what about the evanescent "transmitted wave"? We have seen that it is exponentially decaying in the z direction, but it appears nevertheless to have a propagating wave appearance in the x direction because it is proportional to $\exp\{-k_t yb\}\exp\{i(\omega t - k_t x\sin\theta_i)\}$. Whilst the wave decays very rapidly in the y direction, namely within about a wavelength in the second medium, nevertheless the wave still has a non-zero beam width as it (apparently) propagates in the x direction. This raises a conundrum because all the incident energy is reflected in total internal reflection, so where does the energy come from (albeit small) that appears in the evanescent transmitted wave? The answer is that there is none (once the steady state is set up). Evaluation of the Poynting vector shows it to be purely imaginary. Any energy in the transmitted wave merely shuffles back and forth; there is no net propagation of energy by the evanescent wave (see [11.1-3] for details).

Figure 11.6 ($n_1 < n_2$) and Figure 11.7 ($n_1 > n_2$) plot the phase angle of the reflected waves against the angle of incidence.

Figure 11.4: Reflection and Transmission Coefficients for $n_1 < n_2$. *The relative magnitudes of the reflected and transmitted waves do not faithfully indicate the relative energies the beams convey (see §11.4).*

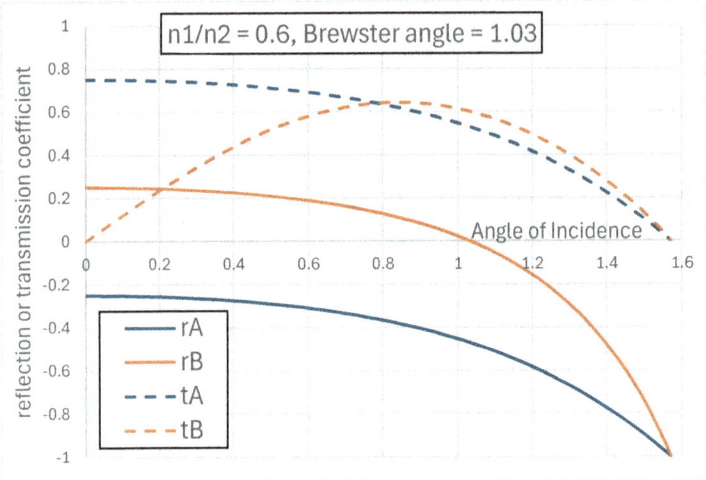

Figure 11.5: Reflection and Transmission Coefficients for $n_1 > n_2$. *The relative magnitudes of the reflected and transmitted waves do not faithfully indicate the relative energies the beams convey (see §11.4). Above the critical angle (total internal reflection) the reflection and transmission coefficients become complex. The plotted value for the reflection coefficient is its modulus.*

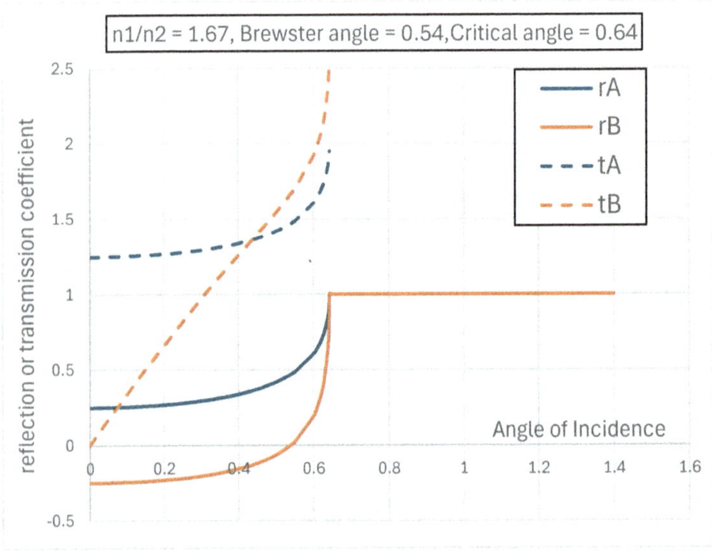

Figure 11.6: Phase Angle of the Reflected Waves for $n_1 < n_2$.

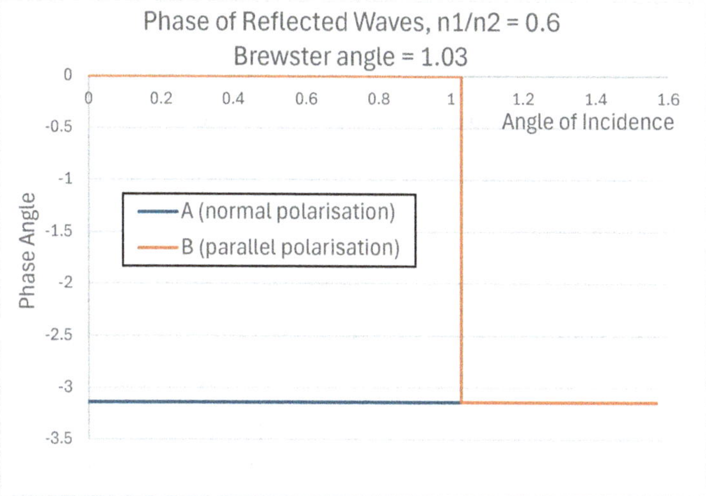

Figure 11.7: Phase Angle of the Reflected Waves for $n_1 > n_2$.

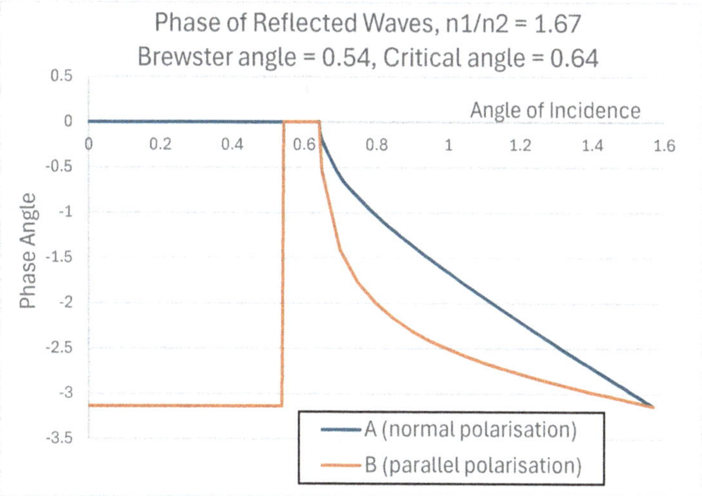

11.7 Absorption of Waves during Propagation

Inevitably, energy is absorbed when an electromagnetic wave travels through a medium. The form of the wave in a medium of refractive index n is,

$$\bar{B} = \bar{B}_0 exp\{-i(\omega t - n\bar{k}\cdot\bar{r})\}; \;\; \bar{E} = \bar{E}_0 exp\{-i(\omega t - n\bar{k}\cdot\bar{r})\} \qquad (11.7.1)$$

where $\bar{k}$ is the wavevector that would apply at angular frequency ω in vacuum. Consequently, attenuation of the wave by energy absorption can be accounted for simply by endowing the refractive index with an imaginary part,

$$n \to \mathcal{N} = n + i\xi \tag{11.7.2}$$

For a wave propagating in the z direction the fields are then,

$$\bar{B} = \bar{B}_0 e^{-\xi kz} exp\{-i(\omega t - nkz)\} \tag{11.7.3a}$$

$$\bar{E} = \bar{E}_0 e^{-\xi kz} exp\{-i(\omega t - nkz)\} \tag{11.7.3b}$$

The wave amplitudes reduce by a factor of $1/e$ in a distance of $\frac{1}{\xi k} = \frac{\lambda}{2\pi\xi}$ where λ is the wavelength in vacuum at angular frequency ω. The Poynting vector (the energy flux) decreases by a factor of $1/e^2$ over this distance.

The derivation of the form of the waves, (11.7.1), applies just as for a real refractive index, which means we can relate the complex refractive index to the relative permittivity by (11.1.3), which, with $\mu_r \approx 1$, reduces to $n = \sqrt{\varepsilon_r}$. Consequently, the relative permittivity has also developed an imaginary part,

$$\varepsilon_r \to \varepsilon_{real} + i\varepsilon_{imag} = (n^2 - \xi^2) + 2in\xi \tag{11.7.4}$$

Because $\bar{D} = \varepsilon_r \varepsilon_0 \bar{E}$ this means, for real $\bar{E}$, that the D-field amplitude develops an imaginary part, which is interpreted as a phase difference wrt the E-field. This phase difference is given by angle $\tan^{-1}\left(\frac{2n\xi}{n^2-\xi^2}\right)$. Recall that the polarisation of the medium is given by $\bar{P} = \bar{D} - \varepsilon_0\bar{E} = (\varepsilon_r - 1)\varepsilon_0\bar{E}$ so that the polarisation also develops a phase difference with the E-field.

Hence the formalism neatly links two apparently different phenomena: attenuation and phase. Whilst we have derived this purely from the mathematical formalism, this link between attenuation and phase must have a physically basis. Polarisation requires charges within the medium to be physically displaced, and the oscillations of a wave mean that these displacements are also oscillating. The finite inertia of the charges introduces the time delay between force and displacement which thus appears as the phase difference between the E-field and the polarisation. But this physical motion of charges within a polarisable medium will not be without energy loss and this is what links the two phenomena. Inertial effects would be expected to be larger at higher frequencies, and this is consistent with more rapid attenuation at higher frequencies as indicated by the smaller decay length, $\frac{1}{\xi k} = \frac{v}{\xi\omega}$.

11.8 Reflection at Metallic Surfaces

So far we have assumed dielectric media with very low electrical conductivity and hence that the boundary between media can be taken to be free of surface charge or surface current. In contrast, in this section we look at reflection from conducting surfaces where these assumptions do not hold.

One of the most obvious characteristics of metals is that they are shiny, which is another way of saying that they are highly reflective. Why? The question is a reasonable one when compared with the reflectivity of dielectrics, e.g., glass. Whilst the reflectivity from a dielectric surface might approach unity when incident at a glancing angle, or beyond the critical angle (Figures 11.4,5), at normal incidence reflectivity is far smaller.

A typical glass used in spectacles might have a refractive index of, say, 1.67, whereas plate glass refractive index is more like 1.5. §11.2 gives the corresponding reflection coefficient at normal incidence to be 0.20 to 0.25, and so only 4% to 6.25% of the incident energy is reflected. We will see here how very poor this is in comparison with a metal surface. This is why a conventional "glass" mirror reflects far more from its silvered backing than its glass surface. It is fortunate that this is so because such a mirror actually produces a double image, one from the front (glass) surface and one from the back (silvered) surface. Thanks to the relative reflectivities only the latter is usually apparent on casual inspection.

The key to the difference is that, being good conductors, the electric field intrinsic to the light (or other electromagnetic radiation) will induce a current in the metal near its surface. Recall from (7.2.1) that the full Maxwell's equations are,

$$\bar{\nabla}\cdot\bar{D} = \rho \qquad \bar{\nabla}\cdot\bar{B} = 0 \qquad \bar{\nabla}\times\bar{H} = \bar{J} + \frac{\partial\bar{D}}{\partial t} \qquad \bar{\nabla}\times\bar{E} = -\frac{\partial\bar{B}}{\partial t} \tag{11.8.1}$$

In this application the current is that induced by the electric field, and is proportional to it, the constant of proportionality being the conductivity, κ, of the metal. Hence,

$$\bar{J} = \kappa\bar{E} \tag{11.8.2}$$

For the units of conductivity see Table 9.2. (We remind the reader here that the universally used term "current density" is strictly a misnomer as $\bar{J}$ is actually the current per unit area, which is the same as the charge flux). The third of the equations (11.8.1) becomes, just within the metal surface, using (11.1.1),

$$\bar{\nabla} \times \bar{B} = \frac{\varepsilon_r \mu_r}{c^2} \frac{\partial \bar{E}}{\partial t} + \mu_r \mu_0 \kappa \bar{E} \tag{11.8.3}$$

The two terms on the RHS of (11.8.3) are respectively the displacement current and the conduction current. The wave equation is modified by the presence of the near-surface conduction current. Following the same approach as previously, take the $curl$ of the last equation in (11.8.1) and substitute (11.8.3) and identity (A1.17) to yield,

$$\nabla^2 \bar{E} = \frac{\varepsilon_r \mu_r}{c^2} \frac{\partial^2 \bar{E}}{\partial t^2} + \mu_r \mu_0 \kappa \frac{\partial \bar{E}}{\partial t} \tag{11.8.4}$$

where we have also appealed to there being no surface charge, so that the first Maxwell equation becomes $\bar{\nabla} \cdot \bar{D} = \bar{\nabla} \cdot \bar{E} = 0$. Following a similar approach shows that the same equation applies for the B-field,

$$\nabla^2 \bar{B} = \frac{\varepsilon_r \mu_r}{c^2} \frac{\partial^2 \bar{B}}{\partial t^2} + \mu_r \mu_0 \kappa \frac{\partial \bar{B}}{\partial t} \tag{11.8.5}$$

If the conduction current is negligible compared with the displacement current these equations revert to those used above for dielectrics. But if the displacement current is negligible compared with the conduction current, because the conductivity is high, then (11.8.4,5) cease to be dominated by a propagating wave and instead approximate to a diffusion equation. We therefore expect rapid attenuation of the wave as it attempts to penetrate the metal. We will see that the latter condition applies in typical cases for metals.

To see the form of the solution of (11.8.4,5) within the metal surface, substitute the usual wave forms, (11.1.4), where $\bar{k}$ is understood to be the wavevector within the metal. The first two Maxwell equations, (11.8.1) give $\bar{k} \cdot \bar{E} = 0$ and $\bar{k} \cdot \bar{B} = 0$ while the last gives $\bar{k} \times \bar{E} = \omega \bar{B}$ and these together mean that the three vectors $\bar{E}, \bar{B}, \bar{k}$ are mutually orthogonal, and that the amplitudes of the two fields are related by $kE_0 = \omega B_0$. Although $v = \omega / |k|$ is the phase velocity of the wave within the metal, as usual, note the use of the modulus in $|k|$ because the wavevector turns out to be complex. The third Maxwell equation, in the form (11.8.3) gives,

$$i \bar{\mathrm{k}} \times \bar{B} = -i \frac{\varepsilon_r \mu_r}{c^2} \omega \bar{E} + \mu_r \mu_0 \kappa \bar{E} \tag{11.8.6}$$

Noting that the right-handed system is in the order $\bar{E}, \bar{B}, \bar{k}$, it follows that $\bar{\mathrm{k}} \times \bar{B}$ is in the opposite sense from $\bar{E}$ and so substitution of $kE_0 = \omega B_0$ into (11.8.6) gives,

$$k^2 = \frac{\varepsilon_r \mu_r}{c^2} \omega^2 + i \mu_r \mu_0 \kappa \omega \tag{11.8.7}$$

Recall that the (complex) refractive index is just the wavevector in the medium divided by the wavevector at the same frequency in vacuum, we have,

$$\mathcal{N}^2 = \varepsilon_r \mu_r + i\frac{\mu_r \mu_0 \kappa c^2}{\omega} \tag{11.8.8}$$

Hence we have recovered the relationship between the refractive index and the dielectric constant, (11.1.3), but with an additional imaginary part similar to (11.7.2). The real and imaginary parts of the complex refractive index, $\mathcal{N} = n + i\xi$, can be found from,

$$n^2 - \xi^2 = \varepsilon_r \mu_r \qquad \text{and} \qquad 2n\xi = \frac{\mu_r \kappa}{\varepsilon_0 \omega} = \frac{\varepsilon_r \mu_r}{\omega \tau} \tag{11.8.9}$$

where we have also used the relaxation time defined as $\tau = \frac{\varepsilon_r \varepsilon_0}{\kappa}$. (11.8.10)

Some sources omit the relative permeability, μ_r, from the whole derivation in this section. For many materials the magnetic susceptibility is extremely small so that $\mu_r \approx 1$, so the omission is justified. For ferromagnetic materials, those materials for which the static μ_r is far larger than unity, the static value is inappropriate at high frequencies. It is invariably the case for ferromagnetic materials that μ_r reduces substantially as the frequency is increased. In particular it is always the case that $\mu_r \approx 1$ at optical frequencies. I have retained the μ_r dependence in the above expressions in case they might be employed for a ferromagnetic material at sufficiently low frequency that the effect of μ_r is significant.

For a wave propagating in the z direction the fields are as (11.7.3a,b),

$$\bar{B} = \bar{B}_0 e^{-\xi k_0 z} exp\{-i(\omega t - nk_0 z)\} \tag{11.7.3a}$$

$$\bar{E} = \bar{E}_0 e^{-\xi k_0 z} exp\{-i(\omega t - nk_0 z)\} \tag{11.7.3b}$$

where $k_0 = \omega / c$ is the wavevector in vacuum. Hence the decay length, or penetration depth, over which the wave amplitudes reduce by a factor of $1/e$ is,

$$\zeta = \frac{1}{\xi k_0} = \frac{2cn\tau}{\varepsilon_r \mu_r} \tag{11.8.11}$$

Solving the quadratic that results from (11.8.9), and noting that the negative root is impossible because n^2 and ξ^2 are both real, we find,

$$n = \sqrt{\frac{\varepsilon_r \mu_r}{2}\left[1 + \sqrt{1 + \frac{1}{(\omega\tau)^2}}\right]} \tag{11.8.12a}$$

$$\xi = \sqrt{\frac{\varepsilon_r \mu_r}{2}\left[-1 + \sqrt{1 + \frac{1}{(\omega\tau)^2}}\right]} \tag{11.8.12b}$$

Based on the above discussion we can take $\mu_r \approx 1$. In the optical range, namely ω values of 2.5×10^{15} to 5.0×10^{15} radians/second (equivalent to frequencies of 400 to 790 THz) the quantity $\omega\tau$ is evaluated in Table 11.1 for some metals and amorphous carbon and glass.

Table 11.1: Electrical Conductivity, Relaxation Time and $\omega\tau$ in the Optical Range for Some Materials

Material	Conductivity S/m	$\tau = \varepsilon_r\varepsilon_0/\kappa$ (s)	$\omega\tau$ (optical)
Silver[(1)]	6.3×10^7	1.4×10^{-18}	0.0035 to 0.0070
Iron[(1)]	1.0×10^7	8.9×10^{-18}	0.022 to 0.044
Mercury[(2)]	1.0×10^6	4.4×10^{-17}	0.11 to 0.22
Amorphous carbon[(3)]	$1 - 2 \times 10^3$	1.1 *to* 2.7×10^{-14}	3 to 130
glass[(3)]	$< 10^{-11}$	>2	$> 5 \times 10^{15}$

[(1)]*using* $\varepsilon_r = 10$, [(2)]*using* $\varepsilon_r = 5$, [(3)]*using* $\varepsilon_r = 2.5$ *to* 3

Mercury has a particularly poor conductivity for a metal, whereas amorphous carbon has a relatively high conductivity for a non-metal. Hence, all the metals satisfy $\omega\tau \ll 1$ whereas non-metals satisfy $\omega\tau \gg 1$ in the optical range. Note that, for metals, accurate values of $\omega\tau$ are difficult to specify because ε_r is rapidly varying over the optical range, reducing rapidly from the red to the blue. The values in Table 11.1 are indicative only.

For sufficiently low conductivity materials such that $\omega\tau \gg 1$ we can approximate (11.8.12a,b) to give,

$$n = \sqrt{\varepsilon_r \mu_r} \qquad \xi = \frac{n}{2\omega\tau} \qquad \zeta = \frac{2c\tau}{n} = 2v\tau \tag{11.8.13}$$

So, for materials with $\omega\tau \gg 1$, the penetration depth is twice the relaxation time times the phase velocity in the material. This is a huge distance for, say, glass, so that attenuation due to conductivity in glass is negligible. (Attenuation in glass is due to other phenomena, including that due to the imaginary part of the dielectric constant arising from polarisation lag, see §11.7).

However, for metals, in contrast, we have $\omega\tau \ll 1$ at optical frequencies and below, e.g., microwaves. In this case we can approximate (11.8.12a,b) to give,

$$n \approx \xi \approx \sqrt{\frac{\varepsilon_r \mu_r}{2\omega\tau}} = \sqrt{\frac{\mu_r \kappa}{2\omega\varepsilon_0}} \qquad \zeta = \frac{c}{\xi\omega} = \sqrt{\frac{2}{\mu_r \mu_0 \kappa\omega}} \tag{11.8.14}$$

Hence, for silver and iron the penetration depth at optical frequencies is between 2 and 8 nanometres, a radically different result than for, say, glass. Consequently metals are opaque. Of interest is the reflectivity of metals, which we now examine.

We can simply lift the result for, say, the reflection coefficient at normal incidence from the analysis of §11.2 provided this is expressed in terms of the refractive index and also accept that this is now a complex quantity, (11.8.8). For both polarisations incident from the vacuum §11.2 derived that $|r_B| = |r_A| = \left|\frac{n-1}{n+1}\right|$ which now becomes,

$$|r| = \left|\frac{\mathcal{N}-1}{\mathcal{N}+1}\right| = \left|\frac{n-1+i\xi}{n+1+i\xi}\right| \tag{11.8.15}$$

Hence the reflectivity (reflecting power) is,

$$R = |r|^2 = \left|\frac{n-1+i\xi}{n+1+i\xi}\right|^2 = \frac{(n-1)^2+\xi^2}{(n+1)^2+\xi^2} \approx \frac{2n^2-2n+1}{2n^2+2n+1} \tag{11.8.16}$$

For silver and iron we have from (11.8.14) that $n \approx \xi \approx 11 \text{ to } 38$. Hence, the reflectivity of metals is,

$$R \approx 1 - \frac{2}{n} \tag{11.8.17}$$

For iron this is about 0.82 whilst for silver it is about 0.95. Hence, for a well silvered glass mirror at normal incidence, the silver backing reflects perhaps 95% of the 96% of the light that is transmitted through the glass, with the glass surface reflecting only 4% of the incident light.

As regards the phase of the reflected beam, because there is never any propagating transmitted wave, the case of metals is analogous to that of dielectrics beyond the critical angle for which we have seen that there is a phase difference which varies with angle of incidence (cf., Figure 11.7). This is also the case for reflection from metals. The reader may consider the derivation as a suitable exercise, or consult [11.1-3].

11.9 References

[1] Jackson, J.D. (1975). *Classical Electromagnetism*. 2nd edition, 1975 John Wiley & Sons. The link is to the edition I use. There are newer editions.

[2] Bleaney, B.I., and Bleaney, B. (1965). *Electricity and Magnetism*. 2nd edition, 1965, Oxford Claredon Press. The link is to the edition I use. There are newer editions.

[3] Born, M., and Wolf, E. (1993). *Principles of Optics: Electromagnetic Theory of Propagation, Interference and Diffraction of Light*. 6th edition with corrections to 1993, Pergamon Press. Link is to the 7th edition.

12. Magnitudes of Electromagnetic Wave Energies and Exposure Safety Limits

The purpose of this chapter is to provide a rough illustrative guide to typical e/m radiant energies, rather than exact values, which are often too variable to be definitive. In most cases the energy fluxes are prodigiously tiny, which only makes it more remarkable that these waves provide such an effective means of mass communication. However, there are some cases where the energy fluxes present more of a challenge to recognised human exposure safety limits, which are also discussed. The particular case of induction hobs is deferred to chapter 17.

The reader is warned that the discussion of safety limits here follows current national or international standards on safety limits for non-ionising radiation for which the acknowledged risk relates simply to heating effects. In due course this may turn out to be inadequate because experimental biology is now identifying that electric fields can have a marked effect on "reprogramming" organisms' development. At the present time the field strength required to cause this emerging biological effect is unclear but may be of the order of 100 V/m.

Contents

Throughout this chapter the key equations to bear in mind are those for the energy density and the Poynting vector (the energy flux), thus,

$$u_{EM} = \frac{1}{2}\bar{E}\cdot\bar{D} + \frac{1}{2}\bar{B}\cdot\bar{H} \qquad (7.4.1)$$

$$\bar{N} = \bar{E}\times\bar{H} \qquad (7.4.6)$$

And the relations between the E and H fields and the D and B fields,

$$\bar{D} = \varepsilon_r\varepsilon_0\bar{E} \qquad \bar{H} = \bar{B}/\mu_r\mu_0 \qquad (7.3.1)$$

(for the idealised case of LIH materials). Also recall that the momentum flux is the energy flux divided by the velocity, $\bar{N}/v$, and that the wave velocity in a medium is,

$$v = \frac{c}{\sqrt{\mu_r\varepsilon_r}} = \frac{1}{\sqrt{\mu_r\varepsilon_r\mu_0\varepsilon_0}} \qquad (7.3.4)$$

And recall that in an electromagnetic wave the amplitudes of the two fields are related by,

$$vB_0 = E_0 \qquad (11.2.2)$$

Averaging over space and for plane waves, the magnitude of the Poynting vector *in vacuo* is,

$$\langle N\rangle = \frac{E_0^2}{2\mu_0 c} \qquad (10.3.1)$$

12.1 Mobile Phones and Safe e/m Radiation Limits

As of 2025, mobile phone batteries can typically deliver 3.0 to 5 amp-hours at a working potential of 3.7 to 4.5 volts. Hence they can deliver a total energy roughly in the range 11 to 22 W-hours, or 40 to 80 kJ. A telephone call typically demands 0.5 to 1.0 W if signal strength is very strong, but double that typically, and perhaps three or four time that minimum for low signal strength. Typically, then, the energy drain is around 5 kJ per hour of a voice call. However, modern mobiles have such a range of functions, many of which are active in the background much of the time, that a charged battery does not last as long as those figures might suggest.

The power that is transmitted by a mobile phone during the "send" part of a call is typically 0.05 to 0.2 W. However, mobiles automatically adjust their transmitted power according to the proximity of the nearest cell phone mast so that a distant mast will demand greater power whilst the power transmitted to a nearby mast may be only a milliwatt. That is not typical, however.

Immediately adjacent to the phone, a transmitted power of 0.2 W corresponds to an E-field amplitude of 77 V/m and an H-field amplitude of 0.20 A/m. (These figures are estimated based on a phone area of 0.025 m^2, accounting for both sides, i.e., 8cm x 16cm x 2, which provides a very crude average of what will be a highly non-uniform transmission pattern).

Is it unsafe to hold such a transmitter next to your ear? People who are versed in radiological protection requirements, in the context of nuclear power or radioactive sources, may indeed initially be alarmed. [12.1] tells us this,

"*At all three operating frequencies: 2.4 GHz, 28 GHz and 100 GHz, the E- and H-fields vanish before passing through the brain layer of the Skin-Bone-Brain layered model.*"

On the positive side this tells us the brain is not affected. On the other hand it also tells us that the skin and skull absorb all the radiation incident upon them. This could be up to half the transmitted power, perhaps 0.1 W. A crude estimate suggests the mass of material doing the absorbing (part of your skull and surrounding tissue) is about 0.25 kg. Hence, in a one hour call your head is absorbing 1.4 kJ/kg. This compares with the allowable effective dose of gamma or beta radiation to the public of 1 mSv = 0.001 J/kg *per year*! Hence, a one hour phone call might involve absorbing radiant energy a million times larger than the ionising radiation safety limit per year.

But this is a completely fraudulent comparison. The two things are not actually comparable at all. Ionising radiation carries a risk because it is ionising, and so can damage DNA or damage cells in other ways. The difference relates to the quantum of energy (hf) involved. In the UK mobiles use a range of frequencies up to 2.6 GHz (4G) or up to 3.8 GHz (5G), with very limited usage of 26 GHz. In the USA mobiles use up to 39 GHz. These frequencies correspond to energy quanta of around 10^{-5} eV to 1.6×10^{-4} eV at most. This is far too small to cause ionisation because ionisation energies are of the order of eV or large fractions of an eV. In contrast, gamma radiation has an energy quantum of at least 10,000 eV, but generally far higher and up to 10 MeV, whilst the kinetic energy of the electrons in beta radiation is at least 0.16 MeV, typically 0.5 MeV, and up to more than 1 MeV. So the differences in energy quanta between mobile transmissions and radioactivity is at least 8 orders of magnitude and typically 10 orders of magnitude, and this is what distinguishes them, very emphatically.

Nevertheless, just as there are restrictions on the doses of ionising radiation to which the public may be exposed, so too there are guidelines on restricting exposure to non-ionising radiofrequency radiation including microwaves, [12.2], which states,

"*After several decades of* RF-EMF *research on numerous potential health effects, the only substantiated effect of* RF-EMF *exposure relevant to human health and safety is heating of exposed tissue.*"

Consequently, guidelines for limiting exposure are based on avoidance of excessive heating (which, in practice, means no noticeable heating). That this is likely to prove a non-event for mobile phone transmissions can be seen as follows.

A crude estimate of the temperature rise of your skull/skin after a one hour call is 0.34^{o}C, based on the above 0.25 kg involved mass, a transmitted power of 0.2 W and assuming a density and a specific heat comparable to water. But even this is unlikely to be realised in practice because, of course, your head is air-cooled on the outside and actively biologically regulated on the inside.

Consequently, unlike ionising radiation where the total dose in J/kg is the concern, safety limits for radio frequency radiation are defined in terms of the rate of energy deposition, i.e., power per kg. The EU has set a radio frequency safety limit for the human head at a Specific Absorption Rate (SAR) of 2 W per kilogram of tissue, [12.3]. The limits recommended in the ICNIRP guidelines, [12.4], draw a distinction between whole body exposure and localised exposure. For exposure of the whole body, a SAR limit to the general public of 0.08 W/kg is recommended, and 0.4 W/kg for occupational exposure. But for exposure localised to the head or torso those limits are 2 W/kg and 10 W/kg. My crude estimates, above, show that mobile phones are within those limits, perhaps around 0.4 W/kg to a small part of the head only.

Readers may have seen videos on the internet of a few mobile phones apparently boiling an egg. These videos are spoofs, of course. You are not going to boil an egg with a radiant power of 0.2 W. (I presume such spoofs are accomplished by removing the magnetron from a microwave cooker and concealing it below the table – that really is dangerous, don't try it at home!).

To indicate how sensitivity cell phone masts must be to pick up the signal from your mobile, consider a mast at a distance of 5 km. The field strength falls inversely with the distance (see chapters 22 and 24), which means the signal power falls as the inverse square of the distance. A transmitted power of 0.2 W from your mobile therefore corresponds to an E-field amplitude of 0.7 mV/m and an H-field amplitude of 1.8 μA/m, reductions of 5 orders of magnitude compared with holding the phone next to your ear, the signal power being reduced by 10 orders of magnitude. And yet the mast can (just) connect with you. The sensitivity of the receiver electronics is quite remarkable.

12.2 Cordless Landline Phones

A typical domestic cordless handset generates about 10 mW of time-averaged power, far less than is typical for a mobile phone. This is not surprising because a cordless handset needs only to communicate with the domestic base station which is meters away rather than kilometres in the case of cell phone masts.

As communication is two-way, there is also the field from the base station to consider. The maximum time-averaged power level for a typical base station is comparable with that of a mobile phone handset, perhaps around 0.2 W. But exposure is less because the base station is not held to the head, and the field strength falls rapidly with distance.

For example, at a distance of 1 metre the power density would be $\frac{0.2}{4\pi} = 0.016$ W/m^2 with an E-field amplitude of 3.5 V/m and a H-field amplitude of around 0.01 A/m. Supposing the cross-section presented by a human is 0.6 m^2 then the incident power at 1 metre is 0.01 W. If the incident power is absorbed within 1 cm, corresponding to about 6 kg mass, then the exposure rate is 0.0017 W/kg, and hence well below the whole-body exposure rate limit of 0.08 W/kg, [12.4].

12.3 The Ubiquitous Decibel

The decibel is a generic logarithmic measure, not specific to a particular physical quantity. It is most familiar in its usage as a measure of sound level. It is a logarithmic measure based on the ratio of the quantity in question to some reference magnitude. Infuriatingly, sources typically fail to state the reference level, without which the dB figure is meaningless.

In the case of electromagnetic waves there are at least two decibel measures in common usage, denoted dBm and dBmV/m. The latter is a measure of the electric field amplitude where the dimensionless ratio is referenced to a field amplitude of 1 mV/m. Hence,

$$\text{dBmV/m} = 20\log_{10}\left(\frac{E_0}{1\text{ mV/m}}\right) \tag{12.3.1}$$

An alternative measure would be to use the energy flux as the basis, which might be referenced to, say, 1 mW/m². This would give a decibel measure, $10\log_{10}\left(\frac{N}{1\text{ mW/m}^2}\right)$ where N is the magnitude of the Poynting vector. However, this is not what is used in practice but rather the measure dBm which is defined in terms of total power, rather than energy flux, and referenced to 1 mW, thus,

$$\text{dBm} = 10\log_{10}\left(\frac{Power}{1\text{ mW}}\right) \tag{12.3.2}$$

In practice the measure provided by a hand-held meter is neither of these, as the "power" is then the power received by the meter. This is not really a measure of signal strength alone because it depends upon the characteristics of the meter, specifically its antenna area and efficiency. In as far as meters themselves are standardised it does, however, function as a convenient relative measure of signal strength. A meter provides a measure of the signal strength at the measured location, of course. Conventional wisdom says that -50 to -60 dBm is a very strong signal, -70 dBm is a acceptable, whilst -80 dBm is a feeble signal that will probably be intermittent and annoying. For the purist, though, dBm is really a measure of a transmitter's total signal strength, not the signal strength at a distant point.

A guide to the signal strength required for decent reception by most devices is an electric field strength of 0.1 mV/m and hence -20 dBmV/m. An alternative field amplitude measure is referenced to 1 $\mu V/m$, denoted dBμ,

$$\text{dB}\mu = 20\log_{10}\left(\frac{E_0}{1\ \mu\text{V/m}}\right) \tag{12.3.3}$$

In terms of this measure a good signal level is 40 dBμ. Note that a signal strength in dBmV/m may be negative when that in dBμ is positive.

Note that factors of 20 occur in the definitions of dB involving field amplitudes, (12.3.1) and (12.3.3), whereas a factor of 10 appears in (12.3.2) where the definition is in terms of power. This is because power is proportional to the square of the field amplitude, so these definitions mean that dBmV/m, dBμ and dBm are all equal apart from additive constants (offsets).

12.4 Mobile Phone Range and WiFi Signal Strength

We have seen in §12.1 that a typical maximum mobile phone transmitted power of 0.2 W corresponds to an E-field amplitude of 77 V/m and an H-field amplitude of 0.20 A/m, and hence an average wave energy flux of about 7.7 W/m^2 immediately adjacent to the mobile (at a distance of roughly 4.5 cm). Because the field amplitude falls inversely with distance (see chapters 22 and 24) it follows that the signal will still be good enough, i.e., at a field strength of at least 0.1 mV/m, at a distance of (77 V/m / 0.1 mV/m)*4.5 cm = 34 km. So, in principle, the nearest cell phone mast could be that far away and still allow connection. In practice, intervening objects will cause absorption or diffraction of the waves, so the true effective range is reduced considerably. However, the above estimate shows that with a mast (say) 3 km away, the mobile emitted signal power might need be only 0.06 W to get connection (though the consumed power will be greater, of course).

Similar remarks apply for home hub WiFi base units (routers). As the distances within a home over which good transmission is required are small compared to cell phone mast distances, the task of the router is simpler – except that signal strength will be attenuated by passage through multiple walls and ceilings, etc.

12.4.1 What Controls the Speed of Your Home WiFi/Internet?

Many readers will have downloaded an app to their mobile which allows them to measure the speed of their WiFi/internet connection in different rooms, successively further away from their router. Such speeds are measured in Mbps, megabits per second. But how does the attenuation of a signal, that is the reduction of the amplitude of the electric field vector, translate into a digital speed in Mbps? The answer depends upon two things.

The first is the bandwidth, B, i.e., the range of frequencies in Hz, which the device and associated communication protocol use. The second is the signal to noise ratio, S/N. Here S is the power density of the signal at the point in question. This is subject to attenuation by walls and other intervening objects. N is the noise power density, which may be fairly constant between rooms.

The speed in Mbps is connected to B and S/N by the Shannon-Hartley noisy channel theorem which states that the speed (or channel capacity) in bps, C, is,

$$C = Blog_2(1 + S/N) \tag{12.4.1}$$

Leaving aside the complication that modern routers do not use a fixed bandwidth but rather deploy a variable, configurable channel width, assume for illustration we had a fixed bandwidth of 80 MHz, which is typical (as of 2026). Suppose your measured WiFi speed was up to 400 Mbps in some rooms but as low as only 20 Mbps in other rooms. The latter corresponds to an S/N ratio of a horrible 0.2, so that the router will spend an inordinate amount of time requesting resends of garbled data. So 20 Mbps will be a very poor connection, perhaps no connection in practice. On the other hand, 400 Mbps corresponds to an S/N ratio of 31, which is good and indicates an excellent connection.

In truth, the technology is not really as this overly simplified account suggests. Modern routers use dynamically switched modulation protocols which involve transmitting 'symbols' consisting of, perhaps, 2, 4, 6 or 8 bits each, being known as QAM (Quadrature Amplitude Modulation). The router constantly optimises which protocol to employ according to prevailing conditions, higher speeds being achieved by switching to larger degrees of QAM. Further details lie within the specialist province of IT comms technology and are beyond the scope of this book or its author's competence.

12.5 Radio and TV Broadcasting

The maximum wave power from AM and FM radio transmitters, in areas accessible to the public, are typically below 0.01 W/m^2. Close to the fence of very powerful transmitters that might be as great as 0.3 W/m^2. For whole body exposure that is roughly an exposure rate (SAR) not exceeding 0.05 W/kg, assuming absorption by 6 kg of body mass, and hence within the ICNIRP limit of 0.08 W/kg for the public. More typically it would be below a SAR of 0.0017 W/kg.

As for digital TV broadcasting, power densities are no higher than 0.04 W/m^2 and probably less than 1 mW/m^2, and so involve lower powers than AM/FM transmissions.

12.6 Cell Phone Mast Power

Scare stories about cell phone masts became a growth industry as 5G was being rolled out. In fact, Ofcom carries out regular surveys in towns and cities around the UK, [12.5]. For example they carried out surveys in 10 different towns in January 2024 alone, and these surveys go back to 2018 at least. The surveys cover 12 different frequencies from 700 MHz to 3.8 GHz, plus a catch-all "other frequencies", and covered 2G, 3G, 4G and 5G masts. (2G has now been discontinued in the UK, and 3G is being discontinued in

2026). I haven't looked through all the data – there's far too much. But the surveys across 6 locations in Cardiff in January 2024 indicated that, summed over all frequencies, all were less than 0.07% of the ICNIRP limit. The survey in Reading the same month showed all locations surveyed to be less than 0.12% of the ICNIRP limit. A survey in London N21 in November 2025 showed all locations surveyed to be less than 0.2% of the ICNIRP limit. There is a wealth of more such data in [12.5].

12.7 Domestic Microwaves

12.7.1 Inside the Microwave

Microwaves usually use a frequency of 2.45 GHz (wavelength 12.2 cm). At this frequency the microwaves will penetrate deep into the volume of the food to be heated (though, inevitably, heating more around the outside). This frequency is not chosen at a water molecule resonant frequency in order to optimise absorption by water molecules, despite many web sites falsely claiming so. (Water molecule vibrational frequencies are far higher than 2.45 GHz. A free isolated water molecule has three vibration frequencies at about 48, 110 and 113 THz. These are modified in the water phase due to interactions with neighbouring molecules, but remain far higher than 2.45 GHz). A microwave which did use a water molecule resonant frequency would be a poor device because it would heat only the surface layer of the food as the absorption would be far too efficient – the outer surface might burn whilst the rest remained cold.

In fact, domestic microwaves are not even monochrome but broadband – and there is nothing magical about the 2.45 GHz deployed. It is merely that which has been allocated for domestic microwave usage.

However, it is indeed principally water molecules which absorb the microwave. The mechanism of absorption depends upon the electric dipole moment of water molecules. These are continually trying to align with the oscillating electric field component of the microwaves, and it is the damping of the resulting motion which results in heating. You have probably noticed that microwaves are less effective on frozen items. This is because, in the solid phase, the water molecules are less mobile (lower velocities means lower friction means lower heating).

The microwaves are generated by a device called a magnetron. Magnetrons were first developed at the university of Birmingham, UK, in 1939. In 1940 the British gave the design of the magnetron to their ally, the USA. These early magnetrons were the secret behind the UK's highly successful radar systems in the second world war. Modern devices such as microwaves use a very different design from those earlier magnetrons, known as resonant cavity magnetrons.

Resonant cavity magnetrons are rather cunning devices. There is a nice explanation of the operation of a magnetron, with diagrams, in [12.6]. A permanent magnet provides a magnetic field along the axis of an annular region. The outside of the annulus is the anode, whilst along the axis lies the heated cathode. Electrons released from the cathode are accelerated across the gap to the anode by a high DC voltage. The electrons follow a curved trajectory due to the magnetic field. The frequency of the microwaves they produce is a result of the geometry of the copper block that serves as the anode. This anode includes slots which cause the generation of microwaves as the electrons sweep past (see also §22.7 which addresses the generation of e/m waves due to charges moving past a corrugated conducting surface). The complex shape of the anode effectively constitutes both capacitive and inductive portions and hence an LCR circuit, which can be regarded as the origin of the particular frequency of microwaves produced, i.e., $f = 1/\sqrt{LC}$ (see §18.3 on LCR circuits for details).

Domestic microwaves have an efficiency of 50% to 65% in turning the input electrical power into microwaves. Hence a 1 kW cooker may produce microwaves of power 650 W. Once emitted from the magnetron the waves are reflected multiple times from the inside surfaces (usually shiny stainless steel) until they are absorbed by the object being heated. Running a microwave empty for longer than a few seconds may damage it as the energy has nowhere to go and will act back on the magnetron.

Estimating the field strengths or energy density within a microwave is quite complicated because it depends upon the dielectric properties, and absorbency, of the food it contains. If there is little in the microwave, or it has low absorbency, the energy builds inside the microwave. A computer simulation reported in [12.7] assumed a 1 kW oven with interior dimensions 42 x 39 x 21 cm, together with a range of different simulated objects being heated. Because the oven dimensions are several times the wavelength (12.2 cm) standing waves are set up due to reflections, so the field intensity has several peaks and troughs. The peaks are likely to be an electric field amplitude of at least 10,000 V/m and probably several times larger. The average energy flux across the section of the microwave is thus at least 66 kW/m^2. As the cross section is 39 x 21 cm^2 = 0.082 m^2 the fluence of energy is at least 5.4 kW, and possibly an order of magnitude higher. How is this possible given that the microwave is drawing only 1 kW of electrical power? The answer is that this 5 to 50 kW fluence is in both directions (or, more accurately, all directions) so it does not represent a net flow. It arises because the waves emanating from the magnetron are reflected multiple times before being absorbed, so the energy fluence within the microwave interior builds up until the rate of absorption balances with that incoming (which is about 0.5 to 0.65 kW given the efficiency of a magnetron). These figures suggest the microwaves are probably reflected 10 to 100 times before being absorbed.

Despite these surprisingly large energy fluences the peak energy density even with an electric field amplitude of 50,000 V/m is only 0.011 J/m^3, and hence the total energy of the microwaves at any instant is only about 0.0002 J. Despite this tiny energy inventory within the microwaves, such a microwave (delivering 650 W) will heat a cup of water (0.25 litres) from 5°C to boiling in 2.5 minutes, a total heat energy of 99,000 J. This is possible because the energy flux is very high despite the energy density being very low – because the former equals the latter times the speed of light. The very low energy density is the reason why the microwave itself never feels hot.

12.7.2 Outside the Microwave

Ideally no microwaves should leak out of the interior with the door shut (and the microwave should not operate with the door open). However, some microwaves are less than perfect and some leakage can happen. The International Electrotechnical Commission (IEC), the International Committee on Electromagnetic Safety (ICES) of the Institute of Electrical and Electronics Engineers (IEEE) and the European Committee for Electrotechnical Standardization (CENELEC), have set a product emission limit of 50 W/m^2 at any point 5 cm away from the external surfaces of the oven, [12.6]. The American FDA sets the same limit, [12.8]. This seems rather lenient to me, as putting your head close to the microwave for a prolonged period would correspond to a specific dose rate in excess of the 2 W/kg limit (perhaps about 4 W/kg). However, moving to a more sensible distance, such as 30 cm (1 foot) decreases this to about 0.13 W/kg and so well inside the limit. In practice you'd expect the microwave's Faraday cage to be far less leaky than these figures suggest. If there is a weak point it is likely to be the door hinge or the door seal. The mechanical integrity of the door seal and hinge are therefore the key to microwave safety.

A simple, if rather crude, way to check if your microwave leaks is to put your mobile phone inside. At the risk of stating the obvious, don't turn the microwave on! Then try calling your phone from another device. If it rings, you have leaks. But don't panic, the leak may still be insignificant in magnitude. You can play around with aluminium foil to identify what spots on the microwave are leaky.

12.8 Domestic Audio Devices

Microphones typically deliver between 5 and 50 mV rms.

Stylus cartridges (from turntables) typically deliver only 0.5 to 10 mV.

Magnetic tape or cassette tape read heads produce very low voltage output, possibly as low as a few μV and certainly not more than hundreds of μV.

Wired earphones or headsets draw their power from the device they are connected to. Wireless earphones or headsets using technology such as Bluetooth consume around 3W and require battery power.

A domestic audio amplifier of sufficient power to satisfy most people might deliver an output of around 15 W rms per stereo channel (depending on the speaker impedance), although it would rarely be turned beyond about one-third that power. Domestic audio amplifiers typical have three stages: the initial voltage gain stage to boost the signal strength followed by the driver stage which is based on current amplification and finally the output stage which delivers what is necessary to the speakers. The total gain across the three stages might be around 1000 (i.e., 60dB).

There was a time that an off-the-shelf audio amplifier could be assumed to work well with a vinyl record turntable, i.e., for a signal derived from a stylus cartridge. This is no longer the case as signals from a stylus cartridge are substantially lower than those from a CD player or microphone, for example. These days you may require a separate pre-amp if you want to play those ancient LPs.

12.9 References

[12.1] Wongkasem, N. (2021). *Electromagnetic pollution alert: Microwave radiation and absorption in human organs and tissues*. Electromagnetic Biology and Medicine, 40(2), 236-253 (2021).

[12.2] International Commission on Non-Ionizing Radiation Protection (2020). *The ICNIRP 2020 Guidelines in Brief*.

[12.3] Scientific Committee on Emerging and Newly Identified Health Risks (SCENIHR) of the European Commission (2009). *Electromagnetic Fields: 2. What are the sources of exposure to radio frequency fields?* in Health Effects of exposure to Electromagnetic Fields (EMF).

[12.4] International Commission on Non-Ionizing Radiation Protection (2020). *ICNIRP Guidelines for Limiting Exposure to Electromagnetic Fields (100 kHz to 300 kHz)*. Health Physics, 118(5), 483-524 (2020).

[12.5] UK Telecommunications Regulator, Ofcom (2024). Electromagnetic field measurements near mobile base stations. 23 February 2024 (updated to 6/2/26 at time of writing).

[12.6] Zerner, T. (2010). *The Physics of Microwave Ovens*

[12.7] T. Santos, T., Costa, L.C, et al (2011). *3D Electromagnetic Field Simulation in Microwave Ovens: a Tool to Control Thermal Runaway*. The Microwave Journal, 15 March 2011. (Available here).

[12.8] US Food and Drug Administration, FDA. (2023). Microwave Ovens, 10 December 2023.

13. Electrostatics: The Effect of Conducting Boundaries (The Method of Images)

Whilst electrostatic fields and their potentials ultimately result from their source charges, the presence of boundaries on which some constraints on the fields or potentials are imposed will modify the resulting fields and potentials. Conducting boundaries are a case in point, being necessarily equipotentials. The method of images is used herein to solve some simple cases.

Contents

13.1 Formulation of the Problem

Under static conditions, the electrostatic scalar potential obeys $\nabla^2\Phi = -\rho/\varepsilon_0$, a Poisson equation, (5.1.2). In the case of a space with no boundaries, the solution for a specified charge density distribution is,

$$\Phi(\bar{r}_f) = \int \frac{\rho(\bar{r})}{4\pi\varepsilon_0|\bar{r}_f - \bar{r}|} d\tau \tag{5.1.6}$$

For a discrete set of point charges this becomes,

$$\Phi(\bar{r}_f) = \sum_i \frac{q_i}{4\pi\varepsilon_0|\bar{r}_f - \bar{r}_i|} \tag{5.1.5}$$

However, when there are conducting boundaries present these solutions are modified so as to ensure that the conducting boundary is an equipotential surface (and hence that the electric field is everywhere normal to the surface). The reason why conducting boundaries must be equipotentials is that any gradient of potential on the boundary surface would cause a current to flow until the gradient was neutralised. In steady state, with no current flowing, no in-surface gradient of potential can persist.

In this chapter we consider some simple cases of a given static point charge, together with certain conducting surfaces, and seek the scalar potential and electric fields which result. Cases in which there are a number of point charges, or a distribution of charge density, can then be found by linear addition of the solutions for each point charge separately.

A key assumption throughout shall be that any solution in which the conducting surfaces are equipotentials, and which tend to the Coulomb field near the point charge, must be the unique solution. This follows from the uniqueness theorem for Laplace's equation which we establish first.

13.2 The Uniqueness Theorem for Laplace's Equation

The theorem states that if a scalar field, Φ, obeys Laplace's equation, $\nabla^2\Phi = 0$ and has Φ specified on a closed surface, then the solution for Φ everywhere within that surface is uniquely defined.

Suppose there are two solutions to $\nabla^2 f = 0$, called f_1 and f_2, which take the same values on a closed boundary. It follows that $F = f_1 - f_2$ is also a solution. Now consider the vector $F\bar{\nabla}F$. The divergence theorem tells us that $\int_V \bar{\nabla}\cdot(F\bar{\nabla}F)d\tau = \oiint_{\delta V} F\bar{\nabla}F \cdot d\bar{S}$. But if we take the surface δV to be that on which the field is specified then these integrals are zero because $F = 0$ on that boundary. But $\bar{\nabla}\cdot(F\bar{\nabla}F) = |\bar{\nabla}F|^2 + F\nabla^2 F = |\bar{\nabla}F|^2$ because $\nabla^2 F = 0$ everywhere. But we then conclude that $\int_V |\bar{\nabla}F|^2 \, d\tau = 0$ but as $|\bar{\nabla}F|^2$ cannot be negative anywhere, then $|\bar{\nabla}F|^2$ must be zero everywhere within the boundary, in which case F is a constant within the boundary, and, since it is zero on the boundary, it must be zero everywhere within it, i.e., the original two solutions are actually identical, $f_1 = f_2$, everywhere within and on the boundary. QED.

13.3 A Single Infinite Plane Conducting Boundary

This problem presents the method of images in its simplest and archetypal application. A point charge, q, is located at $\bar{r}_0 = (d,0,0)$ and there is an infinite conducting plane on $x = 0$. The plane can be regarded as part of a closed surface, so the uniqueness theorem applies, by considering it to be connected to a large hemisphere whose radius is pushed off to infinity. Because the plane is an equipotential, and because the datum of potential is arbitrary, we can consider the whole closed boundary, including the plane, to be at a uniform zero volts.

By considering a notional "image" charge of $-q$ located at $(-d,0,0)$, as if the conducting plane were a mirror which inverts the sign of charge, we get a total scalar potential of,

For $x \geq 0$
$$\Phi = \frac{q}{4\pi\varepsilon_0}\left(\frac{1}{|\bar{r}-\bar{r}_0|} - \frac{1}{|\bar{r}+\bar{r}_0|}\right) \tag{13.3.1}$$

which is applicable only for $x \geq 0$. (On the other side of the plane there is no charge in reality, and the plane is a zero potential boundary so the potential can be taken as zero everywhere with $x \leq 0$, hence no field). On the plane, i.e., for $x = 0$, (13.3.1) gives $\Phi = 0$, as required, and this also applies at infinity. (13.3.1) also has the correct asymptotic form (the Coulomb potential) as the charge is approached, so by the uniqueness theorem, (13.3.1) is the correct solution as it fits the boundary conditions. Writing $\bar{r}_\pm = \bar{r} \pm \bar{r}_0$ the field is,

For $x \geq 0$
$$\bar{E} = \frac{q}{4\pi\varepsilon_0}\left(\frac{\bar{r}_-}{|\bar{r}_-|^3} - \frac{\bar{r}_+}{|\bar{r}_+|^3}\right) \tag{13.3.2}$$

At distances large compared with $2d$ the field will be that of a dipole of dipole moment $2qd$. The force by which the charge is attracted to the conducting plate (and vice-versa) follows immediately because it must equal the force corresponding to charges of $\pm q$ a distance $2d$ apart and so is,

$$\text{Force on charge} = \frac{q^2}{16\pi\varepsilon_0 d^2}\ \text{(attractive)} \tag{13.3.3}$$

The surface charge induced on the plate is $\sigma = \varepsilon_0 E_x$. In terms of the radial distance along the conducting plane, $s = \sqrt{y^2 + z^2}$ the induced charge density reduces as follows,

$$\sigma = -\frac{q}{2\pi d^2}\left(1 + \frac{s^2}{d^2}\right)^{-\frac{3}{2}} \tag{13.3.4}$$

13.4 Two Conducting Planes at Right Angles

Conducting planes line on $x = 0,\ y \geq 0$ and $y = 0,\ x \geq 0$ and a point charge, q, is placed at $\bar{r}_0 = (x_0, y_0, 0)$ where $x_0 > 0,\ y_0 > 0$, so that the charge lies within the 90° wedged-shaped region. The planes are turned into a closed boundary once again by adding a quarter-cylinder, with end caps, whose radius is pushed off to infinity. Notional image charges, $-q$, lie at both $\bar{r}_1 = (x_0, -y_0, 0)$ and at $-\bar{r}_1 = (-x_0, y_0, 0)$. There is also an image-of-an-image charge, $+q$, at $-\bar{r}_0$. Setting $\bar{r}_\pm = \bar{r} \pm \bar{r}_0$ and $\bar{r}_{1\pm} = \bar{r} \pm \bar{r}_1$ the potential is,

Within the wedge: $$\Phi = \frac{q}{4\pi\varepsilon_0}\left(\frac{1}{|\bar{r}_-|} + \frac{1}{|\bar{r}_+|} - \frac{1}{|\bar{r}_{1-}|} - \frac{1}{|\bar{r}_{1+}|}\right) \tag{13.4.1}$$

noting that (13.4.1) gives $\Phi = 0$ on both conducting planes. The field is,

Within the wedge: $$\bar{E} = \frac{q}{4\pi\varepsilon_0}\left(\frac{\bar{r}_-}{|\bar{r}_-|^3} + \frac{\bar{r}_+}{|\bar{r}_+|^3} - \frac{\bar{r}_{1-}}{|\bar{r}_{1-}|^3} - \frac{\bar{r}_{1+}}{|\bar{r}_{1+}|^3}\right) \tag{13.4.2}$$

As the field component parallel to each conducting plane is zero at the plane, it follows that the whole field tends to zero at the origin (the apex of the wedge). The field and potential are zero outside the 90° wedge (the images charges are not real).

The surface charge density on the plates is obtained as the normal component of field (divided by ε_0). The surface charge density, σ, on the $x = 0$ conducting plane varies with the y and z position as,

For $y \geq 0$: $$\sigma = -\frac{q}{2\pi}\left(\frac{x_0}{(x_0^2+(y-y_0)^2+z^2)^{\frac{3}{2}}} - \frac{x_0}{(x_0^2+(y+y_0)^2+z^2)^{\frac{3}{2}}}\right) \tag{13.4.3}$$

This is zero at the origin, $y = 0$ (the apex of the wedge). The vector force acting on the charge, q, is,

$$\bar{F} = \frac{q^2}{16\pi\varepsilon_0}\left\{\left(\frac{x_0}{(x_0^2+y_0^2)^{\frac{3}{2}}} - \frac{1}{x_0^2}\right)\hat{x} + \left(\frac{y_0}{(x_0^2+y_0^2)^{\frac{3}{2}}} - \frac{1}{y_0^2}\right)\hat{y}\right\} \tag{13.4.4}$$

13.5 A Pair of Parallel Conducting Planes

The origin of coordinates is taken at the charge, q, which lies between two parallel planes, one at $x = -d$ and the other at $x = D$. In this case there are an infinite number of images (just as are produced by a pair of parallel mirrors). In terms of the vectors,

$$\bar{r}_{1n} = -\big(2d + 2n(D+d)\big)\hat{x} \quad \bar{r}_{2n} = -2(n+1)(D+d)\hat{x} \tag{13.5.1a}$$

$$\bar{r}_{3n} = \big(2D + 2n(D+d)\big)\hat{x} \quad \bar{r}_{4n} = -\bar{r}_{2n} \tag{13.5.1b}$$

where n is a positive integer or zero. The solution for the potential between the planes is,

$$\Phi = \frac{q}{4\pi\varepsilon_0}\left[\frac{1}{r} + \sum_{n=0}^{\infty}\left(\frac{1}{|\bar{r}-\bar{r}_{2n}|} + \frac{1}{|\bar{r}-\bar{r}_{4n}|} - \frac{1}{|\bar{r}-\bar{r}_{1n}|} - \frac{1}{|\bar{r}-\bar{r}_{3n}|}\right)\right] \tag{13.5.2}$$

It is left as an exercise for the reader to confirm that the potential is zero on both planes $x = -d$ and $x = D$. The electric field is,

$$\bar{E} = \frac{q}{4\pi\varepsilon_0}\left[\frac{\hat{r}}{r^2} + \sum_{n=0}^{\infty}\left(\frac{\bar{r}-\bar{r}_{2n}}{|\bar{r}-\bar{r}_{2n}|^3} + \frac{\bar{r}-\bar{r}_{4n}}{|\bar{r}-\bar{r}_{4n}|^3} - \frac{\bar{r}-\bar{r}_{1n}}{|\bar{r}-\bar{r}_{1n}|^3} - \frac{\bar{r}-\bar{r}_{3n}}{|\bar{r}-\bar{r}_{3n}|^3}\right)\right] \tag{13.5.3}$$

Force and surface charge density can be calculated as before. The potential and the field in the regions $x < -d$ and $x > D$ are zero.

The infinite number of terms in the sums, (13.5.2,3), arises for the same reason that, if you stand between a pair of parallel mirrors, you see images of yourself disappearing to visual infinity. Each reflection reverses the sign of the image charge.

13.6 A Grounded Conducting Sphere and External Point Charge

As we have seen, the "method of images" is well named as regards cases involving plane conducting boundaries. But the name ceases to be so appropriate in general cases, e.g., for curved conducting boundaries. In the general case we seek a placement, and magnitude, of image charge(s) to make the conducting boundary an equipotential.

A conducting sphere of radius a centred on the origin is earthed, and so held at zero potential. It has a charge q placed outside it on the x-axis at $x = d >$

a. A suitable "image" charge in this case is $-\frac{a}{d}q$ placed at $x = b = \frac{a^2}{d}$ (hence inside the sphere). The potential outside the sphere is,

For $r \geq a$ $\quad \Phi = \frac{q}{4\pi\varepsilon_0}\left(\frac{1}{\sqrt{(x-d)^2+y^2+z^2}} - \frac{a}{d}\frac{1}{\sqrt{(x-b)^2+y^2+z^2}}\right)$ (13.6.1)

If correct the potential should be zero on the sphere, $x^2 + y^2 + z^2 = a^2$. On the sphere we have, (13.6.2)

$$\sqrt{(x-d)^2 + y^2 + z^2} = \sqrt{(x-d)^2 + a^2 - x^2} = \sqrt{a^2 + d^2 - 2dx}$$

and also, on the sphere, (13.6.3)

$$\frac{d}{a}\sqrt{(x-b)^2 + y^2 + z^2} = \frac{d}{a}\sqrt{(x-b)^2 + a^2 - x^2} =$$

$$\frac{d}{a}\sqrt{a^2 + b^2 - 2bx} = \frac{d}{a}\sqrt{a^2 + \frac{a^4}{d^2} - 2\frac{a^2}{d}x} = \sqrt{a^2 + d^2 - 2dx}$$

Hence (13.6.3) = (13.6.2) and so $\Phi = 0$ on the sphere as required. The field can be found from $\bar{E} = -\bar{\nabla}\Phi$, as usual. The force of attraction between the charge and the sphere is just that between a charge q and a charge $-\frac{a}{d}q$ at a distance $d - b$ apart, and hence,

$$\text{Force on charge (attractive)} = \frac{q^2}{4\pi\varepsilon_0 a^2} \cdot \left(\frac{a}{d}\right)^3 \left(1 - \frac{a^2}{d^2}\right)^{-2} \quad (13.6.4)$$

13.7 A Grounded Conducting Sphere in a Uniform Field

This case may be simulated by that of §13.6 by adding a charge $-q$ at a distance of d from its centre but diametrically opposite the first charge, i.e., at $x = -d$. If $d \gg a$ the field in the vicinity of the sphere is roughly $\bar{E} = -\frac{2q\hat{x}}{4\pi\varepsilon_0 d^2}$. Allowing $d \to \infty$ whilst increasing q so that q/d^2 remains constant results in a uniform field around the sphere. There are now two image charges, $\pm\frac{a}{d}q$, which become infinite. However, their distance apart, which is $2b = \frac{2a^2}{d}$, shrinks to zero and their dipole moment is $p = \frac{2a^3 q}{d^2} = 4\pi\varepsilon_0 a^3 E$ remains constant as $d \to \infty$. This expression gives the electric dipole moment induced in a conducting sphere when place in a uniform electric field. Taking the polar axis as aligned with the applied external field, $E_0\hat{x}$, this can be written $E_0\hat{x} = E_0(\hat{r}\cos\theta - \hat{\theta}\sin\theta)$. The total field is thus, using (5.2.5) for the dipole field,

$$\bar{E} = E_0\left(\frac{a^3}{r^3}\left[2\hat{r}\cos\theta + \hat{\theta}\sin\theta\right] + \hat{r}\cos\theta - \hat{\theta}\sin\theta\right) \quad (13.7.1)$$

On the surface of the sphere, $r = a$, this is,

$$\bar{E}(on\ sphere) = 3E_0\hat{r}\cos\theta \qquad (13.7.2)$$

Hence the field on the sphere is normal to it, as it must be for an equipotential surface. On the polar axis (here the $\hat{x}$ axis, $\theta = 0$ or $\theta = \pi$) the field is thus $3E_0\hat{x}$. The presence of the conducting sphere thus leads to an intensification of the field on its surface by a factor of 3. (This is analogous to the way in which holes in structural materials intensify stresses). On dividing by ε_0 (13.7.2) also gives the surface charge density induced in the sphere. The reason for the intensification of the field near the sphere is that the equipotential surfaces necessarily become "squashed" closer together near the sphere because the sphere is itself an equipotential, as illustrated in Figure 13.1, and the field is, of course, (minus) the gradient of the potential.

Figure 13.1: Equipotential surfaces for a grounded conducting sphere in a uniform electric field. *Distances in units of the sphere radius, and potentials in units of the remote field times the radius.*

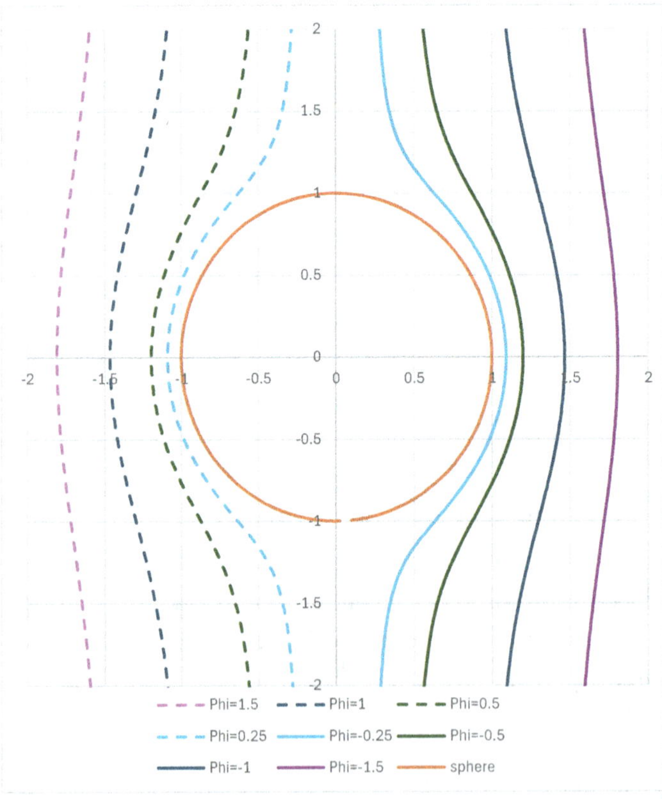

13.8 A Pair of Conducting Planes at Included Angle β

The region within which the field is non-zero is that within the angle β. In this case we do not assume any particular source but consider only the generic behaviour of the potential, the field and the surface charge density on the conducting plates as the corner is approached, $r \to 0$. The detailed theory is expounded in Jackson, [13.1], section 2.11, but can be developed simply as follows.

In cylindrical polars the planes are placed at $\theta = 0$ and $\theta = \beta$. We need the potential to be zero on the planes. We guess that the potential might be expressible in the separable form,

$$\Phi = \Phi_0 r^\alpha f(\theta) \tag{13.8.1}$$

And we note that a possible angular function which gives zero potential on the planes is,

$$f(\theta) = \sin\left(\frac{\pi\theta}{\beta}\right) \tag{13.8.2}$$

In cylindrical polars Laplace's equation becomes,

$$\left[\frac{1}{r}\partial_r(r\partial_r) + \frac{1}{r^2}\partial_\theta^2 + \partial_z^2\right]\Phi = 0 \tag{13.8.3}$$

There is no z dependence, so substituting (13.8.1,2) into (13.8.3) we find that Laplace's equation is obeyed by a potential of this form provided we set the exponent, α, to,

$$\alpha = \frac{\pi}{\beta} \tag{13.8.4}$$

Using $grad(f) = \hat{r}\partial_r f + \hat{\theta}\frac{1}{r}\partial_\theta f + \hat{z}\partial_z f$ in cylindrical polars we find the field is given by,

$$\bar{E} = -\alpha\Phi_0 r^{\alpha-1}\left[\hat{r}f(\theta) + \hat{\theta}\cos(\alpha\theta)\right] \tag{13.8.5}$$

The surface charge density on the plates is just the $\hat{\theta}$ component of the field divided by ε_0 and hence varies with distance from the apex as,

$$\text{LIM } r \to 0 \qquad \sigma \propto r^{\frac{\pi}{\beta}-1} \tag{13.8.1}$$

noting that the cosine function in (13.8.5) means that the surface charge density is equal and opposite on the two plates. The radial component of the field on the plates from (13.8.5) is zero, as it must be. The multiplier Φ_0 is unknown as the source of the field has not been specified.

Hence, for $\beta < \pi$ the charge density is zero at the corner. But when $\beta > \pi$ the surface density of charge is divergent at the corner. In the limit $\beta \to 2\pi$, the two planes are like the two sides of a sharp knife blade (and essentially become just one semi-infinite plate), the singularity in the field and the charge density is most severe, $\propto 1/\sqrt{r}$. This is reminiscent of the singularity in stress near the tip of sharp cracks in structural materials which behave in a linear elastic fashion. The latter accounts for the substantial reduction in a structure's load bearing capacity in the presence of cracks. In the electrostatic case, the divergence in the charge density near a sharp point, and hence the very large field sufficiently close to the tip, will tend to cause discharge.

13.8.1 Lightning Conductors

The large field near such sharp points accounts for the manner in which lightning conductors can "attract" the lightning. However, there has long been a controversy between those advocating lightning conductors with sharp points and those advocating rounded ends on lightning conductors. The advantage of the latter is that, whilst the field is less intense than for a sharp tip very close to the conductor, the field from a rounded end remains fairly large over a longer range than for a sharp point. [13.2] has recently provided empirical evidence in favour of the rounded end.

13.9 Dynamics and the Corrugated Boundary

In all the foregoing static examples, a point charge tends to be attracted to a conducting boundary. Consequently, to maintain the assumed static conditions there must be some restraining force which keeps the charge in position, e.g., by being mechanically fixed.

Alternatively we can envisage, say, a point charge initially moving parallel to a plane, earthed conducting surface. In this case the charge will not remain in its straight trajectory at constant velocity but would be accelerated towards the plate, tending to hit the plate.

An interesting case arises when a conducting boundary is corrugated, i.e., its surface is not planar but sinusoidal. Now envisage a point charge initially moving at speed v in the $\hat{x}$ direction, parallel to the mean position of the surface, at a distance d. A crude heuristic feel for what results is to assume an image charge which also moves along with the real charge in the $\hat{x}$ direction but whose distance from the real charge varies sinusoidally, specifically as $\Delta y = d - a\sin\omega t$ where a is the amplitude of the surface undulations and $\omega = 2\pi v/\lambda$ where λ is the wavelength of the undulations. If the real charge were constrained to continue to move in a straight line without crashing into the surface (or for so long as it does so), then the image charge would be oscillating with an amplitude of $2a$. We will see in chapter 22 that oscillating charges emit electromagnetic radiation. Consequently, we would expect a charge fired parallel to an uneven surface to create radiation. This will be considered again in chapter 22, but we have already seen an example of this phenomenon in the operation of the magnetron (see chapter 12).

13.10 References

[1] Jackson, J.D. (1962). *Classical Electrodynamics*. John Wiley & Sons, 1st edition 1962. Linked is the 3rd edition, 2021.

[2] Moore, C.B., et al (2000). *Lightning Rod Improvement Studies*. Journal of Applied Meteorology and Climatology, 39(5), 593–609, 1 May 2000.

14. Solving Laplace's Equation and the Multipole Expansion

Following chapter 13's presentation of the method of images, this chapter considers two further methods for solving electrostatic problems: separation of variables and, in 2D cases, the complex potential method. The multipole expansion is obtained as a by-product. Because we give examples of specific solutions here, the mathematics is rather more detailed than in many other chapters. A further by-product is an elementary introduction to the Legendre polynomials and the spherical harmonic functions, as well as the cylindrical Bessel functions.

Contents

In this chapter we continue to address electrostatics and so envisage an electric field produced by a distribution of stationary charges (possibly both point charges and continuous distributions of charge). The controlling equation including the source charges is Poisson's equation, (5.1.2), which becomes Laplace's equations, $\nabla^2\Phi = 0$, (5.1.3), at points where there is no charge. In this chapter we consider methods for solving Laplace's equation.

14.1 The Method of Separation of Variables

The method of separating variables is familiar from other branches of physics and mathematics (and we have already deployed it in §13.8). It refers to developing basis solutions, in some coordinate system, as products of functions of each coordinate independently. General solutions can then be expressed as superpositions of such basis solutions. The method is most useful when the chosen coordinate system reflects the symmetry of the boundary conditions to be applied. The solution may then reduce to just one of the basis solutions.

14.1.1 Rectangular Boundaries

Suppose we use a Cartesian coordinate system and assume a solution can be written as $\Phi = X(x)Y(y)Z(z)$. Substituting into $\nabla^2\Phi = 0$ and dividing by Φ gives,

$$\frac{1}{X}\frac{\partial^2 X}{\partial x^2} + \frac{1}{Y}\frac{\partial^2 Y}{\partial y^2} + \frac{1}{Z}\frac{\partial^2 Z}{\partial z^2} = 0 \tag{14.1.1}$$

Note that this has separated the three variables, each term depending upon only one of the coordinates. Since $\frac{1}{X}\frac{\partial^2 X}{\partial x^2}$, which depends upon x only, has to be equal and opposite to $\frac{1}{Y}\frac{\partial^2 Y}{\partial y^2} + \frac{1}{Z}\frac{\partial^2 Z}{\partial z^2}$, which depends upon y and z only, it follows that both must be a constant. The same applies to each term. So we have $\frac{\partial^2 X}{\partial x^2} = -k_x^2 X$, $\frac{\partial^2 Y}{\partial y^2} = -k_y^2 Y$, $\frac{\partial^2 Z}{\partial z^2} = -k_z^2 Z$ where k_x, k_y, k_z are constants and $k_x^2 + k_y^2 + k_z^2 = 0$. The latter means that one or two of the three constants must be purely imaginary (or zero). For example, the case $k_z = 0,\ k_x^2 + k_y^2 = 0, k_x^2 < 0$ (hence $k_x = ik_y$) has a basis solution,

$$X(x) = A_1 e^{k_y x} + A_2 e^{-k_y x}\ \ Y(y) = B_1 \sin(k_y y + \alpha),\ Z(z) = C_1 + C_2 z$$

Suppose the boundary conditions are $\Phi = 0$ on $x = 0$ and on $y = 0$ and on $y = L_y$ whilst on $x = L_x$ the potential is $\Phi = f(y)$ (i.e., we have specified the potential on all the surfaces of a rectangular box). Also, the z component of the field is specified to be zero everywhere (hence $C_2 = 0$).

The most general basis solution that satisfies the conditions $\Phi = 0$ on $x = 0$ and $y = 0$ is of the form,

$$\Phi = (e^{k_y x} - e^{-k_y x}) \sin(k_y y) \tag{14.1.2}$$

To also satisfy $\Phi = 0$ on $y = L_y$ we must choose the parameter k_y to be $n\pi/L_y$ where n is a positive integer (1, 2, 3,...). The most general solution is thus,

$$\Phi = \sum_{n=1}^{\infty} A_n \left(e^{\frac{n\pi x}{L_y}} - e^{-\frac{n\pi x}{L_y}} \right) \sin(n\pi y/L_y) \tag{14.1.3}$$

The final boundary condition, that the potential is some specified function of y, $\Phi = f(y)$, on $x = L_x$ is imposed by requiring that,

$$\sum_{n=1}^{\infty} A_n B_n \sin(n\pi y/L_y) \equiv f(y) \tag{14.1.4}$$

where,
$$B_n = \left(e^{\frac{n\pi L_x}{L_y}} - e^{-\frac{n\pi L_x}{L_y}} \right) \tag{14.1.5}$$

To find coefficients A_n we use what is essentially the usual inverse of a Fourier series (albeit in slightly modified form here) by appealing to the orthogonality of the functions $\sin(n\pi/L_y y)$ for differing n. Specifically,

$$\int_0^{L_y} \sin(n\pi y/L_y) \sin(m\pi y/L_y) dy = \frac{L_y}{2} \delta_{nm} \tag{14.1.6}$$

Hence, multiplying both sides of (14.1.4) by $\sin(m\pi y/L_y)$ and integrating gives,

$$A_n = \frac{2}{L_y B_n} \int_0^{L_y} f(y) \sin(n\pi y/L_y) dy \tag{14.1.7}$$

For example, if the side $x = L_x$ is at a constant potential, $f(y) = \Phi_0$, then we get,

$$A_n = \frac{2[1-(-1)^n]}{n\pi B_n} \Phi_0 \tag{14.1.8}$$

Hence, the solution is,

$$\Phi = \frac{4}{\pi} \Phi_0 \sum_{\text{odd } n=1}^{\infty} \frac{1}{n B_n} \left(e^{\frac{n\pi x}{L_y}} - e^{-\frac{n\pi x}{L_y}} \right) \sin(n\pi y/L_y) \tag{14.1.9}$$

or, $$\Phi = \frac{4}{\pi}\Phi_0 \sum_{\text{odd } n=1}^{\infty} \frac{1}{n}\left(\frac{\sinh\frac{n\pi x}{L_y}}{\sinh\frac{n\pi L_x}{L_y}}\right)\sin(n\pi y/L_y) \tag{14.1.10}$$

where the sums in (14.1.9,10) extend over odd n only (functions of y which are symmetrical about the middle of the rectangle, $y = \frac{L_y}{2}$). Figure 14.1 shows this potential in graphical form.

Figure 14.1: Potential plotted against y coordinate for various x sections for problem of §14.1.1 (from numerical evaluation of (14.1.10), including up to a maximum n of 199).

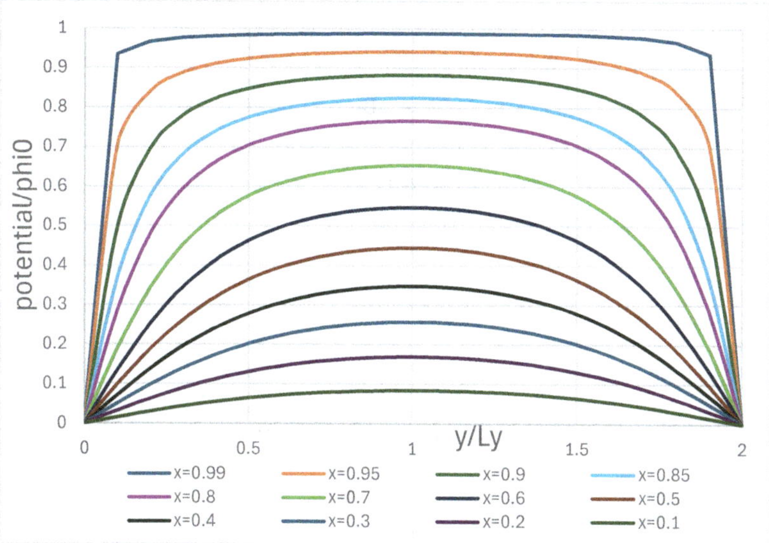

14.1.2 Spherical Boundaries

Separation of variables may be carried out with respect to spherical polar coordinates. Hence we write,

$$\Phi = R(r)\Theta(\theta)\varphi(\phi) \tag{14.1.11}$$

The Laplacian in spherical polars is given in Appendix A2 and Laplace's equation becomes, after substituting (14.1.11) and dividing through by Φ and also multiplying by r^2,

$$\frac{r^2}{\Phi}\nabla^2\Phi = \frac{1}{R}\partial_r(r^2\partial_r R) + \frac{1}{\Theta\sin\theta}\partial_\theta(\sin\theta\,\partial_\theta\Theta) + \frac{1}{\varphi\sin^2\theta}\partial_\phi^2\varphi = 0 \tag{14.1.12}$$

The first term on the RHS depends only on r whilst the second and third terms depend only on θ, ϕ. Hence we conclude there is a constant, k, such that,

$$\frac{1}{R}\partial_r(r^2\partial_r R) = k \qquad \frac{1}{\Theta \sin\theta}\partial_\theta(\sin\theta\partial_\theta\Theta) + \frac{1}{\sin^2\theta}\partial_\phi^2\varphi = -k \qquad (14.1.13)$$

The variables θ, ϕ also separate because the latter equation can be written,

$$\frac{\sin\theta}{\Theta}\partial_\theta(\sin\theta\partial_\theta\Theta) + k\sin^2\theta = -\partial_\phi^2\varphi = m^2 \qquad (14.1.14)$$

for some constant m. The solution for the functions $\varphi(\phi)$ follows immediately,

$$\varphi(\phi) = constant \times e^{\pm im\phi} \qquad (14.1.15)$$

To be single-valued over the angular variable ϕ, which takes values in the range $[0,2\pi)$, we see that m must be an integer (positive, negative or zero). This is also the reason why m must be real, i.e., that in (14.1.14) the constant m^2 is a positive number or zero. The equation in the polar angle therefore becomes,

$$\sin\theta\,\partial_\theta(\sin\theta\partial_\theta\Theta) + (k\sin^2\theta - m^2)\Theta = 0 \qquad (14.1.16)$$

By substituting $\xi = \cos\theta$ this can also be written,

$$\left[\partial_\xi\left((1-\xi^2)\partial_\xi\right) + k - \frac{m^2}{(1-\xi^2)}\right]\Theta = 0 \qquad (14.1.17)$$

(14.1.16) or (14.1.17) are known as Legendre's equation. It is shown in Appendix A7 that finiteness over the whole angular range $\theta\epsilon[0,\pi]$ requires the constant k to be equal to $l(l+1)$ where l is a positive integer, or zero.

The solutions to (14.1.17), denoted $P_l^m(\xi)$, are called the associated Legendre functions of the first kind. (As Legendre's equation is of second order it also has a second, independent, solution known as the associated Legendre functions of the second kind, generally denoted $Q_l^m(\xi)$, but these are singular at $\xi = 1$, i.e., on the polar axis, $\theta = 0$, and so are not usually physically relevant). When $m = 0$ the usual notation is to write $P_l(\xi)$ for $P_l^0(\xi)$, and these $P_l(\xi)$ are known as the Legendre polynomials. As shown in Appendix A7 they are given by,

$$P_l(\xi) = \frac{(-1)^l}{2^l l!}\frac{d^l}{d\xi^l}(1-\xi^2)^l \qquad (14.1.18a)$$

So the Legendre polynomial $P_l(\xi)$ is of order l. In terms of these polynomials the associated Legendre functions of the first kind can be found, for $m \geq 0$, from,

$$P_l^m(\xi) = (-1)^m (1-\xi^2)^{\frac{m}{2}} \frac{d^m P_l(\xi)}{d\xi^m} \tag{14.1.18b}$$

This shows that the P_l^m vanish for $m > l$. Combining (14.1.18a,b) into a single formula gives,

$$P_l^m(\xi) = (-1)^{l+m} \frac{(1-\xi^2)^{\frac{m}{2}}}{2^l l!} \frac{d^{l+m}}{d\xi^{l+m}} (1-\xi^2)^l \tag{14.1.18c}$$

which is now applicable also for $m < 0$. However, this shows that the P_l^m also vanish for $m < -l$. The P_l^m for negative m is equal to $P_l^{|m|}$ except for normalisation. Specifically,

For $m > 0$: $$P_l^{-m}(\xi) = (-1)^m \frac{(l-m)!}{(l+m)!} P_l^m(\xi) \tag{14.1.18d}$$

When solutions (14.1.15) and (14.1.18b) are joined to provide the complete angular solutions in both the azimuthal and polar angles, the combination, after suitable normalisation, is known as a spherical harmonic,

$$Y_l^m(\theta, \phi) = N_{lm} P_l^m(\cos\theta) e^{im\phi} \tag{14.1.19a}$$

The normalisation constant is usually defined as,

$$N_{lm} = (-1)^m \sqrt{\frac{(2l+1)}{4\pi} \frac{(l-m)!}{(l+m)!}} \tag{14.1.19b}$$

This normalisation ensures that the spherical harmonics are orthonormal when integrated over the solid angle $d\Omega = -d(\cos\theta)d\phi$. The first few Legendre polynomials from which the other functions may be derived using (14.1.18b,19) are,

$$P_0(\xi) = 1 \tag{14.1.20a}$$

$$P_1(\xi) = \xi \tag{14.1.20b}$$

$$P_2(\xi) = \frac{1}{2}(3\xi^2 - 1) \tag{14.1.20c}$$

$$P_3(\xi) = \frac{1}{2}(5\xi^3 - 3\xi) \tag{14.1.20d}$$

To complete the solution of Laplace's equation we also require the solution to the radial equation which (14.1.13) now gives as,

$$\partial_r(r^2\partial_r R) = l(l+1)R \tag{14.1.21}$$

the solution to which is readily seen to be,

$$R = Ar^l + \frac{B}{r^{l+1}} \quad (l = 0,1,2,\ldots) \tag{14.1.22}$$

If the boundary conditions are such that the potential is required to be finite at $r = 0$ then we can set $B = 0$. On the other hand if the boundary conditions require the potential to be finite at infinity, then we can set $A = 0$ except for $l = 0$. Example problems using these types of solution are as follows.

14.1.2.1 Cosine potential on surface of a sphere

The sphere $r = a$ has potential specified on it to be $\Phi = \Phi_0 \cos\theta$. If we assume that within the sphere the potential is $\Phi = \Phi_0 \frac{r}{a} \cos\theta = \Phi_0 \frac{r}{a} \xi = \Phi_0 \frac{z}{a}$ then this satisfies Laplace's equation because it is the solution given by (14.1.20b) and (14.1.22) with $l = 1$ when $B = 0$. Since it also satisfies the boundary condition $\Phi = \Phi_0 \cos\theta$ on the surface of the sphere, then by the uniqueness theorem (§13.2) it is the correct solution. This shows that the field inside the sphere is uniform, $E_z = -\frac{\Phi_0}{a}$.

Outside the sphere the angular variation still matches that of $P_1(\xi) = \cos\theta$ and so $l = 1$ and the possible radial dependencies are Ar and B/r^2. The former blows up at infinity and is ruled out. Hence the solution must be $\Phi = B\cos\theta/r^2$ and to match the potential on the sphere we set $B = \Phi_0 a^2$. From (5.2.3) we see this is the field of a dipole with dipole moment $p = 4\pi\varepsilon_0 a^2 \Phi_0$.

14.1.2.2 Specified potential on a sphere varies as $\cos^2\theta$

A sphere $r = a$ has its potential specified as $\Phi = \Phi_0 \cos^2\theta$. Matching the angular variation with Legendre polynomials can only involve $P_0(\xi) = 1$ and $P_2(\xi) = \frac{1}{2}(3\cos^2\theta - 1)$. The first of these is compatible with radial variation $A + B/r$, whereas the second is compatible with radial variation $Cr^2 + D/r^3$. Addressing firstly the field outside the sphere, we must take $C = 0$ to avoid blow-up at infinity, and the datum of potential is arbitrary so we can also take $A = 0$. Hence the solution outside the sphere is of the form,

For $r \geq a$: $$\Phi = \frac{B}{r} + \frac{D}{r^3}\frac{1}{2}(3\cos^2\theta - 1) \tag{14.1.23}$$

Matching the required $\Phi = \Phi_0 \cos^2\theta$ on the sphere gives us $B = \frac{a}{3}\Phi_0$ and $D = \frac{2a^3}{3}\Phi_0$. The first term in (14.1.23) has the form of the potential due to a point charge of a magnitude which would give the same average potential across the sphere, namely $\frac{1}{3}\Phi_0$. The second term is known as a quadrupole term (see §14.1.2.3 below).

What about the field inside the sphere? Now we must discard the terms in the radial solution which are singular at the origin, but the candidate angular functions are the same. Hence the solution is of the form,

For $r \le a$: $$\Phi = A + Cr^2 \frac{1}{2}(3\cos^2\theta - 1) \qquad (14.1.24)$$

Matching the required $\Phi = \Phi_0 \cos^2\theta$ on the sphere gives us $A = \frac{1}{3}\Phi_0$ and $C = \frac{2}{3a^2}\Phi_0$. The field is non-uniform inside the sphere.

14.1.2.3 The Multipole Expansion

Suppose we have a cluster of point charges, q_i, all located within a sphere of radius a and we wish to express the potential, and hence the field, outside of this sphere in a compact manner. The vector position, $\bar{r}$, is defined wrt the centre of the sphere, and the charges are located at $\bar{r}_i$ wrt the same origin. So we are interested in the potential at locations with $r > a$ (and hence $r > r_i$). Because we are not concerned about the field inside the sphere, where the charges reside, we can accept solutions which vary as $1/r^{l+1}$. But we need the potential to drop to zero at large distances so we rule out solutions varying as r^l. The constant term can also be taken to be zero. It follows that we can express the potential due to the cluster of charges in the form,

$$\Phi = \sum_{l=0}^{\infty} \sum_{m=-l}^{l} \frac{A_{lm}}{r^{l+1}} Y_l^m(\theta, \phi) \qquad (14.1.25)$$

Because each term in the sum over l involves increasing powers of the distance r, if we are interested in $r \gg a$ subsequent terms are likely to be rapidly diminishing in significance. This is the basis of the so-called multipole expansion which describes compactly the field due to a cluster of charges. However, it remains to find the coefficients A_{lm} in (14.1.25) in terms of the charges, q_i, and their locations, $\bar{r}_i$. This is most conveniently accomplished by expanding the potential due to a single point charge using the binomial expansion. Hence, for a field point on the polar axis,

$$\frac{q_i}{4\pi\varepsilon_0|\bar{r}-\bar{r}_i|} = \frac{q_i}{4\pi\varepsilon_0 r}\left(1 - 2\frac{r_i\cos\theta_i}{r} + \left(\frac{r_i}{r}\right)^2\right)^{-\frac{1}{2}}$$

$$= \frac{q_i}{4\pi\varepsilon_0 r}\left\{1 + \frac{1}{2}\left(2\frac{r_i\cos\theta_i}{r} - \left(\frac{r_i}{r}\right)^2\right) + \frac{3}{8}\left(2\frac{r_i\cos\theta_i}{r} - \left(\frac{r_i}{r}\right)^2\right)^2 + \right.$$

$$\left.\frac{5}{16}\left(2\frac{r_i\cos\theta_i}{r} - \left(\frac{r_i}{r}\right)^2\right)^3 + \cdots\right\}$$

$$= \frac{q_i}{4\pi\varepsilon_0 r}\left\{\begin{array}{c} 1 + \frac{r_i}{r}\cos\theta_i + \left(\frac{r_i}{r}\right)^2\left[\frac{3}{2}\cos^2\theta_i - \frac{1}{2}\right] + \\ \left(\frac{r_i}{r}\right)^3\left[\frac{5}{2}\cos^3\theta_i - \frac{3}{2}\cos\theta_i\right] + \cdots \end{array}\right\}$$

$$= \frac{q_i}{4\pi\varepsilon_0 r}\sum_{l=0}^{\infty}\left(\frac{r_i}{r}\right)^l P_l(\cos\theta_i) \qquad (14.1.26)$$

We see that the total potential due to a cluster of N charges can therefore be written,

$$\Phi = \sum_{i=1}^{N}\frac{q_i}{4\pi\varepsilon_0 r}\sum_{l=0}^{\infty}\left(\frac{r_i}{r}\right)^l P_l(\cos\theta_i) = \frac{1}{4\pi\varepsilon_0}\sum_{l=0}^{\infty}\frac{Q_l}{r^{l+1}} \qquad (14.1.27a)$$

where, $$Q_l = \sum_{i=1}^{N} q_i r_i^l P_l(\cos\theta_i) \qquad 14.1.27b)$$

and where it is understood that θ_i are the angles between the charge location vector and the field point, i.e., $\cos\theta_i = \hat{r}\cdot\hat{r}_i$.

Q_l is the lth multipole term.

Hence we have succeeded in expressing the potential as a sum over l of terms equal to $1/r^{l+1}$ times a factor, Q_l, that depends only upon the charges and their positions. Note that (14.1.27a,b) involve, separately, the angles of each charge wrt the field point, unlike (14.1.25) which involves only the field point polar angle, so the two expansions are actually quite different, though (14.1.25) can be obtained by projecting out the components of $Y_l^m(\theta,\phi)$ from each term $P_l(\cos\theta_i)$. When this is done we get an alternative form of multipole, namely the spherical tensor components of the multipole, which depend on the azimuthal order, m, as well as the multipole order, l, thus,

$$M_{lm} = \sqrt{\frac{4\pi}{2l+1}}\sum_{i=1}^{N} q_i r_i^l Y_l^m(\theta_s,\phi_s)^* \qquad (14.1.27c)$$

$$\Phi = \frac{1}{4\pi\varepsilon_0}\sum_{l=0}^{\infty}\sum_{m=-l}^{l}\sqrt{\frac{4\pi}{2l+1}}M_{lm}\frac{Y_l^m(\theta_f,\phi_f)}{r^{l+1}} \qquad (14.1.27d)$$

Here θ_s, ϕ_s, r_i refer to the position of the charge q_i, and we have written θ_s to distinguish between the charge's polar angle wrt to the coordinate system and its angle, θ_i, wrt the field point. Similarly, θ_f, ϕ_f, r are the polar coordinates of the field point. The equivalence of the two different forms of multipole expansion requires that,

$$Q_l = \sum_{i=1}^{N} q_i r_i^l P_l(\cos\theta_i) = \sum_{m=-l}^{l} \left(\frac{4\pi}{2l+1}\right) q_i r_i^l Y_l^m(\theta_s, \phi_s)^* Y_l^m(\theta_f, \phi_f)$$

and this follows from the addition theorem for the Legendre functions,

$$P_l(\cos\theta_i) = \sum_{m=-l}^{l} \left(\frac{4\pi}{2l+1}\right) Y_l^m(\theta_s, \phi_s)^* Y_l^m(\theta_f, \phi_f) \qquad (14.1.27e)$$

The zero'th multipole is just the total charge, $Q_0 = \sum_{i=1}^{N} q_i$. The first multipole, $l = 1$, is defined as the vector,

$$\bar{p} = \sum_{i=1}^{N} q_i \bar{r}_i \qquad (14.1.28a)$$

So that,

$$Q_1 = \bar{p} \cdot \hat{r} \qquad (14.1.28b)$$

$\bar{p}$ is called the dipole moment of the cluster of charges because it produces a field equal to that of a dipole of magnitude and direction $\bar{p}$. The next term, $l = 2$, in the power series expansion (14.1.27a) defines the quadrupole moment whose potential falls off as $1/r^3$ and hence whose electric field falls as $1/r^4$. The quadrupole moment is defined as the second rank tensor whose Cartesian components are,

$$q_{ab} = \sum_{i=1}^{N} q_i (3r_{ia} r_{ib} - r_i^2 \delta_{ab}) \qquad (14.1.29a)$$

So that,

$$Q_2 = \frac{1}{2} \sum_{a,b=1}^{3} q_{ab} \hat{r}_a \hat{r}_b \qquad (14.1.29b)$$

(See chapter 27 for the definition and properties of tensors). Here $\hat{r}_a$ is the ath Cartesian component of the unit vector $\hat{r}$ towards the field point, and (14.1.29b) involves a double sum over Cartesian components. The reader will note that (14.1.29a,b) reproduces the P_2 angular dependence, (14.1.20c).

By expressing the dipole moment and quadrupole moment as tensors we render them independent of the coordinate system or the field point (though the components of the dipole moment and quadrupole moment do, of course, depend on the coordinate system, inevitably). In contrast, the multipole terms, $Q_1, Q_2, \ldots$ etc. do depend on the field point.

Higher order multipole terms will involve successively higher ranks of tensors, the third multipole being of the form $Q_3 = \frac{1}{6} \sum_{a,b=1}^{3} q_{abc} \hat{r}_a \hat{r}_b \hat{r}_c$. The utility of the multipole concept is that, in practice, the field due to a complicated cluster of many charges can be captured to adequate accuracy (outside the cluster) using just the total charge plus the dipole and quadrupole moments. At worst perhaps the octupole moment ($l = 3$) may also be added.

14.1.2.4 An Annular Spherical Region

If we consider the annular region between two concentric spheres then, since both $r = 0$ and $r \to \infty$ are excluded, we must include the full radial solution (14.1.22). The most general solution in spherical polars is thus,

$$\Phi = \sum_{l=0}^{\infty} \sum_{m=-l}^{l} \left(A_{lm} r^l + \frac{B_{lm}}{r^{l+1}} \right) P_l^m(\cos\theta) e^{im\phi} \tag{14.1.30}$$

If the two spheres, of radii a and b, have specified potentials, in general given by some functions of the angular variables, say $\Phi_a(\theta, \phi)$ and $\Phi_b(\theta, \phi)$, then we require,

$$\sum_{l=0}^{\infty} \sum_{m=-l}^{l} \left(A_{lm} a^l + \frac{B_{lm}}{a^{l+1}} \right) Y_l^m(\theta, \phi) \equiv \Phi_a(\theta, \phi) \tag{14.1.31}$$

and the similar requirement on $r = b$. To solve for the parameters A_{lm} and B_{lm} we note that the spherical harmonics are orthonormal,

$$\int Y_l^m(\theta, \phi) Y_L^{-M}(\theta, \phi) d\Omega = \delta_{mM} \delta_{lL} \tag{14.1.32}$$

where the integral extends over the whole 4π steradians of the solid angle, $d\Omega = -\mathrm{d}(\cos\theta) d\phi$. Hence, multiplying (14.1.31) by $Y_L^{-M}(\theta, \phi)$ and integrating gives,

$$A_{lm} a^l + \frac{B_{lm}}{a^{l+1}} = \int \Phi_a(\theta, \phi) Y_l^{-m}(\theta, \phi) d\Omega \tag{14.1.33a}$$

And,
$$A_{lm} b^l + \frac{B_{lm}}{b^{l+1}} = \int \Phi_b(\theta, \phi) Y_l^{-m}(\theta, \phi) d\Omega \tag{14.1.33b}$$

From (14.1.33a,b) the parameters A_{lm} and B_{lm} are found for every l, m.

14.1.2.5 The Magnetostatic Multipole Expansion

In analogy with the multipole expansion of the electric scalar potential there is also a multipole expansion of the magnetic vector potential which can be used to find the associated magnetic field. This can be developed by considering the vector potential due to a collection of small current elements, $I_i \delta \bar{L}_i$, as given by (5.5.7b), $\bar{A}_i(\bar{r}) = \frac{\mu_0}{4\pi} \frac{I_i \delta \bar{L}_i}{|\bar{r} - \bar{r}_i|}$, so that the total vector potential is the vector sum of these contributions. Since the derivation of the multipole expansion is essentially just an expansion of $1/|\bar{r} - \bar{r}_i|$ in terms of the Legendre polynomials and inverse powers of r it is essentially the same but with $I_i \delta \bar{L}_i$ replacing q_i. Hence we get,

$$\bar{A} = \frac{\mu_0}{4\pi}\sum_{l=0}^{\infty}\frac{\bar{\mathcal{L}}_l}{r^{l+1}} \quad \text{where} \quad \bar{\mathcal{L}}_l = \sum_{i=1}^{N} I_i \delta\bar{L}_i r_i^l P_l(\cos\theta_i) \tag{14.1.33c}$$

So the lth multipole, $\bar{\mathcal{L}}_l$, is a vector. Strictly, these multipoles are tensorial once we re-express them in terms of their own coordinates, $\theta_s, \phi_s r_i$, rather than in terms of the coordinate $\cos\theta_i = \hat{r}\cdot\hat{r}_i$ relative to the field point. This is accomplished using the addition theorem for the Legendre functions, (14.1.27e), as before, giving the spherical tensorial multipoles,

$$\bar{M}_{lm} = \sqrt{\frac{4\pi}{2l+1}}\sum_{i=1}^{N} I_i \delta\bar{L}_i r_i^l Y_l^m(\theta_s,\phi_s)^* \tag{14.1.33d}$$

$$\bar{A} = \frac{\mu_0}{4\pi}\sum_{l=0}^{\infty}\sum_{m=-l}^{l}\sqrt{\frac{4\pi}{2l+1}}\,\bar{M}_{lm}\frac{Y_l^m(\theta_f,\phi_f)}{r^{l+1}} \tag{14.1.33e}$$

There will, however, be no zero'th multipole ($l = 0$) as there are no magnetic monopoles. This is ensured by requiring all current elements to form complete loops, though there will in general be many discrete complete loops.

An even closer analogue of the electrostatic case can be devised by considering each closed loop of current to be comprised of a pair of fictional equal and opposite magnetic monopoles whose magnitude and relative position reproduce the dipole moment of each current loop. Calling these fictional magnetic monopoles q_i^m we can write the magnetic scalar potential, ϕ_m, in a form that looks formally identical to (14.1.27c,d) but omitting ε_0. From this the magnetic field is given by (5.5.1), $\bar{B} = -\mu_0\bar{\nabla}\phi_m$.

14.1.3 Cylindrical Boundaries

Problems with boundary conditions specified on a cylinder, or concentric cylinders, may have variables separated in cylindrical polars. Thus we write $\Phi = R(r)\Theta(\theta)Z(z)$ and Laplace's equation becomes,

$$\frac{1}{\Phi}\nabla^2\Phi = \frac{1}{rR}\partial_r(r\partial_r R) + \frac{1}{r^2\Theta}\partial_\theta^2 + \frac{1}{Z}\partial_z^2 Z = 0 \tag{14.1.34}$$

From which we get separated expressions involving real constants k, m such that, (14.1.35)

$$\partial_z^2 Z = k^2 Z \qquad \partial_\theta^2\Theta = -m^2\Theta \qquad r^2\partial_r^2 R + r\partial_r R + (k^2r^2 - m^2)R = 0$$

For $k^2 > 0$ the last of these equations is Bessel's equation whose two independent solutions for given k, m are the Bessel functions. Note that if we make the replacement $s = kr$ then the constant k disappears from

Bessel's equation which becomes $s^2\partial_s^2 R + s\partial_s R + (s^2 - m^2)R = 0$. The Bessel functions can then be written $J_m(s) = J_m(kr)$ and $Y_m(s) = Y_m(kr)$. The most general solution in cylindrical polars which is not uniform in the axial direction (i.e., has $k^2 > 0$) is thus of the form, (14.1.36)

$$\Phi = \int dk \sum_{m=0}^{\infty} A_{mk}\big(J_m(kr) + B_{km}Y_m(kr)\big)\cos(m\theta + \alpha_{km})\left(e^{kz} + C_{km}e^{-kz}\right)$$

where A_{km}, B_{km}, C_{km} and α_{km} are constants to be determined by the boundary conditions. The Bessel functions of the first kind, $J_m(kr)$, are finite at the origin whereas the Bessel functions of the second kind, $Y_m(kr)$, are divergent at the origin. Hence solutions which include the origin and are required to be finite there involve only the $J_m(kr)$ but not the $Y_m(kr)$. Both tend to zero at infinity being of order $2/\sqrt{\pi r}$.

Note that it is the requirement for the solution to be periodic in the angular variable θ which means that m must be real (i.e., $m^2 \geq 0$). But $m = 0$ is physical and indicates axisymmetry (the potential being independent of θ). The solution then involves only $J_0(kr)$ and possibly $Y_0(kr)$.

If the situation is essentially 2D, i.e., there is no variation in the axial direction, then $k = 0$ and the radial equation becomes $r^2\partial_r^2 R + r\partial_r R = m^2 R$. This has the solution $R = Ar^m + Br^{-m}$ for $m^2 \neq 0$. However, if the situation is also axisymmetric ($m = 0$) as well as independent of z then the solution is $R = A + Blog(r)$. If the situation is not also axisymmetric the 2D (z independent) potential has the general solution,

$$\Phi = A_0 + B_0 log(r) + \sum_{m=1}^{\infty} A_m(r^m + B_m r^{-m})\cos(m\theta + \alpha_m) \tag{14.1.37}$$

The case of a cylinder with uniform charge density, σ, and no other sources, being axisymmetric and independent of z has the solution $A + Blog(r)$ for the potential and hence the field is,

$$\bar{E} = -\bar{\nabla}\Phi = -\frac{B_0}{r}\hat{r} = \frac{a}{r}\frac{\sigma}{\varepsilon_0}\hat{r} \tag{14.1.38}$$

The final possibility is $k^2 < 0$ for which we write $k = i|k|$. The z-dependence is now oscillatory and the radial equation becomes the modified Bessel equation $s^2\partial_s^2 R + s\partial_s R - (s^2 + m^2)R = 0$ whose solutions are the modified Bessel functions $I_m(s)$ and $K_m(s)$. The solution can then be written, (14.1.39)

$$\Phi = \int dk \sum_{m=0}^{\infty} A_{mk}\big(I_m(kr) + B_{km}K_m(kr)\big)\cos(m\theta + \alpha_{km})\left(e^{i|k|z} + C_{km}e^{-i|k|z}\right)$$

$I_m(s)$ is finite at the origin whereas $K_m(s)$ is divergent at the origin.

14.1.3.1 A Grounded Infinite Cylinder in a Uniform Field

The remote field is assumed to be $\bar{E} = E_0\hat{x} = E_0(\hat{r}\cos\theta - \hat{\theta}\sin\theta)$, in cylindrical polars, which involves only the term in $m = 1$ in the general solution, (14.1.37). The potential is thus of the form $(Ar + Br^{-1})\cos(\theta)$ and is required to be zero on the surface of the cylinder, $r = a$. Hence $B = -Aa^2$. From Appendix A2, $-\bar{E} = \bar{\nabla}\Phi = \hat{r}\partial_r\Phi + \frac{\hat{\theta}}{r}\partial_\theta\Phi$ giving,

$$\bar{E} = A\left[-\left(1 + \frac{a^2}{r^2}\right)\hat{r}\cos\theta + \left(1 - \frac{a^2}{r^2}\right)\hat{\theta}\sin\theta\right] \tag{14.1.40}$$

The remote field shows $A = -E_0$. The field is purely radial on the surface of the cylinder, as it must be, and its magnitude is $2E_0\cos\theta$. On $\theta = 0$ this is $2E_0$, twice the remote field. This is analogous to holes in a structure concentrating mechanical stress, and compares with the field concentration factor of 3 found in §13.7 for a spherical conducting surface in a uniform remote field. The concentration occurs for essentially the same qualitative reason as found there, as illustrated by Figure 13.1.

The surface field of $2E_0\cos\theta$ must equate to σ/ε_0 where σ is the surface charge density. It follows that a grounded conducting cylinder in a uniform electric field has induced on it a surface charge density $\sigma = 2\varepsilon_0 E_0\cos\theta$, and hence the net charge over the whole cylinder is zero (inevitably).

14.2 The Complex Potential Method in 2D

A particularly powerful method exists for solving Laplace's equation in 2D (or when the potential is uniform in the third Cartesian direction). This is provided by the convenient mathematical fact that the 2D Laplace equation is automatically obeyed by any analytic function of the complex variable $z = x + iy$, where this z should not be confused with the third Cartesian coordinate, which will not feature in these 2D problems.

Note that analytic functions of a complex variable are also holomorphic functions, and vice-versa. We shall not dwell on these purely mathematical issues here, for which see any standard text on complex analysis or my own simple account, [14.1]. Without wishing to upset mathematicians with my imprecision, a sufficient feel for what constitutes an analytic function is that it is a function only of z but not of its real or imaginary parts (i.e., x and y) separately, and hence not a function of both z and its complex conjugate, z^*, either. There may be isolated points (e.g., singularities) at which the function ceases to be analytic, often because of a singularity, and these will typically be interpretable in this application with the presence of charges.

Hence, we work here with a complex-valued scalar potential, both the real and imaginary parts of which are functions of $z = x + iy$, and both parts will obey Laplace's equation,

$$\Omega = \Phi + i\Psi \tag{14.2.1}$$

so that $\nabla^2\Phi = 0$ and $\nabla^2\Psi = 0$ are automatic when Ω is analytic (holomorphic). By convention we identify the real part, Φ, with the physical scalar potential. But the utility of this technique does not lie only in automatically ensuring that Laplace's equation is obeyed, though that alone is enormously helpful. It also lies in the fact that the imaginary part, Ψ, is related to the electric field.

Specifically, the lines of constant Ψ are parallel to the field vector at all points. Hence, whilst the lines of constant Φ are equipotential lines, the lines of constant Ψ are known as "flux lines". To establish this recall that $\bar{E} = -\bar{\nabla}\Phi$ so that the field is everywhere perpendicular to the equipotential lines. It therefore remains to establish that the lines of constant Ψ are perpendicular to the equipotential lines, in which case they will be parallel to the field. This occurs because of the Cauchy-Riemann conditions, which are equivalent to the requirement that the complex potential has a unique complex derivative at all points (which in turn is a requirement of being holomorphic). These tell us that,

$$\partial_x\Phi = \partial_y\Psi \quad \text{and} \quad \partial_y\Phi = -\partial_x\Psi \tag{14.2.2}$$

But this means that $\partial_x\Phi\hat{x} + \partial_y\Phi\hat{y}$ is perpendicular to $\partial_x\Psi\hat{x} + \partial_y\Psi\hat{y}$ because the latter is equal to $-\partial_y\Phi\hat{x} + \partial_x\Phi\hat{y}$. But these two vector fields are respectively perpendicular to the equipotential lines and perpendicular to the lines of constant Ψ, from which it follows that the equipotential lines and the lines of constant Ψ are perpendicular. Hence, the lines of constant Ψ are parallel to the field.

Note, however, that the magnitude of the field must still be found from $-\bar{E} = \partial_x\Phi\hat{x} + \partial_y\Phi\hat{y}$. The following section provides an illustration of this powerful method, though it is limited, of course, to 2D situations.

14.2.1 A Conducting Elliptical Cylinder in a Uniform Field

Consider an elliptical cylinder whose axis is the z axis and whose cross section in the x, y plane is an ellipse with semi-major and semi-minor axes a and b in the x and y directions respectively ($a > b$). The cylinder is placed in an electric field which is uniform and oriented in the x direction at large distances from the cylinder. The elliptical cylinder is given by,

$$\frac{x^2}{a^2} + \frac{y^2}{b^2} = 1 \qquad (14.2.3)$$

Elliptical coordinates ξ, η may be defined by,

$$z = x + iy = c\cosh(\xi + i\eta) \qquad (14.2.4)$$

The inversion of the cosh function (positive branch) then gives,

$$\xi + i\eta = log\left(\frac{z}{c} + \sqrt{\frac{z^2}{c^2} - 1}\right) = log\frac{\zeta}{c} \qquad (14.2.5)$$

This defines $\quad \zeta = z + \sqrt{z^2 - c^2} \qquad (14.2.6)$

Surfaces of constant ξ are ellipses because,

$$x + iy = c\cosh(\xi + i\eta) = c\cosh\xi\cos\eta + ic\sinh\xi\sin\eta$$

So that $\quad x = c\cosh\xi\cos\eta, \quad y = c\sinh\xi\sin\eta \qquad (14.2.7)$

Hence $\quad \dfrac{x^2}{(c\cosh\xi)^2} + \dfrac{y^2}{(c\sinh\xi)^2} = 1 \qquad (14.2.8)$

which is an ellipse with semi-major and semi-minor axes $c\cosh\xi$ and $c\sinh\xi$ respectively. The particular $\xi = \xi_0$ reproduces the cylinder, (14.2.3), if we choose ξ_0 such that $c\cosh\xi_0 = a$, from which it follows that to also have $c\sinh\xi_0 = b$ we must choose,

$$c^2 = a^2 - b^2 \qquad (14.2.9)$$

We require a potential which obeys the two boundary conditions,

i. $\Phi = 0$ on the cylinder, and,
ii. $\Phi \to -E_0 x$ as $x \to \pm\infty$

The second condition is to reproduce the remote field, $E_0\hat{x}$.

Hypothesis: The complex potential which solves this problem is,

$$\Omega = -\frac{E_0}{2}\left(\zeta - \frac{A}{c\zeta}\right) \qquad (14.2.10)$$

Due to (14.2.6) this is an analytic function of z except at isolated points which lie inside the cylinder and will not concern us, and hence it is a solution to Laplace's equation in 2D outside the cylinder. Moreover, (14.2.10) respects boundary condition ii. It remains to be seen if we can choose some constant A so that boundary condition i is also obeyed. Recall that the potential is given by $\Phi = \mathcal{R}(\Omega)$ and, from (14.2.5) $\zeta = ce^{\xi}e^{i\eta}$, so that,

$$\Phi = -\frac{E_0}{2}\left(ce^{\xi} - \frac{A}{c^2}e^{-\xi}\right)\cos\eta \qquad (14.2.11)$$

For this to be zero on the cylinder, $\xi = \xi_0$, requires $ce^{\xi_0} - \frac{A}{c^2}e^{-\xi_0} = 0$ hence,

$$A = c^3 e^{2\xi_0} \qquad (14.2.12)$$

But by adding $c\cosh\xi_0 = a$ to $c\sinh\xi_0 = b$ we find that,

$$e^{\xi_0} = \frac{a+b}{c} \qquad (14.2.13)$$

Hence, the desired solution for the complex potential is,

$$\Omega = -\frac{E_0}{2}\left(\zeta - \frac{(a+b)^2}{\zeta}\right) \qquad (14.2.14)$$

The (real) potential is thus,

$$\Phi = \mathcal{R}(\Omega) = -\frac{E_0}{2c}\left[(a^2 - b^2)e^{\xi} - (a+b)^2 e^{-\xi}\right]\cos\eta \qquad (14.2.15)$$

From this the electric field can be found as $\bar{E} = -\bar{\nabla}\Phi$, as usual. In elliptical coordinates (see Appendix A2) this is therefore, (14.2.16)

$$\bar{E} = \frac{E_0}{\sqrt{2(\cosh 2\xi - \cos 2\eta)}}\left\{\left(e^{\xi} + \lambda e^{-\xi}\right)\cos\eta\cdot\hat{\xi} - \left(e^{\xi} - \lambda e^{-\xi}\right)\sin\eta\cdot\hat{\eta}\right\}$$

where $\lambda = \frac{(a+b)^2}{(a^2-b^2)}$. Here the unit vector $\hat{\xi}$ is normal to the ellipses of constant ξ and the unit vector $\hat{\eta}$ is normal to the hyperbolae of constant η (and hence perpendicular to $\hat{\xi}$ at all points). On the positive x axis $\eta = 0$ whilst on the negative x axis $\eta = \pi$ and we get,

$$\bar{E}(y=0) = \frac{E_0}{2\sinh\xi}(e^{\xi} + \lambda e^{-\xi})\hat{x} \qquad (14.2.17)$$

which is consistent with $\bar{E} \to E_0\hat{x}$ as $x \to \pm\infty$. Hence the field on the cylinder at $x = a, y = 0$ is,

$$\bar{E} = \left(1 + \frac{a}{b}\right)E_0\hat{x} \qquad (14.2.18)$$

Hence, for a circular cylinder the field at the cylinder on $y = 0$ is twice the remote field, in agreement with §14.1.3.1. The more eccentric the elliptic cylinder, the greater is the field at the end of its semi-major axis. This result may be expressed in terms of the radius of curvature at the end of the ellipse, namely $\rho = b^2/a$, giving the field at the end to be,

$$\bar{E} = \left(1 + \sqrt{\frac{a}{\rho}}\right) E_0 \hat{x} \tag{14.2.19}$$

and we may use (14.2.19) as an estimate for the field at the end of any prismatic (cylindrical) conductor with a radius of curvature ρ and aligned with that end towards the applied field. Once again there is an analogy with solid mechanics and the stress concentration at the end of an elliptic cylinder in a remotely uniform applied stress, though in that case the stress concentration factor is $\left(1 + 2\sqrt{\frac{a}{\rho}}\right)$ for an applied stress perpendicular to the major axis.

14.3 References

[14.1] Bradford, R.A.W. (2023). *Five Square Roots* (see chapter 2). Principia Publications Unlimited, 1 March 2023.

15. Green's Functions

This mathematical interlude presents Green's functions as a convenient means of expressing the solutions (particular integrals) of linear inhomogeneous differential equations, especially partial differential equations. Physically, a Green's function is the solution for a point source, or, in the time domain, a point source which also persists only for an instant. The Green's functions for the Laplacian operator in three spatial dimensions and the Green's function for the d'Alembertian operator in (1,3) Minkowski spacetime are derived from first principles. In the latter case the retarded and advanced Green's functions are both derived and the unphysical nature of the later stressed. The retarded potential will facilitate the discussion of the generation of e/m waves from accelerating charges in chapter 22.

Contents

15.1 Definition of a Green's Function

A Green's function is the kernel of an integral inverse of a linear, inhomogeneous differential equation. All that means is that given a linear differential operator $\hat{D}$ acting on an independent variable x and which defines a differential equation in the unknown function $y(x)$ in terms of some known function $f(x)$ thus,

$$\hat{D}y(x) = f(x) \tag{15.1.1}$$

then the solution can be written,

$$y(x) = \int G(x, x')f(x')dx' \tag{15.1.2}$$

This generalises to functions of several variables. For example, for a partial differential equation in spatial coordinates $\bar{r} = (x, y, z)$ we might have, say,

$$\hat{D}\Phi(\bar{r}) = f(\bar{r}) \tag{15.1.3}$$

for which the solution is,

$$\Phi(\bar{r}) = \int G(\bar{r}, \bar{r}')f(\bar{r}')d^3r' \tag{15.1.4}$$

Including time dependence there are now four independent variables and the differential operator acts on all four: t, x, y, z. So we have,

$$\hat{D}\Phi(\bar{r}, t) = f(\bar{r}, t) \tag{15.1.5}$$

for which the solution is,

$$\Phi(\bar{r}, t) = \int G(x, x')f(\bar{r}', t')d^3r'dt' \tag{15.1.6}$$

Here it is understood that in this context x is shorthand for $\{x^\mu\}$. Note that (15.1.6) involves a 4D integral.

The solution of a differential equation involves arbitrary constants or functions, chosen to fit the boundary conditions specific to the particular application. The required Green's function, e.g., $G(\bar{r}, \bar{r}')$, is therefore not uniquely specified by the differential equation, i.e., by $\hat{D}$, but may differ according to the required boundary conditions. We do not, therefore, talk of the Green's function for a given equation but a Green's function specific to the boundary conditions. In the derivations below this boundary condition will be that the solution becomes zero at infinity.

15.2 The Dirac Delta Function

Green's functions can be defined succinctly in terms of the Dirac delta function. In just one variable, this function, $\delta(x)$, can be considered heuristically as being zero when $x \neq 0$ but infinite when $x = 0$ such that its integral over any domain which includes the origin is unity. The reader will appreciate that such a wildly discontinuous function has no conventional (Riemann) integral. The spirit of this section will continue to be rather heuristic rather than mathematically rigorous, although rigorous developments do exist, of course, but would be inappropriate here.

The most important property of the delta function is that it picks out a single value of a function from an integral. Thus, for "any" function $y(x)$,

$$y(x) = \int_{-\infty}^{\infty} y(x')\,\delta(x - x')dx' \tag{15.2.1}$$

Historically the origin of the delta function lies in Fourier's integral theorem which states that for "any" function $y(x)$,

$$y(x) = \frac{1}{2\pi}\int_{-\infty}^{\infty} y(x')\left[\int_{-\infty}^{\infty} \cos k(x - x')dk\right]dx' \tag{15.2.2}$$

This immediately gives one representation of the delta function as,

$$\delta(x - x') = \frac{1}{2\pi}\int_{-\infty}^{\infty} \cos\big(k(x - x')\big)dk \tag{15.2.3}$$

Cauchy provided an extension into the complex domain. Taking liberties with the order of integration (which is strictly not always valid) this is,

$$y(x) = \frac{1}{2\pi}\int_{-\infty}^{\infty} y(x')\left[\int_{-\infty}^{\infty} e^{ik(x-x')}\,dk\right]dx' \tag{15.2.4}$$

This provides the best-known and most useful representation of the delta function,

$$\delta(x - x') = \frac{1}{2\pi}\int_{-\infty}^{\infty} e^{ik(x-x')}\,dk \tag{15.2.5}$$

A heuristic rationalisation of the claim that the function defined by (15.2.5) fulfils the key requirement (15.2.1) is as follows. Without loss of generality we can consider $x' = 0$. Consider the limits of the integral in (15.2.5) to be replaced by $\pm K$ where we intend to let $K \to \infty$. Replace the integration variable with $\theta = kx$ so that the integral evaluates to

$$\delta(x - x') = \frac{1}{2\pi}\int_{-K}^{K} e^{ikx}\,dk = \frac{1}{\pi x}\sin Kx \tag{15.2.6}$$

For $x \neq 0$ we now assume the integral covers an integral number of 'wavelengths', i.e., that $Kx = n\pi$, from which we see that (15.2.6) is zero. Alternatively, as $x \to 0$, (15.2.6) becomes K/π (which we are intending to become infinite). As this rapidly drops to zero when x departs from zero, we may approximate for integration purposes that $\delta(x)$ in the vicinity of zero is $K/2\pi$. The integral (15.2.1) then becomes,

$$y(\sim 0)\frac{K}{2\pi}\bigg|_{-\pi/K}^{+\pi/K} = y(0) \tag{15.2.7}$$

as required for $x' = 0$. There are many representations of the delta function but most involve the limit of some sequence of functions. One such is the limit of a Gaussian as the width goes to zero, thus,

$$\delta(x - x') = LIM_{\sigma \to 0}\left[\frac{1}{\sqrt{2\pi}\sigma}e^{-(x-x')^2/2\sigma^2}\right] \tag{15.2.8}$$

But in practice (15.2.5) is the most useful. In three spatial dimensions this becomes,

$$\delta^3(\bar{r} - \bar{r}') = \frac{1}{(2\pi)^3}\int_{-\infty}^{\infty} e^{i\bar{k}\cdot(\bar{r}-\bar{r}')}\, d^3k \tag{15.2.9}$$

and,

$$\delta^3(\bar{r} - \bar{r}') \equiv \delta(x - x')\delta(y - y')\delta(z - z') \tag{15.2.10}$$

Including the time domain we have,

$$\delta^4(x - x') = \frac{1}{(2\pi)^4}\int_{-\infty}^{\infty} e^{-ik_\mu(x-x')^\mu}\, d^4k \tag{15.2.11}$$

where, in this context, $x \equiv \{x^\mu\}$, $k_\mu(x - x')^\mu = k_t(t - t') - \bar{k}\cdot(\bar{r} - \bar{r}')$. The delta function became linked with Dirac's name because of the central use to which Dirac put the function in quantum mechanics.

Note that the units of $\delta^n(x)$ are $(1/\text{Length})^n$.

15.3 Green's Functions in Terms of the Delta Function

Green's functions immediately have an alternative definition in terms of the delta function, for example,

$$\hat{D}G(x,x') = \delta(x - x') \qquad (15.3.1)$$

Or,

$$\hat{D}G(\bar{r},\bar{r}') = \delta^3(\bar{r} - \bar{r}') \qquad (15.3.2)$$

Or,

$$\hat{D}G(x,x') = \delta^4(x - x') \qquad (15.3.3)$$

These follow because, for example,

$$\hat{D}y(x) = \hat{D}\int G(x,x')f(x')dx' = \int \delta(x - x')f(x')dx' = f(x) \qquad (15.3.4)$$

15.4 A Green's Function for Laplace's Operator

We have seen in §5.1 that electrostatic problems can be formulated in terms of Poisson's equation, $\nabla^2\Phi = -\rho/\varepsilon_0$, for which the differential operator is Laplace's operator, ∇^2. Hence we see that in this case the Green's function is the electrostatic potential for a point charge (modulo $-1/\varepsilon_0$). But we already know the solution, (5.1.6), from which we have,

$$G(\bar{r},\bar{r}_f) = \frac{-1}{4\pi|\bar{r}_f - \bar{r}|} \qquad (15.4.1)$$

This illustrates something we should have anticipated, namely that the Green's function depends only on the difference $\bar{r}_f - \bar{r}$, not on the source and field points separately. This is due to the homogeneity of space, i.e., displacing both the source and field points by the same amount can make no difference to the potential and hence the Green's function. So we can write $G(\bar{r},\bar{r}_f) \to G(\bar{r} - \bar{r}_f)$.

Rather than appeal to a solution we already know we illustrate a general procedure for finding a Green's function via its Fourier transform. Different conventions are in use. Here we define the Fourier transform of $G(\bar{r} - \bar{r}_f)$ thus,

$$\tilde{G}(\bar{k}) = \int_{-\infty}^{\infty} e^{-i\bar{k}\cdot\bar{r}}\, G(\bar{r})d^3r \qquad (15.4.2)$$

where,

$$G(\bar{r} - \bar{r}_f) = \frac{1}{(2\pi)^3}\int_{-\infty}^{\infty} e^{i\bar{k}\cdot(\bar{r}-\bar{r}_f)}\tilde{G}(\bar{k})d^3k \qquad (15.4.3)$$

Hence,

$$\nabla^2 G(\bar{r} - \bar{r}_f) = \frac{1}{(2\pi)^3}\int_{-\infty}^{\infty} \nabla^2 e^{i\bar{k}\cdot(\bar{r}-\bar{r}_f)}\tilde{G}(\bar{k}) = \frac{-1}{(2\pi)^3}\int_{-\infty}^{\infty} \mathrm{k}^2 e^{i\bar{k}\cdot(\bar{r}-\bar{r}_f)}\tilde{G}(\bar{k})$$

and this must equal $\delta^3(\bar{r}-\bar{r}') = \frac{1}{(2\pi)^3}\int_{-\infty}^{\infty} e^{i\bar{k}\cdot(\bar{r}-\bar{r}')}\, d^3k$ from which we see that $\tilde{G}(\bar{k}) = -1/k^2$. Hence,

$$G(\bar{r}-\bar{r}_f) = \frac{-1}{(2\pi)^3}\int_{-\infty}^{\infty} \frac{e^{i\bar{k}\cdot(\bar{r}-\bar{r}_f)}}{k^2} d^3k =$$
$$\frac{-1}{(2\pi)^2}\int_{\theta=\pi}^{0}\int_0^{\infty} e^{ikr\cos\theta} d(\cos\theta)dk = \frac{-1}{(2\pi)^2}\int_0^{\infty}\frac{2\sin kr}{kr}dk =$$
$$\frac{-1}{2\pi^2 r}\int_0^{\infty}\frac{\sin\alpha}{\alpha}d\alpha = -\frac{1}{4\pi r} = -\frac{1}{4\pi|\bar{r}-\bar{r}_f|} \qquad \underline{\text{QED}}. \qquad (15.4.4)$$

Here we have used the standard integral $\int_0^{\infty}\frac{\sin\alpha}{\alpha}d\alpha = \frac{\pi}{2}$.

Note that, in this case, the dimensions of the operator, ∇^2, are $1/L^2$ so that $\hat{D}G(\bar{r}-\bar{r}') = \delta^3(\bar{r}-\bar{r}')$ means that the dimensions of $G(\bar{r}-\bar{r}')$ must be $1/L$ in agreement with (15.4.1,4).

We can turn this derivation around. Rather than deriving the Green's function which obeys (15.3.2) given a particular representation of the delta function, we can instead derive the above result by independent means as an alternative representation of the delta function, namely,

$$\delta^3(\bar{r}-\bar{r}_1) = -\frac{1}{4\pi}\nabla^2\frac{1}{|\bar{r}-\bar{r}_1|} \qquad (15.4.5)$$

An independent derivation of (15.4.5) is given in Appendix A5.

15.5 A Green's Function for d'Alembert's Operator

In the case of time dependence, Poisson's equation gives way to the driven wave equation, e.g., (7.3.8a,b), in which the operator is that of d'Alembert, e.g.,

$$\partial^2\Phi = \left(\frac{1}{v^2}\partial_t^2 - \nabla^2\right)\Phi = \frac{\rho}{\varepsilon_r\varepsilon_0} \qquad (7.3.8a)$$

To cover the case of e/m waves propagating in a medium we have written the propagation speed as v, but this will be c, the velocity of light in vacuo when the medium, e.g., air, has negligible effect on the speed.

So we are looking for a 4D Green's function such that,

$$\left(\frac{1}{v^2}\partial_t^2 - \nabla^2\right)G(x-x') = \delta^4(x-x') \qquad (15.5.1)$$

This means we are looking for a $G(x-x')$ with dimensions $1/L^2$. The spacetime Fourier transform is,

$$\tilde{G}(k) = \int_{-\infty}^{\infty} e^{ik_\mu x^\mu}\, G(x)d^4x \qquad (15.5.2)$$

where, $$G(x_s - x_f) = \frac{1}{(2\pi)^4}\int_{-\infty}^{\infty} e^{-ik_\mu (x_s - x_f)^\mu}\tilde{G}(k)d^4k \quad (15.5.3)$$

Here we put $k_\mu x^\mu = \omega t - \bar{k}\cdot\bar{r} = k_0\tau - \bar{k}\cdot\bar{r}$ where $k_0 = \omega/v, \tau = vt$ so that,

$$\left(\frac{1}{v^2}\partial_t^2 - \nabla^2\right)e^{-ik_\mu x^\mu} = -\left(k_0^2 - |\bar{k}|^2\right)e^{-ik_\mu x^\mu} \quad (15.5.4)$$

Hence (15.5.3) gives, (15.5.5)

$$\left(\frac{1}{v^2}\partial_t^2 - \nabla^2\right)G(x_s - x_f) = \frac{-1}{(2\pi)^4}\int_{-\infty}^{\infty}\left(k_0^2 - |\bar{k}|^2\right)e^{-ik_\mu (x_s - x_f)^\mu}\tilde{G}(k)d^4k$$

The limits of the integral are understood to mean integration over the whole of spacetime.

(15.5.5) is required to equal $\delta^4(x - x') = \frac{1}{(2\pi)^4}\int_{-\infty}^{\infty} e^{-ik_\mu (x - x')^\mu}\, d^4k$ so,

$$\tilde{G}(k) = \frac{-1}{(k_0^2 - |\bar{k}|^2)} \quad (15.5.6)$$

Hence,

$$G(x_s - x_f) = \frac{-1}{(2\pi)^4}\int_{-\infty}^{\infty}\frac{e^{-ik_\mu (x_s - x_f)^\mu}}{(k_0^2 - |\bar{k}|^2)}d^4k \quad (15.5.7)$$

The k_0 integral is carried out on a path which avoids the singularities at the zeros of the denominator $k_0^2 - |\bar{k}|^2 = (k_0 - |\bar{k}|)(k_0 + |\bar{k}|)$. The contour is completed on the lower half plane where the exponential drives it to zero and on the contour parallel to the real axis we consider k_0 to have a small positive, constant, imaginary part which therefore brings both poles into the contour. Using Cauchy's theorem, or the residue theorem, the k_0 integral therefore gives,

$$G(x_s - x_f) = \frac{2\pi i}{(2\pi)^4}\int_{-\infty}^{\infty}\frac{\left(e^{i|\bar{k}|\tau} - e^{-i|\bar{k}|\tau}\right)}{2|\bar{k}|}e^{i\bar{k}\cdot(\bar{r}_s - \bar{r}_f)}d^3k \quad (15.5.8)$$

Carrying out the integral over the azimuthal angle gives, (15.5.9)

$$G(x_s - x_f) = \frac{-i}{2(2\pi)^2}\int_0^{\infty}\frac{\left(e^{i|\bar{k}|\tau} - e^{-i|\bar{k}|\tau}\right)}{|\bar{k}|}\int_0^{\pi} e^{i|\bar{k}|r\cos\theta}|\bar{k}|^2 \mathrm{d}|\bar{k}|d(\cos\theta)$$

where $r = |\bar{r}_s - \bar{r}_f|$. Carrying out the θ integral gives, (15.5.10)

$$G(x_s - x_f) = \frac{-i}{2(2\pi)^2}\int_0^{\infty}\frac{\left(e^{i|\bar{k}|\tau} - e^{-i|\bar{k}|\tau}\right)}{i|\bar{k}|^2 r}\left(e^{i|\bar{k}|r} - e^{-i|\bar{k}|r}\right)|\bar{k}|^2 d\,|\bar{k}| = \frac{1}{2(2\pi)^2 r}\int_0^{\infty}\left(e^{i|\bar{k}|\tau} - e^{-i|\bar{k}|\tau}\right)\left(e^{i|\bar{k}|r} - e^{-i|\bar{k}|r}\right)d\,|\bar{k}|$$

$$= \frac{-1}{2(2\pi)^2 r}\int_0^{\infty}\left(e^{i|\bar{k}|(\tau+r)} - e^{i|\bar{k}|(\tau-r)} - e^{-i|\bar{k}|(\tau-r)} + e^{-i|\bar{k}|(\tau+r)}\right)d\,|\bar{k}|$$

The 1st and 3rd terms can be re-expressed by replacing $|\bar{k}| \to -|\bar{k}|$ thus,

$$\int_0^{-\infty}\left(e^{-i|\bar{k}|(\tau+r)} - e^{i|\bar{k}|(\tau-r)}\right)\left(-d|\bar{k}|\right) = \int_{-\infty}^{0}\left(e^{-i|\bar{k}|(\tau+r)} - e^{i|\bar{k}|(\tau-r)}\right)d|\bar{k}|$$

Hence we get,

$$G(x_s - x_f) = \frac{1}{2(2\pi)^2 r}\int_{-\infty}^{\infty}\left(e^{i|\bar{k}|(\tau-r)} - e^{-i|\bar{k}|(\tau+r)}\right)d\,|\bar{k}| \tag{15.5.11}$$

That is, $$G(x_s - x_f) = \frac{1}{4\pi r}\left(\delta(\tau - r) - \delta(\tau + r)\right) \tag{15.5.12}$$

Which is shorthand for,

$$G(x_s - x_f) = \frac{1}{4\pi|\bar{r}_s - \bar{r}_f|}\left(\delta\left(vt - |\bar{r}_s - \bar{r}_f|\right) - \delta\left(vt + |\bar{r}_s - \bar{r}_f|\right)\right) \tag{15.5.13}$$

This is an algebraically complete expression, but, as noted above, it may be necessary to impose appropriate boundary conditions on a general Green's function. In the above expression the term $\delta\left(vt - |\bar{r}_s - \bar{r}_f|\right)$ indicates that to calculate the potential, or field, at the field point, x_f, at time t, one must only involve the source (charge density, in this case) at a distance $r = |\bar{r}_s - \bar{r}_f| = vt$ from the field point. In other words, the influence of the charge at $\bar{r}_s$ propagates to the field point, $\bar{r}_f$, at the finite speed, v, so that only the charge density at a distance of $|\bar{r}_s - \bar{r}_f| = vt$ from the field point will influence the field there at time t. To put that another way, a source at $\bar{r}_s$ will influence the field at $\bar{r}_f$ only through it's charge density a time $|\bar{r}_s - \bar{r}_f|/v$ previously. That interpretation refers to the first delta function in (15.5.13) so we write,

Retarded Green's function: $$G(x_s - x_f) = -\frac{\delta\left(vt - |\bar{r}_s - \bar{r}_f|\right)}{4\pi|\bar{r}_s - \bar{r}_f|} \tag{15.5.14}$$

The second term in (15.5.13) has implications contrary to our usual assumptions about causality. It would imply, if taken seriously, that the effect precedes its cause. That is, it implies that the field at $\bar{r}_f$ is determined by the value that the charge density at $\bar{r}_s$ will take a time $|\bar{r}_s - \bar{r}_f|/v$ *in the future*. We express this as,

Advanced Green's function: $$G(x_s - x_f) = \frac{\delta\left(vt + |\bar{r}_s - \bar{r}_f|\right)}{4\pi|\bar{r}_s - \bar{r}_f|} \tag{15.5.15}$$

In normal usage, the retarded Green's function is assumed in order to retain our usual notions of causality and the advanced Green's function (or "advanced potential") is ignored as unphysical.

In his PhD thesis, [15.1], Richard Feynman took seriously the advanced Green's function and argued that a combined retarded and advanced Green's function was a route to solving the vexing problem of radiation reaction (see chapter 22).

15.6 Solutions for the Time Dependent Potentials

Recall from (15.1.6) that, with time dependence, the solution to a partial differential equation in (1,3) spacetime involves a 4D integral. The solutions for the e/m potentials due to continuous distributions of charge and current are therefore, using (15.5.14) and assuming vacuum/air propagation,

$$\Phi(\bar{r}_f, t_f) = \int \frac{\rho(\bar{r}_s, t_s)\delta(v(t_f - t_s) - |\bar{r}_s - \bar{r}_f|)}{4\pi\varepsilon_0 |\bar{r}_s - \bar{r}_f|} d\tau_s v dt_s \tag{15.6.1}$$

$$\bar{A}(\bar{r}_f, t_f) = \frac{\mu_0}{4\pi} \int \frac{\bar{J}(\bar{r}_s, t_s)\delta(v(t_f - t_s) - |\bar{r}_s - \bar{r}_f|)}{|\bar{r}_s - \bar{r}_f|} d\tau_s v dt_s \tag{15.6.2}$$

The delta functions mean that the sources are evaluated at the retarded time $t_s = t_f - |\bar{r}_s - \bar{r}_f|/v$.

15.7 References

Feynman, R.P. (1942). *Feynman's PhD Thesis: The Principle of Least Action in Quantum Mechanics*. In the book *Feynman's Thesis: A New Approach to Quantum Mechanics* with a Preface by Laurie Brown, 2005, World Scientific.

16. Earnshaw's Theorem and Magnetic Levitation

Earnshaw's Theorem tells us that static charges and static magnets cannot be in stable equilibrium under the action of only static electromagnetic fields (which might arise from other static charges and static magnets or constant currents), excepting only the case of diamagnetic materials. Careful attention must be paid to the conditions under which Earnshaw's Theorem is rigorously true. Relaxing these conditions, however, can produce interesting effects including stable levitation.

Contents

16.1 Earnshaw's Theorem in Electrostatics

In the restricted case of electrostatics, Earnshaw's Theorem states that no system of static point charges can be in stable equilibrium under the action of the electrostatic force alone. An indication of the origin of this result is provided by assuming an arbitrary distribution of point charges and considering the potential at the position of one of the point charges before that charge was introduced. The potential at that point due to all the other charges obeys $\nabla^2\Phi = 0$ and the force on the charge in question (q), when introduced into that potential field, is $\bar{F} = -q\bar{\nabla}\Phi$. For zero force to act, i.e., equilibrium, then $\bar{\nabla}\Phi = 0$ and so all three components of $\bar{\nabla}\Phi$ must be zero, $\partial_x\Phi = 0$, $\partial_y\Phi = 0$, $\partial_z\Phi = 0$. But for this to be a stable equilibrium, if the charge were displaced slightly in the x direction then the resulting force on the charge would need to be in the negative x direction, so as to tend to push the charge back to its equilibrium position. This applies in all three coordinate directions so that stability requires $\partial_x F_x < 0$, $\partial_y F_y < 0$, $\partial_z F_z < 0$. This means that the potential must be a local minimum if the equilibrium is to be stable.

It is tempting to conclude (wrongly) that this requires $\partial_x^2\Phi > 0$, $\partial_y^2\Phi > 0$, $\partial_z^2\Phi > 0$. This would conflict with the sum of the second derivatives of the potential being zero, i.e., $\nabla^2\Phi = 0$, and so cannot be true – apparently proving there cannot be a stable equilibrium. The argument is false, but it does give a simple explanation of roughly where Earnshaw's Theorem comes from as it is 'nearly correct'.

The loophole is that (say) $\partial_x F_x < 0$ can also arise from $\partial_x^2\Phi = 0$ if we also have $\partial_x^3\Phi > 0$ (or, indeed, from $\partial_x^2\Phi = \partial_x^3\Phi = 0$ if $\partial_x^4\Phi > 0$, and so on). Laplace's equation would still hold if this were the case in all three directions.

Here we present a rigorous proof, noting that Earnshaw's Theorem applies only in the case of discrete point charges. "Discrete" means that there is a finite distance between any pair of charges. So we may consider a sphere of finite size centred on any one charge such that it contains no other charges. The proof proceeds by first proving the Average Potential Theorem for Laplace's equation.

16.1.1 The Average Potential Theorem

The Average Potential Theorem states that the average potential over the surface of any sphere which contains no charge is equal to the potential at its centre. The proof proceeds by showing that it holds for the potential due to a point charge placed anywhere outside the sphere. That it holds in general for any solution of Laplace's equation follows immediately because any solution can be expressed as a superposition of point charge solutions.

Consider a sphere of radius r centred on a point $\bar{r}_c$ defined with respect to an origin, O. A point charge of magnitude Q is at the origin, hence giving rise to a potential on, and within, the sphere. Hence the potential at the centre of the sphere is $Q/4\pi\varepsilon_0 r_c$. The vector from the centre to the surface of the sphere is $\bar{r}$. The vector from the charge to the surface of the sphere is therefore $\bar{r}_f = \bar{r}_c + \bar{r}$. The average potential over the surface of the sphere is therefore,

$$\langle\Phi\rangle = \frac{Q}{4\pi\varepsilon_0}\frac{1}{4\pi r^2}\oiint\frac{1}{r_f}dS = \frac{Q}{4\pi\varepsilon_0}\frac{1}{4\pi r^2}\oiint\frac{r^2 d(\cos\theta)d\phi}{\sqrt{r^2+r_c^2+2rr_c\cos\theta}} \qquad (16.1.1)$$

Here we have defined the polar axis to be in the same direction as $\bar{r}_c$. The integration is trivial because,

$$\frac{\partial}{\partial c}\sqrt{r^2+r_c^2+2rr_c c} = \frac{rr_c}{\sqrt{r^2+r_c^2+2rr_c c}} \qquad (16.1.2)$$

Hence,
$$\langle\Phi\rangle = \frac{Q}{(4\pi)^2\varepsilon_0}\frac{2\pi}{rr_c}\sqrt{r^2+r_c^2+2rr_c c}\Big|_{c=-1}^{c=1} \qquad (16.1.3)$$

The source charge is assumed to lie outside the sphere, $r_c > r$, so,

$$\langle\Phi\rangle = \frac{Q}{(4\pi)^2\varepsilon_0}\frac{2\pi}{rr_c}[(r+r_c)-(r_c-r)] = \frac{Q}{4\pi\varepsilon_0 r_c} \qquad (16.1.4)$$

which is the potential at the centre of the sphere, so the average potential over the sphere equals that at the centre. QED.

16.1.2 Electrostatic Proof of Earnshaw's Theorem

The proof of Earnshaw's Theorem now follows immediately. If the potential were a local minimum at the centre of the sphere then it must be greater at all points on a neighbourhood sphere and so the average over such a sphere must exceed that at the centre, which is a contradiction. Therefore there is no such local minimum in the potential and hence no stable equilibrium for a distribution of discrete, static point charges. QED.

16.2 Earnshaw's Theorem in Magnetostatics

16.2.1 Fixed Magnetic Dipoles

There are two versions of Earnshaw's Theorem in magnetostatics according to whether one considers fixed magnetic dipoles or magnetic dipoles induced by an external magnetic field. We initially consider fixed dipoles, meaning a cluster of dipoles, $\bar{m}_I$ for $I \in [1, N]$, each of which has unvarying magnitude and direction. A cheap means of establishing the reasonableness of the theorem is to note that, in the electrostatic case, we may consider the cluster of charges to be composed of equal and opposite pairs of charges a small distance apart, i.e., to consist of a cluster of electric dipoles. Hence, Earnshaw's Theorem must apply also to clusters of electric dipoles.

But there is a direct analogy between electric dipoles and magnetic dipoles. Both may be described in terms of a scalar potential, (5.2.3) and (5.5.4), and these expressions are essentially the same apart from the universal constant, ε_0, defining units. Moreover, the expressions for the force acting on a dipole in terms of the scalar potential are also essentially the same, (5.4.13) and (5.6.6) giving $\bar{F} = -(\bar{p} \cdot \bar{\nabla})\bar{\nabla}\Phi_e$ and $\bar{F} = -\mu_0(\bar{m} \cdot \bar{\nabla})\bar{\nabla}\phi_m$ respectively. Consequently the theorem is essentially already proved for fixed magnetic dipoles by the electric case.

If this seems too slick to be reliable, an alternative proof is as follows. If U is the energy of the system of magnetic dipoles due to their presence within the magnetic field then the force on any dipole is (minus) the gradient of that energy. For stable equilibrium the force acting on any dipole must be zero and the force acting if the dipole is displaced slightly must be such as the push the dipole back to its equilibrium position. In other words U must be a local minimum. Hence U plays the same part as the potential in the electric case. The proof will therefore be as for the electric case if we also have that U obeys Laplace's equation, $\nabla^2 U = 0$. This is established by noting that, from (5.6.5),

$$U = -\sum_{I=1}^{N} \bar{m}_I \cdot \bar{B}(\bar{r}_I) \tag{16.2.1}$$

where $\bar{r}_I$ is the position of the Ith dipole, $\bar{m}_I$. But we recall that the Maxwell equations tells us that, in the absence of current sources between dipoles (i.e., no sources other than the dipoles themselves), and for static conditions so that the displacement current is also zero, then we have $\bar{\nabla} \cdot \bar{B} = 0$ and $\bar{\nabla} \times \bar{B} = 0$. The identity $\nabla^2 \bar{v} = \bar{\nabla}(\bar{\nabla} \cdot \bar{v}) - \bar{\nabla} \times (\bar{\nabla} \times \bar{v})$, then shows that

$\nabla^2 \bar{B} = 0$, which means that each Cartesian component of the magnetic field individually obeys Laplace's equation. As the dipoles are assumed constant (in magnitude and direction) substitution in (16.2.1) then gives $\nabla^2 U = 0$, as required.

The proof of §16.1 showed that a scalar field obeying Laplace's equation cannot have a minimum, and so this also applies here for U. QED.

The reader may still be doubtful because the proof in §16.1 used the potential of a point charge to establish the Average Potential Theorem as a key part of the proof and dipoles do not have the same potential field as a point charge. However, that is irrelevant as the scalar field which obeys Laplace's equation and whose minimum is required for a stable equilibrium is U, not the magnetic scalar potential. The Average Potential Theorem applies also for U in this situation, i.e., the average of U over the surface of a sphere equals its value at the centre, providing the sphere contains no dipoles. The proof is the same as before noting that the Coulomb potential of a point charge is now simply reinterpreted as the Greens function solution for any scalar field obeying Laplace's equation (see chapter 15 for details about Greens functions).

16.2.2 Induced Dipoles

From §5.6.1,2 the energy of a dipole induced by an external field, $\bar{B}_0$, is $-\frac{1}{2}\bar{B}_0 \cdot \bar{m}$. If we imagine the dipole being induced within a small volume, δV, of magnetisable material then we can write, $\bar{m} = \chi_m \bar{H}_0 \delta V$ (see (6.3.9)) so that its energy is $-\chi_m B_0^2 \delta V / 2\mu_0$. The subscript on B_0 here denotes the field excluding that of the induced dipole in question. Hence, summing over all the small volumes which induce dipoles gives the total energy of the dipoles to be,

$$U = -\frac{1}{2\mu_0} \chi_m \sum_{I=1}^{N} B_{0I}^2 \delta V_I \qquad (16.2.2)$$

where B_{0I} is the field at dipole I excluding its own field, and δV_I is the associated small volume of material (to which the magnitude of the dipole is proportional). Hence, in order to explore $\nabla^2 U$ we require $\nabla^2 B_{0I}^2$.

Since $\partial_i^2 B_j^2 = 2(\partial_i B_j)^2 + 2B_j \partial_i^2 B_j$ where i, j are the Cartesian components, it follows that, dropping the $0I$ subscripts,

$$\nabla^2 B^2 = 2 \sum_{i=1}^{3} (\bar{\nabla} B_i) \cdot (\bar{\nabla} B_i) + 2\bar{B} \cdot \nabla^2 \bar{B} \qquad (16.2.3)$$

From §16.2.1 we have $\nabla^2 \bar{B} = 0$ so the second term on the RHS is zero, whereas the first term, being the sum over squares, is necessarily positive or zero. So we have $\nabla^2 B^2 \geq 0$. However, we now argue that, for a stable equilibrium the equality here cannot hold, i.e., that $\sum_{i=1}^{3}(\bar{\nabla} B_i) \cdot (\bar{\nabla} B_i)$ cannot be zero. To be zero would require that every $\partial_i B_j$ were zero for all nine combinations of i, j. But the force on a dipole, (5.6.6), is, in the ith coordinate direction,

$$F_i = \sum_{j=1}^{3} m_j \partial_j B_i \propto \sum_{j=1}^{3} B_j \partial_j B_i \tag{16.2.4}$$

where the last expression applies because the dipole is induced by the field. Hence,

$$\partial_i F_i \propto \partial_i \sum_{j=1}^{3} B_j \partial_j B_i = \sum_{j=1}^{3} (\partial_i B_j)(\partial_j B_i) + \sum_{j=1}^{3} B_j \partial_j \partial_i B_i \tag{16.2.5}$$

The first term in (16.2.5) must be zero if all the $\partial_i B_j$ are zero. If we sum (16.2.5) over i the second term becomes zero because $\bar{\nabla} \cdot \bar{B} = 0$ is a Maxwell equation. Hence we conclude that $\sum_{i=1}^{3} \partial_i F_i = 0$. But stable equilibrium requires that $\partial_i F_i < 0$ for each i so $\sum_{i=1}^{3} \partial_i F_i$ cannot be zero for stability. Hence the assumption that $\sum_{i=1}^{3}(\bar{\nabla} B_i) \cdot (\bar{\nabla} B_i)$ is zero conflicts with stability. Hence we conclude that (16.2.3) requires the strict inequality,

$$\nabla^2 B^2 > 0 \tag{16.2.6}$$

Provided the magnetic susceptibility, χ_m, is positive – which means that the dipoles are induced in either ferromagnetic or paramagnetic materials – then (16.2.2) gives,

$$\nabla^2 U < 0 \tag{16.2.7}$$

But this contradicts the requirement for a stable equilibrium which requires that U is a minimum and hence that each $\partial_i^2 U \geq 0$ so that $\nabla^2 U \geq 0$.

The loophole in Earnshaw's Theorem for induced dipoles is the requirement that the materials in question be either ferromagnetic or paramagnetic, and hence have positive magnetic susceptibility. In contrast, diamagnetic materials have negative magnetic susceptibility (i.e., a relative permeability less than 1). In this case (16.2.2) together with (16.2.6) gives $\nabla^2 U > 0$ which is consistent with stability. This loophole permits diamagnetic materials to develop induced magnetism which leads to stable equilibrium in an applied magnetic field. This loophole has been exploited in experiments, as we will see.

16.3 Pseudo-Levitation with Mechanical Constraint

The rest of this chapter consists of cases of experimental equipment, or commercial technology, which appear to be in conflict with Earnshaw's theorem in the magnetic case. All of them fall into one of three categories, defined by a violation of some pre-condition for Earnshaw's theorem to hold, namely that,

(i) levitation is not provided stably in all three dimensions, or,

(ii) conditions are not static, or,

(iii) diamagnetic materials are involved.

The most obvious "cheat" is the first of these cases, in which stable levitation is provided in only one or two dimensions, whilst unstable motion in the third direction is prevented by mechanical restraints or limits to movement.

16.4 Magnetic Ion Traps

Magnetic ion traps have become a standard tool for experimentalists. Being able to keep the objects of study – ions, in this case – pinned down to a small location has obvious experimental advantages (compared to, say, an ion beam in which said ions exit your apparatus in a tiny fraction of a second). Such ion traps come in many forms. Broadly there are those in which the ion itself is not stationary (e.g., Penning traps or storage rings) and those in which the field is not stationary (e.g., the radiofrequency Paul trap).

There are misleading discussions regarding magnetic ion traps on the internet. We know from §16.2.1 and §16.2.2 that, in free space, $\nabla^2 B^2 > 0$ so the magnitude of the magnetic field can have a minimum but not a maximum. Discussions may then observe that mobile ions with permanent magnetic dipoles which are not free to rotate will tend to move towards such a minimum in the field magnitude if they happen to be anti-aligned with the field (because their energy, $-\bar{B}_0 \cdot \bar{m}$, is then positive and can be minimised by moving to the minimum of the field). True, but that fails to observe also that, under strictly stationary conditions, the ion will not thereby achieve a stable equilibrium, as we have seen. Freedom to rotate will cause alignment with the field so that $-\bar{B}_0 \cdot \bar{m}$ becomes negative and the minimum in B^2 becomes a maximum in the energy wrt displacement.

Some sort of dynamic condition must be used to induce the ion to remain in a roughly prescribed location.

In the Penning trap, for example, the ions are caused to execute simple harmonic motion (SHM) between a pair of electrodes at positive potential (assuming the ions are positively charged). This is stable in the axial direction (i.e., the direction of the SHM) but this electric field will not cause lateral stability. To address this a large axial magnetic field is also applied, which causes any tendency to move laterally to be converted into circular motion about the axial direction (akin to motion in a cyclotron). In short, the ions are caused to execute a complex motion that keeps them within a small region (e.g., a few cm). Other trap designs are very different in detail but all involve dynamic conditions, including cases where the field is oscillatory. See [16.1] for other examples.

16.5 Magnetic Levitation Stabilised by Rotation

If the magnet which is being levitated is rotating, stabilisation can arise under a suitably shaped external magnetic field due to a combination of centrifugal effects and gyroscopic precession. The earliest example of this was a demonstration device, the Levitron, a video of which can be found on the internet, see [16.2]. For the theory of spin stabilised levitation see [16.3].

16.6 The Levitation used by MagLev Trains

Various technologies exist, some of which combine the fields required for levitation with those used for propulsion – in others these are separate. In this context the levitation is known as electromagnetic suspension or electrodynamic suspension. Details of the stabilisation methodology vary. In one, stabilisation is achieved by continually adjusting the current to electromagnets in a manner determined by a feedback loop so as to cancel the instability, i.e., controlled dynamic stabilisation.

16.7 The Levitating Frog

Undoubtedly the most famous example of genuine static levitation caused by diamagnetism is that of the levitating frog, [16.4]. It is the water content that provides the dominant diamagnetism, as the levitation of many different items demonstrates in the video of [16.4], including levitation of water droplets and an ice cube. However, because diamagnetism is so weak, extremely strong magnetic fields are required for such levitation (e.g., 16 Tesla). Andre Geim and Michael Berry won an Ignobel prize in 2000 for not only levitating a frog but also a sumo wrestler, [16.5].

16.8 References

[16.1] Knoop, M., Madsen, N., and Thompson, R.C. (2013). *Physics with Trapped Charged Particles*. 28 November 2013.

[16.2] Harrigans, R. (1983). *The Levitron*. Video available here.

[16.3] Simon, M.D., Heflinger, L.O., and Ridgway, S.L. (1997). *Spin stabilized magnetic levitation*. Am. J. Phys. 65(4), 286–292 (1997).

[16.4] Beautiful Minds, Series Two (2012). *Andre Geim levitates a frog* (video). BBC Four, 5 April 2012.

[16.5] Nadis, S. *Ig Nobel glory for levitating frogs and collapsing toilets*. Nature **407**, 665 (2000).

17. The Skin Effect and Induction Hobs

At sufficiently high AC frequencies, the current in wires and other conductors is not uniformly distributed over their cross section but is concentrated near the surface. This is the skin effect. From the inclusion of induction hobs in this chapter you may anticipate that the skin effect is relevant to their operation.

Contents

17.1 The Skin Effect in a Solid Cylinder, Wire or Cable

17.1.1 Static Applied EMF (Uniform Current)

A current passing down a wire creates a magnetic field. (4.2.21) gives the magnitude of the magnetic flux density due to a long straight wire carrying a constant (DC) current and how it varies with distance from the wire. However, this applies only outside the metal of the wire itself. As a gentle introduction to the AC case we firstly consider how the magnetic field varies within the thickness of the wire when the applied emf is steady (DC), and hence the current is also constant. We assume that both are uniform across the thickness of the wire, though this will not be the case when AC is considered.

Figure 17.1 shows the geometry of the wire, and we consider the more general case of a hollow cylindrical conductor. (For a wire we would have $R_i = 0$). We have seen in §11.8 that, for frequencies up to, and including, optical frequencies (<10^{15} Hz), and for metals which have high conductivity and $\varepsilon_r \approx 1$, the displacement current may be neglected in favour of the conduction current when an electric field is present. Hence the third Maxwell equation, (7.2.1), becomes $\overline{\nabla} \times \overline{H} = \overline{J} = \kappa \overline{E}$ where κ is the conductivity (either isotropic or that in the axial direction of the wire).

As the electric field is in the z direction, and because the fields cannot vary in the z or θ directions by symmetry, it follows from $\overline{\nabla} \times \overline{H}$ expressed in cylindrical polars that the only non-zero component of $\overline{H}$ is H_θ. Using Stokes's theorem on a circular path across the wire cross-section and concentric with it, $\overline{\nabla} \times \overline{H} = \overline{J}$ gives,

For $r < R_i$ $\quad H_\theta(r) = 0 \quad$ (17.1.1a)

For $R_i \leq r \leq R_0$ $\quad H_\theta(r) = \frac{r^2 - R_i^2}{2r} J \quad$ (17.1.1b)

For $r > R_0$ $\quad H_\theta(r) = \frac{R_0^2 - R_i^2}{2r} J = \frac{I}{2\pi r} \quad$ (17.1.1c)

where I is the total current. The last expression agrees with (4.2.21) whilst (17.1.1b) shows that the magnetic field varies across the thickness of the wire or cylinder from zero at the bore.

Figure 17.1: Geometry for §17.1 and §17.2

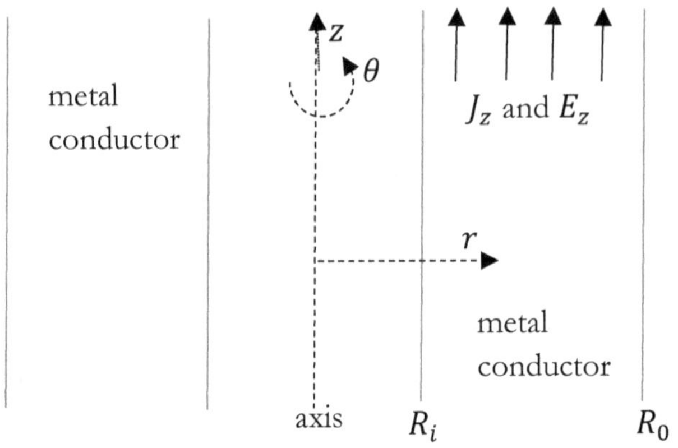

17.1.2 AC Skin Effect (Wire or Solid Cylindrical Conductor)

The skin effect is specific to AC currents at sufficiently high frequencies. It is the phenomenon whereby the current does not flow uniformly over the cross section of the wire or other conductor but is concentrated in a narrow region close to its outer surface. The higher the frequency, the narrower the region to which the current is confined, and the more intense is the current density in this near-surface "skin" as a consequence. This section derives the skin effect from the Maxwell equations.

We consider an applied alternating emf, driving an alternating current. If the current density remained uniform over the cross-section of the conductor then we could simply replace J in (17.1.1b,c) with $J = J_0 e^{i\omega t}$. This would then indicate that the magnetic field also alternates. The fourth Maxwell equation, $\bar{\nabla} \times \bar{E} = -\frac{\partial \bar{B}}{\partial t}$, then requires that the electric field, $\bar{E}$, varies across the cross section. But this conflicts with Ohm's law, namely $\bar{J} = \kappa \bar{E}$, as we assumed the current density was uniform. It follows that the current density and the electric field must both vary over the cross section of the conductor when AC is applied. It is this which gives rise to the skin effect.

We continue to assume that the current and electric field are axial and hence that the magnetic field is in the circumferential (θ) direction. Assuming the conduction current dominates the displacement current then the third Maxwell equation becomes $\bar{\nabla} \times \bar{H} = \bar{J} = \kappa \bar{E}$, which is,

$$\frac{1}{r}\partial_r(rH_\theta) = \kappa E_z \tag{17.1.2}$$

All fields are varying sinusoidally, with a factor $e^{i\omega t}$, and hence $\bar{\nabla} \times \bar{E} = -\frac{\partial \bar{B}}{\partial t} = -i\omega B_{\theta 0} e^{i\omega t} \hat{\theta} = -i\omega\mu H_{\theta 0} e^{i\omega t} \hat{\theta}$, which becomes,

$$\partial_r(E_z) = i\omega\mu H_{\theta 0} e^{i\omega t} \tag{17.1.3}$$

Here $\mu = \mu_r \mu_0$, the magnetic permeability of the metal.

Hence, $\frac{1}{r}\partial_r\left(r\partial_r(E_{z0})\right) = i\omega\mu\kappa E_{z0}$ which gives,

$$r^2\partial_r^2 E_{z0} + r\partial_r E_{z0} - i\omega\mu\kappa r^2 E_{z0} = 0 \tag{17.1.4}$$

This is Bessel's equation of order zero and hence the general solution is a linear superposition of $\mathrm{J}_0(kr)$ and $\mathrm{I}_0(kr)$, the Bessel functions of the first and second kinds, where,

$$k^2 = -i\omega\mu\kappa \tag{17.1.5}$$

Hence, $$k = \frac{(1-i)}{\sqrt{2}}\sqrt{\omega\mu\kappa} = (1-i)\sqrt{\pi f \mu\kappa} = \frac{(1-i)}{\delta} \tag{17.1.6}$$

where we have replaced the angular frequency with the frequency, $f = \omega/2\pi$. The Bessel functions are therefore to be evaluated at a complex valued argument, which induces a varying phase across the section. For a wire or solid cylinder we can rule out $\mathrm{I}_0(kr)$ as this would be divergent on the axis. Hence the solution for E_{z0}, and hence also the distribution of current, J_z (not to be confused with the Bessel function J_0) is some constant times $\mathrm{J}_0(kr)$. Evaluation of $\mathrm{J}_0(kr)$ with the complex k given by (17.1.6) increases rather steeply for larger r. It is therefore convenient for numerical evaluation to normalise by the Bessel function at the surface of the wire, $\mathrm{J}_0(kR_o)$, so the quantity we calculate is $\mathrm{J}_0(kr)/\mathrm{J}_0(kR_o)$ which becomes 1 on the outer surface but is less than 1 beneath the surface.

Numerical evaluation using the power series expansion for the Bessel function (see for example [17.1]) produces the results displayed in Figures 17.2-4. Curves are labelled by R_o/δ where $\delta = 1/\sqrt{\pi f \mu\kappa}$. In the present context this quantity is called the skin depth, but it is the same as the penetration depth of an electromagnetic wave incident upon a conducting medium, i.e., ζ, in (11.8.14). The current (and hence the electric field) is strongly concentrated near the outer surface. The current density in the centre of the wire is virtually zero for $\frac{R_o}{\delta} > 5$.

Figure 17.4 is perhaps the best illustration of how the skin effect concentrates the current density near the outer surface, as all the curves relate to the same total current.

Figure 17.2: Distribution of either J_z or E_z across the section of a wire for various R_0/δ, normalised to the same surface value of unity. *This is plotted as if for wires of different diameters but the same skin depth, δ.*

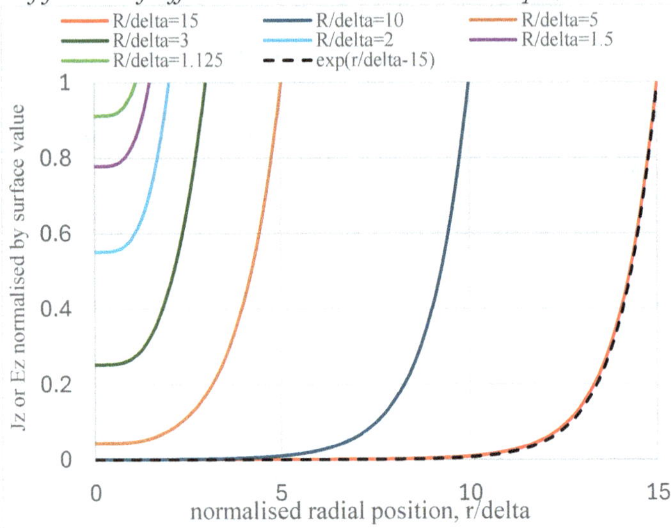

Figure 17.3: Distribution of either J_z or E_z across the section of a wire for various R_0/δ, normalised to the same surface value of unity. *This is plotted as if for wires of the same diameter but differing skin depth, δ (e.g., due to differing AC frequencies)*

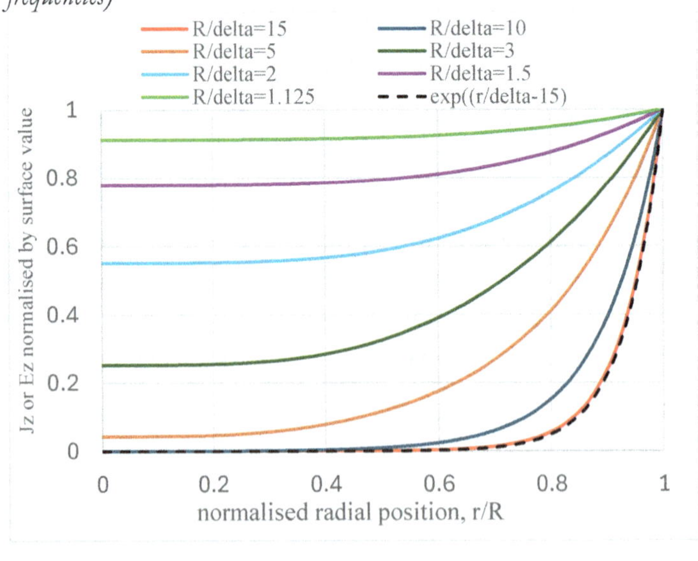

Figure 17.4: Distribution of either J_z or E_z across the section of a wire for various R_0/δ, normalised to the same total current or emf. *This is plotted as if for wires of the same diameter but differing skin depth, δ (e.g., due to differing AC frequencies)*

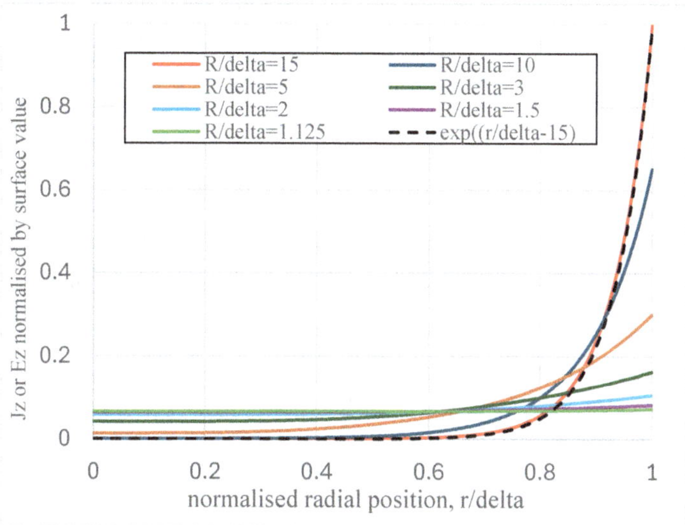

17.1.3 Skin Effect in a Hollow Cable or Coaxial Cable

Consider firstly a hollow wire or cylinder with wall thickness t. Confine attention to sufficiently high frequencies such that the skin depth is a small fraction of the metal thickness, $\delta \ll t$. Imagine that the hollow core was filled in to make the cylinder solid, with unchanged outer diameter. The current and magnetic and electric fields would be essentially zero within the added material. Removing the added material therefore makes no difference to the electromagnetic situation. It follows that the current density at the bore remains virtually zero when the cylinder is hollow. We conclude that, for a hollow wire or tube, the current density is concentrated by the skin effect only near the outer surface but not near the inner surface, provided that $\delta \ll t$.

For a coaxial cable, however, the situation is different. Suppose the inner conductor has radius a and the outer conductor has inner and outer radii b and c respectively ($a < b < c$). The inner and outer conductors carry equal and opposite current. By Stokes's theorem, the magnetic field, H_θ, in the interspace, ($a < r < b$), is $\frac{I}{2\pi r}$. It therefore falls from its peak value of $\frac{I}{2\pi a}$

at the surface of the inner conductor to $\frac{I}{2\pi b}$ at the inner surface of the outer conductor.

But also by Stokes's theorem, the net current linked becomes zero when the path of the integral lies on the outer surface of the outer conductor, so that H_θ becomes zero there. Consequently, the radial variation of H_θ is shown qualitatively by Figure 17.5. But recall that the third Maxwell equation, assuming the conduction current dominates the displacement current, gives,

$$\frac{1}{r}\partial_r(rH_\theta) = \partial_r H_\theta + \frac{H_\theta}{r} = J \tag{17.1.7}$$

But the current density must be negative in the outer conductor because it is the return path in the circuit. Thus the clearly negative term $\partial_r H_\theta$ must dominate the positive term $\frac{H_\theta}{r}$. This condition is met if H_θ is proportional to $e^{-(r-b)/\delta}$ in the outer conductor and this term dominates the derivative, since then $\partial_r H_\theta \approx -\frac{H_\theta}{\delta}$ which dominates $\frac{H_\theta}{r}$ in the assumed regime that $\delta \ll a < b$. (We do not claim this is an exact solution – it is not, this is only a qualitative argument). The current density in the outer conductor, from (17.1.7), is then also roughly proportional to $e^{-(r-b)/\delta}$. The result is that the magnetic field and the current density vary with radius as shown schematically by Figure 17.5.

Hence, for a coaxial cable the skin effect concentrates the current density near the outer surface of the inner conductor, but also near the inner surface of the outer conductor. In contrast, the material near the outer surface of the outer conductor carries negligible current at sufficiently high frequencies.

17.1.4 Litz Wire

The skin effect tends to increase losses because of the increased current density in a narrow region, recalling that losses are related to $\frac{1}{2}RI^2$. Another way of looking at this is that the resistance is effectively increased because the useful current-carrying cross section of the wire is reduced. This is wasteful of material as the central region of the wire is doing nothing. To avoid the worst of these problems, radio frequency circuitry often deploys Litz wire, which consists of multiple strands of copper braided together, with each strand being insulated, typically using a lacquer. If each individual strand has a thickness comparable with, or smaller than, the skin depth, then the phenomenon has little effect on the wire as a whole.

Figure 17.5: Schematic Illustration of the Variation of the Magnetic Field with Radius for a Coaxial Cable

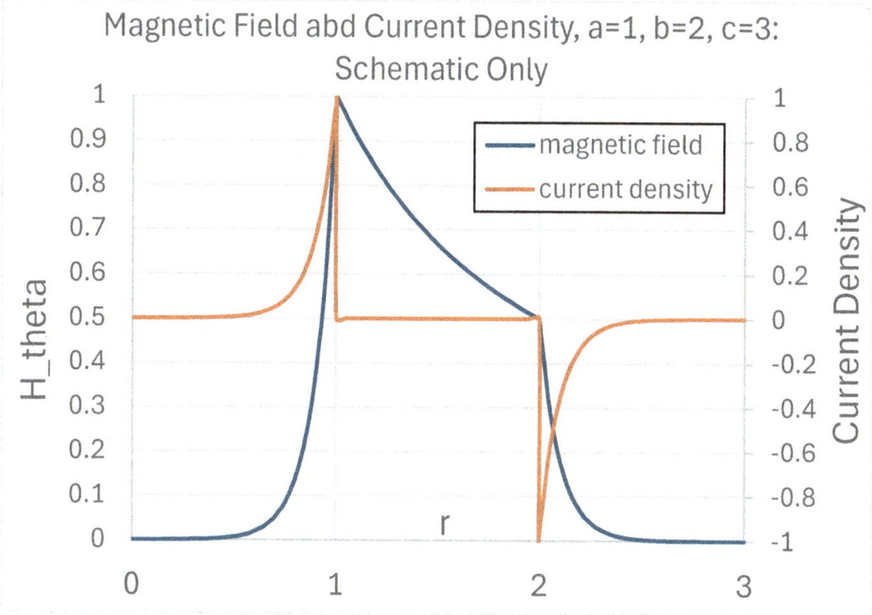

17.2 Induction Hobs

As the name implies, induction cooking is based on heat generation from currents induced by an alternating magnetic field. Unlike conventional electrical hobs the heat is not generated within the hob unit but only within the pan. Consequently, the hob remains cold whilst the contents of the pan are rapidly raised to boiling point. An ice cube can remain frozen on the hob adjacent to the boiling pan, initially to the mystification of the scientifically uneducated.

As I write, induction hobs which are readily available for the domestic market are intended to work only with pans made of ferromagnetic material (cast iron, carbon steel and ferritic stainless steel) but not with, say, aluminium, copper or austenitic stainless steel pans. Until I thought more seriously about the theory of such hobs I had thought that this was because the high permeability of ferromagnetic materials caused a much larger magnetic flux density in these materials than in non-ferromagnetic pans. But this is not right, though it may be a contributory factor, as we will see. But first, the basic construction of induction hobs.

The hob consists of a copper coil, usually composed of Litz wire, embedded within a heat-resistant ceramic with a low coefficient of thermal expansion. The coil is supplied with an alternating current at a frequency between 20 and 40 kHz, typically 24 kHz. The external appearance is a cool, smooth ceramic surface, easy to clean and not prone to burnt-on deposits as the hob never gets very hot. The coil produces an alternating magnetic field immediately above the hob, and by Farady's law of induction, i.e., the Maxwell equation $\bar{\nabla} \times \bar{E} = -\frac{\partial \bar{B}}{\partial t}$, an alternating electric field is generated above the hob. Hence, if an electrically conducting material, e.g., a saucepan, is placed on the hob, alternating currents are generated within the metal of the pan. In this context such currents are called "eddy currents". Heat is then generated simply by the usual process of resistance heating (sometimes called Joule heating).

But the amount of heat generated depends sensitively on the currents induced, and these depend upon the magnetic flux density within the metal of the pan…and on skin depth! And this is where it gets interesting.

Consider an extremely simplified model of the process, as illustrated by Figure 17.6. Here I idealise the magnetic field as vertical (B_z). An equally highly simplified account of the induced electric field might then be an x component, E_x, which varies in the y direction only, and perhaps also a y component, E_y, which varies in the x direction only, such that the Maxwell equation $\bar{\nabla} \times \bar{E} = -\frac{\partial \bar{B}}{\partial t}$ becomes $-i\omega B_z = \partial_x E_y - \partial_y E_x$, where ω is the angular frequency of the supplied excitation current in the coils.

Figure 17.6: A Simple Model of the Induction Hob

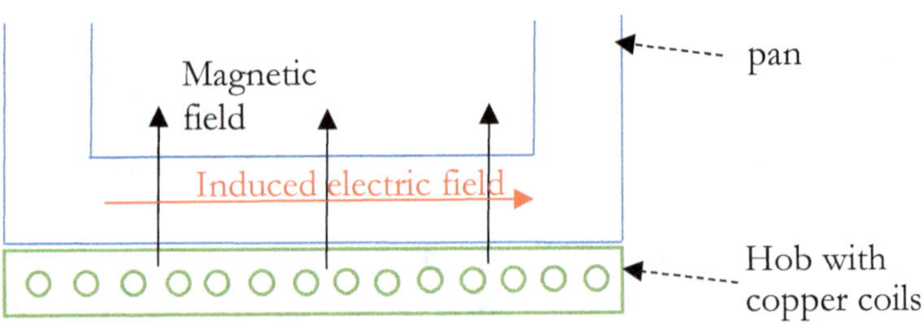

In this crude model, the B_z in the base of the pan is equal to the B_z generated within the hob – despite the former having a large relative permeability, μ_r, compared to the latter's $\mu_r = 1$. This is because B_z is the normal component at the interface, and so is continuous (see §6.7). It was this observation that initially bewildered me as I had expected the pan to benefit from its magnetic material via an enhanced magnetic flux density. Of course, the pan does indeed benefit from its magnetic material, but not via B_z but rather via the skin effect.

Recall that the skin depth is $\delta = 1/\sqrt{\pi f \mu \kappa}$ where $\mu = \mu_r \mu_0$, the magnetic permeability of the metal of the pan. Consequently, ferromagnetic materials have a shallower skin depth than non-ferromagnetic materials, other things being equal. Other things are not equal, however, and the likes of copper and aluminium partly compensate for their $\mu_r \approx 1$ by having high conductivity. Table 17.1 gives the skin depth for a number of metals relevant in the context of saucepans, at the typical hob coil frequency of 24 kHz. The ferromagnetic materials have a substantially smaller skin depth at this frequency, and hence produce more heating.

Table 17.1: Skin Depths at 24 kHz (mm)

Metal	Conductivity, κ (S/m)	μ_r	Skin depth, δ (mm)
Cast iron	1×10^7	100	0.10
Carbon steel	0.6×10^7	100	0.13
Ferritic stainless steel	0.2×10^7	1000	0.07
Austenitic stainless steel	0.15×10^7	1.05	2.6
Copper	6×10^7	1	0.4
Aluminium	3.5×10^7	1	0.55

However, at higher frequencies the skin depth reduces for all materials. Why do induction hobs not use a higher frequency and thereby provide a facility that would work even for copper and aluminium pans? Actually, this has been done, but not in designs that are generally available (at the time of writing, February 2026) but this may change in future. Part of the problem is that to reduce the skin depth for copper to ~0.1mm, typical of ferromagnetic materials at 24 kHz, the frequency would need to be 16 times higher, hence nearly 0.4 MHz. The reason why commercially available domestic induction hobs have so far been restricted to frequencies not exceeding 40 kHz is because of the lack of sufficiently cheap semiconductor components capable of delivering higher frequencies at high power levels. Hence, the goal of all-metal induction hobs depends upon the development of such semiconductor technology.

However, ferromagnetic pans will always perform better on induction hobs than non-ferromagnetic pans. Their skin depth will always be smaller. But there is another effect, which goes back to the magnetic flux density within the pan.

The magnetic field produce by the coils is not really vertical, at least not at all positions. It will approximate a dipole-type field, and thus will have a substantial horizontal component. The currents induced will be in varying directions as the magnetic field lines curve, thus producing loops of current, whence the term "eddy currents" originates. But the horizontal component of the B-field, being tangential at the interface with the pan, is not continuous. Even if possible surface currents are ignored, it is (say) B_x/μ which is continuous, i.e., the components B_x and B_y are increased by a factor of μ_r when entering the pan compared to that generated by the coils. Hence, ferromagnetic pans do indeed have an enhanced magnetic flux density compared to other pans, albeit only the horizontal components. So the induced electric fields will be greater in ferromagnetic pans and this is another reason why ferromagnetic pans will always perform better than others with induction hobs.

17.3 References

[1] M.Abramowitz and I.A.Stegun (1965). Handbook of Mathematical Functions. Dover, NY.

18. Circuit Electricity

It would be too glaring an omission to include nothing about circuit electricity in a book about electromagnetism. On the other hand, circuit electricity is a vast topic. So vast, indeed, that even a work dedicated solely to it would have to be in at least five volumes, addressing separately high power applications (electrical engineering), domestic electrical installations (the craftsman electrician), analogue circuit design, information technology hardware (electronic engineering) and telecommunications hardware. In this chapter we merely dip a toe in these waters, initially explaining what circuit electricity actually is from the perspective of field theory, and then using transmission lines and simple transistor circuits as examples. Transmission lines will be addressed further in chapters 21 and 23.

Contents

In this chapter we address a few topics in circuit electricity. The scope of potential topics that could be addressed under this heading is vast and could fill several texts on its own. There are several distinct professional disciplines involved, including electrical engineering, electronic engineering and telecommunications. In fact, the design of electronic equipment now covers a number of subdisciplines as information technology has become so dominant and so sophisticated in the last few decades. We cannot do justice to such a large scope, and in any case our concern is with field theory which is immediately left behind in the circuit theory of electricity.

However, the complete omission of any mention of circuit electricity would be odd. And there is scientific value in emphasising how the methodology of circuit analysis arises from the field description. To this end we start with how Kirchhoff's laws arise from Maxwell's equations and from there the basics of simple circuit analysis theory via Thévenin's Theorem and illustrated by the LCR circuit.

Chapters 19 and 20 will continue the theme of reminding readers that, indispensable though circuit theory is in practice, to forget that the underlying reality is the electromagnetic field is to commit such follies as believing that energy flows in wires. That fallacy is encouraged by circuit theory but is false.

18.1 What is Circuit Theory?

In circuit theory there is no mention of the electromagnetic fields, whether that be $\bar{E}, \bar{B}, \bar{D}$ or $\bar{H}$. Instead an electric circuit is taken to consist of lumped components connected by conducting wires. What may be happening outside the wires or components is not involved in the analysis of the circuit. The key variables are the current, I, and the potential, V, at points in the circuit. These quantities are addressed only at a few discrete points for each component, typically 2, 3 or 4 points (or a great many more for integrated circuits), known as terminals.

Each component is treated as a "black box", for which no discussion of internal states is necessary, instead the required response of the component is taken to be adequately described by a few simple characteristic properties. We have already met the most important of these lumped-component characteristics: resistance (R), capacitance (C) and inductance (L), see for example §5.4.5, §5.6.4 and §9.1.8. Semiconductor components, or the older valve technology, require other characteristics to define their behaviour, as we will see.

18.2 Kirchhoff's Laws

Consider some electrical circuit, consisting of linear 'wires' connecting lumped circuit components. It can be as complicated as you like, and can contain active as well as passive components. (An active component provides energy to a circuit, hence batteries, generators or photovoltaic elements, etc. Passive components can only transmit or dissipate energy).

Now consider some closed surface which cuts through a certain number of wires. If a wire is conveying current out of the closed surface that is denoted a positive current. If a wire is conveying current into the closed surface that is denoted a negative current. Kirchhoff's First Law states that,

$$\sum_i I_i = 0 \tag{18.2.1}$$

In particular, (18.2.1) will apply at any junction of wires. Now consider any closed path in the circuit and let ΔV_j be a contiguous sequence of potential differences around it, ending at the starting point. Then Kirchhoff's Second Law is,

$$\sum_j \Delta V_j = 0 \tag{18.2.2}$$

These are so familiar to us that we are likely to take them for granted. But the underlying reality of all electrical phenomena is the electromagnetic field, which we claim is fully specified by Maxwell's equations. In that case, how do (18.2.1,2) arise from Maxwell's equations? The use of closed surfaces and closed paths in the above description rather gives it away.

Consider firstly steady conditions in which all electric and magnetic fields are unvarying. Then we have, from (7.2.1), $\overline{\nabla} \times \bar{H} = \bar{J}$ and $\overline{\nabla} \times \bar{E} = 0$ and the former gives $\overline{\nabla} \cdot \bar{J} = 0$. Hence by Gauss's and Stokes's theorems,

$$\oiint \bar{J} \cdot d\bar{S} = 0 \quad \text{and} \quad \oint \bar{E} \cdot d\bar{s} = 0 \tag{18.2.3}$$

where these integrals are over any closed surface and any closed loop respectively. But these integrals are just the same thing as (18.2.1,2) but expressed in continuum form rather than discrete form (recalling that the electric field is (minus) the potential gradient).

This would complete the proof were it not for the fact that we will deploy Kirchhoff's Laws also in the dynamic case when potentials, currents and fields are time varying, such as AC circuits, discharging capacitors, etc. This initially appears suspect because the Maxwell equations are now $\overline{\nabla} \times \bar{H} = \bar{J} + \frac{\partial \bar{D}}{\partial t}$ and $\overline{\nabla} \times \bar{E} = -\frac{\partial \bar{B}}{\partial t}$. Using also $\overline{\nabla} \cdot \bar{D} = \rho$ the former gives $\overline{\nabla} \cdot \bar{J} = -\frac{\partial \rho}{\partial t}$. Hence we find, instead of (18.2.3),

$$\oiint \bar{J} \cdot d\bar{S} = -\frac{\partial Q}{\partial t} \qquad \text{and} \qquad \oint \bar{E} \cdot d\bar{s} = -\frac{\partial \varphi}{\partial t} \tag{18.2.4}$$

where Q is the charge within the closed surface of the first integral, and φ is the magnetic flux linked within the closed path of the second integral. However, in practice we can continue to use Kirchhoff's Laws in the form (18.2.1,2) providing certain idealised conditions are realised sufficiently closely in practice. This involves assuming that the wires of the circuit have negligible capacitance and inductance, although they may have resistance. Furthermore it is assumed that all magnetic and electric fields are confined to within the lumped components, the fields being negligible outside them. Under these circumstances Q and φ will be negligible if we confine attention to the wires, so that Kirchhoff's Laws in the form of (18.2.1,2) can be retained as long as the current and potential in question relate only to points along the wires, not within the components.

The trouble with this argument is that, as we will see in the next two chapters, it is a fiction to think that the fields outside a current-carrying wire can be ignored. However, more careful analysis shows that Kirchhoff's Laws do continue to remain adequate in practice provided that the characteristic time for fields to propagate across the circuit is very short compared to the characteristic time over which currents and potentials vary. In other words, if τ is the timescale of significant change in any currents or potentials, such as the periodic time for AC, then Kirchhoff's Laws will be an adequate basis for analysing the circuit as long as $\tau \gg x/c$ where x is the greatest linear dimension of the circuit. We will see below that this requirement is challenged by very long transmission lines, which is why transmission lines stand at the boundary between circuit theory and field theory. Transmission lines can therefore be analysed using circuit theory (§18.5) or by treating them as waveguides (chapter 23).

18.3 LCR Series Circuit

Consider the circuit of Figure 18.1, which consists of an inductor (L), a capacitor (C) and a resistor (R) in series.

Figure 18.1: The LCR Series Circuit

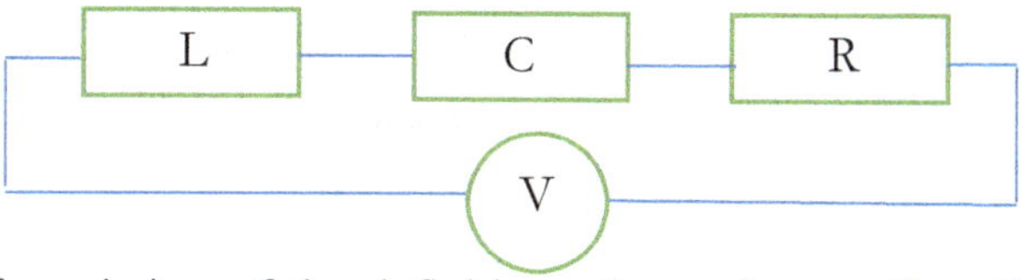

Differentiation of the definition of capacitance, $Q = CV_c$, (5.4.5a), gives us the current through a capacitor, $I = C\frac{dV_c}{dt}$. Similarly, (5.6.15) gives us the potential across an inductor and resistor in series, $V_{RL} = RI + L\frac{dI}{dt}$. Differentiating the latter and combining with the former gives the rate-of-change of potential across an inductor, a resistor and a capacitor in series,

$$L\frac{d^2I}{dt^2} + R\frac{dI}{dt} + \frac{1}{C}I = \frac{dV}{dt} \tag{18.3.1}$$

18.3.1 No Applied EMF

If the voltage source in Figure 18.1 is short-circuited then the RHS of (18.3.1) is replaced by zero. The general solution of the resulting homogeneous differential equation is obtained by substitution of,

$$I = I_0 e^{\Omega t} \tag{18.3.2a}$$

giving,
$$L\Omega^2 + R\Omega + \frac{1}{C} = 0 \tag{18.3.2b}$$

Hence,
$$\Omega = -\frac{R}{2L} \pm \sqrt{\left(\frac{R}{2L}\right)^2 - \frac{1}{LC}} \tag{18.3.2c}$$

There are two cases to consider. If $\left(\frac{R}{2L}\right)^2 \geq \frac{1}{LC}$ then the solutions for Ω are both real and negative, so any initial current or potential will decay exponentially, the longer time constant (smaller $|\Omega|$) being given by the positive sign in (18.3.2c). This corresponds to the charge on the capacitor and the magnetic field in the inductor both decaying away. In the special case that there is no capacitor, omitting the $1/LC$ term in (18.3.2c) reproduces the time constant for an inductor discharging through a resistor found in (5.6.19), namely $\tau = -\frac{1}{\Omega} = \frac{L}{R}$. Similarly, the case of a capacitor discharging through a resistor can be found from (18.3.2c) in the limit that $L \to 0$ (by expanding the root using the binomial theorem), i.e., $\tau = -\frac{1}{\Omega} = RC$, in agreement with (5.4.6f).

The second possibility is $\left(\frac{R}{2L}\right)^2 < \frac{1}{LC}$ in which case we can write the roots of Ω as,

$$\Omega = -\frac{1}{\tau} \pm i\omega_n \text{ where } \tau = \frac{2L}{R} \text{ and } \omega_n = \sqrt{\frac{1}{LC} - \left(\frac{R}{2L}\right)^2} \tag{18.3.3}$$

The solution now has an oscillatory part with natural angular frequency ω_n, but is still exponentially decaying, with a time constant of $\frac{2L}{R} > \frac{RC}{2}$. If the resistance is sufficiently small that $\left(\frac{R}{2L}\right)^2 \ll \frac{1}{LC}$ then the periodic time of the oscillation is $\approx \sqrt{LC}$ whilst the decay time constant is $\gg \sqrt{LC}$, so there would be many oscillations before the current dies away.

18.3.2 Applied Sinusoidal Voltage

§18.3.1 consider an undriven circuit discharging naturally. We now consider a driven circuit in which the voltage source in Figure 18.1 is $V = V_0 e^{i\omega t}$. The most general solution of the resulting inhomogeneous differential equation, (18.3.1), consists of a particular integral plus the general solution of the closed-circuit homogeneous equation. But the latter is always exponentially decaying if $R \neq 0$, as we have seen in §18.3.1. And in practice the resistance is always non-zero because inductors never have zero resistance in reality. Consequently, as long as we are interested only in times long compared with the decay time from §18.3.1 we can dispense with the part of the solution corresponding to the homogeneous equation. To find a particular integral, substitute both $V = V_0 e^{i\omega t}$ and $I = I_0 e^{i\omega t}$ with assumed real frequency, ω, into (18.3.1) which gives,

$$\left(-L\omega^2 + iR\omega + \frac{1}{C}\right) I_0 = i\omega V_0 \tag{18.3.4}$$

But from (18.3.3) we can write

$$\left(-L\omega^2 + iR\omega + \frac{1}{C}\right) = L\left(i\omega + \frac{1}{\tau} - i\omega_n\right)\left(i\omega + \frac{1}{\tau} + i\omega_n\right) \tag{18.3.5}$$

The frequency-dependent impedance, $Z(\omega)$, is defined as the complex-valued equivalent of resistance which incorporates the effects of the capacitance and inductance. Hence,

$$V_0 = Z(\omega) I_0 \tag{18.3.6}$$

Substitution of (18.3.6) into (18.3.5) and using (18.3.3) gives,

$$Z(\omega) = R + i\left(\omega L - \frac{1}{\omega C}\right) \tag{18.3.7}$$

The imaginary part of the impedance, $\left(\omega L - \frac{1}{\omega C}\right)$, is called the reactance. Note that the minus sign in $\left(\omega L - \frac{1}{\omega C}\right)$ indicates that the inductive and capacitive parts of the reactance are in antiphase and will tend to partly cancel each other. This will be seen in chapter 21 to be of great importance in electricity generation and transmission.

For a given applied voltage amplitude the resulting amplitude of current is greatest when $|Z(\omega)|$ is a minimum. But from (18.3.7),

$$|Z(\omega)| = \sqrt{R^2 + \left(\omega L - \frac{1}{\omega C}\right)^2} \tag{18.3.8}$$

The minimum of $|Z(\omega)|$ clearly occurs when the reactance is zero, i.e., when $\omega L - \frac{1}{\omega C} = 0$, i.e., at an angular frequency of,

$$\omega = \omega_R = \frac{1}{\sqrt{LC}} \tag{18.3.9}$$

or equivalently with a time constant, $\tau = \frac{1}{\omega} = \sqrt{LC}$. We see that this is the geometrical mean of the decay time constants for capacitor discharge, $\tau_C = \frac{1}{\omega_C} = RC$, (5.4.6d), and inductor discharge, $\tau_I = \frac{1}{\omega_I} = \frac{L}{R}$, (5.6.19). Thus the peak current occurs at the frequency which makes the reactance zero, at which condition the current is in phase with the voltage. Normalising (18.3.8) by its peak value we get,

$$\frac{|Z(\omega)|}{|Z_{peak}|} = \sqrt{1 + \left(\frac{\omega}{\omega_I} - \frac{\omega_C}{\omega}\right)^2} \tag{18.3.10}$$

For the case $\omega_I = \omega_C$ this is plotted in Figure 18.2. The peak amplitude is clear from the graph. This behaviour is generally called a resonance. However it should be noted that the frequency of this resonance in the driven circuit, (18.3.9), does not equal the natural frequency of the shorted circuit, which is given by (18.3.3) to be $\omega_n = \sqrt{\frac{1}{LC} - \left(\frac{R}{2L}\right)^2}$.

In fact, the resonance at frequency (18.3.9) in the driven circuit still applies even if the shorted circuit has no natural frequency, i.e., if $\left(\frac{R}{2L}\right)^2 \geq \frac{1}{LC}$ so that the shorted circuit displays only exponential decay without oscillation. In this case, as is readily shown, the impedance is still given by (18.3.7).

Figure 18.2: LCR Series Circuit Resonance

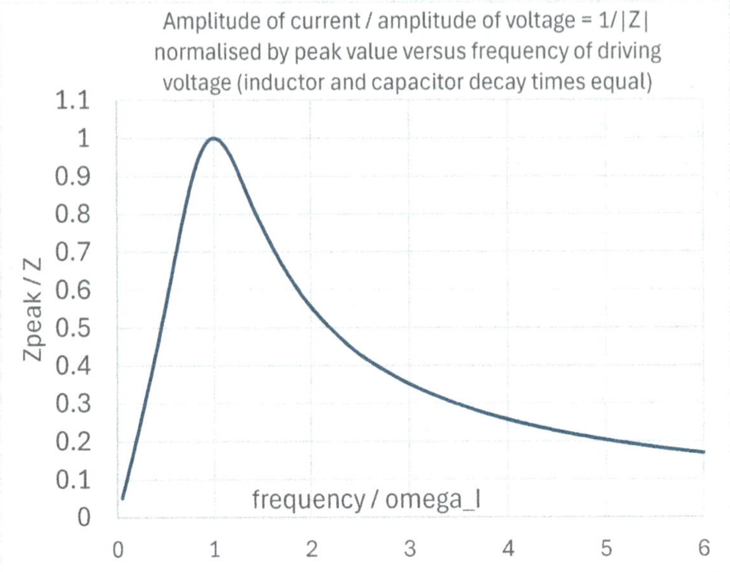

18.4 Thévenin's Theorem

Thévenin's Theorem states that any single port electrical circuit containing any configuration of impedances, voltage sources and current sources can be replaced at its port with a voltage source in series with an impedance. Here a "port" means a pair of terminals giving access to the circuit. And "replaced" means the behaviour of the single voltage-source-plus-impedance would be indistinguishable from the behaviour of the original circuit.

The proof is straightforward. Let the open circuit voltage at the port be V_0. Now imagine shorting out all voltage sources and replacing any current sources with open circuits. The impedance at the port is then measured to be Z_0. We want to find the current, I, that would flow through an arbitrary impedance, Z, placed across the port and show that it is the same as the current that would flow if the circuit were replaced with an impedance of Z_0 and the resulting closed circuit subject to an emf of V_0. The latter current is clearly $I_{equiv} = V_0/(Z + Z_0)$. We need to show that $I = I_{equiv}$.

This is established by initially imagining inserting a voltage source in parallel with Z with magnitude V_0 but of the opposite sense to the circuit's open circuit voltage. As the two cancel, no current flows through Z. We then imagine inserting another voltage source with magnitude V_0 but with the same sense as the circuit's open circuit voltage. This is merely to apply V_0 to a pair of impedances Z and Z_0 in parallel and so produces a current $V_0/(Z+Z_0)$. The final step simply acknowledges that the two notional voltage sources cancel out and might as well not be there, so this must also equal I, the current that would flow through Z if connected across the port of the original circuit. Hence the circuit can be replaced by Z_0 in parallel with a voltage source V_0. QED.

The importance of Thévenin's Theorem in designing and understanding circuits can hardly be overstated. Connecting two complex circuits together at a single port, the combined response can be understood in terms of the equivalent impedance and voltage source of each.

18.5 Transmission lines

Referring to the conditions under which Kirchhoff's laws remain valid, as discussed in §18.2, they are no longer valid in the case of transmission lines. The inductance and capacitance of these long lines cannot be ignored. Moreover, the lines are themselves the "component" in question, at least for over tens or hundreds of miles of length. We must now talk of L being the inductance per unit length of line, and C the capacitance per unit length of line. Consider what happens when we do this.

For simplicity we assume a line with negligible resistance losses. The change in potential along length δx of the line is then $\delta V=-L\delta x\frac{\partial I}{\partial t}$ whereas the change in current is $\delta I=-C\delta x\frac{\partial V}{\partial t}$. Note that we must be dealing now with partial derivatives because we have introduced variations in space (x) as well as in time (t). Thus we have,

$$\frac{\partial V}{\partial x}=-L\frac{\partial I}{\partial t} \quad \text{and} \quad \frac{\partial I}{\partial x}=-C\frac{\partial V}{\partial t} \tag{18.5.1}$$

Substitution of one of those equations into the other and we get the wave equation,

$$(\partial_x^2-LC\partial_t^2)V=0 \quad \text{and} \quad (\partial_x^2-LC\partial_t^2)I=0 \tag{18.5.2}$$

The phase velocity of the wave is seen to be $v=1/\sqrt{LC}$. This phase velocity is numerically equal to the angular frequency at which unit length of the line has minimum impedance, (18.3.9). ($v=1/\sqrt{LC}$ is a velocity here, rather than an angular frequency because L and C are per unit length in this case but not in (18.3.9)).

As an illustration consider a pair of parallel cables acting as a transmission line. (Most high power transmission is actually 3-phase, as we shall see in more detail in chapter 21, so this is only a rough guide). (5.4.11) gives the capacitance per unit length to be $\frac{\pi\varepsilon_0}{log\left(\frac{D-a}{a}\right)}$ where D is the distance between the axes of the two cables and a is their radius. (5.7.4) gives the inductance per unit length to be $\frac{\mu_0}{\pi} log\left(\frac{D-a}{a}\right)$. Hence $LC = \varepsilon_0\mu_0$ and so from (10.1.8) the phase velocity is $v = c$, the speed of light in vacuum. (Do not confuse this velocity with the drift speed of the electrons in the cables, which is extremely sluggish, see Appendix A8).

Hence, at 50 Hz, the frequency of UK mains electricity, the wavelength of the V, I waves is 6,000 km. Over a hundred or two hundred kilometres we can therefore expect the circuit analysis based on V and I, rather than the underlying fields, to remain sufficient. Nevertheless, we are dealing with wave propagation whichever dependent variables we use, whether V and I or $\bar{E}$ and $\bar{B}$. Because of this it is found that the theory of transmission lines can be addressed either in terms of circuit theory, as here, or by treating the transmission lines as a waveguide (for which see chapter 23). Hence, transmission lines provide a halfway house between circuit theory and field theory.

The effective impedance of a transmission line follows from the above analysis if we assume a wave $V = V_0e^{i(\omega t-kx)}$ and $I = I_0e^{i(\omega t-kx)}$ so that the effective impedance is $Z = V/I$. Using (18.5.1) $-ikV = -Li\omega I$ so that

$$Z = L\frac{\omega}{k} = Lv = \sqrt{\frac{L}{C}} \tag{18.5.3}$$

As an example, using the inductance and capacitance per unit length from (5.4.11) and (5.7.4) quoted above for a pair of parallel cables with conductor diameter 30mm spaced 4m apart we get $Z = 670\Omega$. More realistically the impedance of 3-phase overhead transmission lines in the UK is about 400 Ω.

Note that because (18.5.3) involves the ratio of the inductance per unit length to the capacitance per unit length, it is not itself an impedance per unit length but is an impedance which is independent of length. This may initially seem rather baffling if one is used to impedance being associated with resistance, which would be proportional to length. The explanation is that (18.5.3) has been derived from assumed lossless propagation of the voltage and current, i.e., in the limit of negligible resistance, so the usual source of length dependence is absent. In reality there will also be a resistance per unit length, R, and the formula for impedance will include contributions from all three of L, C, R and become length dependent.

In addition, it has been assumed in deriving (18.5.3) that the waves of voltage and current propagate in the forward direction only. In reality there will be a wave propagating in the reverse direction also, originating from reflection at the load with which the line terminates (e.g., the primary windings of the transformer at the substation). However, this reflected wave will be absent, and so (18.5.3) valid, if the impedance of the load matches that of the transmission line, (18.5.3). This is proved next.

18.5.1 Transmission Line Wave Reflection

Here we show that there is no reflected wave from the load if the load has the same impedance as the transmission line, $Z_{load} = Z$.

The sum of an incident wave and a reflected wave is given by, (18.5.4)

$$V = V_1 e^{i(\omega t - kx)} + V_2 e^{i(\omega t + kx)} \text{ and } I = I_1 e^{i(\omega t - kx)} + I_2 e^{i(\omega t + kx)}$$

where the subscript 1 denotes the incident wave and subscript 2 the reflected wave, and recall that the amplitudes will be complex in general. By requiring that equations (18.5.1) hold at all points along the line we find that $V_1 = ZI_1$ and $V_2 = ZI_2$. Note that V_1 and V_2 both refer to potential differences across the pair of cables comprising the transmission line. If we place the load at $x = 0$ then the potential difference across the load is $V_{load} = V(0) = (V_1 + V_2)e^{i\omega t} = Z(I_1 + I_2)e^{i\omega t}$. But the current through the load is the incident current minus that reflected, $I = (I_1 - I_2)e^{i\omega t}$, so the potential across the load is $V_{load} = Z_{load} I = Z_{load}(I_1 - I_2)e^{i\omega t}$. But this must equal $Z(I_1 + I_2)e^{i\omega t}$ so $Z_{load}(I_1 - I_2) = Z(I_1 + I_2)$, which gives,

$$I_2 = \frac{Z_{load} - Z}{Z_{load} + Z} I_1 \quad (18.5.6)$$

> Hence the reflected wave vanishes ($I_2 = 0$) iff the impedance of the load matches that of the transmission line, $Z_{load} = Z$.

We will see in chapter 21 that this impedance matching is the desirable state of affairs, leading to the most efficient transmission of power.

18.6 Semiconductor Circuits

18.6.1 The Transistor

A presentation of semiconductor technology at any depth is beyond the scope of this book. However, the importance of this technology in the modern world cannot be over-emphasized. Prior to the early 1960s when transistors first began to be mass produced, electronic equipment – be it TVs, radios, audio devices or radio transmitters – all required valve technology in their crucial amplification circuitry. But thermionic valves require high voltages and mains power. Only with the advent of low voltage, low power transistors did such things as the transistor radio become available, indeed ubiquitous by the mid-1960s, and the size of radio receivers shrank from being virtually pieces of furniture to small enough for your pocket.

The early transistor components were just single transistors, with 3 terminal wires. But by the late 1960s or early 1970s, integrated circuits were also being mass produced. These semiconductor "chip" devices had large numbers of transistors, and other semiconductor components, all placed on the same small "chip" of silicon. Such integrated circuits have large numbers of terminal connectors. The most recent such chips (at 2026) have numbers of individual transistors of around several trillion. We will say no more about these crucial IT chip technologies which is a specialist area even within electronic engineering.

However, a few examples of basic transistor circuits will be given below. It is important to appreciate that there are many different types of transistor. We will confine examples below to the earliest mass-produced type. These are the Bipolar Junction Transistors (BJT). The substrate material is usually silicon but may be germanium or gallium arsenide. These BJTs come in two broad types: pnp and npn. This notation refers to how the pure silicon is "doped" by trace amounts of other elements, such as boron, gallium, phosphorus or arsenic. Boron and gallium are electron acceptors and doping with these elements produces a so-called p layer, i.e., the trace elements mop up some electrons and so leave positively charged "holes" in the silicon substrate. In contrast, phosphorus and arsenic are electron donors, providing free electrons into the silicon substrate and causing an n type layer.

In BJT types of transistor the three terminals are called the base, the emitter and the collector (see Figure 18.3).

The other broad class of transistors are the Field Effect Transistors (FETs). These also have two broad subclasses, the Junction-Field Effect Transistors (JFETs) and the Metal Oxide Semiconductor Field Effect Transistors (MOSFETs). In FETs the three terminals are called the gate, the source and the drain, corresponding to the BJT's base, emitter and collector respectively.

The simplest semiconductor device is a diode. This has a single p-n junction, i.e., the silicon substrate is doped to be p type on one side of an interface but doped to be n type on the other side. Consequently, electrons will move fairly freely from the n region into the p region, but are reluctant to move in the reverse direction. (Equivalently, current flows easily from p to n type, but not the reverse). By this means a diode is formed, which passes current when forward biased but prohibits current when reverse biased. Like all semiconductor devices this is a simplified picture of a diode's characteristics and applies only within a strictly limited range of voltages and/or currents. The true behaviour is actually highly non-linear. A sufficiently large reverse biased voltage will overcome the p-n barrier and large reverse currents can be made to flow beyond the "breakdown voltage". Even smaller reverse voltages will cause a small leakage current. And when forward biased, the current is only large if the voltage is sufficient. Nevertheless, the simplified picture is adequate to understand how circuits operate, providing the device is kept within its declared operating range.

A single transistor consists of a sandwich of either an n layer between two p layers (pnp) or the reverse (npn). The latter case is illustrated in Figure 18.3.

Figure 18.3: Common-Emitter Configuration of an NPN Type Transistor

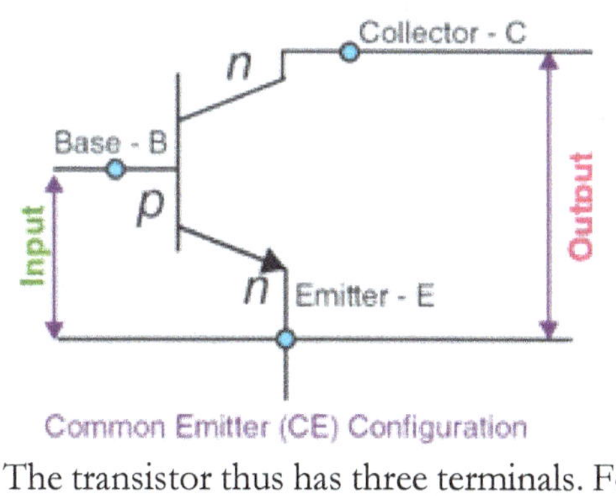

The transistor thus has three terminals. For npn type the "base" is connected to the p type region whilst the "collector" and the "emitter" are connected to the two different n type regions.

Any given transistor can be used in a number of different ways, for example according to which of the three terminals is the common reference for potential (called in obvious terminology common-base, common-collector or common-emitter configurations). Figure 18.3 shows an npn type transistor in common-emitter configuration in which the emitter is the common reference zero potential.

The characteristics of a given transistor can be expressed in terms of how the current in any terminal, or the potential difference between any pair of terminals, varies with two other of these variables. For this purpose it is useful to regard some variables as inputs and some as outputs, but which is which will depend upon the configuration of the circuit (whether it is common-base, common-collector or common-emitter configuration). For the common-emitter case Figures 18.4a-c show the characteristics in diagrammatic form.

Figure 18.4a is the input characteristic and shows the base current as it varies with base-emitter (BE) voltage for a fixed common-emitter (CE) voltage. A key qualitative feature is that there is no base current until the BE voltage exceeds some minimum value greater than zero.

Figure 18.4b is the output characteristic and shows the collector current as it varies with collector-emitter (CE) voltage for a fixed base current. The key qualitative feature here is that the collector current is zero unless there is some base current. For sufficiently small collector currents or CE voltage the relationship between the two is linear (the "ohmic region"). Beyond a certain collector current the collector current no longer increases with further increase in the CE voltage. The collector current is said to be "saturated". At CE voltages beyond this point, with the collector current saturated, the transistor is said to be "switched on". This is the active region in which a transistor is intended to be operated. But, most importantly…

Figure 18.4c shows that the saturation current in the collector varies linearly with the base current (for fixed CE voltage), at least within a certain range of currents. In particular, no collector current can flow unless there is a base current, and, moreover, the saturated collector current is far larger than the relatively small base current. The ratio of those two currents is the current gain and can typically be about 100. This is the central feature of a transistor: a small base current controls the larger collector current, and is proportional to it. This is the basis of a transistor as an amplifier.

The other main function of a transistor is to act as a switch. We have seen that the transistor is "switched on" when a base current is flowing, where by "switched on" is meant that a collector current can flow. If the CE voltage is also sufficient, the saturated collector current flows and the transistor is "fully switched on". In contrast, if there is no base current then there can be no collector current and the transistor is "switched off". Hence, these characteristics are the basis of the transistor as a switch. This in turn is the basis of multi-transistor circuits acting as digital storage of data, with the on/off status of each transistor representing one bit.

Figure 18.4a: Transistor Input Characteristics for the Common-Emitter Configuration

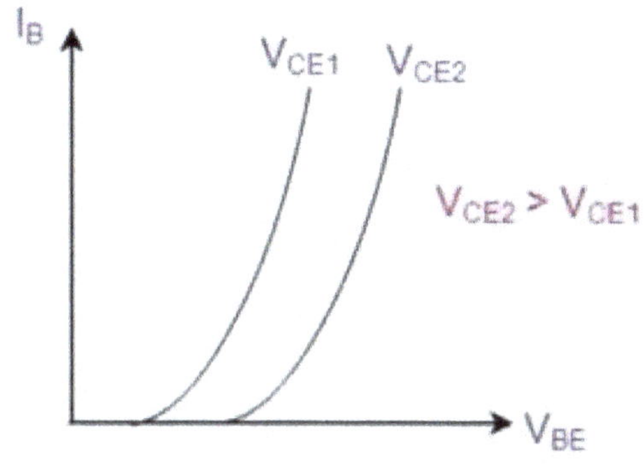

Figure 18.4b: Transistor Output Characteristics for the Common-Emitter Configuration

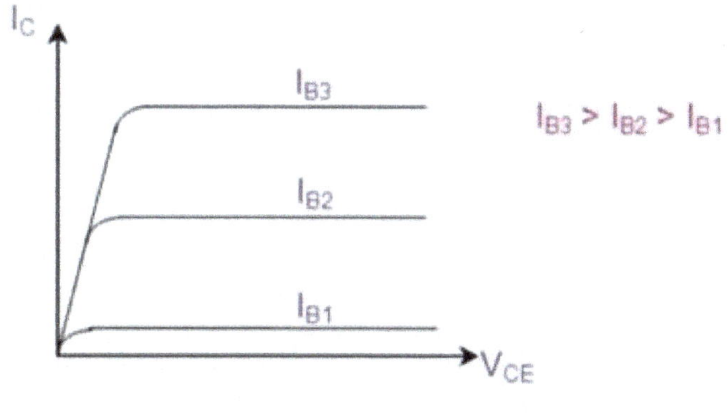

Figure 18.4c: Transistor Input Characteristics for the Common-Emitter Configuration

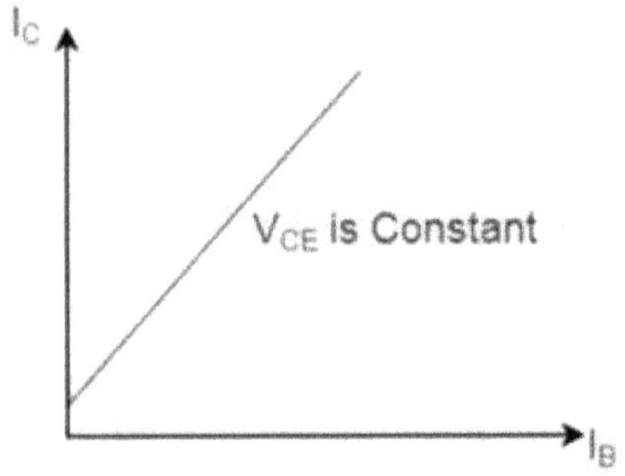

A few examples of simple transistor circuits follow.

18.6.2 Amplifier Circuit

The circuit is shown in Figure 18.5. This is a common-emitter amplifier using an npn transistor. It is a voltage amplifier (as opposed to a current amplifier).

Typical values for the resistors and capacitors might be,

R1 = 110 kΩ, R2 = 10 kΩ, R_C = 10 kΩ, R_E = 1 kΩ, Cin, Cout, $C_E \sim 0.1\ \mu F$.

The supplied voltage might be, say, 20 V. The emitter, base and collector voltages might then be around 1 V, 1.6V and 10 V respectively.

The input capacitor, Cin, blocks any DC part of the signal, allowing only the AC signal through to be amplified. Similarly, the output capacitor, Cout, blocks any spurious DC generated by the transistor from being output, allowing only the amplified AC signal through.

The resistors R1 and R2 are used to provide the base of the transistor with a suitable potential and to provide a small base current which is necessary for the transistor to operate.

The resistor R_E provides negative feedback and stabilizes the operating point of the transistor (the base voltage and current), improving linearity.

The resistor R_C with R_E define the voltage gain, which is $-\frac{R_C}{R_E}$ (and hence a voltage gain of $\times\ 10$ using the resistances suggested above).

Figure 18.5: A Simple, Low Power Common-Emitter Amplifier Circuit.

Credit [Aflafla1](#) June 2014, Original file [here](#)

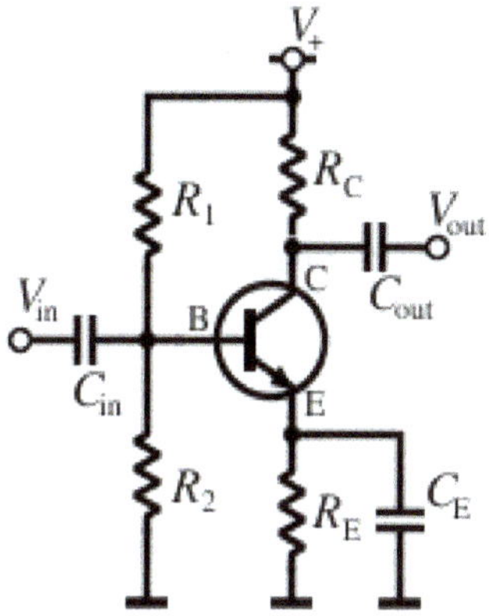

18.6.3 The Flip-Flop, or Latch, Circuit

A latch is a device which has two alternative stable states. (A flip-flop is a latch operating in the time domain whose state continually flips from one state to the other and back again). Figure 18.6 shows a simple latch circuit with two transistors. The key to this being a two-state device is that the circuit is designed so that if one transistor is on the other is off. The two states thus relate to which transistor is on.

To understand the key feature, namely that one and only one of the transistors will be on, assume initially that there is current flowing from A1 to E2. This is the base current of the transistor on the right. This transistor is therefore on. Hence, assuming the supply voltage, Vcc, is sufficient, there will be a comparatively large collector current flowing through the right-hand transistor. To put that another way, the effective collector-emitter resistance of the right-hand transistor is low when the transistor is "on", and so the resistor between A2 and E1 has been virtually shorted-out (there is virtually no voltage over it). Hence there is no base current flowing into the transistor on the left, and that transistor is off. This means that the left-hand transistor has no collector current either. This in turn means that any current reaching A1 must be flowing towards E2 – which is consistent with our starting assumption. Hence this state is stable.

But the circuit is symmetrical and we could equally have argued that the left-hand transistor was on if the right-hand transistor was off. Hence the circuit has two stable states and could be in either one. QED.

One only needs to introduce some means of "kicking" the circuit to induce a change from one state to the other. Here we have the basis of rewriteable (but volatile) digital memory.

Figure 18.6: A Simple Latch Circuit with Two Transistors

Credit Nolanjshettle, own work, CC BY-SA 3.0

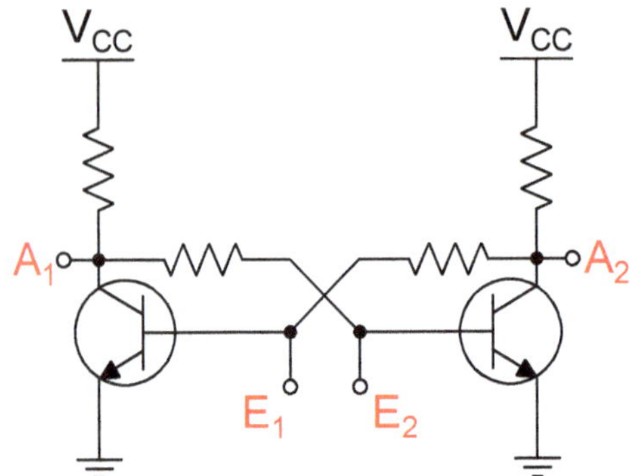

18.6.4 Example Logic Gates

In logic circuits we talk about "high" voltage representing logic 1, or "true", whilst "low" voltage represents logic 0, or "false". Because transistors work at quite low voltage, "high" voltage here means only a few volts, whilst "low" voltage means virtually zero volts.

A NOT gate consists of one input and one output and has the effect of reversing logic 1 to 0 and vice-versa. A very simple NOT gate is shown in Figure 18.7.

The operation should be clear now. If the input is "high" (logic 1) then the base is supplied with current and so the transistor is "on", meaning that the collector-emitter resistance is low and so the output voltage is "low" (logic 0). But if the input voltage is changed to "low" (logic 0) then the base has no current, the transistor is "off", the collector-emitter resistance is high and the output voltage is "high" (logic 1). The resistance R1 is chosen so that in the latter case the voltage at the output is high enough to reach the criterion for logic 1.

Figure 18.7: A NOT Gate

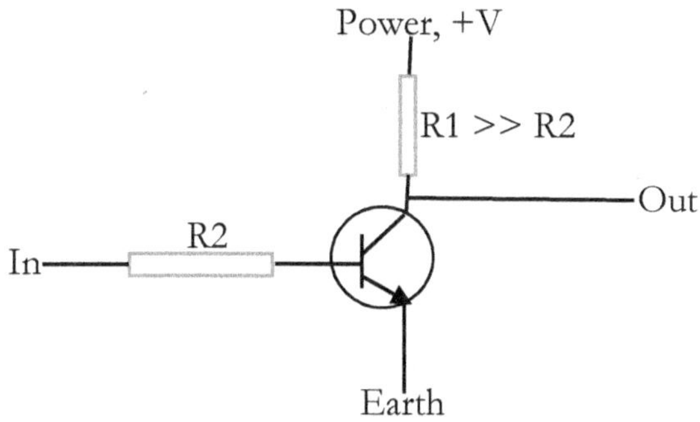

As an example of a two-input, single output gate, consider the AND gate. This is defined such the logic 1 is obtained at the output iff both inputs are logic 1. A simple implementation is shown in Figure 18.8.

Figure 18.8: An AND Gate

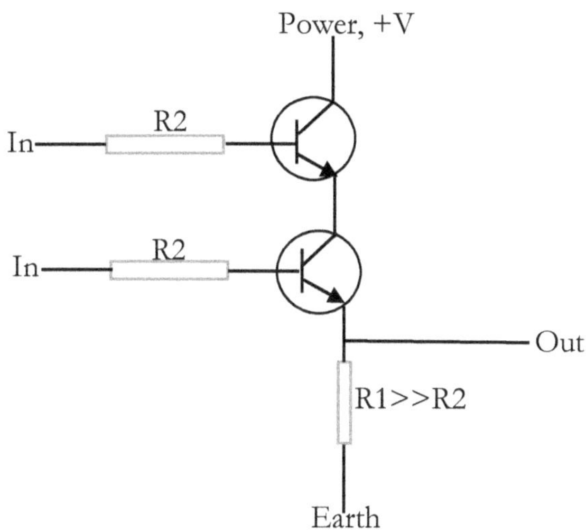

Again it should be clear how this works now. As the emitter of the first transistor is connected to the collector of the second transistor, current can flow from the power source to ground only if both transistors are "on", and hence only if the bases of both transistors are supplied with current, and hence only if both inputs are at logic 1. Only then does the output become "high" (logic 1). If either, or both, inputs are low then at least one of the transistors is "off" and thus has very high collector-emitter resistance, so no current flows through R1 and so the output is essentially zero volts (logic 0).

19. Energy Does Not Flow in Wires, (1): The Fields and Energy Flows due to a Steady Current in a Long Straight Wire or Cylinder

You might think there cannot be much to say about such a trivial case as a steady current flowing in a long straight wire. But the complete solution for the fields is more subtle than you probably imagine. This is the archetypal case which illustrates one of the most surprising conclusions to come out of a study of the complete solution for the fields, rather than using the sort of circuit theory that is familiar from school level accounts. Energy does not flow in wires. It flows in the fields surrounding the wire. Energy enters the wire itself only to be dissipated as heat.

Contents

In this chapter we solve Maxwell's equations for the case of a steady current flowing in a straight wire. Such a wire can deliver electrical power to some remotely attached device. Our purpose here is to clarify where that flowing power lies, and, specifically to show that it does not lie within the wire. Other surprising features of what seems so simple an arrangement also emerge, such as the surface charge on the wire which is inevitable and which plays a crucial role in enabling the power flow.

19.1 The Magnetic Field

For generality we consider a hollow conducting cylinder with inner and outer radii R_i and R_o, see Figure 19.1. We regain the case of a wire or solid cylinder when $R_i = 0$. We wish to study the case of a steady current which has uniform current density, $\bar{J}$, within the material of the cylinder (henceforth "the wire"). Given that, for isotropic materials, $\bar{J} = \kappa\bar{E}$, where κ is the conductivity, it follows that the electric field within the wire is also uniform and aligned with the wire axis (taken here to be the z axis).

A current passing down a wire creates a magnetic field. We have seen in §11.8 that, for frequencies up to, and including, optical frequencies ($<10^{15}$ Hz), and for metals which have $\varepsilon_r \approx 1$, the displacement current may be neglected in comparison with the conduction current when an electric field is present. Hence the third Maxwell equation in (7.2.1), becomes $\bar{\nabla} \times \bar{H} = \bar{J} = \kappa\bar{E}$.

For a long straight wire the magnetic field cannot vary in the z or θ directions by symmetry. Consequently, the Maxwell equation $\bar{\nabla} \cdot \bar{B} = 0$ in cylindrical polars (Appendix A2) reduces to the requirement that rB_r must be a constant. If the wire is solid then, considering the centre ($r = 0$) this constant must be zero. Hence $B_r = 0$ everywhere. For a hollow cylinder this argument shows that $B_r = 0$ inside the hollow ($r < R_i$) and as B_r is continuous at the boundary with the metal it follows that $B_r = 0$ within the metal. Continuity at the outer boundary then implies $B_r = H_r = 0$ everywhere.

Inside the wire $\bar{\nabla} \times \bar{H} = \bar{J} = \gamma\bar{E}$ is in the z direction. Expressing $\bar{\nabla} \times \bar{H}$ in cylindrical polars (Appendix A2) the θ component is zero iff H_z is uniform (i.e., independent of r) within the wire. Within the hollow ($r < R_i$) and outside the wire ($r > R_o$) there is no current (and there is no displacement current because the situation is static) hence $\bar{\nabla} \times \bar{H} = 0$ and so again H_z is uniform. But the field must go to zero at infinity and hence everywhere outside the wire $H_z = 0$.

Figure 19.1: Geometry for §19.1 and §19.2

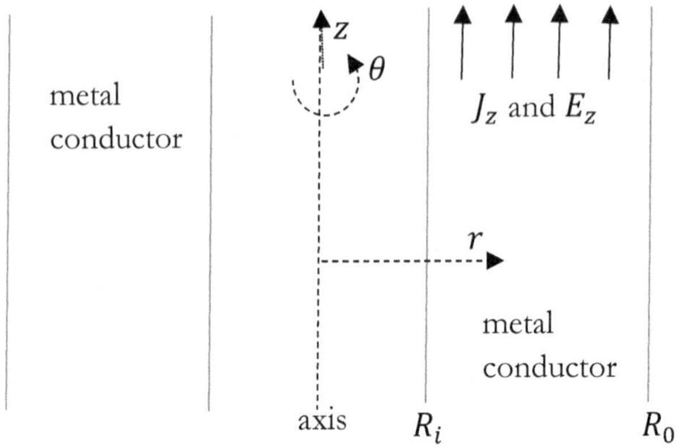

We will see that we cannot rule out the possibility of surface currents, because of the presence of surface charges, but any such currents can have no θ component. This is sufficient to ensure that, as H_z must be continuous, we can conclude that $H_z = 0$ everywhere (because H_z would be discontinuous by the surface current density in the θ direction, see §6.7).

We are left with H_θ only, and this is a function of r only. Using Stokes's theorem on a circular path across the wire cross-section and concentric with it, $\overline{\nabla} \times \overline{H} = \overline{J}$ within the wire and $\overline{\nabla} \times \overline{H} = 0$ elsewhere then gives,

For $r < R_i$ $\qquad H_\theta(r) = 0 \qquad$ (19.1.1a)

For $R_i \leq r \leq R_0$ $\qquad H_\theta(r) = \frac{r^2 - R_i^2}{2r} J \qquad$ (19.1.1b)

For $r > R_0$ $\qquad H_\theta(r) = \frac{R_0^2 - R_i^2}{2r} J = \frac{I}{2\pi r} \qquad$ (19.1.1c)

where I is the total current. The last expression thus agrees with (4.2.21) whilst (19.1.1b) shows that the magnetic field varies across the thickness of the wire or cylinder, being zero at the inner surface (or centre, if solid).

19.2 The Electric Field

We have an applied electric field inside the wire which is in the z direction, and there is no symmetry requirement for the radial component of the electric field to be zero, though the circumferential component must be zero. Hence we need to address two components, E_z and E_r.

The charge density is zero everywhere so the two remaining Maxwell equations are $\overline{\nabla} \cdot \overline{D} = 0$ and $\overline{\nabla} \times \overline{E} = 0$ in the static case. Because E_z cannot vary in the z direction, $\overline{\nabla} \cdot \overline{D} = 0$ reduces to $\partial_r(rD_r) = 0$ hence $D_r = \frac{A}{r}$ for some constant A. At $r = 0$ we can only avoid a divergent field if $A = 0$. Hence $E_r = 0$ throughout the hollow (or just at the centre if solid).

So why does this not imply that $E_r = 0$ everywhere? Because surface charges are necessarily induced. This is the feature of the solution, and the physical reality, that is omitted from standard current electricity theory.

Now $\varepsilon_r \varepsilon_0 E_r$ is discontinuous by the surface charge density, which we denote σ_i on the inner surface and σ_o on the outer surface. Hence we get,

$0 < r < R_i$: $$E_r = 0 \qquad (19.2.1a)$$

$R_i < r < R_o$: $$E_r = \frac{A}{r} = \frac{R_i}{r} \frac{\sigma_i}{\varepsilon_r \varepsilon_0} \qquad (19.2.1b)$$

$R_o < r$: $$E_r = \frac{A'}{r} = \frac{R_o \sigma_o + R_i \sigma_i}{\varepsilon_r \varepsilon_0 r} \qquad (19.2.1c)$$

But, for a steady state solution the radial field E_r within the metal must be zero because otherwise it would drive a constant radial current – which has nowhere to go other than to build up surface charge without limit. Hence we conclude that, other than in a very short transient period,

$$E_r = 0 \text{ and } \sigma_i = 0 \qquad (19.2.1d)$$

The same objection does not apply outside the metal, $r > R_0$, but non-zero outer surface charge is required to drive non-zero E_r there.

It remains to satisfy $\overline{\nabla} \times \overline{E} = 0$ and hence to find E_z outside the wire and shed light on the surface charge densities that determine E_r. The radial and axial components of $\overline{\nabla} \times \overline{E}$ are zero by the symmetry requirements, so $\overline{\nabla} \times \overline{E} = 0$ reduces to $\partial_r E_z - \partial_z E_r = 0$. If $\partial_z E_r = 0$ then $\partial_r E_z = 0$ and we would conclude that E_z was the same at all radii, which contradicts the fields reducing to zero an infinity. Hence we must concede that E_r is indeed non-zero and that E_r varies in the z direction, from which (19.2.1c) implies that the surface charge density, σ_o, varies in the z direction.

For $R_i < r < R_o$ we therefore have $\partial_r E_z = 0$ hence $E_z = E_0$ the applied field inside the wire.

For $r > R_o$ we have $\partial_r E_z = \frac{\partial_z A'}{r}$ hence $E_z = (\partial_z A') log(r)$ + a constant. The axial component of the E field must be continuous over the material boundary and hence we conclude that the solution for $r > R_o$ must be,

$$E_z = E_0 + (\partial_z A') log\left(\frac{r}{R_o}\right) \tag{19.2.2}$$

The reader will spot the problem. This solution implies that the field blows up at infinity. In practice, of course, the return path for the current – the other half of the complete circuit – will be encountered at some finite distance and hence invalidate our assumption that the wire is isolated. This return path can be included without breaking the cylindrical symmetry which is crucial to the above analysis by taking the wire to be enclosed within a concentric cylinder of radius $R_c > R_o$ which provides the return path for the current. If R_c is sufficiently large, or the cylinder sufficiently thick, or the cylinder material has sufficiently large conductivity, then we can take the electric field within in it, necessary to drive the return current, as negligible. Hence continuity now requires a boundary condition $E_z = 0$ at $r = R_c$ and our final, finite, solution is,

For $R_o < r < R_c$:

$$E_z = E_0 \left[1 - \frac{log\left(\frac{r}{R_o}\right)}{log\left(\frac{R_c}{R_o}\right)}\right] \tag{19.2.3}$$

This means that,

$$\partial_z A' = -\frac{E_0}{log\left(\frac{R_c}{R_o}\right)} \tag{19.2.3b}$$

so the complete solution for the radial field outside the wire is,

$R_o < r$:

$$E_r = E_{r0} - \frac{E_0}{log\left(\frac{R_c}{R_o}\right)} \cdot \frac{z}{r} \tag{19.2.4}$$

where E_{r0} is an integration constant. The clue to the magnitude of this constant is provided by considering the axial energy flow. If we imagine the wire to terminate at some large positive $z = z_0$ at which point all the energy flow has been dissipated in resistance heating then the energy flow must drop to zero at that point, so we can write (19.2.4) as,

$R_o < r, z \leq z_0$:

$$E_r = \frac{E_0}{log\left(\frac{R_c}{R_o}\right)} \cdot \frac{(z_0 - z)}{r} \tag{19.2.4b}$$

Equivalently,

$$A' = \frac{(z_0 - z)}{log\left(\frac{R_c}{R_o}\right)} E_0 \tag{19.2.4c}$$

Note that this means $E_r > 0$ so the axial component of the Poynting vector (i.e., the energy flow) is upwards in Figure 19.1, i.e., in the same direction as the current as one would expect.

Recall also from (19.2.1c) that the surface charge density on the outside of the wire is proportional to A', and hence that the surface charge density also varies linearly along the length of the wire. The radial field, E_r, and the surface charge density, σ_o, are largest at the smallest z coordinate, i.e., at the point where the potential driving the axial current is applied (the energy source, which must also be the source of the surface charge).

Outside the outer cylinder, $r > R_c + t$, where t is the thickness of the outer cylinder, the electric and magnetic fields are zero. The radial electric field falls discontinuously to zero within the outer cylinder material, due to surface charges. Through the thickness of the outer cylinder the magnetic B field falls from $\frac{I}{2\pi R_c}$ to zero as the net current contained falls from I to zero. The solution is therefore finite everywhere.

19.3 The Energy Flows

The circumferential component of the Poynting vector, $\bar{N} = \bar{E} \times \bar{H}$, is zero because $H_r = H_z = 0$. No energy flows around the wire circumferentially.

The radial component of the Poynting vector, $\bar{N} = \bar{E} \times \bar{H}$, is $-E_z H_\theta$. As the z component of the electric field is positive (upwards) this points radially inwards, into the wire. Just outside the outer surface of the wire the radial component of the energy flux is thus, using (19.1.1c),

$$N_r = -\frac{E_0 I}{2\pi r} \tag{19.3.1}$$

The power entering length L of the wire is thus $\frac{E_0 I}{2\pi r} 2\pi r L = E_0 I L = VI$ where V is the potential drop over length L. Hence, we see that the energy entering the wire from the external fields per second per unit length equals the energy dissipated as Joule heating due to the wire's resistance over the same time and length. Hence, ***all*** the energy entering the wire is dissipated.

The axial component of the Poynting vector is $E_r H_\theta$. The radial component of the electric field is outwards (positive) so that the axial energy flux is upwards, i.e., in the same sense as the current and the applied electric field inside the wire, consistent with (19.2.4b). From (19.1.1c) and (19.2.1c) the axial energy flux outside the wire is thus,

$$N_z = \frac{I}{2\pi r} \cdot \frac{A'}{r} \tag{19.3.2}$$

The total power flowing axially in the annular space between the wire and the outer cylinder is thus,

$$P_z = \int_{R_0}^{R_C} \frac{I}{2\pi r} \cdot \frac{A'}{r} \cdot 2\pi r dr = IA' log\left(\frac{R_C}{R_0}\right) \tag{19.3.3}$$

Hence, using (19.2.4c), the axial power (flow of energy) is,

$$P_z = IE_0(z_0 - z) \tag{19.3.4}$$

The rate at which this axial power flow changes along a length L of the wire is thus,

$$\frac{\partial P_z}{\partial z} = -E_0 I = -\frac{VI}{L} \tag{19.3.5}$$

Hence, the axial flow of power diminishes by VI over length L, where V is the potential drop over that length. This exactly balances with the flow of power radially inwards and which equals the resistive Joule heating. Hence, ***all*** the energy flows parallel to the wire but outside the wire (i.e., through the air) but, per unit length, a proportion turns to flow radially into the wire where it is dissipated as Joule heating by the wire's electrical resistance.

In practice, the very long length of wire which leads to all the electrical energy being dissipated in resistance heating would be replaced by some load doing useful work (or useful heating). What the analysis shows is that, until that load is encountered, ***all*** the energy that reaches it has flowed through the air outside the wire.

19.4 Numerical Example

Consider a copper wire of radius $R_0 = 1$mm and a return path of radius $R_C = 10$mm. Suppose it carries a current of 1A, hence a current density in the wire of 3.18×10^5 A/m². The conductivity of copper is 58 MS/m so the applied axial field inside the copper is $E_0 = 0.0055$V/m. So, for a wire of length 1m, the driving voltage is just 0.0055V.

From (19.2.1c) and (19.2.4c) the surface charge density is $\sigma_o = \frac{(z_0 - z)\varepsilon_0 E_0}{R_0 log(R_C/R_0)}$ which, assuming a length of wire $z_0 = 1$m, and evaluating the charge density at $z = 0$ gives $\sigma_o = 2.11 \times 10^{-11}$C/m² and the total surface charge on the wire is therefore 6.6×10^{-14}C.

The radial field, E_r, just outside the surface of the wire, at $z = 0$, is thus 2.38V/m, or $434E_0$. It is worth noting that the radial electric field is so much larger than the applied axial field which drives the current and so is the ultimate cause of the radial field.

The magnetic field just outside the surface of the wire is $H_\theta = \frac{I}{2\pi R_0} = 159\text{A/m}$. Hence the Poynting vector, i.e., the axial energy flux, immediately outside the surface of the wire is 379W/m². This reduces with radius as $(R_o/r)^2$ and so the total power flowing through the air in the annular region between R_0 and R_c at the power source ($z = 0$) is,

$$\int_{R_0}^{R_C} 379 \left(\frac{R_0}{r}\right)^2 2\pi r dr = 0.0055$$

which is exactly VA, the energy supplied by the power source to drive the current, as it should be.

The radial electric field is not merely a small refinement but is crucial, being far larger than the applied axial field. It is the radial field which is responsible for driving the whole of the electrical power, not the axial field which only drives the current.

19.5 What We Have Learnt

From this archetypal example we have learnt five unexpected lessons,

(1) Energy does not flow in wires but in the space around the outside of the wire, and parallel to it.

(2) The component of the energy flow (the Poynting vector) which is perpendicular to the wire is radially inwards, into the surface of the wire from outside, and equals the energy dissipated by resistance (Joule heating). Thus the energy flows parallel to, but outside, the wire until part of it turns radially inwards which is that dissipated as heat.

(3) The axial component of the electric field on the outside of the wire, which must be non-zero by continuity with the inside, is sustained by the necessary presence of a surface charge density on the outer surface which varies axially. This axial component of the electric field outside the wire is essential to drive energy into the wire metal and hence to cause resistance dissipation.

(4) The radial component of the electric field outside the wire is also sustained by the surface charge on the wire and is essential in driving the axial flow of energy parallel to the current but outside the wire. Without this radial component of the electric field outside the wire, and hence without the surface charge density on the wire, there would be no energy conveyed by an electric circuit and the entirety of electrical technology would be void.

(5) To obtain the complete solution for the fields everywhere which is physical (finite) the whole electrical circuit must be analysed, including the return path for the current. The topologically closed nature of an electrical circuit is essential for the associated fields to be finite.

20. Energy Does Not Flow in Wires, (2): The Fields and Energy Flows due to a Steady Current in a Circular Loop

In the previous chapter the analysis of the fields due to a steady current in a long straight wire or cylinder led to several important learning points, exemplified by "energy does not flow in wires". Here we consolidate these learning points by considering another archetypal situation: a circular loop of thin wire bearing a steady current. Even for this simple geometry we will need to have recourse to approximations, but these suffice to illustrate the main features once again.

Contents

20.1 The Magnetic Field

The geometry of the circuit to be assumed here, a circular loop of radius a, is shown in Figure 20.1. The wire is assumed very thin for the purposes of deriving the magnetic field.

Figure 20.1: Geometry for Magnetic Field Solution

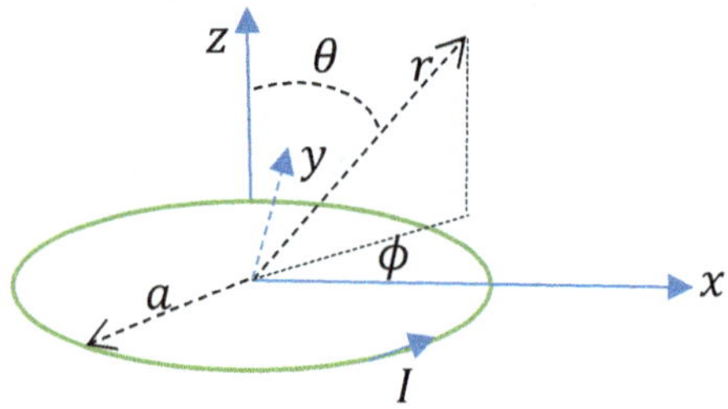

[20.1] gives the solution for the magnetic field in spherical polar coordinates, Cartesian coordinates and cylindrical polar coordinates. Here we opt for spherical polars, as shown in Figure 20.1,

$$x = r\sin\theta\cos\phi, y = r\sin\theta\sin\phi, z = r\cos\theta$$

In this coordinate system the B_ϕ component is zero everywhere and the other two components are independent of ϕ. [20.1] gives the solution in terms of parameters, α, β, k, defined as follows, (20.1.1)

$$\alpha^2 = a^2 + r^2 - 2ar\sin\theta, \beta^2 = a^2 + r^2 + 2ar\sin\theta, k^2 = 1 - \frac{\alpha^2}{\beta^2}$$

The parameter k^2 is the argument of the complete elliptic integrals in terms of which the solution can be found, giving,

$$B_r = \mu_0 I a^2 \frac{\cos\theta}{\pi\alpha^2\beta} E(k^2) \tag{20.1.2a}$$

$$B_\theta = \mu_0 I \frac{1}{2\pi\alpha^2\beta\sin\theta}\left[(r^2 + a^2\cos 2\theta)E(k^2) - \alpha^2 K(k^2)\right] \tag{20.1.2b}$$

Here $K(k^2), E(k^2)$ are the compete elliptic integrals of the first and second kind respectively, given by,

$$K(\xi) = \int_0^{\frac{\pi}{2}} (1 - \xi\sin^2\varphi)^{-\frac{1}{2}} d\varphi; \;\; E(\xi) = \int_0^{\frac{\pi}{2}} (1 - \xi\sin^2\varphi)^{\frac{1}{2}} d\varphi \tag{20.1.3}$$

where it is understood that the argument, ξ, is in the range $[0,1)$, noting that $K(\xi)$ blows up at $\xi = 1$. By construction of the parameter k^2 it does lie in this range except on the ring of current where $\alpha = 0$ and $k^2 = 1$. On the axis ($\theta = 0$) $B_\theta = 0$ and the solution becomes a vertical field (B_r) given by,

$$B_r = \frac{\mu_0 I a^2}{2(z^2+a^2)^{\frac{3}{2}}} \tag{20.1.4}$$

Sufficiently far from the loop, $r \gg a$, the solution (20.1.2a,b) becomes the field of a magnetic dipole, (5.5.5), with dipole moment $m = \pi a^2 I$, as it should.

20.1.1 Derivation of the Solution

The solution (20.1.2a,b) may be derived from the general expression (5.5.7) for the magnetic vector potential in terms of an arbitrary current density,

$$\bar{A}(\bar{r}_f) = \frac{\mu_0}{4\pi}\left(\int \frac{1}{r_{sf}} \bar{J}(\bar{r}_s)\, d\tau_s\right) \tag{5.5.7}$$

where $r_{sf} = \left|\bar{r}_f - \bar{r}_s\right|$ is the distance from source point to field point, and the integration is over the source. The current density in this case is a line source, namely on the ring. Due to symmetry we can consider a field point with $y = 0$ in which case we can write $\bar{r}_f = (r\sin\theta\,, 0, r\cos\theta)$ in terms of the usual spherical polar coordinates, and $\bar{r}_s = (a\cos\phi\,, a\sin\phi\,, 0)$ so that

$$\begin{aligned} r_{sf}^2 &= (r\sin\theta - a\cos\phi)^2 + (a\sin\phi)^2 + (r\cos\theta)^2 \\ &= r^2 + a^2 - 2ax\cos\phi \end{aligned} \tag{20.1.5}$$

Note that r, x are field point coordinates whereas ϕ is the source point coordinate. The vector current element on the ring, taken to be in the same sense as increasing ϕ, is $\bar{J}(\bar{r}_s)d\tau_s = I\hat{\phi}ad\phi = Ia(\hat{y}\cos\phi - \hat{x}\sin\phi)d\phi$. When integrated over the source (i.e., over ϕ) the $\hat{x}$ component of the vector potential will vanish to leave only the $\hat{y}$ component, which is actually the polar ϕ component because we have assumed the field point has $y = 0$. Hence we get,

$$A_\phi = \frac{\mu_0 I a}{4\pi}\int_0^{2\pi} \frac{\cos\phi d\phi}{\sqrt{r^2+a^2-2ax\cos\phi}} \tag{20.1.6}$$

This integral may be recast in terms of the complete elliptic integrals of the first and second kind defined by (20.1.3). Shifting the origin of the integral in (20.1.6) to $\phi = \pi$ we get,

$$A_\phi = -\frac{\mu_0 I a}{4\pi}\int_{-\pi}^{\pi} \frac{\cos\phi d\phi}{\sqrt{r^2+a^2+2ax\cos\phi}} \tag{20.1.8}$$

Substituting $\phi = 2\gamma$ so that $\cos\phi = 1 - 2\sin^2\gamma$ this becomes,

$$A_\phi = -\frac{\mu_0 I a}{2\pi}\int_{-\pi/2}^{\pi/2} \frac{(1-2\sin^2\gamma)\, d\gamma}{\sqrt{r^2+a^2+2ax-4ax\sin^2\gamma}} \tag{20.1.9}$$

Noting that the integrand is symmetrical about zero, and using the parameters defined in (20.1.1) this is just,

$$A_\phi = -\frac{\mu_0 Ia}{\pi\beta}\int_0^{\pi/2}\frac{(1-2\sin^2\gamma)\,d\gamma}{\sqrt{1-k^2\sin^2\gamma}} \tag{20.1.10}$$

But we can write,

$$\frac{k^2\sin^2\gamma}{\sqrt{1-k^2\sin^2\gamma}} = (1-k^2\sin^2\gamma)^{-\frac{1}{2}} - (1-k^2\sin^2\gamma)^{\frac{1}{2}} \tag{20.1.11}$$

So (20.1.10) becomes, (20.1.12)

$$A_\phi = -\frac{\mu_0 Ia}{\pi\beta}\int_0^{\pi/2}\left\{\left(1-\frac{2}{k^2}\right)(1-k^2\sin^2\gamma)^{-\frac{1}{2}} + \frac{2}{k^2}(1-k^2\sin^2\gamma)^{\frac{1}{2}}\right\}d\gamma$$

i.e., $$A_\phi = \frac{\mu_0 Ia}{\pi\beta}\left[\left(\frac{2}{k^2}-1\right)K(k^2) - \frac{2}{k^2}E(k^2)\right] \tag{20.1.13}$$

This is the solution for the vector potential in terms of the standard complete elliptic integrals, (20.1.3), where k, β are given by (20.1.1). It now remains to find the field components from (20.1.13) using $\bar{B} = \bar{\nabla}\times\bar{A}$. In spherical polars these are,

$$B_r = \frac{1}{r\sin\theta}\partial_\theta(A_\phi\sin\theta) \qquad B_\theta = -\frac{1}{r}\partial_r(rA_\phi) \qquad B_\phi = 0 \tag{20.1.14}$$

To evaluate derivatives of (20.1.13) we shall need expressions for the derivatives of the complete elliptic integrals. These are given by the identities,

$$\frac{dE(m)}{dm} = \frac{1}{2m}(E(m)-K(m)) \qquad \frac{dK(m)}{dm} = \frac{1}{2m}\left(\frac{E(m)}{1-m} - K(m)\right) \tag{20.1.15}$$

These can be established from, for example,

$$\frac{dE(m)}{dm} = -\int_0^{\frac{\pi}{2}}\frac{1}{2}(1-m\sin^2\alpha)^{-\frac{1}{2}}\sin^2\alpha\, d\alpha \tag{20.1.16}$$

and then using (20.1.11). Substitution of (20.1.13) into (20.1.14) and using (20.1.15) leads, after heavy algebra, to the fields as given in (20.1.2a,b), noting that the coefficient of K in B_r cancels to zero.

Figure 20.2 plots the B field lines (where the ring is at $r/a = 1$). The field is clockwise along the lines if the current on this section is into the page. The magnitude of the B field on these field lines is plotted in Figure 20.3. The vertical B field on the axis is plotted in Figure 20.4 and across the plane of the loop, and inside the loop, in Figure 20.5.

***Figure 20.2: B Field Lines on* $y = 0$**

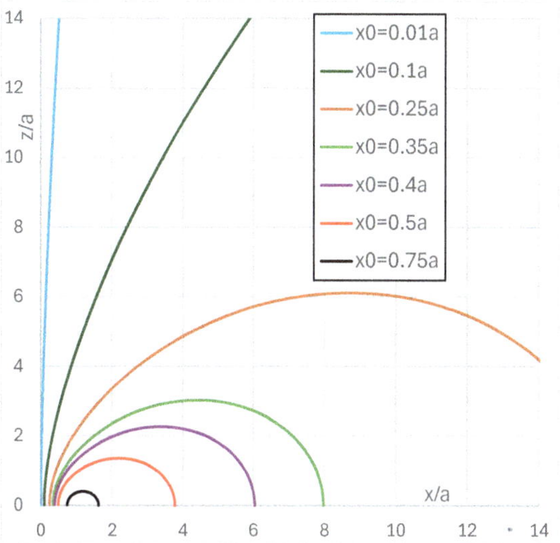

Figure 20.3: Magnitude of the B Field along the Field Lines of Fig.20.2
(normalised by $\mu_0 I / a$)

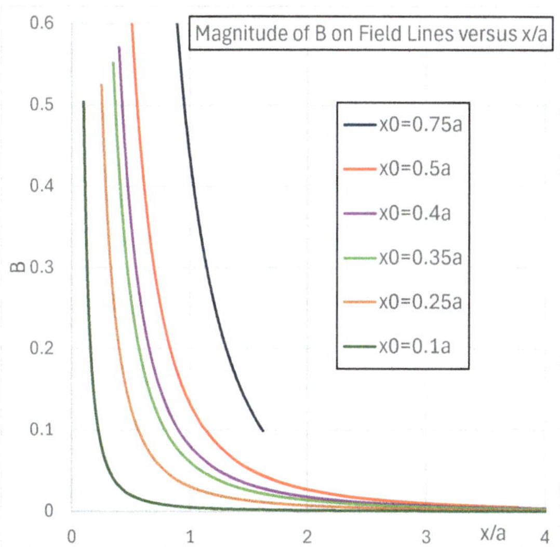

Figure 20.4: B Field on the Axis is Vertical and Given by (20.1.4) *(normalised by $\mu_0 I/a$)*

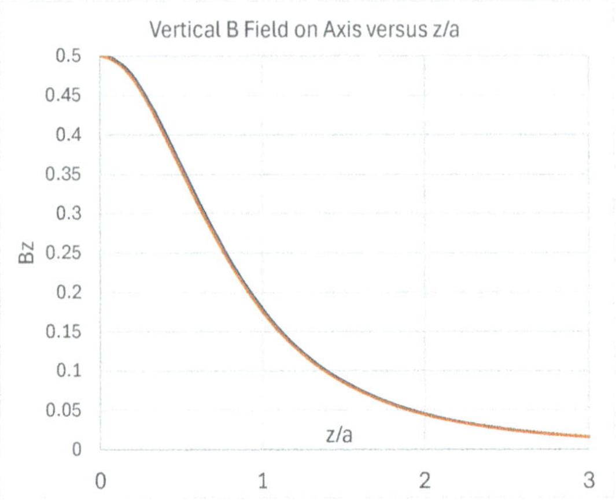

Figure 20.5: B Field in the Plane of the Loop is Vertical *(plotted below within the loop, normalised by $\mu_0 I/a$).*

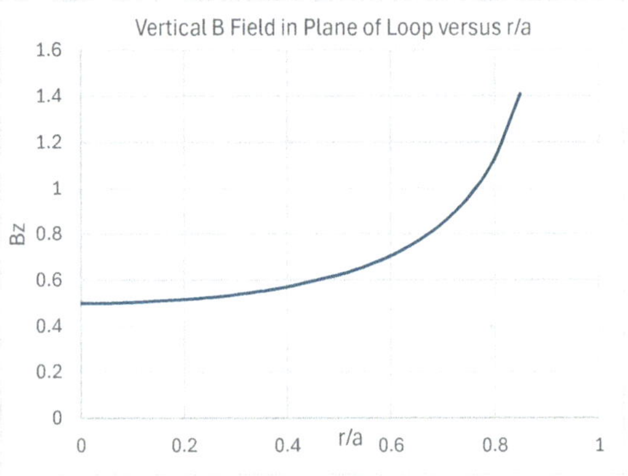

20.2 The Electric Field

Hernandes and Assis, [20.2], derived expressions for the electric field due to a steady current flowing within a solid or hollow toroid. It is necessary to consider a toroid rather than a ring of current if one wishes to derive finite surface charge densities and avoid singular electric fields. The geometry considered in [20.2] is shown in Figure 20.6. The major and minor radii of the toroid are R_0 and r_0 as shown. Consistent with §20.1 the axis of symmetry is the z axis. The current is assumed flowing in the sense of increasing ϕ, as before. The applied potential gradient is such that, on or within the toroid, equi-potentials lie across sections at a constant ϕ. Note that as $r_0 \to 0$ we can identify R_0 with the radius of the ring of current in §20.1. Also note that Figure 20.6 shows the position of the battery which drives the current.

Figure 20.6: Geometry Analysed by Hernandes and Assis, [20.2] *(with acknowledgements to those authors for the image).*

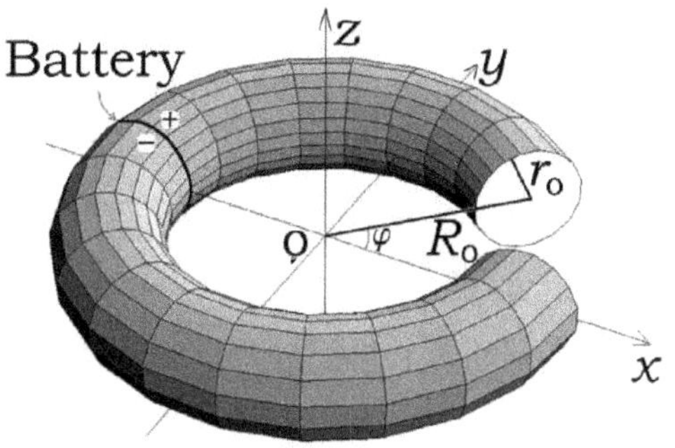

Hernandes and Assis work in toroidal coordinates defined by,

$$x = \rho \cos \phi, \quad y = \rho \sin \phi \tag{20.2.1}$$

$$\rho = \sqrt{x^2 + y^2} = \frac{a \sinh \eta}{\cosh \eta - \cos \xi} \tag{20.2.2}$$

$$z = \frac{a \sin \xi}{\cosh \eta - \cos \xi} \tag{20.2.3}$$

Surfaces of constant ξ are spheres centred on the z axis, and given by,

$$\rho^2 + (z - a \cot \xi)^2 = \left(\frac{a}{\sin \xi}\right)^2 \tag{20.2.4}$$

Hence the centre of these spheres is at $z = a \cot \xi$ and they have radius $\frac{a}{\sin \xi}$. In cross-section (taken at a given ϕ) these spheres are shown as the red circles in Figure 20.7. Surfaces of constant η are toroids, shown in section as the blue circles in Figure 20.7. These toroids are given by,

$$z^2 + (\rho - a \coth \eta)^2 = \left(\frac{a}{\sinh \eta}\right)^2 \tag{20.2.5}$$

The centre of the cross-section of the toroids (i.e., the blue circles) is therefore at $\rho = a \coth \eta$ and their inner radius is $\frac{a}{\sinh \eta}$.

The toroidal circuit is specified by a given $\eta = \eta_0$. From Figure 20.6 we can thus identify $R_0 = a \coth \eta_0$ and $r_0 = \frac{a}{\sinh \eta_0}$. We shall ultimately be interested in the case of a "thin" toroid, approximating to a ring, and hence with $r_0 \ll R_0$ which occurs for $\eta_0 \gg 1$ (because $\frac{R_0}{r_0} = \cosh \eta_0$).

Figure 20.7: Illustration of Toroidal Coordinates. *Red circles are at constant ξ. Blue circles are at constant η. The latter generate toroids when rotated about the axis of symmetry (the vertical, z, axis). The toroidal circuit is defined by a specified $\eta = \eta_0$.* *Attribution WillowW (original); Pbroks13 (redraw) - Own work based on: Apollonian circles.png Created: 6 June 2008. Copyright notice CC BY-SA 3.0, licenced under the terms of the* ***GNU Free Documentation License****, Version 1.2 or any later version published by the Free Software Foundation, a copy of the license is included in the section entitled GNU Free Documentation License.*

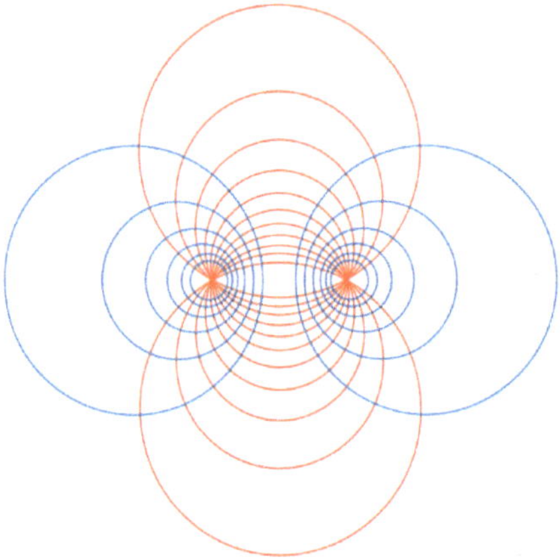

All the constant ξ spheres (red) pass through the limit ring of the toroids (blue), of radius a, obtained as $\eta \to \infty$.

Because this is an electrostatic problem, it can be formulated in terms of the scalar potential, and so as a solution to Laplace's equation, (5.1.3). [20.2] develops the solution for the scalar potential firstly by expressing its boundary condition, the potential specified on/in the toroidal circuit, as a Fourier series in the azimuthal angle, ϕ. After removing a factor of $\sqrt{\cosh \eta - \cos \xi}$, Laplace's equation turns out to be separable in the three toroidal coordinates, η, ξ, ϕ. This leads, as is usual in cases of separability, to basis solutions in terms of two integer parameters (essentially eigenvalues) which the authors denote q and p. The first is just the Fourier order, and produces terms involving $\sin q\phi$. The p parameter produces terms of the form $\cos \xi \cdot P^q_{p-1/2}(\cosh \eta)$ where $P^q_{p-1/2}$ is an elliptic Legendre polynomial. The exact solution is then expressed as a double infinite sum over both integers q and p. This is a messy thing to deal with. (Note that, lamentably, the toroids of constant η are not equipotentials, as one might have hoped).

[20.2] makes an approximation appropriate for a "thin" toroid, $r_0 \ll R_0$, which allows the infinite sum over p to be replaced by a single term. The authors then make a second approximation appropriate when interested in the field near to, but outside, the toroidal circuit, so that $\eta \sim \eta_0 \gg 1$. [20.2] finally notes that, except near the battery or source, the potential varies linearly with ϕ to a good approximation. We now assume that the boundary conditions at the battery are equivalent to applying an equal and opposite potential to its two poles, so that the potential is zero diametrically opposite.

We express the solution for the electric field in terms of the arc length along the toroidal circuit between azimuthal angles $\pm\phi_0$, i.e., $L = 2a\phi_0$. After introducing further approximations, [20.2] gives a result equivalent to,

$$\bar{E} = E_0 \left(-\frac{a\phi}{\rho' log\left(\frac{L}{r_0}\right)} \hat{\eta} + \frac{log\left(\frac{L}{\rho'}\right)}{log\left(\frac{L}{r_0}\right)} \hat{\phi} \right) \tag{20.2.6a}$$

where, $$\rho' = \sqrt{z^2 + (\rho - a)^2} \tag{20.2.6b}$$

Hence, the electric field near to, but outside, the toroidal circuit has a component normal to its surface ($\hat{\eta}$) plus a component parallel to its azimuthal length ($\hat{\phi}$). As regards the latter, we have used the opposite sign to [20.2] because we have the current (and the applied potential gradient) in the positive sense of $\hat{\phi}$ (anticlockwise in the x, y plane). The normal component of the field points outwards at positive ϕ because the unit vector $\hat{\eta}$ points inwards.

The expression (20.2.6a) depends upon an arbitrary arc length, L, of the circuit. [20.2] notes that, near the surface of the toroid, the solution for the electric field coincides with the solution for a straight cylinder of length L. It is reasonable, then, to set $L = 2\pi a$ for the toroidal case. Hence we see that the solution (20.2.6a,b) is the same as that for a long straight cylinder given by our (19.2.3,4b) if we make the replacements,

Straight wire $R_c \to L = 2\pi a$; Straight wire $r \to \rho'$

Straight wire $z - z_0 \to a\phi$; Straight wire $R_0 \to r_0$

The comments regarding energy flows made in §19.3 will therefore also carry over to this case.

By extension we expected the solution for the fields and energy flows sufficiently close to any circuit to be indicated, qualitatively, by the solutions here and in chapter 19.

Further away from the toroidal circuit, [20.2] gives the form of the equipotentials based on the exact (double sum) solution as illustrated in Figures 20.8a,b.

Figure 20.8a: Equipotentials in the* $\mathbf{z} = \mathbf{0}$ *plane from [20.2]. *(With acknowledgements to those authors for the image).*

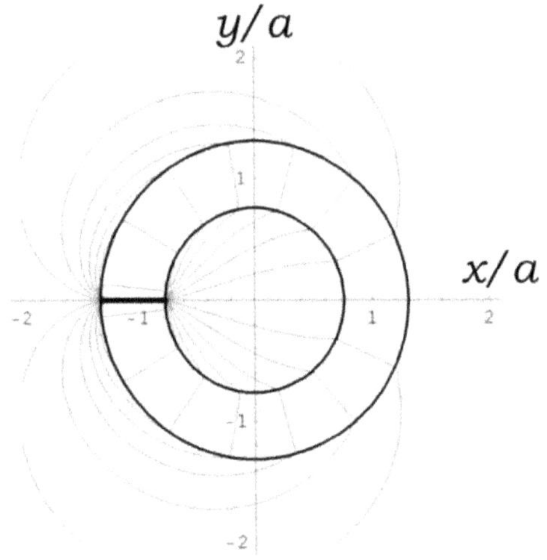

Figure 20.8b: Equipotentials in the* $\mathbf{x} = \mathbf{0}$ *plane from [20.2]. *(With acknowledgements to those authors for the image).*

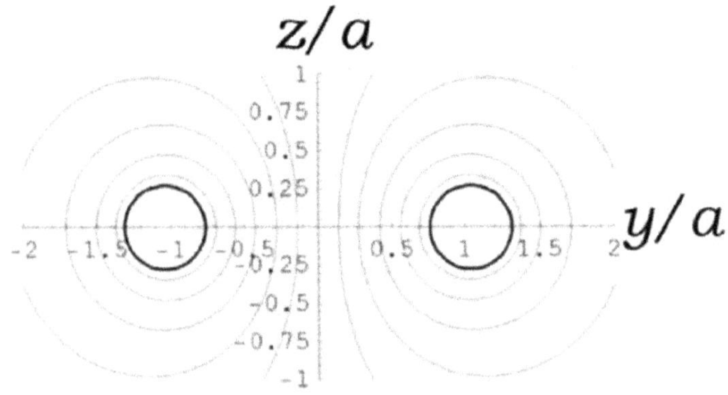

20.3 References

[20.1] Simpson, J.C., et al. (2001). *Simple Analytic Expressions for the Magnetic Field of a Circular Current Loop*. NASA/TM-2013-217919, February 2001.

[20.2] Hernandes, J.A., and Assis, A.K.T. (2004). *Surface Charges and External Electric Field in a Toroid Carrying a Steady Current*. Brazilian Journal of Physics, 34(4B) 1738-1744, December, 2004.

21. Power Generation and Transmission

Chapter 12 addressed some low-power, domestic appliances and their e/m fields. Here we look at high power electrical generation and transmission. Unless otherwise stated, data will be that applicable to the UK. Discussion of energy issues tempts one to stray into policy issues and I have not entirely avoided that temptation. Some key features of the major generating technologies are discussed. In particular, some technical issues with wind and solar generation are addressed, exposing some factors that are not widely disseminated in popular accounts of these renewables. The opportunity is taken to present the basic theory of three-phase electrical generation and supply. I shed some light on the tricky but very important issue of reactive power and megavars.

Contents

21.1 Steam Turbogenerators

The title of this section does not relate to outmoded 19th century technology as the word "steam" might suggest to some. Every nuclear power station, every coal fired power station and every oil-fired power station uses steam turbogenerators to generate the electricity which they produce. Whatever the source of heat, a nuclear reactor or conventional combustion, this heat is used to raise steam which is fed into a steam turbogenerator. Throughout the 20th century, prior to its last decade, nearly all public supplies of electricity in England and Wales were generated from steam turbogenerators. The same is likely true in other countries except where hydroelectric supplies were plentiful (e.g., in Canada or the north of Scotland). Only from about the mid-1990s did other sources, such as wind power and solar power, make significant contributions to electricity generation in England and Wales. Gas generation also started to make large contributions to generation in England and Wales from the early 1990s but via CCGT (Combined Cycle Gas Turbine) plant which does involve steam turbogenerators but in conjunction with gas turbines.

The story of steam turbogenerators through the 20th century in England and Wales is one of ever-increasing size in terms of generated power. As this happened, small, local generation gave way gradually to large, remote sources supplying electricity via a national grid. In 1900 a large turbogenerator was rated at 1 MW, but more plentiful were machines rated at only 250 kW. By the end of the second world war, turbogenerators rated up to 45 MW had become the norm, and by the mid-1950s machines rated up to 90 MW were just as plentiful. By 1960, 180 MW machines had become the norm for new orders. But it was after 1965 that huge machines became *de rigueur* in new plant, initially rated at 360 MW but rapidly standardised at 500 MW and 660 MW. Prior to the millennium, almost all new coal, oil or nuclear power stations ordered after 1965 used the new 500 MW or 660 MW sets. Consequently, the number of units being ordered fell markedly through the 1960s, whilst at the same time the total generating capacity of the ordered plant increased substantially (see Figure 21.1).

The European Pressurised Reactor (EPR) station under construction at Hinkley Point C (and that planned for Sizewell C) will use two steam turbogenerators each rated at 1,600 MWe, similar to those now in service at Flamanville, France, and Olkiluoto, Finland.

Figure 21.1: Steam Turbogenerators Ordered 1946 – 1984, England & Wales

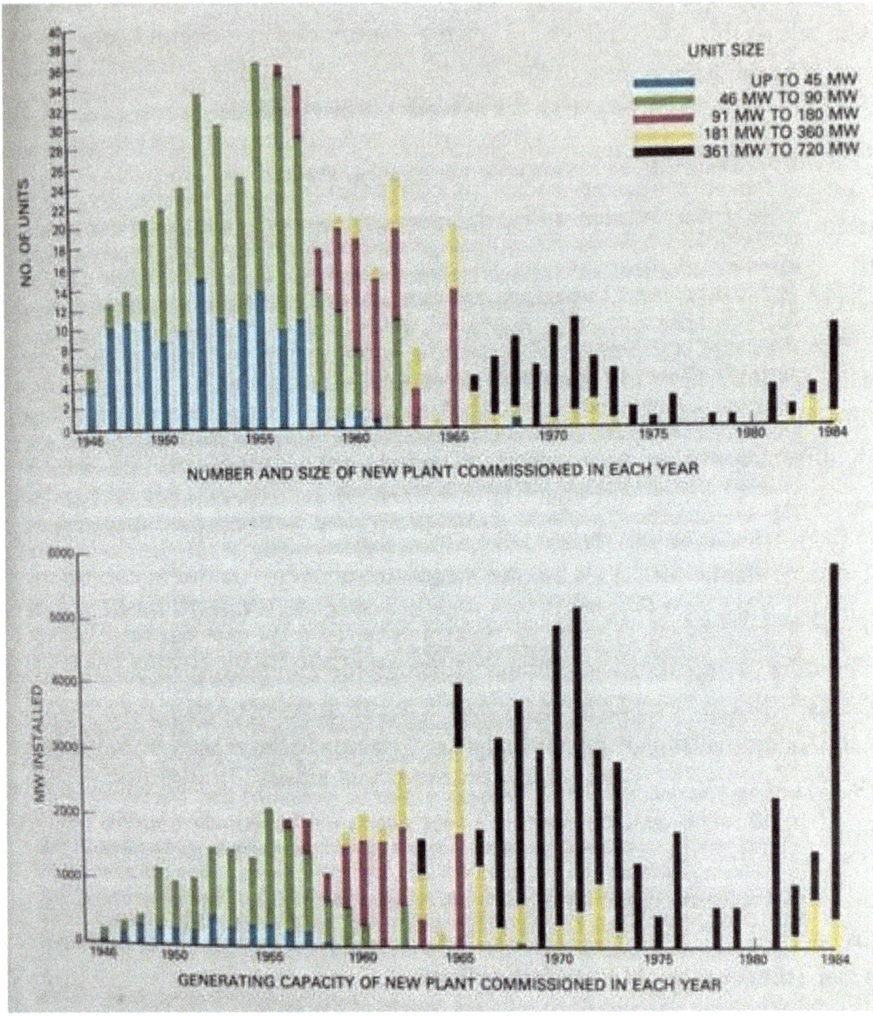

The rotational speed of the rotor in the UK's 500MW and 660MW designs is 3,000 rpm, consistent with the grid frequency of 50 Hz. These are known as "full speed" turbines. Half speed tubogenerators run at 1,500 rpm so that in order to feed into a 50 Hz grid they employ double the number of poles in the generator. The EPR stations' turbogenerators will be half-speed machines, the advantage being substantial reduction in centrifugal stressing.

The kinetic energy inherent in a running turbogenerator is very large. The total rotor mass, including both the turbine and the generator, is roughly the order of 500 tonnes or more for a 660 MW machine, and probably more than 700 tonnes for the 1,600 MW machines in the EPR stations (design data not published at the time of writing). For the 660 MW machines, the turbine rotor diameter is up to about 5 metres in the low pressure (LP) turbine (though far smaller in the high and intermediate pressure turbines). For the 1,600 MW machines the rotor diameter is up to about 8 metres in the LP section.

We are interested here only in the electrical end of the business, not the design of the steam turbine which drives the generator. These large steam turbogenerators deliver three-phase AC into the grid, which also transmits in three-phase. Generators work through electromagnetic induction, i.e., by virtue of $\overline{\nabla} \times \bar{E} = -\frac{\partial \bar{B}}{\partial t}$. An emf is generated when conducting loops experience a changing magnetic flux linkage. A magnetic field must, therefore, be provided. Permanent magnets will obviously not do when the power levels are as indicated above. Instead, coils attached to the rotor have current passed through them in order that they produce the required magnetic field. This magnetic field then rotates with the rotor, sweeping past coils in the stator windings which surround the rotor. It is the stator, the static coils, which generate the electricity.

This describes generators of the rotating field type. It is possible to reverse the role of the stator and rotor windings, having the rotor windings generate the electricity whilst the stator provides the magnetic field. These are called rotating armature designs. However, these are not a practical possibility for large power levels because the output power from the rotor would need to be passed through slip rings, which is not a technological possibility at such high currents and voltages. In contrast, rotating field designs require only the lower power required to create the magnetic field to be passed through slip rings.

Three-phase generators use three separate sets of windings, or coils, spaced 120 degrees apart around the stator. Each set of windings produces an alternating current (AC) that is out of phase with the other two by 120°, resulting in three-phase output. Each of these three phases actually consists of a pair of coils diametrally opposite each other, one being wound clockwise and the other anticlockwise. In a full speed (3,000rpm) machine the rotor has two poles, so the north pole is passing one of these stator coils as the south pole is passing the other. The two stator coils therefore produce equal emfs at the same time. In a half speed machine the rotor has four poles at 90° with a corresponding change to the stator configuration, thus generating electricity at a frequency which is double the machine speed.

To reduce resistance losses, the windings of the generator are made of thick section copper. Nevertheless, it is inevitable that resistance losses occur in view of the very high currents involved. In addition, losses occur due to eddy currents resulting from the alternating magnetic field. As a result, a great deal of heat is produced by the generator, which therefore needs to be cooled. In the UK's 500 MW and 660 MW designs, pressurised hydrogen at 4 Bar was the chosen coolant for the rotor and the outer surfaces of the stator. The stator windings were of hollow section within which cooling water flowed. The larger 1600 MW machines at the Flamanville EPR station use similar hydrogen and water cooling.

To supply current to the rotor, a smaller generator called the exciter is also installed on the same extended shaft. Yet another, even smaller generator, the pilot exciter, is also driven by the same shaft and provides the field current for the main exciter. All are three-phase, the pilot exciter also providing power for control equipment. For a typical 660 MW design the currents and voltages involved are: output (stator) 23,500 V and 19,076A; exciter output (main rotor) 560 V and 4,500 A; pilot exciter output 200 V and 260 A. To reproduce the 660 MW output from the above stator figures do recall that this is three-phase (so multiply the product of volts and amps by $\sqrt{3}$, see below) and also recall that the potential and the current will not be in phase, so multiply also by the Power Factor (defined below) of 0.85.

21.2 Combined Cycle Gas Turbines

Between the end of the second world war and the early 1990s, the state-owned Central Electricity Generating Board (CEGB) had a monopoly on electricity generation in England and Wales. In this period the CEGB held to a policy of avoiding gross electricity generation from gas. The CEGB did operate gas turbine plant of the open cycle type (OCGT), which consist of a gas turbine coupled directly to a generator. But these OCGT units were requested to run only to "peak lop" sudden spikes in demand, or to backfill for sudden losses of other generators. The great advantage of OCGTs for this purpose is that they can be run-up to full power relatively quickly. From a cold start an OCGT can reach full electrical output in perhaps 15 to 30 minutes. Where plant has been requested to warm-up in advance, the run-up time of an OCGT may be reduced to a few minutes. In the CEGB days,

OCGT units ran only for as long as it took to bring higher merit plant onto the bars – or until the spike in demand subsided. (For an explanation of what "merit" means in this context see the next section).

The gas turbines in OCGT and CCGT plant are essentially the same as jet aeroengines. Indeed, the first generations of gas turbines for electricity generation were aeroengines converted for the purpose. OCGT generation has an efficiency of around 35% - 40%. The hot exhaust gases from the gas turbine are wasted heat.

Following the break-up and privatisation of the CEGB in the early 1990s, the new companies invested widely in CCGT plant (known colloquially as "the dash for gas"). Figure 21.2 shows a diagrammatic layout of a CCGT power station. CCGTs are effectively two types of generation in one. One part of the plant consists of a gas turbine coupled directly to a generator, i.e., an OCGT unit. However, instead of wasting the heat in the exhaust gases, these are used to raise steam which is then fed into a steam turbogenerator. This raises the efficiency of the plant overall to about 50% and hence is commercially highly beneficial. (On the other hand, the steam raising plant and steam turbogenerator increases the capital cost compared with a simple OCGT – but still much cheaper than a large coal or nuclear power station).

Figure 21.2: Diagrammatic Layout of a CCGT

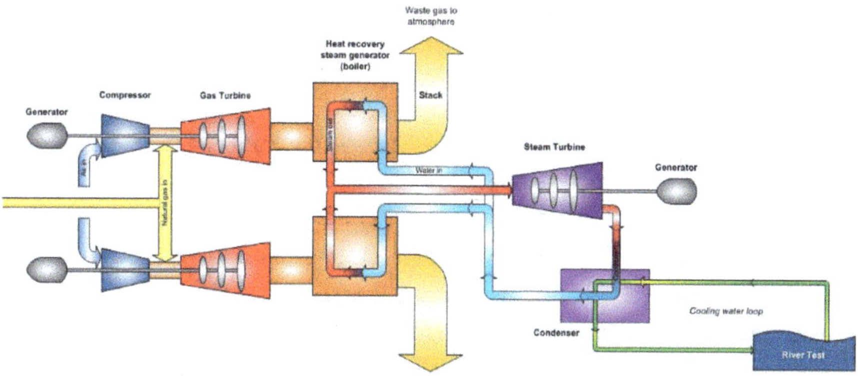

Note in Figure 21.2 that separate generators are typically provided for the OCGT part of the unit and the steam part of the unit. This permits a CCGT to benefit from the quick start times of an OCGT and also, a little while longer, from the higher output of the combined cycle generation from both sets of generators. Full power from a CCGT may be obtained in 30 to 60 minutes from cold, or between 5 to 20 minutes from warm/hot.

The power rating of individual gas turbines in CCGT plant is typically 150 to 300 MW. A common arrangement is two gas turbines feeding waste heat into one steam unit with a power rating for the whole unit of perhaps 440 MW. CCGT stations would typically consist of three, four or five such units and hence with total capacity up to perhaps 2.2 GW.

As regards the generators themselves, the main difference from those discussed in §21.1 is that a CCGT station includes a larger number of smaller generators, perhaps rated at only 150 to 300 MW each, but the design is broadly similar.

21.3 Policy Issues on Electricity Generation in England and Wales, and the Use of Gas Before the 1990s

21.3.1 Keeping the Lights On

Under the Electricity Act 1989, the Secretary of State for Energy (or the equivalent) has a duty to secure that all reasonable demands for electricity are met. However, this is a high-level policy duty, not an operational one – and I leave the reader to make his own assessment of the degree of security implied by a politician's duty.

The National Grid Electricity System Operator (NESO), formerly part of the National Grid company, has the duty to balance supply and demand continuously and has a statutory duty to operate the system in an efficient and economic way. But NESO does not have a legal duty to ensure there is always enough generation capacity available. It balances what exists.

The main mechanism that is supposed to ensure demand can always be met is a commercial one, the "Capacity Market" system. Under this system, generation companies are paid to provide availability in future years. This is now the main tool to ensure adequacy of supply. But it is dependent on independent companies finding the terms and conditions of continuing to generate electricity to their commercial liking. No companies are obliged to continue to do so; they might find entirely different markets more lucrative.

In short, since the demise of the CEGB in 1990 the responsibility for "keeping the lights on" has been diffuse rather than clearly resting on a single body or person. As anyone knows who has ever managed anything, responsibility shared is responsibility neglected.

21.3.2 Is Gas Generation Wise?

I do not refer in this section to the issue which causes many people to oppose all forms of electricity production from fossil fuels, that is the emission of carbon-dioxide and its relationship to climate change. Climate change issues are beyond the scope of this book. I refer here to a scientifically less contentious issue: the efficient use of energy. I refer to a radical change in the use of gas that came about due to privatisation of the electricity industry in 1990 and thereafter.

Up to 1990, as a state-owned monopoly, the CEGB was not constrained by competition and the need to maximise profit for shareholders. As a result, the CEGB was free to exercise a *de facto* electricity policy. This was manifested in practice by the Merit Order. This was a simple list of existing generating stations, of all types, in "merit" order, where "merit" was defined primarily in terms of operating costs. When extra generation was required, the plant that was brought on load was the highest merit plant not currently running but available.

Operating costs are defined as the costs of running the plant minus the costs of not running the plant. In particular this excludes capital build costs, but it also excludes all fixed, unavoidable costs associated with maintaining plant availability (including dedicated staff salaries, for example). That resulted in all the CEGB's nuclear stations being at the top of the merit order, having the lowest operating costs. After that came primarily coal fired stations, essentially in order of size or modernity, as this reflected efficiency and hence operating costs. After those came the oil-fired plant, which sank down the merit order dramatically after the OPEC-driven oil price hikes in the 1970s.

What is noticeable by its absence is gas generation. The OCGT plant operated by the CEGB was never regarded as a sustained generation source, being there only to peak-lop and to be quickly replaced by other generation as soon as possible.

But why?

The CEGB could have generated base load electricity from gas had they so wished. This was to happen after privatisation in 1990. But cheap and plentiful gas had been available from the North Sea for 20 years before that. Why did the CEGB not generate more from gas? They could have made a good profit, as the privatised companies were to do after 1990.

We in the industry at the time understood the reason to be a policy decision based on what was best for the nation, as opposed from maximising profit. As a nationalised monopoly the CEGB could make such a decision. The company recognised, on the basis of the efficient use of fuel, that gas was best left for heating – not primarily electricity production. The CEGB essentially recognised the iniquity of gas generation for simply thermodynamic reasons.

The essence of the matter is that gas can be used far more efficiently for heating (domestic or industrial) than for electricity generation. You might be tempted to say the same of coal – which was the fuel from which most electricity was generated before 1990. But there are several crucial differences between gas and coal for domestic use. Firstly, gas is convenient to transport down pipes. Coal is rather different. It comes in sacks and has to be delivered door to door on the back of a lorry. (My recollection of the 1960s is that the coal would be dumped at your back gate, leaving the 10 year old lad of the house (me) responsible for moving perhaps 15 cwt of coal into our shed before someone complained they couldn't get out of the yard).

But then there was the smog. I was raised in Manchester in the 50s and 60s when central heating was some exotic foreign thing that no one had. Everyone had an open coal fire, generally as their only source of space heating. In many cases their only way of heating their water also, via a back-boiler. The result was pea-souper smogs of a type that you need to have experienced to appreciate. Appallingly unhealthy for one thing, but they also made travel very difficult. On more than one occasion I was sent home from school early because traffic was at a virtual stand-still. I walked the 10 miles home – not in itself a big issue, but visibility was so bad one risked bumping into trees, walls, other people, etc. Smokeless zones duly followed and were entirely sensible. The switch to smokeless fuel (coke) helped, but coke was a by-product of coal gas production, an industry which died when the North Sea was tapped for its methane. No way would we want to go back to widespread burning of ordinary coal in open fires by most people.

In contrast, as we know well, gas is a clean and convenient fuel for domestic space heating and water heating, as well as for cooking, and equally so for industrial heating purposes. And it is easily transported. The issue, though, is the relative efficiency of using gas in this way compared to its usage for power generation.

Straight gas turbine generation would have been about 35%-40% efficient, i.e., only about 40% of the calorific value of the gas would appear as electrical energy. For CCGT plant achieved efficiencies are about 50%, though higher values up to 60% are sometimes claimed.

At one time I would have conceded that these CCGT efficiencies beat the best coal stations into a cocked hat, typical coal generation efficiencies in the past being around 34%. However, the use of supercritical technology in coal generation has now pushed efficiencies to the same level as existing CCGTs, around 50%. However, these improvements in coal plant efficiencies have come about since the UK abandoned new coal plant in 1990. (Bear in mind that greater efficiency means lower carbon emissions per megawatt-hour).

Another issue of which to be aware, as regards efficiency issues, is that in practice, as things now are in 2026, CCGT plant often is not given the opportunity to realise its full potential efficiency. The reason is that their high efficiency results from adding the steam raising cycle to the directly coupled generation. But much CCGT plant is used to cover periods when the wind drops and the output from the UK's wind turbines decreases. This often involves periods of time too short to permit the steam cycle to get up to full efficiency, or for steam to even be raised at all. In short, the large installed capacity of wind generation drives down the efficiency of the CCGT plant which is required to back-fill when the wind drops – thus resulting in more carbon dioxide emissions from CCGT plant, per unit generated, than would otherwise be the case.

However, it is not the comparative efficiency of electricity generation from coal and gas that is apposite. The iniquity of which I speak is the relative inefficiency of using gas for electricity generation compared to its use in domestic (or industrial) heating. Modern domestic gas-powered condensing boilers have efficiencies of 90% - 94%, far higher than even the most efficient generators of electricity from gas. And even old-fashioned, non-condensing boilers had efficiencies of 75% - 78%, way higher than CCGT efficiencies.

There is a reason for this which mere technological advances can never overcome. It is, in principle, possible to convert gas's calorific value (i.e., chemical energy) to heat with 100% efficiency. It is not, however, possible even in principle to convert the heat produced by burning gas to electrical energy with 100% efficiency. Heat engines have a fundamental limit on their maximum possible efficiency defined by the range of temperatures between which they work. We are up against the second law of thermodynamics.

Being a science-driven organisation, the CEGB was well aware of this issue and gave it priority over a crude "dash for gas for cash" policy. But after privatisation there was no home for such thinking and commercial advantage was everything.

The percentage of total electricity generation in the UK by fuel type since 1990 is shown in Figure 21.3, for gas, nuclear, renewables and thermal generation other than gas. Renewables (mostly wind) became the largest total supplier of electricity in 2024, but gas generation is still comparable. Gas generation will inevitably remain substantial whilst renewables are dominant and there is no other source capable of stepping in when the wind drops. Nuclear might eventually perform that role, but as of 2026 that is a long way off and may depend upon the commercial success of nuclear hydrogen production. For now, England and Wales are locked into the thermodynamically questionable use of gas for electricity production as a by-product of the politically driven rise of renewables.

Figure 21.3: UK Generation by Fuel Type (data from [21.1]).

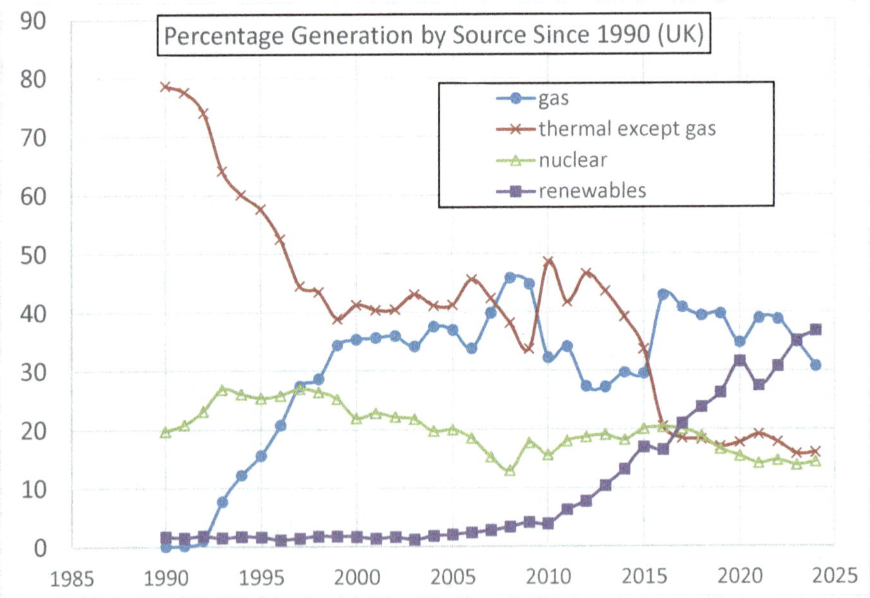

21.4 Wind Turbines

The generator of a wind turbine is housed in the nacelle. In common with the generators discussed above, the basic operation consists of either windings or permanent magnets attached to the rotor and windings comprising the stator.

The most obvious difference is that wind turbines operate at varying rotational speeds – and rotational speeds which are obviously far slower than the 3,000 rpm needed for a synchronous machine. The generator rotor speed is stepped up by gearing with respect to the turbine shaft speed, but is still variable.

In all designs the rotor provides the magnetic field while the electricity is generated in the stator windings, as with turbogenerators though at far smaller power levels. There are two main designs in use. In a Permanent Magnet Synchronous Generator (PMSG) the rotor contains permanent magnets, which provide the magnetic field. In contrast, in a Doubly Fed Induction Generator (DFIG) the rotor carries windings which constitute electromagnets. The rotor's electromagnets are powered from the grid via slip rings and brushes, and usually use a three-phase supply.

Feeding the power from wind generators into the national grid is a good deal more involved than is the case for synchronous turbogenerators, the main problem being that the frequency of the AC produced is variable. Arrangements will differ according to design specifics but will always entail sophisticated power electronics. A likely arrangement is that the variable frequency AC is first rectified to DC and then an electronic inverter is used to produce AC at 50 Hz from the DC. To be more precise, the inverter monitors the grid and converts the DC to precisely the instantaneous grid frequency – and phase. But that is not all. As we will see below, efficient power transmission does not involve voltage and current which are in phase but rather they must be out of phase by a certain amount. The inverter therefore creates the desired phase difference and matches it to the grid. The need for such sophisticated power electronics adds to the capital cost of wind generation. It may also cause grid stability problems relative to power from turbogenerators (see §21.10).

In practice, even with modern designs, the turbine captures only around 40% of the kinetic energy of the wind (and 59% is a theoretical upper limit).

Typical designs reach their maximum power at wind speeds of 15 m/s (33 mph). It generally takes a wind speed of about 3 m/s (7 mph) for any generation to initiate, and the turbine must be protected from overspeed when wind speeds exceed about 25 m/s (55 mph). Larger turbines are actively controlled to face the wind.

As of 2025 the total installed capacity of wind generators in the UK is about 32 GW, about half being from onshore and half from offshore turbines. UK government plans are currently to almost double the latter. Offshore turbines are individually rated at typically 6 to 12 MW, whilst onshore wind turbines are smaller with capacities generally in the range 0.5 to 5 MW. The total number of individual wind turbines is about 12,000, suggesting a strong bias towards the lower end of the individual power ratings. Installed capacity of wind turbines is misleading, however, as the wind does not always blow at its optimal strength (or at all).

The load factor for the UK's installed wind turbines (that is the actual generation divided by the capacity) has a seasonal variation from 20% to 35% for onshore turbines, or 25% to 50% for offshore turbines. That compares with the historical load factors for the UK's gas-cooled nuclear reactor power stations of between 60% and 80% (average 70%), and typical PWR nuclear stations' load factors in excess of 85%. Since the mid-1990s the load factor for the UK's coal and gas generation plant says nothing about the plants' reliability as these stations have been increasingly constrained-off due to the priority given to wind generation.

21.5 Photovoltaics

Photovoltaic generation in the UK is rather the poor relation of wind power, accounting for 14% of renewable generation in 2024, or 5% of total generation. The reason is obvious: lack of strong sunlight in the UK for most of the time. PV panels generate DC, of course, so a similar electronic inverter system is required to feed into the 50Hz AC grid.

21.6 Intermittency and Storage

The major problem with wind and solar generation is their intermittency: the wind does not always blow and the sun does not always shine. If electricity is to remain available during these periods then alternative sources must be available, and must be capable of running up to power quickly. The rapid rise of wind power in the UK from 2005 to the present (2026), and the complete phasing-out of coal generation, has only been possible as a result of the large installed capacity of CCGT generation...and this was simply happenstance as it resulted from the "dash for gas" following privatisation of the industry in the early 1990s. These CCGT plants are now ageing and will soon be passed their commercial life. Some have already been closed. This will present a problem if electrical supplies are to be continuously available and the political determination to phase out all fossil fuel generation, including gas, continues.

It should be noted that however great the installed capacity of wind power, it alone can never solve the intermittency problem. Analysis of periods of low wind generation in [21.2] shows that battery storage can never practically be the solution. Nor is pumped storage the solution due to lack of suitable locations in the UK. At present it appears that there are two possible future scenarios. One involves new nuclear build coupled to hydrogen production, the intention of which would be to have the capability to divert power quickly into the grid without having to cycle the nuclear plant (which has particular problems). But as of 2026 this is a long way off, lead times for new nuclear plant being frustratingly long – and the associated technology for hydrogen production being currently non-existent in commercial form. The only other plan is Demand Side Reduction (DSR), in other words, encouraging consumers to "switch off" during periods of low generation by raising the price to prohibitive levels at these times, [21.3]. The heavily politicised industry is not being very forthright with the public about this policy direction.

21.7 Three-Phase Electricity

Three-phase electricity involves three conducting lines, not just the more familiar two lines. Each line carries AC. Figure 21.4 shows how the current varies for each of the three phases (or lines). As the name implies, these alternating currents are 120° out of phase. The so-called "phase voltages", that is the potential of a given line compared with neutral (Earth), are similarly 120° out of phase.

We can also talk of the "line voltage", which is the potential of one line wrt the potential of another. If the potential of phase 1 wrt Earth is $V_0 \cos \omega t$ then the potential of phase 2 is $V_0 \cos\left(\omega t - \frac{2\pi}{3}\right)$ and the potential of phase

3 is $V_0 \cos\left(\omega t - \frac{4\pi}{3}\right)$. The line voltage of phase 1 wrt phase 2 is thus $V_0 \left[\cos \omega t - \cos\left(\omega t - \frac{2\pi}{3}\right)\right]$. The use of familiar trig formulae gives,

$$\text{Line 1 voltage wrt line 2} = V_0 \left[\frac{3}{2}\cos \omega t + \frac{\sqrt{3}}{2}\sin \omega t\right] \qquad (21.7.1)$$

The line 2 wrt line 3 and line 3 wrt line 1 voltages will be similar but each at 120° phase wrt the others. The important thing about the line voltage, (21.7.1), is that its amplitude is not V_0 but $\sqrt{3}V_0$, where the $\sqrt{3}$ is found from (21.7.1) as $\sqrt{\left(\frac{3}{2}\right)^2 + \left(\frac{\sqrt{3}}{2}\right)^2}$.

Figure 21.4: Waveform of Three-Phase Current or (Phase) Voltage

J JMesserly, original SiriusA File:3-fas-spänningar.svg, Public Domain.

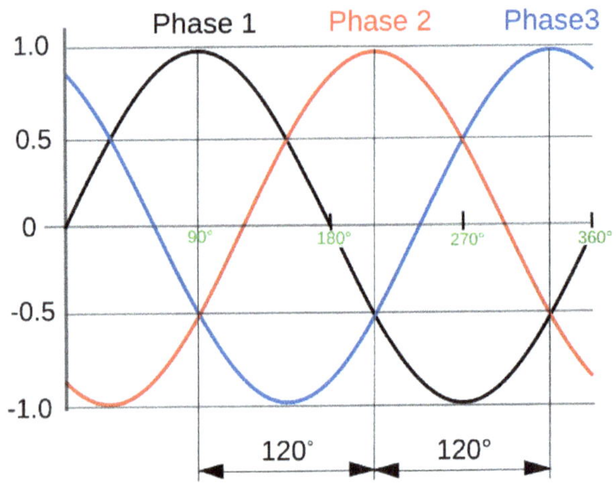

Note that a consequence of having three-phase current is that the net current (i.e., the algebraic sum of the three line currents) is zero. Hence, the return path for the current in line 1 is lines 2 and 3.

In the UK, the statement that our domestic mains supply is at 230 V refers to the phase voltage, i.e., that of each line wrt neutral. The line voltage of UK domestic supply is thus $\sqrt{3} \times 230 = 398$ V (usually rendered as 400 V).

Why use 3-phase in transmission of electrical power? It is a matter of efficient use of cabling, which is expensive. The electrical power being conveyed by each of the three pairs of cables is the same as the power that would be conveyed by a single cable pair bearing the same current. Hence, three times the power capacity is obtained for a 50% increase in cabling.

Typical overhead transmission lines in the UK involve two separate circuits, each consisting of three cables carrying three-phase electricity, plus a neutral line. Because the net current involved in three-phase electricity is zero, the neutral line need not be capable of carrying large currents, thus saving on costs.

When using the power transmitted as three-phase, a circuit including the load may be completed either by connecting across any pair of the three phases (400V) or by connecting between any one of the phases and the neutral line (230V). These configurations are known as delta and star (or Y) respectively. However, some loads may themselves be designed to use three-phase, and in that case all is well if source and load are configured the same way. If they are configured differently, then a transformer will be needed to (say) accept delta configuration into the primary windings whilst outputting in star configuration from the secondary windings.

21.8 UK Transmission Lines

The UK introduced Overhead Lines (OHLs) in the 1920s. Modern high voltage transmission OHLs in the UK are operated at voltages ranging from 275 to 400kV. Most routes have two circuits each, with three phases per circuit. A neutral or earth line usually runs between the tops of the towers.

In principle each of the two circuits in a 400kV supergrid line can carry up to 2.5GW. In practice, the complete OHL, both circuits together, would be limited to perhaps 3.5GW. Conveniently, this means a single OHL with two three-phase circuits can carry the whole output from the new power station under construction at Hinkley Point C (rated at 3.2GW).

Standard OHL conductor sizes range from 24.7 to 41.04mm diameter. The conductors are made of stranded aluminium with a core of steel for strength. Copper is not used despite its better conductivity due to both cost and weight. Aluminium wins due to far better strength-to-weight ratio and conductance-to-weight ratio. Weight is a major factor due to the long spans between transmission towers – and the greater the weight of the cables, the stronger and more expensive the towers would need to be.

21.9 MegaVARs, Reactive Power and Power Factors

Other than professional electrical engineers, even people in the electricity generation industry tend to twitch when megaVARs are mentioned. They know that megaVARs are very important but have only the vaguest notion why. We must start by looking back at the LCR circuit analysed in §18.3.

Recall that a sinusoidal source applied to a circuit with inductive (L) and capacitive (C) elements, as well as resistance (R), has a relationship between the resulting voltage and current amplitudes give by $V_0 = Z(\omega)I_0$, where the frequency dependent $Z(\omega)$ is a complex impedance. The real part of $Z(\omega)$ is just R, which relates to the part of the response of the circuit in which the current and the potential are in-phase. But the imaginary part of $Z(\omega)$ is that part of the potential which is 90° out of phase with the current. The imaginary part of $Z(\omega)$, called the reactance, is $\left(\omega L - \frac{1}{\omega C}\right)$.

The ratio of the magnitudes of the voltage and current, is thus,

$$\frac{|V_0|}{|I_0|} = |Z(\omega)| = \sqrt{R^2 + \left(\omega L - \frac{1}{\omega C}\right)^2} \tag{18.3.8}$$

In the context of transmission lines what concerns us is delivering power efficiently to a load which will, in general, have inductance and capacitance as well as resistance. So we need to ask what happens to the transmitted power when the load itself creates a phase difference (say ϕ) between the voltage and current?

The instantaneous power being delivered at time t is simply $V(t)I(t)$, where, for cogency, we will deal with real voltage and current quantities. As these are supposed to be out of phase we set $V = \cos\omega t, I = \cos(\omega t + \phi)$. The average power delivered over some time period τ, assumed to be a large integral multiple of the periodic time, is then,

$$\text{Average Power Delivered} = \frac{|V_0 I_0|}{\tau}\int_0^\tau \cos\omega t\cos(\omega t + \phi)\,dt \tag{21.9.1}$$

which is readily found to be $\frac{1}{2}|V_0 I_0|\cos\phi$. Hence, when there is no phase difference between the voltage and current the average power delivered to the load is $\frac{1}{2}|V_0 I_0|$ as would be expected. Note that this is the same as the product of the rms voltage and the rms current, i.e., $\frac{|V_0|}{\sqrt{2}} \times \frac{|I_0|}{\sqrt{2}}$. But where a phase difference occurs between the voltage and the current, the power delivered to the load is reduced by a factor of $\cos\phi$. In the extreme, if the load had zero resistance and was purely inductive and capacitive, then the phase difference would be $\phi = \pi/2$ and no power would, on average, be delivered to the load at all.

The impact of these observations on transmission efficiency is that, where there is a phase difference the power delivered is diminished by the factor $\cos\phi$ but the transmission lines are, nevertheless, still carrying the full current amplitude, I_0. The resistive losses along the transmission line are therefore the same as they would be had the maximum power of $\frac{1}{2}|V_0 I_0|$ been delivered. The actual situation is therefore inefficient because the same losses are incurred whilst delivering less power to the load.

The quantity $\frac{1}{2}|V_0 I_0|$ is known as the apparent power. It is the power that would flow along the transmission lines transiently before the load was first encountered (or if the load were pushed off to an infinite distance). What happens, then, to the excess power being initially transmitted, i.e., the excess given by $\frac{1}{2}|V_0 I_0|(1-\cos\phi)$? This "reactive power" merely slops back and forth without being consumed, i.e., it does no work and raises no heat. If the load is predominantly inductive, the reactive work is temporarily stored as magnetic energy in the load and then released back into the transmission line whence it came. The same applies if the load is predominantly capacitive, except the temporary storage is as an electric field. This slopping back and forth occurs at twice the line frequency, i.e., 100 Hz.

We will see in chapter 23 that when transmission lines are analysed from the perspective of being waveguides, the full power $\frac{1}{2}|V_0 I_0|$ corresponds to the incident wave whilst the power supplied to the load, $\frac{1}{2}|V_0 I_0|\cos\phi$, corresponds to the transmitted wave. However, the reactive power, the difference between the two, corresponds to a reflected wave. It is the reactive part of the load's impedance which causes the reflection. From this perspective, the most efficient delivery of power is when no power is reflected back along the line.

The "power factor" of the load is the ratio of the real power delivered to the apparent power, i.e., the power factor is just $\cos\phi$.

What generators ideally attempt is to cancel the reactive impedance of the load by generating with, in effect, an equal and opposite reactive impedance. More simply put, power stations deliberately supply voltage and current which are slightly out of phase, a phase which is judged to be about equal and opposite to that created by the load. The 90° out of phase component of the generation has, in principle, the same units as power, namely volts × amps, but because this "reactive power" is not real power the units in this case are not written as Watts but rather as VARs (Volts× Amps Reactive). In practice we are dealing with MegaVARs.

So this is the first reason why MegaVARs are so important. They are necessary in order to make the power transmission as efficient as possible.

But there is also a second reason, which in practice is even more important: they contribute to grid stability. To see this simply, consider what happens if the generator fails to balance a predominantly inductive load. Less real power is absorbed and this leads to a depression of line voltage. The provision by the generator of balancing (capacitive) megaVARs is therefore a key ingredient in maintaining line voltage stability.

Note that the reason why supplying capacitance on the system can balance an inductive load, i.e., bring the total reactance close to zero, is because the reactance is $\left(\omega L - \frac{1}{\omega C}\right)$ so capacitance and inductance contribute with opposite signs.

Note that power stations do not deliver power direct to the consumer but rather to a substation. So it is the substation that is the load in this case. Further downstream we may have a large industrial consumer for whom the direct supply of power is from the substation, so another stage in the balancing of reactive impedance applies between the substation and the consumer.

But generators or suppliers will not accommodate loads with an arbitrary amount of reactive impedance. That would present too great a challenge to balance. In practice there is a lower limit on the acceptable load power factor, e.g., 0.9 or 0.95. Large industrial consumers may need to employ Power Factor Correction (PFC) to get their requirements into the acceptable range. So, for example, large electric motors will be inductive due to their coils. PFC may then involve switchable banks of capacitors to cancel the reactance down to the acceptable range.

Finally, we have alluded above to power stations deliberately generating out-of-phase voltage and current – but how is this done? In the case of traditional turbogenerators it is accomplished by overexcitation or underexcitation. Whilst the mechanical work being done by the turbine controls the real power delivery, the reactive power is controlled by the excitation current supplied to the field coils (the rotor). We will pass over the details but the essence of it is that increasing the excitation current pushes up the supplied voltage and cancels the depressing effect on voltage of an inductive load. Hence, increasing excitation current acts like adding capacitance. Similarly, decreasing the excitation current will help balance a capacitive load. In practice these adjustments to the excitation current are implemented automatically by the AVR (Automatic Voltage Regulator). By this means a direct 'awareness' of the reactive nature of the load is not needed; instead the system responds to correct any impact on the supplied voltage.

There is a rather misleading terminology in use, namely references to "consuming reactive power". This actually refers to the balancing of an inductive load using capacitance, or vice-versa, not the actual consumption of real power at all.

21.10 Renewables and Grid Stability

An increasing proportion of renewable generation within a grid system challenges its stability in several different ways. The most obvious is the intermittency of wind and solar generation. This challenges stability of supply because the frequent need to run-up alternative sources very quickly provides equally frequent opportunities for something to go wrong. However, there are three other challenges to grid stability which result from large quantities of renewables.

The first is the issue of the megaVARs. Unlike conventional turbogenerators, renewables cannot deliver megaVARS via variable exciter current (because these sources are not synchronised to the grid). Renewables can provide megaVARs but only 'artificially' via the electronic inverters that connect them to the grid. Hence, renewable megaVARs arise only via control algorithms. Renewables lack the intrinsic "voltage stiffness" of conventional turbogenerators. Moreover, inverter systems lack the deep reserves of potential reactive support that are intrinsic to turbogenerators, being limited in their reactive range, especially at high power.

The second challenge to stability again works via the reactive power issue. Wind turbines are generally in remote places, including many offshore. This requires longer lengths of transmission line, and the transmission line itself imposes reactive impedance. So, large quantities of wind power cause more megaVARs to be needed on the system, whilst providing less.

The third issue which impacts adversely on grid stability concerns inertia. Recall the huge mass and size of turbogenerators, and consider the combined kinetic energy of hundreds of such machines at 3,000 rpm. Now consider what happens if there is a sudden loss of a power station off the grid. The sheer inertia of these turbogenerators can continue to keep the grid voltage within the tolerable range for the precious minutes needed to bring on alterative sources. In contrast, there is no effective inertia associated with renewables because they are not directly grid connected.

21.11 Health Effects of E/M Fields from Transmission Lines?

There has been some public concern over the possibility of adverse health effects due to living close to high power transmission lines. Rather than attempt a digression that would either be inadequately brief, or adequate but far too medically detailed for this book, I refer the reader to [21.4-6]. I repeat the warning of chapter 12, namely that there is emerging evidence from experimental biology that electric fields can have a marked effect on "reprogramming" organisms' development. However, [21.4-6] are based on epidemiological studies which would account for any and all mechanisms of harm, which may reassure you.

21.12 References

[21.1] UK Government (2025). *Energy Trends: UK electricity*, generation data to end 2024.

[21.2] Bradford, R. (2022). *Could Wind Power Alone Be Enough?* 18 October 2021.

[21.3] Ofgem. (2020). *Potential consumer impacts following the implementation of Market-Wide Half-Hourly Settlement*. Office of Gas and Electricity Markets.

[21.4] UK Health Security Agency (2013). *Guidance on Electric and magnetic fields: health effects of exposure*. 1 July 2013.

[21.5] A.F.McKinlay, et al. (2004). *Advice on Limiting Exposure to Electromagnetic Fields (0-300 GHz)*. Documents of the National Radiological Protection Board: Volume 15, No. 2.

[21.6] NRPB (2001). *ELF Electromagnetic Fields and the Risk of Cancer: Report of an Advisory Group on Non-Ionising Radiation*. Documents of the National Radiological Protection Board: Volume 12, No. 1.

22. Oscillating Dipoles and Accelerating Charges as Sources of E/M Radiation

Intuitively it is reasonable to expect that an oscillating charge or dipole would produce e/m radiation. The reason is that, as the source oscillates it will move alternately closer to, then further away from, a given field point. The electric field there will therefore oscillate. But an oscillating electric field causes an oscillating magnetic field due to $\bar{\nabla} \times \bar{B} = \frac{1}{c^2}\frac{\partial \bar{E}}{\partial t}$. *With both fields oscillating it is reasonable to expect that at least part of the overall e/m field will be a propagating e/m wave. However, specific scenarios must be considered in order to confirm this rigorously and to derive the full solution including amplitudes, as herein. In particular it will be found that oscillation is not necessary to produce radiant energy. The concept of retarded potentials, introduced in chapter 15, is central to these analyses.*

Contents

22.1 The Oscillating Electric Dipole

In this section we consider an ideal electric dipole such that, in the static case, there would be only a dipole field but no higher multipole fields, nor the field of a net charge. However, we consider that the dipole has a time-varying magnitude (p) but is fixed in direction, namely oriented in the z direction. Later we will assume a sinusoidally varying dipole moment but initially we develop the potentials and fields for an arbitrary time variation in p provided its orientation is fixed.

The solutions for the e/m potentials in terms of time dependent distributions of charge and current have been derived from the Green's functions of the d'Alembertian operator in chapter 15, specifically, for vacuum/air propagation,

$$\Phi(\bar{r}_f, t_f) = \int \frac{\rho(\bar{r}_s, t_s)\delta(c(t_f - t_s) - |\bar{r}_s - \bar{r}_f|)}{4\pi\varepsilon_0 |\bar{r}_f - \bar{r}_s|} d\tau_s c dt_s \tag{22.1.1}$$

$$\bar{A}(\bar{r}_f, t_f) = \frac{\mu_0}{4\pi} \int \frac{\bar{J}(\bar{r}_s, t_s)\delta(c(t_f - t_s) - |\bar{r}_s - \bar{r}_f|)}{|\bar{r}_f - \bar{r}_s|} d\tau_s c dt_s \tag{22.1.2}$$

The effect of the delta functions is to ensure that the sources at position and time $\bar{r}_s, t_s$ only contribute to the potentials, hence fields, at position and time $\bar{r}_f, t_f$ when the source time takes the specific "retarded" value,

$$t_s = t_{retarded} = t_f - |\bar{r}_s - \bar{r}_f|/c \tag{22.1.3}$$

Physically, the difference $t_f - t_s$ is the time taken for the influence of the source to propagate to the field point at the speed of light. Carrying out the time integrals gives,

$$\Phi(\bar{r}_f, t_f) = \int \frac{\rho(\bar{r}_s, t_f - |\bar{r}_s - \bar{r}_f|/c)}{4\pi\varepsilon_0 |\bar{r}_f - \bar{r}_s|} d\tau_s \tag{22.1.4}$$

$$\bar{A}(\bar{r}_f, t_f) = \frac{\mu_0}{4\pi} \int \frac{\bar{J}(\bar{r}_s, t_f - |\bar{r}_s - \bar{r}_f|/c)}{|\bar{r}_f - \bar{r}_s|} d\tau_s \tag{22.1.5}$$

These differ from the expressions for the potentials due to static sources, (5.1.6), (5.5.7), because of the use of the retarded times in the sources.

It is convenient to solve for the vector potential first. The product $\bar{J} d\tau_s$ for a varying dipole can be replaced with $I\bar{L} = \dot{\bar{p}} = \dot{p}\hat{z}$, where p is the time dependent dipole moment and $\dot{p}$ its derivative. We locate the dipole at the origin, noting that we are thereby idealising it as a point source. (22.1.5) then becomes,

$$\bar{A}(\bar{r}_f, t_f) = \frac{\mu_0}{4\pi}\frac{\dot{\bar{p}}(t_f - r_f/c)}{r_f} \tag{22.1.6}$$

The reader is warned that deducing (22.1.6) from (22.1.2) has involved a hidden assumption that the source moves only at non-relativistic speeds. This assumption is hidden in the idealisation that the dipole is a point entity. More realistically, the dipole will comprise equal and opposite charges at some finite distance L. The development here will therefore apply only at distances such that $r \gg L$. But there is a further limitation when considering oscillating dipoles. It is implicitly assumed here that the finite distance between the charges will not cause any significant phase difference, or retarded time difference, between the two charges. This requires that the frequency of oscillation respect $\omega \ll c/L$, i.e., that $L \ll c/\omega$. In §22.2.1 we will derive the relativistic potentials for moving sources, the so-called Liénard-Wiechert potentials, which avoids this limitation.

We can drop the subscripts on the field point now since $\bar{r}_s, t_s$ have been removed from the (22.1.6). This immediately tells us that the vector potential is oriented along the z axis and we can write its components in spherical polars as,

$$A_r = \frac{\mu_0}{4\pi}\frac{\dot{p}(t-r/c)}{r}\cos\theta \qquad A_\theta = -\frac{\mu_0}{4\pi}\frac{\dot{p}(t-r/c)}{r}\sin\theta \tag{22.1.7}$$

We denote the second derivative of p as $\ddot{p}$, and we have $\partial_t p(t - r/c) = \dot{p}$ and $\partial_r p(t - r/c) = -\dot{p}/c$, and it will be understood that $\dot{p}$ and $\ddot{p}$ are to be evaluated at the retarded time, $t - r/c$.

Recalling from Appendix A2 that the divergence in spherical polars when there is no azimuthal variation is $\frac{1}{r^2}\frac{\partial}{\partial r}(r^2 v_r) + \frac{1}{r\sin\theta}\frac{\partial}{\partial\theta}(v_\theta \sin\theta)$ we find,

$$\bar{\nabla}\cdot\bar{A} = -\frac{\mu_0}{4\pi}\left(\frac{[\dot{p}]}{r^2} + \frac{[\ddot{p}]}{rc}\right)\cos\theta \tag{22.1.8}$$

The square brackets indicate the function is evaluated at the retarded time. A shortcut to finding the scalar potential is to recall the Lorenz gauge condition,

$$\bar{\nabla}\cdot\bar{A} + \frac{1}{c^2}\partial_t\Phi = 0 \tag{10.4.6}$$

Note that this is not optional in this case but essential as it is already implicit in the solution (22.1.1,2) which was derived from the Green's function for the d'Alembertian operator, ∂^2, and without the Lorenz gauge assumption the wave equations $\partial^2\Phi = \partial^2\bar{A} = 0$ would not apply (see §10.4.1). From this we can immediately write down,

$$\Phi = \frac{1}{4\pi\varepsilon_0}\left(\frac{[p]}{r^2} + \frac{[\dot{p}]}{rc}\right)\cos\theta \qquad (22.1.9)$$

Hence, we have found the solution for the potentials in terms of the retarded dipole moment and its first and second derivatives, (22.1.7,9). To find the fields is a simple matter of substituting these expressions into,

$$\bar{B} = \bar{\nabla} \times \bar{A} \qquad (5.5.6)$$

$$\bar{E} = -\frac{\partial\bar{A}}{\partial t} - \bar{\nabla}\Phi \qquad (7.2.2)$$

and using the expressions for $grad$ and $curl$ in spherical polars from Appendix A2. The result is,

$$\bar{B} = \frac{\mu_0}{4\pi}\left(\frac{[\dot{p}]}{r^2} + \frac{[\ddot{p}]}{rc}\right)\hat{\phi}\sin\theta \qquad (22.1.10a)$$

$$\bar{E} = \frac{1}{4\pi\varepsilon_0}\left\{2\left(\frac{[p]}{r^3} + \frac{[\dot{p}]}{r^2c}\right)\hat{r}\cos\theta + \left(\frac{[p]}{r^3} + \frac{[\dot{p}]}{r^2c} + \frac{[\ddot{p}]}{rc^2}\right)\hat{\theta}\sin\theta\right\} \qquad (22.1.10b)$$

The electric and magnetic fields are therefore orthogonal. The fields may be considered as in three parts according to their r dependence, thus,

- The terms in the electric field in $\frac{[p]}{r^3}$ are seen to be the same as the dipole field of (5.2.5) except the dipole moment is replaced by its retarded value, as we would expect.

- The term in $\frac{[\dot{p}]}{r^2}$ in the magnetic field can be seen to be just that due to a current element $Id\bar{L} = \dot{\bar{p}}$ given by the Biot-Savart field (4.2.7), noting that $Id\bar{L} \times \hat{r} = \hat{\phi}[\dot{p}]\sin\theta$. Substitution of this in $\bar{\nabla} \times \bar{B} = \frac{1}{c^2}\frac{\partial\bar{E}}{\partial t}$ requires the corresponding terms in $\frac{[\dot{p}]}{r^2}$ in the electric field (as well as the dipole field). These terms are sometimes called the "induction fields".

- This leaves the terms in $\frac{[\ddot{p}]}{r}$ in both fields. These are the e/m radiation fields and are the only parts of the fields that propagate energy away to infinity. For any other pair of terms their product decreases at least as fast as $1/r^3$ and hence their integral over a sphere becomes zero as the sphere is taken infinitely large.

Separating out the radiation terms we write,

$$\bar{B}_{rad} = \frac{\mu_0[\ddot{p}]}{4\pi cr}\hat{\phi}\sin\theta \qquad (22.1.11a)$$

$$\bar{E}_{rad} = \frac{[\ddot{p}]}{4\pi\varepsilon_0 c^2 r}\hat{\theta}\sin\theta \qquad (22.1.11b)$$

A useful exercise is to confirm that the complete fields (22.1.10a,b) obey all Maxwell's equations in free space. Note that the radiation fields alone, i.e., the above terms in $[\ddot{p}]$, do not separately respect all four Maxwell equations because the induction fields will give terms in $[\ddot{p}]$ on differentiation of terms in $[\dot{p}]$.

The cross product of (22.1.11a,b) gives the Poynting vector to be,

$$\bar{N} = \bar{E}\times\bar{H} = \frac{1}{(4\pi)^2\varepsilon_0 c^3}\left(\frac{[\ddot{p}]\sin\theta}{r}\right)^2\hat{r} \qquad (22.1.12)$$

This shows that the radiated power has a $\sin^2\theta$ angular dependence. The reader is reminded that this is a solution only for non-relativistic source speeds.

Integration over a sphere readily gives the total power being radiated away to infinity,

$$\text{Power} = \frac{[\ddot{p}]^2}{6\pi\varepsilon_0 c^3} \qquad (22.1.13)$$

So far we have left the time variation of the dipole moment completely open. But if we assume a sinusoidal variation $p = p_o\cos\omega t$ then the average of $[\ddot{p}]^2$ over a whole number of cycles is $p_0^2\omega^4/2$ and the average radiated power is then,

$$\text{Average power} = \frac{p_0^2\omega^4}{12\pi\varepsilon_0 c^3} \qquad (22.1.14)$$

This shows how sensitive is the radiated power to the frequency of oscillation, as well as its amplitude.

22.2 A Point Charge in Arbitrary Motion

22.2.1 Liénard-Wiechert Potentials

The solutions for the e/m potentials in terms of time dependent distributions of charge and current have been derived from the Green's functions of the d'Alembertian operator in chapter 15 and reproduced in this chapter specifically for vacuum/air propagation in (22.1.1,2). The two features by which these expressions differ from those applicable under static conditions, namely,

$$\Phi(\bar{r}_f) = \int \frac{\rho(\bar{r}_s)}{4\pi\varepsilon_0|\bar{r}_f - \bar{r}_s|} d\tau_s \tag{5.1.6}$$

$$\bar{A}(\bar{r}_f) = \frac{\mu_0}{4\pi}\left(\int \frac{1}{|\bar{r}_f - \bar{r}_s|} \bar{J}(\bar{r}_s)\, d\tau_s\right) \tag{5.5.7}$$

are (i) the replacement of the spatial integral with a spacetime integral, and, (ii) the presence of the delta functions which mean that, after the time integral is carried out, the sources are evaluated at the retarded time $t_s = t_f - |\bar{r}_s - \bar{r}_f|/v$.

We are interested here in a point charge following some arbitrary path, which in general may involve accelerations and even higher order time derivatives. This is described by specifying $\bar{r}_s$ as a function of time, $\bar{r}_s = \bar{f}(t_s)$. The charge density is therefore a spatial delta function,

$$\rho(\bar{r}_s, t_s) = q\delta^3\left(\bar{r}_s - \bar{f}(t_s)\right) \tag{22.2.1}$$

Substitution in (22.1.1) gives,

$$\Phi(\bar{r}_f, t_f) = \int \frac{q\delta^3(\bar{r}_s - \bar{f}(t_s))\delta(c(t_f - t_s) - |\bar{r}_s - \bar{r}_f|)}{4\pi\varepsilon_0|\bar{r}_f - \bar{r}_s|} d\tau_s c dt_s \tag{22.2.2}$$

Carrying out the spatial integral gives,

$$\Phi(\bar{r}_f, t_f) = \int \frac{q\delta(c(-t_f + t_s) + |\bar{f}(t_s) - \bar{r}_f|)}{4\pi\varepsilon_0|\bar{r}_f - \bar{f}(t_s)|} c dt_s \tag{22.2.3}$$

It is straightforward to prove, e.g., by substitution, the general theorem,

$$\int f(x)\delta(g(x))dx = \left.\frac{f}{\frac{dg}{dx}}\right|_{g=0} \tag{22.2.4}$$

We therefore note that,

$$\frac{d}{dt_s}\left(c(-t_f + t_s) + |\bar{f}(t_s) - \bar{r}_f|\right)$$

$$= c + \frac{(f_x - x)f_x' + (f_y - y)f_y' + (f_z - z)f_z'}{|\bar{r}_f - \bar{f}(t_s)|}$$

$$= c - \frac{\left(\bar{r}_f - \bar{f}(t_s)\right)\cdot\bar{f}'}{|\bar{r}_f - \bar{f}(t_s)|} = c - \hat{r}_r \cdot \bar{f}' = c - \hat{r}_r \cdot \bar{u}_r \tag{22.2.5}$$

where we have put $\bar{r}_r = \bar{r}_f - \bar{f}(t_s)$, the retarded distance from source to field point, and $\hat{r}_r = \bar{r}_r / r_r$, and $\bar{u}_r = \bar{f}'(t_s)$ is the instantaneous velocity of the charge at the retarded time.

The condition in (22.2.4) that the RHS is evaluated at $g = 0$ is just the definition of retarded time, t_s,

$$t_s = t_f - \left|\bar{r}_f - \bar{f}(t_s)\right|/c \tag{22.2.6}$$

Hence, we finally arrive at,

$$\Phi\left(\bar{r}_f, t_f\right) = \frac{q}{4\pi\varepsilon_0 r_r(1-\hat{r}_r\cdot\bar{u}_r/c)} = \frac{q}{4\pi\varepsilon_0(r_r-\bar{r}_r\cdot\bar{u}_r/c)} \tag{22.2.7}$$

The expression for the vector potential follows in the same way, noting that the current due to the moving point charge is,

$$\bar{J}(\bar{r}_s, t_s) = q\bar{u}\delta^3\left(\bar{r}_s - \bar{f}(t_s)\right) \tag{22.2.8}$$

So we get,

$$\bar{A}\left(\bar{r}_f, t_f\right) = \frac{\mu_0 q \bar{u}_r}{4\pi(r_r-\bar{r}_r\cdot\bar{u}_r/c)} \tag{22.2.9}$$

The potentials (22.2.7) and (22.2.9) are known as the Liénard-Wiechert potentials. Both the source position vector, $\bar{r}_r$, and the velocity of the charge, $\bar{u}_r$, are retarded quantities.

Note that we have adopted the usual sign convention here such that the velocity $\bar{u}_r$ is positive when the charge is moving towards the field point.

I challenge the reader to devise a convincing physical argument to explain the origin of the denominators of $(r_r - \bar{r}_r \cdot \bar{u}_r/c)$ in (22.2.7,9), rather than simply the retarded distance, r_r, alone as one might have expected. Some authors attempt to do so, but I have not seen one that convinces. Its mathematical origin in the integration over the delta function in (22.2.3) is clear, however. That the additional term, $-\bar{r}_r \cdot \bar{u}_r/c$, depends upon u_r/c betrays it to be a relativistic correction (in contrast to the non-relativistic development of §22.1).

22.2.2 The Fields

In principle the derivation of the fields from the potentials, (22.2.7,9), is a straightforward matter of substitution in (5.5.6) and (7.2.2), i.e., $\bar{B} = \bar{\nabla} \times \bar{A}$ and $\bar{E} = -\frac{\partial \bar{A}}{\partial t} - \bar{\nabla}\Phi$. However there are some technical difficulties due to the quantities being evaluated at the retarded time. Dropping subscripts and understanding all terms to be retarded, we have, succinctly,

$$\Phi = \frac{q}{4\pi\varepsilon_0 s} \qquad \bar{A} = \frac{\mu_0 q\bar{u}}{4\pi s} \qquad s = r - \bar{r}\cdot\bar{u}/c \tag{22.2.10}$$

Because of the sign convention that the velocity $\bar{u}$ is positive when the charge is moving towards the field point we have $\bar{u} = -\dot{\bar{r}}$. We now proceed by deriving a number of identities which allow conversion from differential expressions in t to differential expressions in the retarded time which we now denote T,

$$T = t - r/c \tag{22.2.11}$$

Firstly we note that,

$$\frac{\partial r^2}{\partial T} = 2r\frac{\partial r}{\partial T} = 2x\frac{\partial x}{\partial T} + 2y\frac{\partial y}{\partial T} + 2z\frac{\partial z}{\partial T} = -2\bar{r}\cdot\bar{u}$$

So $$\frac{\partial r}{\partial T} = -\frac{\bar{r}\cdot\bar{u}}{r} \tag{22.2.12}$$

Next we note that $r = c(t - T)$ gives,

$$\frac{\partial r}{\partial t} = \frac{\partial r}{\partial T}\frac{\partial T}{\partial t} = -\left(\frac{\bar{r}\cdot\bar{u}}{r}\right)\frac{\partial T}{\partial t} = c\left(1 - \frac{\partial T}{\partial t}\right)$$

Hence, $$\frac{\partial T}{\partial t} = \frac{1}{1-\left(\frac{\bar{r}\cdot\bar{u}}{rc}\right)} = \frac{r}{s} \tag{22.2.13}$$

Hence, $$\frac{\partial}{\partial t} = \frac{r}{s}\frac{\partial}{\partial T} \tag{22.2.14}$$

That takes care of the translation of explicit time derivatives in terms of the retarded time, but we also need to translate the spatial derivatives in $\bar{\nabla}$ (in which t is held constant) into a form of spatial derivative which we write as $\bar{\nabla}_T$ which holds T constant. This can be illustrated by considering ∂_x acting on a function $f(\bar{r},t) = F(\bar{r},T)$ where T is understood to be a function of $\bar{r},t$. We have,

$$\partial_x f(\bar{r},t) = \partial_x F(\bar{r},T) + \frac{\partial T}{\partial x}\frac{\partial F}{\partial T} \tag{22.2.15}$$

Hence, $$\bar{\nabla} = \bar{\nabla}_T + (\bar{\nabla}T)\frac{\partial}{\partial T} \tag{22.2.16}$$

To derive $\bar{\nabla}_T$ we firstly note that (22.2.16) gives,

$$\bar{\nabla}r = \bar{\nabla}_T r + (\bar{\nabla}T)\frac{\partial r}{\partial T} = \hat{r} + (\bar{\nabla}T)\frac{\partial r}{\partial T} \tag{22.2.17}$$

Here we have used $\overline{\nabla}_T r = \hat{r}$ because $\overline{\nabla}_T$ is precisely that version of the $grad$ operator which legitimises ignoring that r is a retarded quantity and so allows us to use, for a function of r only, that $\overline{\nabla}_T \equiv \hat{r}\partial_r$ and hence that $\overline{\nabla}_T r = \hat{r}$. But we also have that,

$$\overline{\nabla} r = \overline{\nabla} c(t - T) = -c\overline{\nabla} T \tag{22.2.18}$$

Equating (22.2.17,18) and using (22.2.12) gives,

$$\overline{\nabla} T = \frac{-\hat{r}}{c + \frac{\partial r}{\partial T}} = \frac{-\hat{r}}{c - \frac{\bar{r}\cdot\bar{u}}{r}} = -\frac{\bar{r}}{cs} \tag{22.2.19}$$

So, finally, (22.2.16) becomes,

$$\overline{\nabla} = \overline{\nabla}_T - \frac{\bar{r}}{cs}\frac{\partial}{\partial T} \tag{22.2.20}$$

This is the sought-for translation that will now permit us to evaluate $\overline{\nabla}$ in either $grad$ or $curl$ expressions by taking the $\overline{\nabla}_T$ part to operate as normal by ignoring the quantities being retarded, and similarly $\frac{\partial}{\partial T}$ directly creates velocity and acceleration terms which are retarded. In the first instance we therefore get,

$$\bar{E} = -\left(\overline{\nabla}_T - \frac{\bar{r}}{cs}\frac{\partial}{\partial T}\right)\frac{q}{4\pi\varepsilon_0 s} - \frac{r}{s}\frac{\partial}{\partial T}\left(\frac{\mu_0 q\bar{u}}{4\pi s}\right)$$

$$\Rightarrow \bar{E} = -\frac{q}{4\pi\varepsilon_0}\left\{\overline{\nabla}_T\left(\frac{1}{s}\right) - \frac{\bar{r}}{cs}\frac{\partial}{\partial T}\left(\frac{1}{s}\right) + \frac{r}{sc^2}\frac{\partial}{\partial T}\left(\frac{\bar{u}}{s}\right)\right\}$$

$$\Rightarrow \bar{E} = -\frac{q}{4\pi\varepsilon_0}\left\{-\frac{1}{s^2}\overline{\nabla}_T s + \frac{\bar{r}}{cs^3}\frac{\partial s}{\partial T} + \frac{r}{sc^2}\left(\frac{\dot{\bar{u}}}{s} - \frac{\bar{u}}{s^2}\frac{\partial s}{\partial T}\right)\right\} \tag{22.2.21}$$

Noting that,

$$\overline{\nabla}_T s = \hat{r} - \bar{u}/c \quad \text{and} \quad \frac{\partial s}{\partial T} = -\frac{\bar{r}\cdot\bar{u}}{r} - \frac{\bar{r}\cdot\dot{\bar{u}}}{c} + \frac{u^2}{c} \tag{22.2.22}$$

$$\bar{E} = -\frac{q}{4\pi\varepsilon_0}\left\{-\frac{1}{s^2}\left(\hat{r} - \frac{\bar{u}}{c}\right) + \frac{r\dot{\bar{u}}}{c^2 s^2} + \left(\frac{\bar{r}}{cs^3} - \frac{r\bar{u}}{c^2 s^3}\right)\left(-\frac{\bar{r}\cdot\bar{u}}{r} - \frac{\bar{r}\cdot\dot{\bar{u}}}{c} + \frac{u^2}{c}\right)\right\}$$

$$\Rightarrow \bar{E} = \frac{q}{4\pi\varepsilon_0 s^3}\left\{s\left(\hat{r} - \frac{\bar{u}}{c}\right) - \frac{rs\dot{\bar{u}}}{c^2} + \left(\frac{\bar{r}}{c} - \frac{r\bar{u}}{c^2}\right)\left(\frac{\bar{r}\cdot\bar{u}}{r} + \frac{\bar{r}\cdot\dot{\bar{u}}}{c} - \frac{u^2}{c}\right)\right\} \tag{22.2.23}$$

It is convenient to separate the terms in acceleration, and also substituting for s,

$$\bar{E}_{acc} = \frac{q}{4\pi\varepsilon_0 s^3}\left[\left(\frac{\bar{r}}{c} - \frac{r\bar{u}}{c^2}\right)\frac{\bar{r}\cdot\dot{\bar{u}}}{c} - \frac{r\dot{\bar{u}}}{c^2}\left(r - \frac{\bar{r}\cdot\bar{u}}{c}\right)\right] \tag{22.2.24}$$

Using the vector identity (A1.3) in the form,

$$\bar{a} \times (\bar{b} \times \bar{c}) = (\bar{a} \cdot \bar{c})\bar{b} - (\bar{a} \cdot \bar{b})\bar{c}$$

one can expand $\bar{r} \times \left(\left(\bar{r} - \frac{r\bar{u}}{c}\right) \times \dot{\bar{u}}\right)$ to show that it equals c^2 times the term in brackets in (22.2.24), so,

$$\bar{E}_{acc} = \frac{q}{4\pi\varepsilon_0 c^2 s^3}\left[\bar{r} \times \left(\left(\bar{r} - \frac{r\bar{u}}{c}\right) \times \dot{\bar{u}}\right)\right] \quad (22.2.25)$$

The remaining terms simplify to,

$$\bar{E}_{rest} = \frac{q}{4\pi\varepsilon_0 s^3}\left(1 - \frac{u^2}{c^2}\right)\left(\bar{r} - \frac{r\bar{u}}{c}\right) \quad (22.2.26)$$

Overall, $\bar{E} = \bar{E}_{acc} + \bar{E}_{rest}$. A similar analysis of $\bar{B} = \bar{\nabla} \times \bar{A}$ leads to,

$$c\bar{B} = \hat{r} \times \bar{E} \quad \text{and} \quad c\bar{B}_{acc} = \hat{r} \times \bar{E}_{acc} \quad (22.2.27)$$

The presence of the relativistic gamma factor, $\gamma = \left(1 - \frac{u^2}{c^2}\right)^{-\frac{1}{2}}$, in these fields betrays the relativistic nature of the solution (see chapters 26, 27).

Practical examples of e/m radiation emitted by accelerated (or decelerated) charges are synchrotron radiation (circular motion, generally at relativistic speeds) and bremsstrahlung (radiation due to slowing or stopping an electron beam, this being the usual technological source of X-rays).

22.2.3 Non-Relativistic Power Radiated

For speeds small compared with c and so with $s \approx r$ and noting that $\bar{r} \times (\bar{r} \times \dot{\bar{u}}) = r^2\dot{u}\hat{\theta}\sin\theta$ where the direction of acceleration is the polar axis, the radiation fields become, approximately,

$$\bar{E}_{acc} = \frac{q\dot{u}\hat{\theta}}{4\pi\varepsilon_0 c^2 r}\sin\theta \quad (22.2.28a)$$

$$\bar{B}_{acc} = \frac{q\dot{u}\hat{\phi}}{4\pi\varepsilon_0 c^3 r}\sin\theta = \frac{\mu_0 q\dot{u}\hat{\phi}}{4\pi c r}\sin\theta \quad (22.2.28b)$$

These are the same as the radiation fields due to an oscillating electric dipole, (22.1.11a,b) except that $q\dot{u}$ replaces $[\ddot{p}]$. The total power integrated over a large sphere is correspondingly the same,

$$\text{Power} = \frac{q^2\dot{u}^2}{6\pi\varepsilon_0 c^3} \quad (22.2.29)$$

In particular we note that, in this non-relativistic case, the angular distribution of the power is $\sin^2\theta$. We will see below that this is radically altered by relativistic motion.

22.3 Linear Acceleration for a Finite Period

Consider a point charge, initially stationary, which undergoes a constant acceleration, a, in the z direction for a finite period, τ. Thereafter its velocity remains constant at $a\tau^2\hat{z}/2$ (assuming non-relativistic speed). We consider this simple case to emphasise that radiated power does not always come in the form of oscillatory e/m waves.

The radiated power does travel radially, of course, as the Poynting vector $\bar{N} = \bar{E}_{acc} \times \bar{B}_{acc}/\mu_0$ confirms. But the fields (22.2.28a,b) are zero except within the time period τ, and during that period they are constant, not oscillatory. Nevertheless they correspond to a burst of power over this finite duration which then travels as a spherically symmetric pulse of radiation, without oscillation.

22.3.1 Relativistic Angular Distribution of Radiated Power

We envisage here the same scenario as above, specifically with acceleration in a constant direction and parallel to the velocity. (22.2.25,27) then become, putting $a = \dot{u}$ (the acceleration), and recalling that the acceleration and the velocity defines the polar axis, and that $\bar{r}$ is the field point wrt the source,

$$\bar{E}_{acc} = \frac{qar^2\hat{\theta}}{4\pi\varepsilon_0 c^2 s^3}\sin\theta \qquad (22.3.1a)$$

$$\bar{B}_{acc} = \frac{\mu_0 qar^2\hat{\phi}}{4\pi cs^3}\sin\theta \qquad (22.3.1b)$$

Also putting $\beta = u/c$ then,

$$s = r(1 - \hat{r}\cdot\bar{u}/c) = r(1 - \beta\cos\theta) \qquad (22.3.2)$$

Hence the radiation fields become,

$$\bar{E}_{acc} = \frac{qa\hat{\theta}}{4\pi\varepsilon_0 c^2 r}\frac{\sin\theta}{(1-\beta\cos\theta)^3} \qquad (22.3.3a)$$

$$\bar{B}_{acc} = \frac{\mu_0 qa\hat{\phi}}{4\pi cr}\frac{\sin\theta}{(1-\beta\cos\theta)^3} \qquad (22.3.3b)$$

The Poynting vector is thus,

$$\bar{N} = \frac{q^2a^2\hat{r}}{(4\pi)^2\varepsilon_0 c^3 r^2}\frac{\sin^2\theta}{(1-\beta\cos\theta)^6} \qquad (22.3.4)$$

So we regain the non-relativistic distribution of radiant power, i.e., $\propto \sin^2\theta$, as derived in §22.2.3 when $\beta \ll 1$.

But it only makes physical sense to consider the period of acceleration, τ, to be defined in terms of the retarded time, T, i.e., the time referenced to the charge itself. (If it were interpreted as an interval of the delayed time, t, that would make the time interval dependent on the charge's spatial location). Hence, to adjust the Poynting vector (22.3.4) for the retarded time we must factor it by $\frac{\partial t}{\partial T}$ which (22.2.13) gives as $\frac{s}{r} = 1 - \beta\cos\theta$. Hence we get the adjusted Poynting vector to be,

$$\overline{N} = \frac{q^2 a^2 \hat{r}}{(4\pi)^2 \varepsilon_0 c^3 r^2} \frac{\sin^2\theta}{(1-\beta\cos\theta)^5} \tag{22.3.5}$$

This gives the angular distribution of the radiant power in the relativistic case. In the extreme relativistic limit that $u \to c$, i.e., $\beta \to 1$, the denominator of $1 - \beta\cos\theta$ causes the power flux to be very strongly peak near, but not at, the forward direction, i.e., near the charge's motion and acceleration.

This is illustrated graphically in Figure 22.1 and as polar plots in Figures 22.2 to 22.4 for the non-relativistic, relativistic and extreme relativistic cases respectively. In the non-relativistic case the power is radiated sideways wrt the direction of motion. But in the relativistic cases the lobes of peak radiation swing round to (nearly) face the forward direction, like the headlamp beams of a car. The exact forward direction, however, sees no radiation, because of the factor of $\sin^2\theta$.

The total radiated power is found by integrating (22.3.5) over all angles around a sphere, i.e.,

$$\text{Power} = \int_{sphere} \overline{N} \cdot \hat{r}\, r^2 d\Omega \tag{22.3.6}$$

The exact integral is found to be,

$$\text{Power} = \frac{q^2 a^2}{8\pi\varepsilon_0 c^3 \beta^3} \left\{ A\left(\frac{\beta^2-1}{4}\right) + \frac{2}{3}B - \frac{1}{2}C \right\} \tag{22.3.7}$$

where,

$$A = \left[\frac{1}{(1-\beta)^4} - \frac{1}{(1+\beta)^4}\right] \quad B = \left[\frac{1}{(1-\beta)^3} - \frac{1}{(1+\beta)^3}\right] \quad C = \left[\frac{1}{(1-\beta)^2} - \frac{1}{(1+\beta)^2}\right]$$

In the non-relativistic limit, $\beta \to 0$, the { } in (22.3.7) becomes $\frac{4}{3}\beta^3$ to leading order so the power becomes,

$$\text{Power (non-relativistic)} = \frac{q^2 a^2}{6\pi\varepsilon_0 c^3} \tag{22.3.8}$$

in agreement with (22.2.29). In the extreme relativistic limit, $\beta \to 1$, we find in sharp contrast,

$$\text{Power (extreme relativistic)} = \frac{q^2 a^2}{48\pi\varepsilon_0 c^3 (1-\beta)^3} = \frac{\gamma^6 q^2 a^2}{6\pi\varepsilon_0 c^3} \tag{22.3.9}$$

where $\gamma = 1/\sqrt{1-\beta^2}$ is the relativistic gamma factor. So we see that extreme relativistic speeds increase the radiated power by a potentially massive factor of γ^6. At 99% of light speed this increases the radiant power by more than five orders of magnitude.

Figure 22.1: Angular Distribution of Radiant Power due to Linearly Accelerated Charge at up to Relativistic Speeds (Graphical) *normalised by power density at* $\theta = \pi/2$.

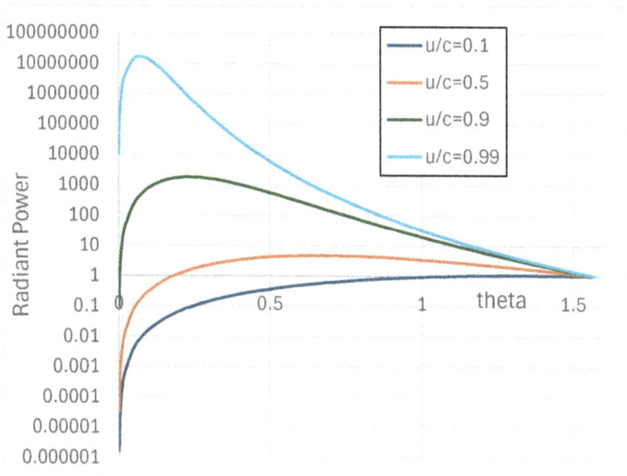

Figure 22.2: Angular Distribution of Radiant Power due to Linearly Accelerated Charge, Non-Relativistic (Actual Shape) *normalised by power density at* $\theta = \pi/2$.

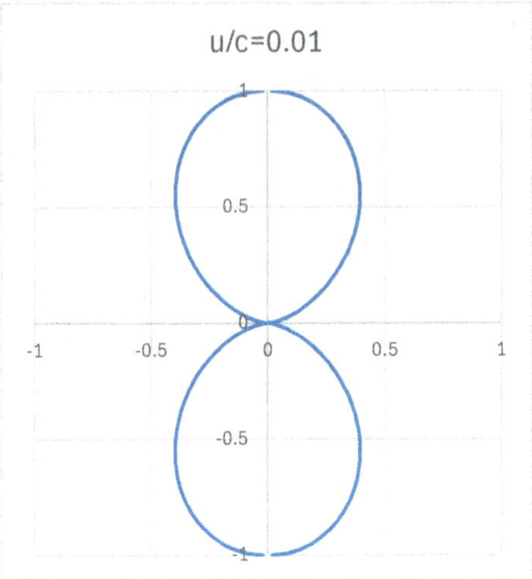

Figure 22.3: Angular Distribution of Radiant Power due to Linearly Accelerated Charge, Relativistic u/c=0.9 (Actual Shape) *normalised by power density at* $\theta = \pi/2$. *Power density is given by the radial coordinate in this plot.*

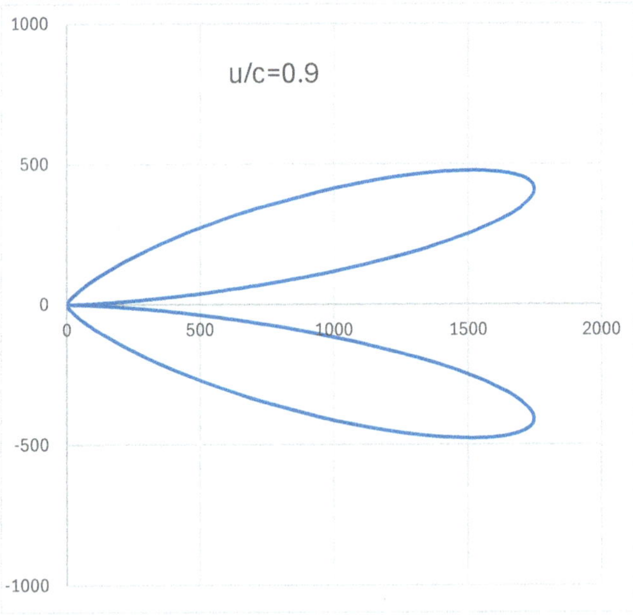

Figure 22.4: Angular Distribution of Radiant Power due to Linearly Accelerated Charge, Extreme Relativistic u/c=0.99 (Actual Shape) *normalised by power density at* $\theta = \pi/2$. *Power density is given by the radial coordinate in this plot.*

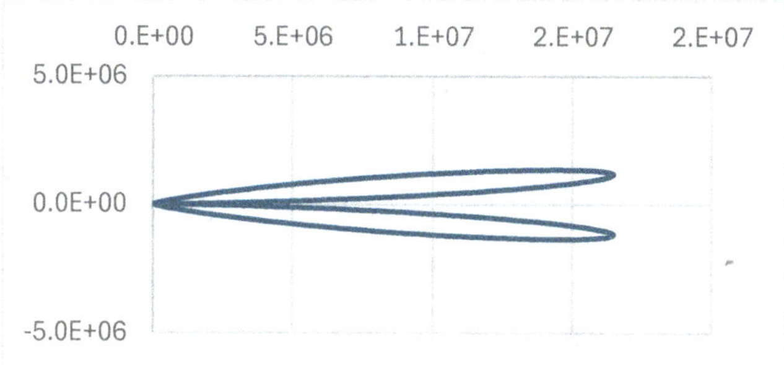

22.4 Instantaneous Circular Motion

The general solution (22.2.25,27) can be used to evaluate the fields and energy flux in the case when the instantaneous acceleration is perpendicular to the velocity, i.e., the motion is locally a circular arc. Details are not given here but can be an exercise for the reader (see Jackson, [22.1], if stuck).

When relativistic, the exact angular distribution of radiant power differs from (22.3.5) but is qualitatively similar with a very large increase in power radiated near the 'forward' direction, similar to Figures 22.3,4. However, the 'forward' direction is now sweeping out a circular arc, so the beams of radiation are like the headlights of a car driving around a circular bend.

However, in the extreme relativistic case, the total radiated power is less than in the case of linear acceleration as the calculation shows the relativistic gamma factor appears to the power 4 rather than to power 6, specifically,

$$\text{Power (extreme relativistic)} = \frac{\gamma^4 q^2 a^2}{6\pi\varepsilon_0 c^3} \tag{22.4.1}$$

22.5 Does a Charge Uniformly Accelerated Forever Radiate?

This is a famous conundrum, avoidance of which is the reason why in the above discussion the acceleration was assumed to occur only for a finite duration. In contrast, if we assume that the charge has been accelerating forever, and will continue to accelerate forever, there is a problem. General relativity allows us to transform to the frame of reference in which the charge is stationary but instead experiences a gravitational field. But if the charge is radiating energy to infinity, how is this possible when the charge is stationary and merely in a gravitational field? A charge which is stationary on your lab bench in the Earth's gravitational field does not radiate. Many authors have attempted to address this rather pathological case, with conflicting results. Pauli was of the view that there is no radiation in this case, whereas Drukey came to the opposite conclusion. The matter has been discussed by Bondi and Gold, [22.2], who opine that there is radiation and the relativity argument fails because the gravitation betrays itself by inevitable inhomogeneity.

22.6 Where Does the Energy Come From? Radiation Reaction

It was implicitly assumed in the above treatments that the charge or dipole could be maintained in some specified motion by some external agency. But as energy is being radiated away, it follows that the source must be providing that energy. In other words there must be a backreaction on the source tending to slow down its motion – unless its motion is sustained by force from an external agency which continuously supplies the balancing energy.

Conversely, if the charge is not subject to any external force, then its motion will clearly be damped in order to supply the energy radiated away.

The issue of the backreaction or damping force on the source is a fraught one which unavoidably encounters the fundamental nature of the source itself, i.e., the nature of a charged particle, because it involves the effect of the source's field on itself. We alluded to Feynman's attempt to get to grips with this issue in his thesis, [22.4].

Some feel for the matter is easily accomplished simply by considering energies, rather than forces. As an illustration consider a classical (i.e., non-quantum) model of a hydrogen atom. Assuming the electron is in a circular orbit of the Bohr radius (0.53×10^{-10}m), held there by the Coulomb force, then the angular speed of the electron is $\omega = \sqrt{\frac{q^2}{4\pi\varepsilon_0 mr^3}}$ so $\omega = 4.1 \times 10^{16}$ and the radiated power, using the non-relativistic formula, (22.3.8), is 4.5×10^{-8} W compared with the kinetic energy of 2.2×10^{-18} J. This illustrates the problem that led to the development of the Bohr model of the atom, and thence to quantum mechanics, as classical physics suggests the electron would spiral into the nucleus on a timescale of a small fraction of a nanosecond. The solution to the damping force problem in this case is that there is no problem because the classical theory of electromagnetism does not apply and quantum mechanics holds that there is no radiation (provided the electron is in the ground state).

A crude model for the magnitude of the damping force can be obtained by equating the total radiated power with the product of a damping force and the charge's velocity, this product being the work done by the charge in overcoming the radiation reaction. This leads to a damping force given by,

$$\bar{F}_{damping} = \frac{q^2}{6\pi\varepsilon_0 c^3}\ddot{\bar{v}} \tag{22.6.1}$$

where $\bar{v}$ is the charge's velocity. This damping force, known as the Abraham-Lorentz model, involves the third temporal derivative of the charge's position, i.e., the rate-of-change of its acceleration, which is unusual. The equation of motion of the charge would then become,

$$m\dot{\bar{v}} = \bar{F}_{em} - \bar{F}_{damping} \tag{22.6.2}$$

However, this model is pathological and will produce runaway solutions is treated literally. It can be used as a rough guide only if the damping force is treated as a small perturbation on the undamped solution. For further discussion of what classical developments can yield for radiation damping see Jackson, [22.1], chapter 17. The bottom line, though, is no thoroughly satisfactory solution exists.

22.7 A Charge Moving Parallel to a Corrugated Conducting Surface

In §12.7 we saw how a magnetron generates microwaves due to the passage of electrons past a periodically shaped anode. Here we look at a rather heuristic description of the mechanism behind the phenomenon. We envisage an electron passing at speed u parallel to a conducting surface. But the conducting surface has a regular undulation in its surface given by $y = y_0 \sin\frac{2\pi x}{L}$.

The electron is at a distance $y = D$ from the surface. In reality the electron would be attracted to the surface due to the induced positive charges there. However, we assume that by some contrivance the electron is caused to maintain its velocity uniform, proceeding in a straight line in the x direction at a constant $y = D$. If the undulations are of small amplitude compared to their spatial period, L, then the effect of the conducting surface on the e/m field may be approximated by a moving image charge which will be at $y = 2y_0 \sin\frac{2\pi x}{L} - D$. But as the charge is moving we have $x = ut$ so that the image charge is undergoing sinusoidal oscillations $y(t) = 2y_0 \sin\frac{2\pi ut}{L} - D$.

Consequently, e/m radiation is expected to be produced. Indeed there is, as Smith and Purcell first reported in 1953, having observed visible light produced in their apparatus, [22.3]. The image charge is a fiction, of course. What is really happening is that the positive surface charge density induced by the electron moves along with the electron, and thus moves up and down the undulations. This real charge therefore undergoes real oscillations. (In Smith and Purcell's original work a grating was used rather than a corrugated surface).

This very heuristic analysis ignores the fact that it would take some finite time for the image charge to be set up, by which time the charge will have moved on. In other words there is 'retarded time' factor at work, as is inevitable with any moving source. We have ignored this for simplicity of explaining the qualitative phenomenon.

The frequency of the radiation would be expected to correspond to that of the electron passing each undulation, hence $f = \frac{u}{L}$. The wavelength of the radiation would then be $\lambda = cL/u$. As Smith and Purcell showed in their original work, [22.3], the wavelength actually diminishes with viewing angle. Moreover, because they used a grating there are other "diffraction orders" so their formula for the radiation frequency was $\lambda = \frac{cL}{nu}\left(1 - \frac{u}{c}\cos\theta\right)$ for a sequence $n = 1,2,3 \ldots$. While our analysis here is very heuristic, it correctly predicts the qualitative phenomenon of Smith-Purcell radiation.

22.8 References

[22.1] J.D.Jackson (1975). Classical Electrodynamics. John Wiley & Sons. (See §14.3).

[22.2] H.Bondi and T.Gold (1955). The field of a uniformly accelerated charge, with special reference to the problem of gravitational acceleration. Royal Society Proceedings Series A, Vol.229, p416-424 (24 May 1955).

[22.3] G.Doucas (2025). Simon-Purcell Radiation: Basic Theory and Applications. Oxford University Press.

[22.4] Feynman, R.P. (1942). *Feynman's PhD Thesis: The Principle of Least Action in Quantum Mechanics*. In the book *Feynman's Thesis: A New Approach to Quantum Mechanics* with a Preface by Laurie Brown, 2005, World Scientific.

23. Waveguides

Here we present the basic theory of e/m waveguides, firstly with conducting boundaries where transmission is through the air. Solutions are developed for straight waveguides with cross sections that are rectangular or circular. The crucial difference is emphasised between singly-connected and multiply-connected geometries (i.e., with two or more conductors). Transmission lines were considered in §21.8 but we mention in passing that transmission lines can also be considered to be multiple-conductor waveguides. Optic fibres for communication purposes are also waveguides for e/m radiation, albeit in the optical (actually infrared) frequency regime. The structure and data rate capacity of optic fibres are discussed and the basic theory of dielectric waveguides is derived. Application to optic fibre explains why some fibres, especially the most efficient, are "single mode" waveguides. Single mode transmission is seen to arise only when the frequency is below a "cutoff" frequency characteristic of the particular fibre. Above the cutoff frequency multiple transmission modes become excited. Operation below the cutoff frequency therefore results in the greatest efficiency and the longest transmission distances before signal boosting becomes necessary.

Contents

23.1 Conducting Waveguides, Basic Theory

The simple theory developed here will be based on the idealisation that the conducting boundaries which form the waveguide have effectively infinite conductivity. Consequently, within the metal material itself the electric field is zero. In particular this means that $\bar{\nabla} \times \bar{E} = -\frac{\partial \bar{B}}{\partial t} = 0$ so that the only possible magnetic field within the metal is static. Boundary conditions under static conditions were specified in §6.7 but now we need to rethink them for the case of time dependence, as required here.

The easy case is the magnetic field because $\bar{\nabla} \cdot \bar{B} = 0$ is always true and this leads via Stoke's theorem to the boundary condition that the component of the magnetic field normal to the conducting surface is zero,

$$B_n = 0 \tag{23.1.1a}$$

For the electric field, consider integrating $\bar{\nabla} \times \bar{E} = -\frac{\partial \bar{B}}{\partial t}$ over a narrow rectangular area whose long side runs parallel to the conductor surface, partly inside and partly outside the metal, as illustrated in Figure 23.1.

Figure 23.1: Integration contour to derive (23.1.1b)

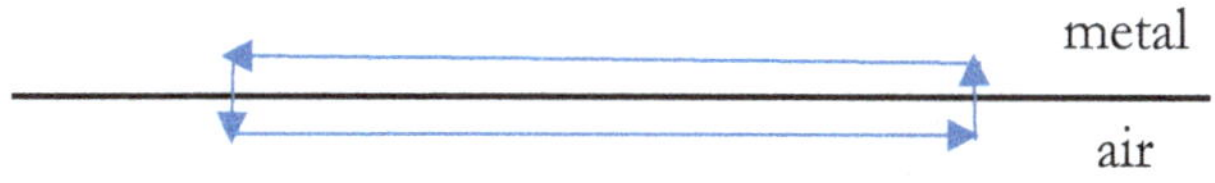

As the width of this contour shrinks to zero the area integral over $-\frac{\partial \bar{B}}{\partial t}$ becomes zero, which implies by Stoke's theorem that the line integral of $\bar{E}$ is zero. It follows that the tangential components of the electric field are continuous, and hence zero on the air side as they are within the metal, i.e.,

$$E_t = 0 \tag{23.1.1b}$$

where the subscript t refers to any tangential direction.

We now assume a prismatic conductor, that is a long straight conductor with a uniform cross section which may be any shape, including possibly the case of a multiply connected conductor (i.e., two or more separate conductors). The axial direction will be taken to be the Cartesian z direction so the cross section of the waveguide is the x, y plane. The purpose of the waveguide is to propagate e/m waves, and hence energy, along the axial (z) direction.

The rest of this first section considers the conditions under which both the electric and magnetic fields can be transverse to the direction of net energy transfer, i.e., transverse to z (TEM waves). We show the following,

23.1.1 TEM Waves Cannot be Plane Waves

A plane wave in this context is one which does not vary across the x, y plane, i.e., the cross-sections of the waveguide are surfaces of constant phase. By the TEM assumption $B_z = E_z = 0$. The boundary condition (23.1.1b) means that there must be some points on the closed cross-section such that $E_x = 0$ and also some points where $E_y = 0$. But, by assumption, there is no variation across the x, y plane so these are sufficient to conclude that $E_x = E_y = 0$ everywhere and so there is no electric field at all and so no magnetic field. Hence any TEM wave cannot be a plane wave. Note that this proof applies whether the waveguide is simply connected or multiply connected.

23.1.2 Non-Plane TEM Waves

We now consider the case where both the electric and magnetic fields are transverse to z (i.e., TEM) but variation of the fields across the x, y plane are permitted. Because the TEM assumption tells us that $\frac{\partial B_z}{\partial t} = \frac{\partial E_z}{\partial t} = 0$ it follows from the Maxwell equations $\overline{\nabla} \times \bar{E} = -\frac{\partial \bar{B}}{\partial t}$ and $\overline{\nabla} \times \bar{H} = \frac{\partial \bar{D}}{\partial t}$ in free space that,

$$\frac{\partial E_y}{\partial x} - \frac{\partial E_x}{\partial y} = 0 \quad \text{and} \quad \frac{\partial B_y}{\partial x} - \frac{\partial B_x}{\partial y} = 0 \tag{23.1.2}$$

These imply the existence of scalar functions U and V which are independent of z and such that $\bar{E} = \overline{\nabla} U$ and $\bar{B} = \overline{\nabla} V$ so that $\overline{\nabla} \cdot \bar{B} = 0$ and $\overline{\nabla} \cdot \bar{E} = 0$ within the waveguide imply $\nabla^2 U = \nabla^2 V = 0$. The scalar function U therefore obeys Laplace's equation and is subject to the boundary condition (23.1.1b) which implies that U is constant on the conducting surface. Consequently U is identical to the electrostatic potential that can arise subject to the conducting boundary conditions, Φ.

If we consider a waveguide with a simply connected closed boundary, it follows there is no non-trivial solution because the electrostatic field inside a closed, simply connected surface at uniform potential is zero. Hence,

There are no TEM waves inside a simply connected prismatic waveguide

The only possibility for TEM waves therefore lies with multiply connected conducting surfaces, such as an inner conductor (e.g., coaxial) or a waveguide which guides waves which lie between two or more separated conductors (e.g., parallel transmission lines). This analysis shows that the electrical potential is constant around the cross-section of a given conductor (but will in general be different between different conductors at the same z). This justifies the use of circuit theory in such situations, and in particular the analysis of transmission lines using circuit theory in §18.5.

23.2 Simply Connected (Hollow) Conducting Waveguides

We continue to consider waveguides consisting of perfectly conducting boundaries but now confine attention to the case of long straight prismatic waveguides with uniform hollow simply connected closed cross sections. Initially we consider an arbitrary cross section subject to that restriction. We anticipate solutions of the form,

$$\bar{E} = \bar{E}_0 e^{i(k_g z - \omega t)} \qquad \bar{B} = \bar{B}_0 e^{i(k_g z - \omega t)} \tag{23.2.1}$$

where the amplitudes $\bar{E}_0$ and $\bar{B}_0$ are independent of z, t but will in general depend upon x, y. Here k_g is the wavenumber within the waveguide, anticipating that this will differ from $k_0 = \omega/c$, the wavenumber in free space. Recall that, in free space, all field components obey the wave equation, (10.1.5,11), $\left(\nabla^2 - \frac{1}{c^2}\frac{\partial^2}{\partial t^2}\right)\bar{E} = 0$ and $\left(\nabla^2 - \frac{1}{c^2}\frac{\partial^2}{\partial t^2}\right)\bar{B} = 0$. For the z components with (23.2.1) this gives,

$$\left(\partial_x^2 + \partial_y^2\right)E_z = -\left(k_0^2 - k_g^2\right)E_z \tag{23.2.2a}$$

$$\left(\partial_x^2 + \partial_y^2\right)B_z = -\left(k_0^2 - k_g^2\right)B_z \tag{23.2.2b}$$

We know from §23.1 that there is no non-trivial solution for $E_z = B_z = 0$. However, we may seek a solution for $B_z = 0$ with non-zero E_z (transverse magnetic case, denoted TM). Alternatively, we may seek a solution for $E_z = 0$ with non-zero B_z (transverse electric case, denoted TE). To confirm that solutions for the x and y components will be possible in these cases, and for arbitrary shaped waveguides, we proceed as follows.

The x component of the Maxwell equation $\bar{\nabla} \times \bar{E} = -\frac{\partial \bar{B}}{\partial t}$ and the y component of the Maxwell equation $\bar{\nabla} \times \bar{H} = \frac{\partial \bar{D}}{\partial t}$ give,

$$\partial_y E_z - ik_g E_y = ick_0 B_x \qquad -\partial_x B_z + ik_g B_x = -ik_0 E_y/c \tag{23.2.3}$$

These can be used to find B_x and E_y in terms of derivatives of the z components,

$$B_x = \frac{i}{k_0^2 - k_g^2}\left[-\frac{k_0}{c}\partial_y E_z + k_g \partial_x B_z\right] \tag{23.2.4a}$$

$$E_y = \frac{i}{k_0^2 - k_g^2}\left[k_g \partial_y E_z - k_0 c \partial_x B_z\right] \tag{23.2.4b}$$

In the same way, from the y component of the Maxwell equation $\overline{\nabla}\times\bar{E} = -\frac{\partial\bar{B}}{\partial t}$ and the x component of the Maxwell equation $\overline{\nabla}\times\bar{H} = \frac{\partial\bar{D}}{\partial t}$ we find,

$$E_x = \frac{i}{k_0^2 - k_g^2}\left[k_g \partial_x E_z + k_0 c \partial_y B_z\right] \tag{23.2.4c}$$

$$B_y = \frac{i}{k_0^2 - k_g^2}\left[\frac{k_0}{c}\partial_x E_z + k_g \partial_y B_z\right] \tag{23.2.4d}$$

Hence we see that there are non-trivial solutions given by (23.2.4a-d) for both the case $B_z = 0$ (TM) and also for $E_z = 0$ (TE). The general solution would be a linear combination of the two. We now proceed to specific cases, starting with the rectangular waveguide.

23.3 Rectangular Waveguide

23.3.1 TM Modes ($B_z = 0$)

The waveguide cross section is defined by boundaries at $x = 0, x = a, y = 0$, and $y = b$. Bearing in mind that (23.1.1b) requires that $E_z = 0$ on the whole boundary, the general solution for E_z is any linear combination of the basis solutions,

$$E_z^{pq} = \sin\left(\frac{\pi p x}{a}\right)\sin\left(\frac{\pi q y}{b}\right)e^{i\left(k_g^{pq} z - \omega t\right)} \tag{23.3.1}$$

where p and q are positive integers (>0). Satisfaction of the wave equation, (23.2.2a), determines the wavenumber in the waveguide, k_g,

$$k_g^{pq} = \sqrt{k_0^2 - \pi^2\left[\left(\frac{p}{a}\right)^2 + \left(\frac{q}{b}\right)^2\right]} \tag{23.3.2}$$

Here the superscripts on k_g^{pq} emphasise that the wavelength within the waveguide depends upon the mode, defined by p and q, which is in question. The largest wavenumber TM mode that can be transmitted is k_g^{11} (i.e., $\frac{2\pi}{k_g^{11}}$ is the smallest transmittable TM mode wavelength within the waveguide).

(23.3.2) makes clear that only a finite number of modes can be excited for a given frequency, according to the argument of the root in (23.3.2) remaining positive. The cutoff wavenumber below which no TM mode transmission will occur through the waveguide is $k_0 = \pi\sqrt{\left(\frac{1}{a}\right)^2 + \left(\frac{1}{b}\right)^2}$ or equivalently a cutoff frequency of,

$$f_{cutoff} = \frac{c}{2}\sqrt{\left(\frac{1}{a}\right)^2 + \left(\frac{1}{b}\right)^2} \tag{23.3.3}$$

If the frequency is lower than the cutoff, no TM mode transmission will occur. The solutions for the other field components follow from (23.2.4a-d). For completeness we write them explicitly,

$$E_x^{pq} = \left(\frac{\pi p}{a}\right)\left(\frac{ik_g}{k_0^2-\left(k_g^{pq}\right)^2}\right)\cos\left(\frac{\pi px}{a}\right)\sin\left(\frac{\pi qy}{b}\right)e^{i\left(k_g^{pq}z-\omega t\right)} \tag{23.3.4a}$$

$$E_y^{pq} = \left(\frac{\pi q}{b}\right)\left(\frac{ik_g}{k_0^2-\left(k_g^{pq}\right)^2}\right)\sin\left(\frac{\pi px}{a}\right)\cos\left(\frac{\pi qy}{b}\right)e^{i\left(k_g^{pq}z-\omega t\right)} \tag{23.3.4b}$$

$$B_x^{pq} = \left(\frac{\pi q}{cb}\right)\left(\frac{-ik_0}{k_0^2-\left(k_g^{pq}\right)^2}\right)\sin\left(\frac{\pi px}{a}\right)\cos\left(\frac{\pi qy}{b}\right)e^{i\left(k_g^{pq}z-\omega t\right)} \tag{23.3.4c}$$

$$B_y^{pq} = \left(\frac{\pi p}{ca}\right)\left(\frac{ik_0}{k_0^2-\left(k_g^{pq}\right)^2}\right)\cos\left(\frac{\pi px}{a}\right)\sin\left(\frac{\pi qy}{b}\right)e^{i\left(k_g^{pq}z-\omega t\right)} \tag{23.3.4d}$$

As an example consider a wave of frequency 10^{10}Hz and hence with $k_0 = 2.094$ cm^{-1}. Suppose we wish this to be transmitted down a rectangular waveguide with $a = 2$ cm, $b = 3$ cm. We find that the lowest TM mode, k_g^{11}, can be transmitted with a waveguide wavelength of 6.935 cm. However, no higher TM modes can be excited as they would all correspond to imaginary k_g^{pq}. For a frequency below 0.9×10^{10}Hz no TM mode transmission will occur.

23.3.2 TE Modes ($E_z = 0$)

What are the boundary conditions on B_z in this case? On the boundaries $x = 0$ and $x = a$, boundary condition (23.1.1a) means that $B_x = 0$. But, because $E_z = 0$, (23.2.4a) then tells us that $\partial_x B_z = 0$ on these boundaries. Similarly $\partial_y B_z = 0$ on the boundaries $y = 0$ and $y = b$. Hence, the solution for B_z is.

$$B_z^{pq} = \cos\left(\frac{\pi px}{a}\right)\cos\left(\frac{\pi qy}{b}\right)e^{i\left(k_g^{pq}z-\omega t\right)} \tag{23.3.5}$$

In this case p and q are positive integers or zero, excepting only the case that both are zero (as this would reduce to a TEM mode which is not possible in a simply connected waveguide). Satisfaction of the wave equation (23.2.2b) again requires that the wavenumber in the waveguide is given by (23.3.2). However we may now include the lower modes 01 and 10. In the above example of a 2cm by 3cm waveguide, the cutoff frequency would drop from 0.9×10^{10}Hz for TM modes to 0.5×10^{10}Hz for TE modes. For example, for a free space wavelength of 5cm (frequency 0.6×10^{10}Hz) only one mode would be excited, the TE_{01} mode with guide wavelength 9.05cm.

Hence, for this example 2cm by 3cm waveguide, neither TM nor TE modes would be available for transmission below 5 GHz. As the general solution is a superposition of TM and TE modes, it follows that a conducting waveguide has a minimum cutoff frequency below which no transmission is possible at all. For the rectangular guide with $a < b$ the mode that is excited at the lowest frequency is the TE$_{01}$ mode. Below the frequency $f_{cutoff} = \frac{c}{2b}$ there is no transmission. Transmission in a single mode can be ensured for frequencies between the TE$_{01}$ cutoff and the TE$_{10}$ cutoff. Only the TE$_{01}$ mode is then excited and this can have efficiency advantages. We will see later that for dielectric waveguides (of which optic fibre is the most important example) the conditions for single mode transmission are very different.

For completeness the solutions for the other field components follow from (23.2.4a-d) and are,

$$E_x^{pq} = -\left(\frac{\pi q}{b}\right)\left(\frac{ik_0 c}{k_0^2-\left(k_g^{pq}\right)^2}\right)\cos\left(\frac{\pi px}{a}\right)\sin\left(\frac{\pi qy}{b}\right)e^{i\left(k_g^{pq}z-\omega t\right)} \qquad (23.3.6a)$$

$$E_y^{pq} = \left(\frac{\pi p}{a}\right)\left(\frac{ik_0 c}{k_0^2-\left(k_g^{pq}\right)^2}\right)\sin\left(\frac{\pi px}{a}\right)\cos\left(\frac{\pi qy}{b}\right)e^{i\left(k_g^{pq}z-\omega t\right)} \qquad (23.3.6b)$$

$$B_x^{pq} = -\left(\frac{\pi p}{a}\right)\left(\frac{ik_g}{k_0^2-\left(k_g^{pq}\right)^2}\right)\sin\left(\frac{\pi px}{a}\right)\cos\left(\frac{\pi qy}{b}\right)e^{i\left(k_g^{pq}z-\omega t\right)} \qquad (23.3.6c)$$

$$B_y^{pq} = -\left(\frac{\pi q}{b}\right)\left(\frac{ik_g}{k_0^2-\left(k_g^{pq}\right)^2}\right)\cos\left(\frac{\pi px}{a}\right)\sin\left(\frac{\pi qy}{b}\right)e^{i\left(k_g^{pq}z-\omega t\right)} \qquad (23.3.6d)$$

Hence, the x and y components of the fields have the same functional form for TM and TE modes, differing only in their amplitudes (and sign).

23.4 Circular Waveguide

23.4.1 TM Modes ($B_z = 0$)

Here we again consider a straight conducting waveguide, oriented along the z axis, but now with a circular cross section, $r = a$. In cylindrical polars the wave equation (23.2.2a) becomes,

$$\left[\frac{1}{r}\partial_r(r\partial_r) + \frac{1}{r^2}\partial_\theta^2 + \Delta^2\right]E_z = 0 \qquad \text{where, } \Delta^2 = k_0^2 - k_g^2 \qquad (23.4.1)$$

Multiplying by r^2 and changing the independent variable to the dimensionless $s = \Delta r$ we can then separate variables in the standard manner (see §14.1) from which we conclude the angular variation is $e^{im\theta}$ and periodicity requires that m is an integer (including zero). (23.4.1) then becomes Bessel's equation of order m,

$$[s^2\partial_s^2 + s\partial_s + (s^2 - m^2)]E_z(s) = 0 \tag{23.4.2}$$

The general solution is a linear combination of $J_m(s)$ and $Y_m(s)$, the Bessel functions of the first and second kinds. However, $Y_m(s)$ is divergent at the origin and so can be ruled out for this simply connected waveguide. (If we considered instead a twin conductor waveguide consisting of a pair of coaxial cylinders, we would have to include the $Y_m(s)$ term in the annular region transmitting the signal).

For TM modes the boundary condition is that $E_z = 0$ for $r = a$. Hence, the criterion which determines the allowed wavenumbers, k_g, is,

$$J_m(\Delta a) = 0 \tag{23.4.3}$$

The first three zeros of the first three Bessel functions of the first kind are given in Table 23.1

Table 23.1: Zeros of $J_m(s)$

n	$J_{m=0}$	$J_{m=1}$	$J_{m=2}$
1	2.40482	3.83171	5.13562
2	5.52007	7.01559	8.41724
3	8.65372	10.17347	11.61984

Calling these zeros s_{nm}, the allowed wavenumbers are,

$$k_g^{nm} = \sqrt{k_0^2 - \left(\frac{s_{nm}}{a}\right)^2} \tag{23.4.4}$$

The cutoff frequency is thus given by,

$$k_0 = \frac{2\pi}{c}f = \frac{2.40482}{a} \tag{23.4.5}$$

As an illustration consider a waveguide with the same cross section area as the 2cm x 3cm waveguide considered in §23.3, and hence with $a = 1.38$cm. The cutoff frequency for TM modes is thus 0.83×10^{10}Hz. This compares with 0.9×10^{10}Hz for the rectangular waveguide.

Consider an input wave with frequency 1.5×10^{10}Hz ($k_0 = 3.14\ \text{cm}^{-1}$). The nm TM modes 10 and 11 can be excited, with guide wavelengths of 2.40 and 4.26 cm respectively. The 01 mode is axisymmetric whereas the 11 mode has angular variation $e^{i\theta}$. No other modes can be excited at this frequency.

The r and θ components of the fields can again be found from (23.2.4a-d) by first converting these expressions to ones involving the derivatives ∂_r and ∂_θ. This will involve derivatives of the Bessel functions. This is left as an exercise for the reader.

23.4.2 TE Modes ($E_z = 0$)

In cylindrical polars the wave equation for B_z, (23.2.2b), is the same as that for E_z, (23.4.2), so again the solutions are $J_m(s)$. However, the boundary condition, (23.1.1a), implies $\partial_r B_z = 0$ on $r = a$. So the criterion which determines the guide wavenumber is no longer (23.4.3) but rather,

$$J_m'(\Delta a) = 0 \qquad (23.4.6)$$

where the dash denotes the derivative. The first three zeros in the derivatives of the first three Bessel functions are given in Table 23.2.

Table 23.2: (Approximate) Zeros of the Derivatives of $J_m(s)$

n	$J_{m=0}'$	$J_{m=1}'$	$J_{m=2}'$
1	3.84	1.84	3.05
2	7.0	5.33	6.7
3	10.15	8.53	10.0

Note that there is also a zero in J_0' at $s = 0$ which would imply $\Delta = 0$ and hence that B_z was constant over the cross section. But then (23.2.4a) implies that $\bar{E} = 0$ and hence this solution degenerates to the trivial case of no fields at all. Hence we have omitted this zero from Table 23.2.

The allowed wavenumbers in the wave guide are again given by (23.4.4) but with s_{nm} given by Table 23.2. Considering the same illustrative case used in §23.4.1, i.e., $a = 1.38$cm, which, at a frequency of 1.5×10^{10}Hz could excite two TM modes, we see that three TE modes can also be excited, for guide wavelengths of 2.21, 2.81 and 4.28cm. These are modes 11, 12 and 10 respectively. The first two have angular dependencies of $e^{i\theta}$ and $e^{2i\theta}$ respectively. The third is axisymmetric. The lowest mode is not axisymmetric in this case. The lowest cutoff frequency for the TE modes, namely mode TE11, is given by $k_0 = \frac{2\pi}{c} f = \frac{1.84}{a}$, i.e., 0.64×10^{10}Hz which is lower than that for the lowest TM mode (as was the case for the rectangular waveguide).

The r and θ components of the fields can again be found from (23.2.4a-d) by first converting these expressions to ones involving the derivatives ∂_r and ∂_θ. This will involve derivatives of the Bessel functions. Readers who enjoy algebra can evaluate them explicitly.

23.5 Phase Velocity and Group Velocity

In all the above waveguide modes the time variation is given by $e^{i(k_g z-\omega t)}$, (23.2.1). Hence the phase velocity of the signal is,

$$v_{ph}=\frac{\omega}{k_g}>c \tag{23.5.1}$$

That the phase velocity exceeds c is because (23.3.2) and (23.4.4) show that $k_g<k_0=\omega/c$. However the group velocity is given by $v_{gr}=\frac{\partial\omega}{\partial k_g}$ and in all cases we find that $k_g^2=k_0^2-X=\frac{\omega^2}{c^2}-X$ where X is some constant for our present purposes. So we find,

$$v_{gr}=\frac{\partial\omega}{\partial k_g}=c^2\frac{k_g}{\omega}=\frac{k_g}{k_0}c<c \tag{23.5.2}$$

So the group velocity, which is the maximum speed that energy and hence information can be transmitted, is less than the speed of light in *vacuo*, as we would expect.

Hence, despite involving e/m radiation propagating in air (much the same as vacuum) the signal speed down a conducting waveguide is not c, but less than c. This may be rationalised by considering the path of the e/m waves down the waveguide to be non-axial and to involve successive reflections from its inside surface. The path length is then greater than the axial distance travelled by the signal. Assuming propagation at speed c along this zig-zag path implies the cosine of the angle the path makes with the axial is $\frac{k_g}{k_0}$.

23.6 Electromagnetic Cavities

A complete enclosure will support only "quantised" modes of standing waves. For example, a rectangular cavity of dimensions $a\times b\times c$ supports TE modes of the form,

$$B_z^{pqs}=\cos\left(\frac{\pi px}{a}\right)\cos\left(\frac{\pi qy}{b}\right)\sin\left(\frac{\pi sz}{c}\right)e^{i\left(k_g^{pqs}z-\omega t\right)} \tag{23.6.1}$$

where p,q are integers ≥ 0 and s is an integer >0, and,

$$k_g^{pq}=\sqrt{k_0^2-\pi^2\left[\left(\frac{p}{a}\right)^2+\left(\frac{q}{b}\right)^2+\left(\frac{s}{c}\right)^2\right]} \tag{23.6.2}$$

In practice such a cavity will not usually be truly completely enclosed in that one or two openings will be required to inject, or to sample, the field. Such resonant cavities have a number of technological applications.

23.7 Data Transmission Rates Basics

Digital information technology (IT) is beyond the scope of this book. However, optical fibres are a form of waveguide for optical signals, and light is an electromagnetic wave. As optical fibres are now the dominant technology for long distance digital communication by cable, something must be included on optical fibres from the waveguide perspective in this chapter. However, to explain why optical fibres have such advantages over copper cables, and have therefore become dominant in IT technology, it is necessary to digress briefly into the basics of data transmission rates, bandwidths, etc.

Two things are required in order to understand the limitations of copper as a carrier of high speed data. The first is the relationship between frequency bandwidth in Hz and data transmission rate in bps (bits per second). This was addressed briefly in §12.4.1 but is worth repeating. The second is the relationship between signal frequency and signal attenuation in copper. For the first of these we reprise two fundamental theorems.

The Shannon-Hartley theorem gives an upper limit on the data rate, R, in bps in terms of the bandwidth, B, in Hz and the signal-to-noise ratio, ξ,

Shannon-Hartley: $$R \leq B\log_2(1+\xi) \quad (23.7.1)$$

However, data may be transmitted in the form of a set of M symbols in which case there is the Nyquist-Shannon theorem which also limits the possible data rate in terms of the bandwidth and the number of symbols used,

Nyquist-Shannon: $$R \leq 2B\log_2(M) \quad (23.7.2)$$

Both theorems apply which means that there is an upper limit to the number of symbols that can be used before the limitation due to the (23.7.1) is violated. Physically this is because distinguishing between too many symbols becomes unreliable at some point according to the noise level.

Details aside, the theoretical maximum data transmission rate is proportional to the frequency bandwidth in Hz. This fact, and the above theorems, apply whether we are talking about copper or optical fibre, or anything else.

But in the case of copper there is a practical limitation due to attenuation of the signal over long lengths of cable, essentially due to resistance. For example, at 100 Mbps you may be limited to around 100m. Achievable data speeds in standard copper hardware and software protocols over one or two kilometres may be only 20 Mbps, reducing to only 1 or 2 Mbps at 5 kilometres. Longer distances can be achieved at lower frequencies but only at the expense of a slower connection in terms of bps. In practice, internet connection over hundreds of kilometres, let alone thousands, requires amplifiers or repeaters at a great many places along such lengths, boosting the signal but increasing the expense of the system.

Actually the limitation of copper is not only attenuation but also other factors including crosstalk (interference between different channels on the same line) and other factors which can distort the signal.

The advantage of optic fibre communication lies in both the above factors: (i) the frequency of "light" is enormously higher than the ~10 MHz or so relevant in copper, and, (ii) at the frequencies of "light" used its attenuation in modern optic fibre is dramatically less than that of copper at ~10 MHz.

The word "light" is in quotes here because the most common wavelengths used in "optical" IT comms are actually 850, 1300 and 1550 nm, which are in the infrared range (frequencies of about 2 to 3.5 $\times 10^{14}$ Hz). Signal boosting may still be needed with optic fibre but we are looking at distances of 50km to 100km before that is necessary, dramatically reducing the number of boosters required (and possibly none).

Existing long-distance data systems achieve optic fibre data rates of around 100 to 400 Gbps, so tens of thousands of times faster than copper. But the capacity is vastly greater still even with normal commercially available fibres which have been demonstrated to be capable of ~400 Tbps, i.e., $\sim 4 \times 10^{14}$bps.

Another benefit of optical fibre is that it is immune to the electrical interference that can affect copper cables.

23.8 Optic Fibres as Waveguides

23.8.1 Optic Fibre Composition

Optic fibres for communications purposes are not composed of uniform material over their cross section. Instead they consist of a core surrounded by cladding, the two materials having very slightly different refractive indices, which we will denote $n_1 = \sqrt{\mu_1 \varepsilon_1}$ in the core and $n_2 = \sqrt{\mu_2 \varepsilon_2}$ in the cladding. There may be a clear boundary between the two at some radius, or there may be a gradation of refractive index as a continuous function of radius. The latter has some advantages in efficiency. However, to facilitate analysis we shall assume a clear boundary at some radius a.

Optic fibre for IT applications typically has a diameter of 0.125mm. The core diameter, $2a$, is typically 0.0085mm for single mode fibre or 0.05 to 0.0625mm for multimode fibre. Hence signal transmission occurs in only a small minority of the fibre cross-section (the core).

The material is generally silica doped with, for example, germanium dioxide, aluminium oxide, fluorine, boron trioxide or rare earth metals. Differing amounts of doping cause the variation in refractive index. For sake of argument in what follows we shall assume a cladding refractive index of 1.444 at the frequencies generally in use. However, the core will have a slightly larger refractive index so that at the interface total internal reflection will occur provide the angle of incidence exceeds the critical angle (see §11.5.1).

From the perspective of geometrical optics, this total internal reflection is what causes the light (actually infrared) to propagate along the fibre with little attenuation.

The reader might wonder why there need be two regions of differing refractive index since total internal reflection would arise at the glass/air boundary anyway. The reason relates to the number of waveguide modes that can be excited. For a homogeneous fibre the number of active modes would be extremely large, as we will see, and this detracts badly from the efficiency of transmission. Fibres supporting low attenuation over large distances require only a few modes to be active. In fact, the highest quality communication is achieved using "single mode" fibres, and to achieve this requires the percentage difference $(n_1 - n_2)/n_1$ to be less than 1%.

This small difference in refractive index means the critical angle is large. This means that light must be injected into the fibre within a narrow angle of acceptance near the axial in order to exceed the critical angle and hence for transmission to occur.

The efficiency of single mode fibres is further enhanced by the use of laser light, and for this purpose LED lasers are generally used.

Figure 23.2 illustrates how single mode fibres using laser generated light can achieve good transmission over larger distance than multimode fibres using standard LED light.

Figure 23.2: Illustrating the transmission distances at various data speeds attained for single mode versus multimode fibres (and LED versus laser excitation). *Note that this data refers to relatively early technology and does not reflect the higher data rates now available.*

The following graph shows the expected data rates and transmission distances for the various mode and indexed fibers at different wavelengths.

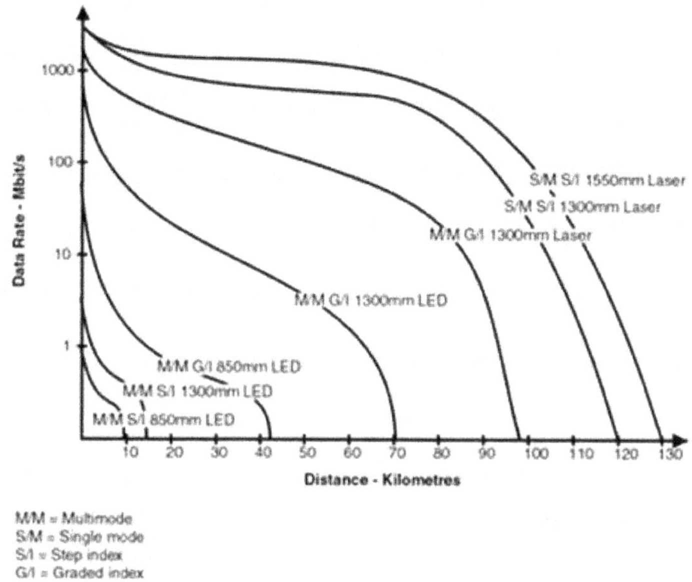

We next present an analysis of e/m wave propagation in dielectrics with a boundary between two different media, as in optic fibres. This follows the similar analyses for conducting waveguides in §23.3 and §23.4 the main differences being that, (i) we now have reduced wave speeds in media with refractive index $(n) > 1$, and, (ii) the boundary conditions are different. The reader is reminded of the analysis of §11.6 relating to situations of total internal reflection, and specifically the crucial role of the evanescent wave in the analysis.

In particular, having stressed above the importance of single mode fibres, this analysis elucidates the conditions under which single mode transmission occurs in optic fibres – very different conditions from those in conducting waveguides.

23.8.2 Mode Analysis

23.8.2.1 Formulation of Equations

Despite the qualitative understanding of light propagation along optic fibres based on geometrical optics and total internal reflection, this does not provide any means of analysing what frequencies the fibre can transmit. The analogy with conducting waveguides and far lower (e.g., microwave) frequencies warns us that there are likely to be restrictions on the possible frequencies. Unlike conducting waveguides, however, we will find that there is a mode in optic fibres which transmits at arbitrarily low frequencies, though this is joined by other modes above a certain cutoff frequency, and by an increasing number of other modes at even higher frequencies.

Moreover, we have already alluded to single mode fibres being the most efficient and we need some means of calculating under what conditions there is indeed only a single mode, and what the cutoff frequency may be above which the fibre becomes multimode. So we need to solve Maxwells equations with the appropriate boundary conditions.

Again we anticipate solutions of the form,

$$\bar{E} = \bar{E}_0 e^{i(k_g z - \omega t)} \qquad \bar{B} = \bar{B}_0 e^{i(k_g z - \omega t)} \tag{23.8.1}$$

where z is the direction of the fibre length and the amplitudes $\bar{E}_0$ and $\bar{B}_0$ are independent of z, t but will in general depend upon x, y. Here k_g is the wavenumber within the waveguide. We initially develop the equations in Cartesian coordinates and then convert to cylindrical polars in order to apply the boundary conditions.

The quantity k_{10} is defined as the wavenumber for free plane wave propagation in a medium of refractive index n_1,

$$k_{10} = \frac{\omega}{v_1} \quad \text{and similarly,} \quad k_{20} = \frac{\omega}{v_2} \tag{23.8.2}$$

where the wave speeds in these media are,

$$v_1 = \frac{c}{n_1} \quad \text{and} \quad v_2 = \frac{c}{n_2} \tag{23.8.3}$$

Recall that, within any one medium, all field components obey the wave equation provided that we use the local wave speed, i.e.,

$$\left(\nabla^2 - \frac{1}{v_1^2}\frac{\partial^2}{\partial t^2}\right)\bar{E} = 0 \text{ and } \left(\nabla^2 - \frac{1}{v_1^2}\frac{\partial^2}{\partial t^2}\right)\bar{B} = 0 \tag{23.8.4}$$

and similarly in the cladding using v_2. Substituting (23.8.1) into (23.8.4) gives, for the z components in the core,

$$\left(\partial_x^2 + \partial_y^2\right)E_{1z} = -\left(k_{10}^2 - k_g^2\right)E_{1z} \tag{23.8.5a}$$

$$\left(\partial_x^2 + \partial_y^2\right)B_{1z} = -\left(k_{10}^2 - k_g^2\right)B_{1z} \tag{23.8.5b}$$

and similarly for the cladding. Considering the x and y components of the Maxwell equations $\overline{\nabla} \times \bar{E} = -\frac{\partial \bar{B}}{\partial t}$ and $\overline{\nabla} \times \bar{H} = \frac{\partial \bar{D}}{\partial t}$, noting that the latter can be written $\overline{\nabla} \times \bar{B} = \frac{1}{v_1^2}\frac{\partial \bar{E}}{\partial t}$ in the core or $\overline{\nabla} \times \bar{B} = \frac{1}{v_2^2}\frac{\partial \bar{E}}{\partial t}$ in the cladding, we can derive the equivalent of (23.2.4a-d), expressing the x and y components in terms of the z components, namely, for the core,

$$B_{1x} = \frac{i}{k_{10}^2 - k_g^2}\left[-\alpha_1 \partial_y E_{1z} + k_g \partial_x B_{1z}\right] \tag{23.8.6a}$$

$$E_{1y} = \frac{i}{k_{10}^2 - k_g^2}\left[k_g \partial_y E_{1z} - \omega \partial_x B_{1z}\right] \tag{23.8.6b}$$

$$E_{1x} = \frac{i}{k_{10}^2 - k_g^2}\left[k_g \partial_x E_{1z} + \omega \partial_y B_{1z}\right] \tag{23.8.6c}$$

$$B_{1y} = \frac{i}{k_{10}^2 - k_g^2}\left[\alpha_1 \partial_x E_{1z} + k_g \partial_y B_{1z}\right] \tag{23.8.6d}$$

and similarly in the cladding, changing all subscripts 1 to 2. Here we have defined,

$$\alpha_1 = \frac{k_{10}}{v_1} = \frac{\omega}{v_1^2} \quad \text{and} \quad \alpha_2 = \frac{k_{20}}{v_2} = \frac{\omega}{v_2^2} \tag{23.8.7}$$

We now convert (23.8.6a-d) to cylindrical polars using the transform,

$$\begin{pmatrix} u_x \\ u_y \end{pmatrix} = \begin{pmatrix} \cos\theta & -\sin\theta \\ \sin\theta & \cos\theta \end{pmatrix}\begin{pmatrix} u_r \\ u_\theta \end{pmatrix} \tag{23.8.8}$$

where the 2D vector $\bar{u}$ may be $\bar{E}$ or $\bar{B}$ or $\begin{pmatrix} \partial_r \\ \frac{1}{r}\partial_\theta \end{pmatrix}$ transforming to $\begin{pmatrix} \partial_x \\ \partial_y \end{pmatrix}$.

Hence we find that (23.8.6a-d) become,

$$B_{1r} = \frac{i}{k_{10}^2 - k_g^2}\left[-\frac{\alpha_1}{r}\partial_\theta E_{1z} + k_g \partial_r B_{1z}\right] \tag{23.8.9a}$$

$$B_{1\theta} = \frac{i}{k_{10}^2 - k_g^2}\left[\alpha_1 \partial_r E_{1z} + \frac{k_g}{r}\partial_\theta B_{1z}\right] \tag{23.8.9b}$$

$$E_{1r} = \frac{i}{k_{10}^2 - k_g^2}\left[k_g \partial_r E_{1z} + \frac{\omega}{r}\partial_\theta B_{1z}\right] \tag{23.8.9c}$$

$$E_{1\theta} = \frac{i}{k_{10}^2 - k_g^2}\left[\frac{k_g}{r}\partial_\theta E_{1z} - \omega \partial_r B_{1z}\right] \tag{23.8.9d}$$

and similarly in the cladding, changing all subscripts 1 to 2. In the core the wave equations (23.8.5a,b) for the z components become in cylindrical polars,

$$\left[\frac{1}{r}\partial_r(r\partial_r)+\frac{1}{r^2}\partial_\theta^2+\Delta_1^2\right]E_{1z}=0 \text{ where, } \Delta_1^2=k_{10}^2-k_g^2 \tag{23.8.10a}$$

$$\left[\frac{1}{r}\partial_r(r\partial_r)+\frac{1}{r^2}\partial_\theta^2+\Delta_1^2\right]B_{1z}=0 \tag{23.8.10b}$$

Separating variables the θ dependence must be $e^{im\theta}$ where m is an integer or zero. Both equations then become Bessel's equation in $s=\Delta_1 r$,

$$[s^2\partial_s^2+s\partial_s+(s^2-m^2)]E_{1z}(s)=0 \tag{23.8.11}$$

The general solution is a linear combination of $J_m(s)$ and $Y_m(s)$, the Bessel functions of the first and second kinds. However, $Y_m(s)$ is divergent at the origin and so can be ruled out in the core. A solution in the core would therefore be, for $r\le a$,

$$E_{1z}=C_{1E}J_m(\Delta_1 r)e^{im\theta} \quad \text{and} \quad B_{1z}=C_{1B}J_m(\Delta_1 r)e^{im\theta} \tag{23.8.12}$$

for some constants $C_{1E,B}$.

Now the solution in the cladding can be written,

$$\left[\frac{1}{r}\partial_r(r\partial_r)+\frac{1}{r^2}\partial_\theta^2-\Delta_2^2\right]E_{2z}=0 \text{ where, } \Delta_2^2=-k_{20}^2+k_g^2 \tag{23.8.13a}$$

$$\left[\frac{1}{r}\partial_r(r\partial_r)+\frac{1}{r^2}\partial_\theta^2-\Delta_2^2\right]B_{2z}=0 \tag{23.8.13b}$$

These are formally identical to (23.8.10a,b) merely with Δ_2^2 defined with the opposite sign from Δ_1^2, i.e., $\Delta_2^2=k_g^2-k_{20}^2$. The reason for this change in definition is so that, assuming Δ_2 is real, we will find that the solution in the cladding decays exponentially with radius. This is essential if the signal is to be kept away from the outer surface of the fibre (which would present an opportunity for signal leakage). But also, if the signal did propagate freely to the outer surface it would not be possible to constrain the signal to a single mode because the fibre would then support a very large number of modes, with serious impact on its efficiency. In contrast, by assuming Δ_1 and Δ_2 are both real the guide wavenumber is tightly constrained to lie in the interval,

$$\frac{\omega n_2}{c}<k_g<\frac{\omega n_1}{c} \tag{23.8.14}$$

Using $s=\Delta_2 r$ in the cladding equs.(23.8.13a,b) are seen to become the modified Bessel equation in standard form,

$$[s^2\partial_s^2+s\partial_s-(s^2+m^2)]E_z(s)=0 \quad \text{(ditto for } B_z(s)) \tag{23.8.15}$$

The two independent solutions for a given order (m) are called the modified Bessel functions of the first and second kind, conventionally denoted $I_m(s)$ and $K_m(s)$. $I_m(s)$ is finite at the origin but diverges for large arguments, $s \to \infty$, as $e^s/\sqrt{s}$. $K_m(s)$ is divergent at the origin but rapidly reduces towards zero for large arguments, as $e^{-s}/\sqrt{s}$. In the cladding the origin does not occur but we require a decreasing function at larger r so we need K_m and can exclude I_m. The solutions in the cladding can be written, for $r \geq a$,

$$E_{2z} = C_{2E} K_m(\Delta_2 r) e^{im\theta} \quad \text{and} \quad B_{2z} = C_{2B} K_m(\Delta_2 r) e^{im\theta} \tag{23.8.16}$$

for some constants $C_{2E,B}$. We have written these with the same order parameter, m, as in the core, anticipating the boundary conditions will force this to be so.

23.8.2.2 Boundary Conditions

Now we must impose the boundary conditions where the core meets the cladding at $r = a$. The boundary conditions appropriate for a dielectric boundary are as follows,

- Using Stoke's theorem on a suitable contour together with the Maxwell equation $\bar{\nabla} \times \bar{E} = -\frac{\partial \bar{B}}{\partial t}$ always requires the tangential components of $\bar{E}$ to be continuous. (This does not require $\bar{B}$ to be static).
- In the case of a conducting boundary the Maxwell equation $\bar{\nabla} \times \bar{H} = \bar{J} + \frac{\partial \bar{D}}{\partial t}$ with Stokes's theorem imply the tangential components of $\bar{B}$ are discontinuous by $\mu_0 \mu_r$ times the surface current. But in the case of a dielectric there will be no current. So, unlike the case of a conducting boundary, for a dielectric boundary the tangential components of $\bar{B}$ are also continuous.
- Gauss's theorem with Maxwell equation $\bar{\nabla} \cdot \bar{B} \equiv 0$ means that the normal component of $\bar{B}$ is also continuous.
- Maxwell equation $\bar{\nabla} \cdot \bar{D} = \rho$ with Gauss's theorem means that the normal component of $\bar{E}$ is discontinuous by $1/\varepsilon_0 \varepsilon_r$ times the surface charge density. As any charge density is not known, E_r does not provide a boundary condition constraint.

The above discussion leads to the following being continuous over the boundary between the core and the cladding in optic fibre,

$$B_{1r}(a) = B_{2r}(a), B_{1z}(a) = B_{2z}(a),\ B_{1\theta}(a) = B_{2\theta}(a) \tag{23.8.17a}$$

$$E_{1z}(a) = E_{2z}(a),\ E_{1\theta}(a) = E_{2\theta}(a) \tag{23.8.17b}$$

23.8.2.3 The TE Mode

We seek a solution with,

$$E_{1z} = 0 \quad \text{and} \quad B_{1z} = C_{1B}J_m(\Delta_1 r)e^{im\theta} \tag{23.8.18a}$$

$$E_{2z} = 0 \quad \text{and} \quad B_{2z} = C_{2B}K_m(\Delta_2 r)e^{im\theta} \tag{23.8.18b}$$

The other components, from (23.8.9a-d) are thus,

$$B_{1r} = \xi_1 k_g \partial_r B_{1z} \qquad B_{1\theta} = \xi_1 \frac{k_g}{r} \partial_\theta B_{1z} \tag{23.8.19a}$$

$$E_{1r} = \xi_1 \frac{\omega}{r} \partial_\theta B_{1z} \qquad E_{1\theta} = -\xi_1 \omega \partial_r B_{1z} \tag{23.8.19b}$$

Where we have put,

$$\xi_1 = \frac{i}{k_{10}^2 - k_g^2} = \frac{i}{\Delta_1^2} \quad \text{and} \quad \xi_2 = -\frac{i}{k_{20}^2 - k_g^2} = \frac{i}{\Delta_2^2} \tag{23.8.20}$$

The solution for the radial and circumferential components in the cladding is thus obtained from (23.8.19a,b) by replacing subscripts 1 with 2 and changing the overall sign.

The requirement $B_{1z}(a) = B_{2z}(a)$ gives,

$$C_{1B}J_m(\Delta_1 a) = C_{2B}K_m(\Delta_2 a) \tag{23.8.21a}$$

The requirement $B_{1\theta}(a) = B_{2\theta}(a)$ gives,

$$m\xi_1 C_{1B}J_m(\Delta_1 a) = -m\xi_2 C_{2B}K_m(\Delta_2 a) \tag{23.8.21b}$$

But (23.8.21a) and (23.8.21b) are in conflict and this can only be avoided if $m = 0$, i.e., that the sought-for TE mode is axisymmetric. (The alternative that $J_m(\Delta_1 a) = K_m(\Delta_2 a) = 0$ is not possible because the functions K_m have no zeros for real m). (23.8.21a) then becomes,

$$C_{1B}J_0(\Delta_1 a) = C_{2B}K_0(\Delta_2 a) \tag{23.8.22}$$

So the only non-zero field components in the TE mode are B_z, B_r, E_θ. Continuity of E_θ provides the same condition as continuity of B_r, i.e.,

$$\xi_1 C_{1B}\Delta_1 \left.\frac{dJ_0(s)}{ds}\right|_{s=\Delta_1 a} = -\xi_2 C_{2B}\Delta_2 \left.\frac{dK_0(s)}{ds}\right|_{s=\Delta_2 a} \tag{23.8.23}$$

The following are identities for the derivatives of the Bessel functions,

$$\frac{dJ_m}{ds} = \frac{m}{s}J_m - J_{m+1} \qquad \frac{dK_m}{ds} = \frac{m}{s}K_m - K_{m+1} \tag{23.8.24}$$

In particular, $\frac{dJ_0}{ds} = -J_1$ and $\frac{dK_0}{ds} = -K_1$ so (23.8.23) becomes,

$$\xi_1 C_{1B}\Delta_1 J_1(\Delta_1 a) = -\xi_2 C_{2B}\Delta_2 K_1(\Delta_2 a) \tag{23.8.25}$$

So, finally, for the TE mode the complete boundary conditions reduce to (23.8.22) plus (23.8.25). We can choose one constant arbitrarily, say $C_{1B} = 1$, because this merely scales the overall amplitude of the wave. The two boundary conditions then permit elimination of C_{2B} to leave the condition which defines the guide wavenumber, k_g, for this axisymmetric TE mode,

$$-\frac{J_1(\Delta_1 a)}{\Delta_1 J_0(\Delta_1 a)} = \frac{K_1(\Delta_2 a)}{\Delta_2 K_0(\Delta_2 a)} \tag{23.8.26}$$

Recall from (23.8.10a) and (23.8.13a), i.e., $\Delta_1^2 = k_{10}^2 - k_g^2$ and $\Delta_2^2 = k_g^2 - k_{20}^2$ that,

$$\Delta_1^2 + \Delta_2^2 = \frac{\omega^2}{c^2}(n_1^2 - n_2^2) \tag{23.8.27}$$

So (23.8.26,27) allow the one remaining degree of freedom, k_g, to be found (if, indeed, there is a solution). From (23.8.14) the minimum possible ak_g is $\frac{a\omega n_2}{c} = ak_{20}$ so the maximum possible $a\Delta_1$ is $a\sqrt{k_{10}^2 - k_{20}^2} = \frac{a\omega}{c}\sqrt{n_1^2 - n_2^2}$.

The minimum possible $a\Delta_1$ at which we may start our search for a solution to (23.8.26) is zero. But for small $a\Delta_1$ the two sides of (23.8.26) are of opposite sign, so there is no solution. This changes suddenly as $a\Delta_1$ increases past the smallest zero of $J_0(\Delta_1 a)$ at $a\Delta_1 = 2.40482$ when the LHS falls rapidly from positive infinity and a solution is found, as illustrated by Figure 23.3 for the case of a core radius 4.25μm, $n_1 = 1.452$, $n_2 = 1.444$ and a free space wavelength 1550nm ($\omega = 1.2161 \times 10^{15}$ s^{-1}). This Figure is plotted with ak_g as the x-axis so $a\Delta_1$ is increasing towards the left. It can be seen that the solution occurs for $a\Delta_1$ just above the singularity due to the zero of $J_0(\Delta_1 a)$. Because the whole possible range of ak_g is plotted in Figure 23.3, it can be seen that there is only one solution in this case. Consequently, there is only one TE transmission mode with $E_z = 0$ at this frequency for this fibre.

Because any solution to (23.8.26) can occur only beyond the first zero of $J_0(\Delta_1 a)$, and because the maximum possible $a\Delta_1$ is $\frac{a\omega}{c}\sqrt{n_1^2 - n_2^2}$, it follows that there is no solution to (23.8.26) and no TE mode can occur within this fibre unless the frequency exceeds the cutoff of,

$$\omega_{cutoff} = \frac{2.40482c}{a\sqrt{n_1^2 - n_2^2}} \tag{23.8.28}$$

For our illustration this is $\omega_{cutoff} = 1.115 \times 10^{15}$ s^{-1} (1690nm) and we were careful to use a frequency just above this.

The reader should not assume that there is no mode at all at lower frequencies, however. We shall shortly see that there is, and it is not axisymmetric and has non-zero axial components of both the electric and magnetic fields, the so-called HE mode.

Figure 23.3: Illustration of the solution to (23.8.26) for core radius 4.25μm, $n_1 = 1.452$, $n_2 = 1.444$, free space wavelength 1550nm ($\omega = 1.2161 \times 10^{15}$ s^{-1}). *The black line is the RHS of the equation; the blue lines are the LHS. The x-axis is ak_g and covers the whole possible range from ak_{20} to ak_{10}. There is thus only one TE mode solution, at $ak_g = 24.885145$ at this frequency. But this is not a single mode fibre at this frequency because, as we will see below, there is also another active mode, the so-called HE mode. The discontinuity in the LHS occurs at the lowest zero of $J_0(\Delta_1 a)$, i.e., at $\Delta_1 a = 2.40482$ which occurs at $ak_g = 24.899316$.*

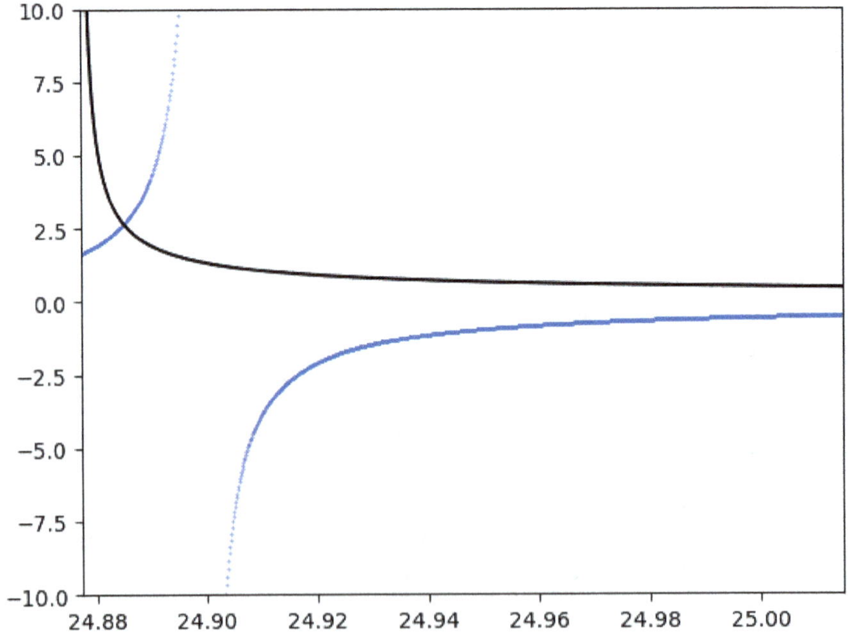

Extending the same argument we expect a second TE mode to arise if the frequency exceeds the cutoff for the second zero of $J_0(\Delta_1 a)$, and hence above a frequency $\omega = \frac{5.52007c}{a\sqrt{n_1^2 - n_2^2}}$ which is 2.55996 x 10^{15} s^{-1}. Choosing an input frequency just above this, corresponding to a free wavelength of 711nm we do indeed now find that two transmission modes arise, as illustrated by Figure 23.4.

The generalisation is clear: an increasing number of axisymmetric TE modes arise as higher frequencies are used for a given fibre, each new mode appearing as the frequency exceeds the next zero of $J_0(\Delta_1 a)$, the nth such mode be denoted TE_{0n}.

Figure 23.4: Illustration of the solution to (23.8.26) for the same optic fibre as Fig.23.3 but now with an input free space wavelength 711nm ($\omega = 2.651133 \times 10^{15}$ s^{-1}). *The black line is the RHS of the equation; the blue lines are the LHS. The x-axis is ak_g and covers the whole possible range from ak_{20} to ak_{10}. There are now two solutions, at $ak_g = 54.2418$ and $k_g = 54.4378$, showing that at this higher frequency this fibre can carry a TE mode signal in two different modes (denoted TE modes 01 and 02). The discontinuities in the LHS occur at the two lowest zeros of $J_0(\Delta_1 a)$, i.e., at $\Delta_1 a = 2.40482$ and $\Delta_1 a = 5.52007$ which occur at $ak_g = 54.48076$ and at $ak_g = 54.25371$ respectively.*

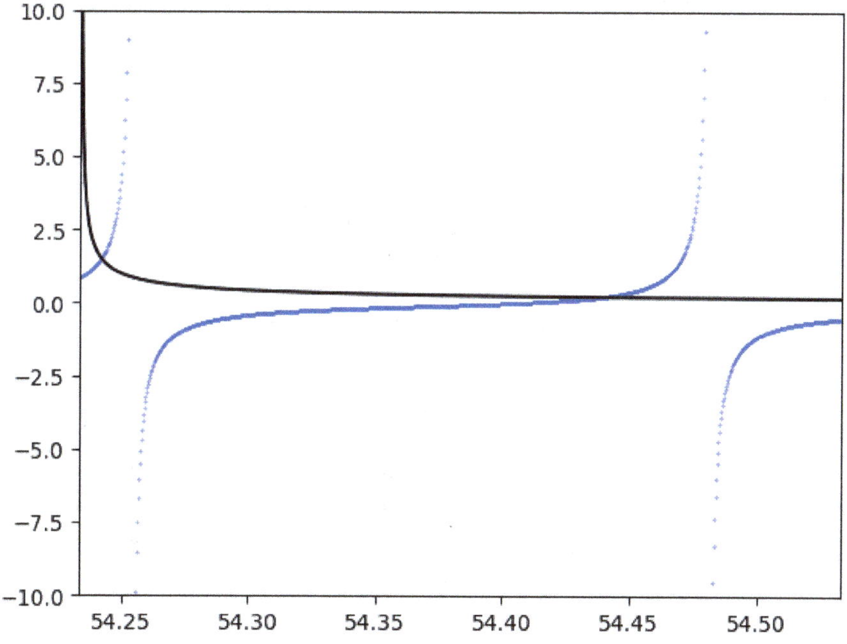

23.8.2.4 The TM Mode

In a similar manner to the TE mode, above, we now seek a TM mode solution with,

$$B_{1z} = 0 \quad \text{and} \quad E_{1z} = C_{1E} J_m(\Delta_1 r) e^{im\theta} \tag{23.8.18a}$$

$$B_{2z} = 0 \quad \text{and} \quad E_{2z} = C_{2E} K_m(\Delta_2 r) e^{im\theta} \tag{23.8.18b}$$

The other components, from (23.8.9a-d), are thus,

$$B_{1r} = -\xi_1 \frac{\alpha_1}{r} \partial_\theta E_{1z} \qquad B_{1\theta} = \xi_1 \alpha_1 \partial_r E_{1z} \tag{23.8.29a}$$

$$E_{1r} = \xi_1 k_g \partial_r E_{1z} \qquad E_{1\theta} = \xi_1 \frac{k_g}{r} \partial_\theta E_{1z} \tag{23.8.29b}$$

with similar expressions in the cladding in terms of $\xi_1 \to -\xi_2$. The boundary conditions are as specified by (23.8.17a,b). The requirements $E_{1z}(a) = E_{2z}(a)$ and $E_{1\theta}(a) = E_{2\theta}(a)$ give us,

$$C_{1E} J_m(\Delta_1 a) = C_{2E} K_m(\Delta_2 a) \tag{23.8.30a}$$

$$m\xi_1 C_{1E} J_m(\Delta_1 a) = -m\xi_2 C_{2E} K_m(\Delta_2 a) \tag{23.8.30b}$$

But these are in conflict unless $m = 0$, so we conclude that TM mode we seek must be axisymmetric, as was the case for the TE mode. (23.8.30a) then becomes,

$$C_{1E} J_0(\Delta_1 a) = C_{2E} K_0(\Delta_2 a) \tag{23.8.31}$$

So the only non-zero field components in the TE mode are E_z, E_r, B_θ. Continuity of B_θ requires,

$$\xi_1 \alpha_1 C_{1E} \Delta_1 \left. \frac{dJ_0(s)}{ds} \right|_{s=\Delta_1 a} = -\xi_2 \alpha_2 C_{2E} \Delta_2 \left. \frac{dK_0(s)}{ds} \right|_{s=\Delta_2 a} \tag{23.8.32}$$

where $\alpha_{1,2}$ are given by (23.8.7). Using identities $\frac{dJ_0}{ds} = -J_1$ and $\frac{dK_0}{ds} = -K_1$ these become,

$$\xi_1 \alpha_1 C_{1E} \Delta_1 J_1(\Delta_1 a) = -\xi_2 \alpha_2 C_{2E} \Delta_2 K_1(\Delta_2 a) \tag{23.8.33}$$

So, finally, for the TM mode the complete boundary conditions reduce to (23.8.31) plus (23.8.33). We can choose one constant arbitrarily, say $C_{1E} = 1$, because this merely scales the overall amplitude of the wave. The two boundary conditions then permit elimination of C_{2E} to leave the condition which defines the guide wavenumber, k_g, for this axisymmetric TM mode,

$$-\frac{\alpha_1 J_1(\Delta_1 a)}{\Delta_1 J_0(\Delta_1 a)} = \frac{\alpha_2 K_1(\Delta_2 a)}{\Delta_2 K_0(\Delta_2 a)} \tag{23.8.34a}$$

For a given fibre, that is for given $n_{1,2}$, at a given frequency, we know from (23.8.27) that $\Delta_1^2 + \Delta_2^2 = \frac{\omega^2}{c^2}(n_1^2 - n_2^2)$ and from (23.8.7) that $\alpha_1 = \frac{\omega}{v_1^2}$ and $\alpha_2 = \frac{\omega}{v_2^2}$. So (23.8.34) allows the only remaining variable, the guide wavenumber, k_g, to be found, subject to frequency, and the number of solutions for k_g will depend on frequency, as before. It can be rewritten as,

$$-\frac{J_1(\Delta_1 a)}{\Delta_1 J_0(\Delta_1 a)} = \left(\frac{n_2^2}{n_1^2}\right)\frac{K_1(\Delta_2 a)}{\Delta_2 K_0(\Delta_2 a)} \qquad (23.8.34b)$$

For typical optic fibres n_1 and n_2 differ by around 1% or less (in our numerical example above they differ by only 0.55%). Consequently the solutions of (23.8.34b) will be very close to those of the corresponding TE mode as (23.8.34b) is then very close to (23.8.26). In particular, the cutoff frequency below which the optic fibre will not carry a TM signal is the same, (23.8.28), and the frequencies at which further TM modes appear are again the same, being given by the zeros of $J_0(\Delta_1 a)$.

However, this virtually identical behaviour of the TE and TM modes is specific to the case when n_1 and n_2 differ very little, as is the case with optic fibres for IT purposes. But for other dielectric waveguides we might have $n_1 \gg n_2$ in which case the TM modes will have appreciably different threshold frequencies from the TE modes.

23.8.2.5 The HE Modes

Any modes other than the axisymmetric TE and TM modes in our optic fibre must have the axial components of both the electric and magnetic fields non-zero and be non-axisymmetric. These are called the HE or EH modes according to whether the magnetic or electric fields dominate respectively. Hence, in (23.8.12) and (23.8.16) all four constant coefficients are non-zero, and so from (23.8.9a-d) we are dealing with all six field components being non-zero. The boundary conditions from §23.8.2.2 are that all field components other than E_r are continuous. Continuity of E_z and B_z require,

$$C_{1E} J_m(\Delta_1 a) = C_{2E} K_m(\Delta_2 a) \qquad (23.8.35a)$$

$$C_{1B} J_m(\Delta_1 a) = C_{2B} K_m(\Delta_2 a) \qquad (23.8.35a)$$

which therefore requires,

$$\frac{C_{2E}}{C_{1E}} = \frac{C_{2B}}{C_{1B}} = \frac{J_m(\Delta_1 a)}{K_m(\Delta_2 a)} \qquad (23.8.36)$$

From (23.8.9a-d) it is readily derived that,

$$k_g E_{1\theta} + \omega B_{1r} = \frac{m}{r} E_{1z} \text{ and } \quad k_g E_{2\theta} + \omega B_{2r} = \frac{m}{r} E_{2z} \tag{23.8.37}$$

Consequently, the continuity of E_z and E_θ implies continuity of B_r. It therefore remains only to impose continuity of E_θ and B_θ. This is simple to impose by substitution of (23.8.12) and (23.8.16) into (23.8.9a-d). We thus derive two boundary condition equations which, using (23.8.36), can both be expressed in terms of the constants C_{1E} and C_{1B} alone and are homogeneous. The condition that these two equations are consistent, i.e., that they have a non-trivial solution, reduces to,

$$\left[\frac{J_m'}{\Delta_1 J_m} + \frac{K_m'}{\Delta_2 K_m}\right]\left[\frac{J_m'}{\Delta_1 J_m} + \left(\frac{n_2}{n_1}\right)^2 \frac{K_m'}{\Delta_2 K_m}\right] = \left(\frac{mk_g}{an_1 k_0}\right)^2 \left(\frac{1}{\Delta_1^2} + \frac{1}{\Delta_2^2}\right)^2 \tag{23.8.38}$$

Here $k_0 = \frac{2\pi}{\lambda_0} = \frac{\omega}{c}$ and other terms are defined in (23.8.2,3,7,10a,13a,20). It is understood that the J Bessel functions are evaluated at $\Delta_1 a$ whereas the K Bessel functions are evaluated at $\Delta_2 a$, and dashes denote the derivative functions. The latter can be expressed in terms of the base functions using (23.8.24).

(23.8.38) is the condition which defines the guide wavenumber, k_g, in the general case, for any mode.

For the axisymmetric case, $m = 0$, this condition reduces to either (23.8.26), i.e., the TE mode (when the first bracket on the LHS of (23.8.38) is zero), or to (23.8.34b), i.e., the TM mode (when the second bracket on the LHS of (23.8.38) is zero).

Turning now to the case $m = 1$ we find (e.g., by numerical evaluation) that (23.8.38) has a solution at arbitrarily small frequencies, ω. In each case the wavelength in the waveguide lies between λ_0/n_1 and λ_0/n_2. For example, the input wavelengths (λ_0) 1800, 2000 and 3000nm, all of which are below the cutoff frequency, (23.8.28), for our example fibre with a core radius 4.25μm, $\mathrm{n}_1 = 1.452$, $\mathrm{n}_2 = 1.444$, have solutions with waveguide wavelengths of 1243, 1382 and 2075.6nm respectively.

For any solution for k_g the above equations permit the complete solution for the fields to be found. The solution below the cutoff frequency is found to have $C_{1B} > C_{1E}$ and hence is termed an HE mode.

For any mode, the numerical subscripts mn denote the angular order (i.e., the angular dependency $e^{im\theta}$) and the radial order, i.e., the nth solution for k_g for the stated m.

23.8.2.6 Summary of Optic Fibre Transmission Modes

The solution below the cutoff has $m = 1$, as we have seen, and it is the first solution of the HE kind, being the only one below the cutoff frequency. It is therefore denoted HE_{11}. This is also called the "fundamental mode" of the fibre.

As illustrated by Figure 23.5, other HE and EH modes occur at higher frequencies. The HE_{21} mode typically has a very similar wavelength-frequency curve as the TM_{01} and TE_{01}, whereas the first EH mode, EH_{11}, does not appear until a higher frequency is reached.

Telecommunications engineers usually employ a different notation for the modes (denoted LP), preferring to use an approximation in which the modes are linearly polarised in the x or y directions.

Operation of a given optic fibre at a frequency below the cutoff frequency given by the first zero of $J_0(\Delta_1 a)$ at $\Delta_1 a = 2.40482$ as given by (23.8.28), ensures that only the HE_{11} mode can be excited. This is the single mode condition and provides the most efficient transmission. Operation just a little above the cutoff frequency will generally cause four modes to become active, HE11, HE21, TE01 and TM01. At higher frequencies more and more modes become excitable.

Note that, although we have throughout referred to these as unique modes, for all cases where $m > 0$ there are clearly a pair of degenerate modes, i.e., for every $e^{im\theta}$ dependence we can equally choose $e^{-im\theta}$.

When the frequency is sufficiently far above cutoff, an estimate of the number of modes supported is given by $V^2/2$ where $V = ak_0\sqrt{n_1^2 - n_2^2}$.

Figure 23.5: Illustrative positions of the various modes on a plot of normalised waveguide wavenumber against normalised frequency. *I acknowledge P.Hudson, Brunel University as the source of this graphic, here β is my k_g and the abscissa is thus k_g/k_0 which lies between n_2 and n_1 (which are defined in reverse in this plot).*

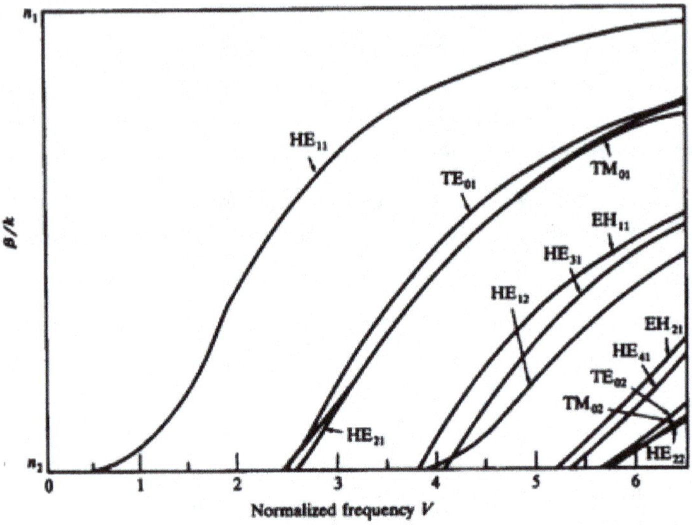

24. Spherical E/M Waves and Their Sources

Here we define what is meant by a spherical wave and find their general form for the electromagnetic field. Specific cases are discussed including standing waves (quantised modes) in cavities and central point sources emitting power to infinity. It is shown that an electric multipole field whose magnitude is sinusoidally oscillating creates a TM spherical wave of the same l,m order, whereas a magnetic multipole field whose magnitude is sinusoidally oscillating creates a TE spherical wave of the same l,m order.

Contents

24.1 What is a Spherical Wave?

A plane wave is a wave whose wavefronts (surfaces of uniform phase) are planes. A spherical wave is a wave whose wavefronts are spheres. A spherical wave centred on the origin of spherical coordinates has uniform phase on the spheres $r =$ constant.

We have already met such a wave in chapter 22, though it may not have been obvious. The solution (22.1.11a,b) for the radiation fields of a varying electric dipole is a spherical wave when the dipole varies sinusoidally, i.e., as $e^{-i\omega t_s}$, in terms of the source time, t_s. Recall from (22.1.3) that this time appears as the retarded time when the time at the field point is used, i.e.,

$$t_s = t_f - \frac{|\bar{r}_s - \bar{r}_f|}{c} \equiv t - r/c \text{ for brevity} \tag{24.1.1}$$

Hence the solution (22.1.11a,b) for the fields have the form of waves with the wave-like factor $e^{i\omega(t-r/c)}$. The radiation fields are thus of the form,

$$\bar{B}_{rad} \text{ or } \bar{E}_{rad} = \frac{\bar{F}(\theta,\phi)}{r} e^{-i\omega(t-r/c)} \tag{24.1.2}$$

which is a spherical wave. A term of this form will always dominate at large radii in any spherical wave if the wave is assumed to propagate energy to infinity from a point source at the origin. The reason is that the Poynting vector, averaged over time and integrated over the sphere of radius r, i.e., the average energy propagating across the sphere, is then a constant independent of r for sufficiently large r, i.e., it is the energy propagating to infinity from the source. Any other radial dependency would either have zero energy propagating to infinity or increasing energy crossing ever larger spheres, which would necessarily imply energy sources distributed throughout the volume and extending to infinity.

However, (24.1.2) is only the dominant form at large r, even when the only source is confined to a finite region and this source provides a non-zero net power injection. Even in such cases, at smaller r, other terms will arise. And then there are the cases of spherical waves without a net energy source. The question to be addressed is the exact solution for spherical waves in general.

24.2 Formulation of the Problem of E/M Spherical Waves

Other than at places where there is a source, and assuming free space without a medium, the equations which must be obeyed are the Maxwell equations in the source-free form,

$$\bar{\nabla}\cdot\bar{E}=0 \qquad \bar{\nabla}\cdot\bar{B}=0 \tag{24.2.1a}$$

$$\bar{\nabla}\times\bar{B}=\frac{1}{c^2}\frac{\partial\bar{E}}{\partial t} \qquad \bar{\nabla}\times\bar{E}=-\frac{\partial\bar{B}}{\partial t} \tag{24.2.1b}$$

We know, from chapters 5 and 10, that these equations imply that all six components of the e/m field obey the wave equation,

$$\nabla^2\bar{B}=\frac{1}{c^2}\frac{\partial^2\bar{B}}{\partial t^2} \qquad \nabla^2\bar{E}=\frac{1}{c^2}\frac{\partial^2\bar{E}}{\partial t^2} \tag{24.2.2}$$

Assuming wave-like time dependence $e^{-i\omega t}$ equs.(24.2.1b) become,

$$\bar{\nabla}\times\bar{B}=\frac{-i\omega}{c^2}\bar{E} \qquad \bar{\nabla}\times\bar{E}=i\omega\bar{B} \tag{24.2.3}$$

whilst the wave equations, (24.2.2), become,

$$\nabla^2\bar{B}+k^2\bar{B}=0 \qquad \nabla^2\bar{E}+k^2\bar{E}=0 \tag{24.2.4}$$

where $k=\omega/c$. An issue which arises in solving these wave equations for both the electric and magnetic fields is that the two fields are not independent but constrained to obey (24.2.3). A strategy to address this is to focus on solving for $\bar{B}$ and then use the first of (24.2.3) to find $\bar{E}$. However, the solution of the wave equation for $\bar{B}$ must also respect $\bar{\nabla}\cdot\bar{B}=0$. A neat means of imposing that constraint will emerge shortly. To start, as we are interested in spherical waves, express ∇^2 in spherical polars so that, using Appendix A2, we get,

$$\left(\frac{1}{r^2}\partial_r(r^2\partial_r)+\frac{1}{r^2\sin\theta}\partial_\theta(\sin\theta\partial_\theta)+\frac{1}{r^2\sin^2\theta}\partial_\phi^2\right)\bar{B}+k^2\bar{B}=0 \tag{24.2.5}$$

Readers familiar with the non-relativistic quantum theory of the hydrogen atom via the Schrodinger equation will anticipate the approach to the solution to (24.2.5) and we have already addressed this in §14.1.2 but do so again here in brief. Define the vector operator,

$$\bar{L}=-i\bar{r}\times\bar{\nabla} \tag{24.2.6}$$

In quantum mechanics this is the angular momentum operator (which is why the factor of $-i$ has been included, to be consistent with normal practice). Using $\bar{\nabla}$ in polars,

$$\bar{\nabla}=\hat{r}\partial_r+\frac{\hat{\theta}}{r}\partial_\theta+\frac{\hat{\phi}}{r\sin\theta}\partial_\phi \tag{24.2.7}$$

we get,

$$\bar{L}=-i\left(\hat{\phi}\partial_\theta-\frac{\hat{\theta}}{\sin\theta}\partial_\phi\right) \tag{24.2.8}$$

An exercise for the reader is to derive the following identities,

$$\partial_\phi \hat{\phi} = -(\hat{\theta}\cos\theta + \hat{r}\sin\theta) \qquad \partial_\theta \hat{\phi} = 0 \tag{24.2.9a}$$

$$\partial_\phi \hat{\theta} = \hat{\phi}\cos\phi \qquad \hat{\phi}\cdot\partial_\theta\hat{\theta} = 0 \tag{24.2.9b}$$

With the aid of these we find that,

$$-L^2 = -\bar{L}\cdot\bar{L} = \frac{1}{\sin\theta}\partial_\theta(\sin\theta\,\partial_\theta) + \frac{1}{\sin^2\theta}\partial_\phi^2 \tag{24.2.10}$$

So that (24.2.5) becomes,

$$\left(\frac{1}{r^2}\partial_r(r^2\partial_r) - \frac{1}{r^2}L^2\right)\bar{B} + k^2\bar{B} = 0 \tag{24.2.11}$$

Consider initially how a scalar field solution to this equation is formulated (as addressed already in §14.1.2). Separation of variables in the form $R(r)\Theta(\theta,\phi)$ leads to the eigenvalue equation,

$$L^2\Theta(\theta,\phi) = l(l+1)\Theta(\theta,\phi) \tag{24.2.12}$$

and the corresponding radial equation,

$$[\partial_r(r^2\partial_r) - l(l+1) + k^2r^2]R = 0 \tag{24.2.13}$$

Here we have assumed that the eigenvalues of the squared angular momentum operator, L^2, are $l(l+1)$ where $l = 0,1,2,3,\ldots$ which will be familiar to most readers from quantum mechanics. (For the proof see Appendix A7). The angular functions are therefore the spherical harmonics which we met in §14.1.2,

$$\Theta(\theta,\phi) = Y_l^m(\theta,\phi) \tag{24.2.14}$$

where m takes integer values from $-l$ to $+l$ inclusive.

Re-expressing the radial equation, (24.2.13), in terms of the independent variable $s = kr$ we see that it becomes the spherical Bessel equation,

$$[s^2\partial_s^2 + 2s\partial_s + s^2 - l(l+1)]R(s) = 0 \tag{24.2.15}$$

This differs from the standard Bessel equation which occurs in problems with cylindrical symmetry by the factor of 2 in the first order derivative term. Also, the constant term (the angular eigenvalue), rather than being an integer-squared is here $l(l+1)$. The standard solutions to the spherical Bessel equation are conventionally denoted $j_l(s)$ and $y_l(s)$, spherical Bessel functions of the first and second kind respectively. So, for a notional scalar field, ψ, obeying (24.2.11) the general solution would be an arbitrary linear combination of the base solutions $j_l(kr)Y_l^m(\theta,\phi)$ and $y_l(kr)Y_l^m(\theta,\phi)$,

$$\psi = \sum_{l,m}\left[C_{lm}^j j_l(kr) + C_{lm}^y y_l(kr)\right] Y_l^m(\theta,\phi)e^{-i\omega t} \tag{24.2.16}$$

Now we need to face the real issue which is the solution for a vector field which also respects the divergence-free condition. The trick is to consider vector functions of the form $\bar{L}\chi$ where χ is a scalar solution, (24.2.16). There are three crucial features of this vector function which make this a convenient approach,

(i) $\bar{L}\psi$ is a vector solution of the wave equation because $\bar{L}$ commutes with $\nabla^2 + k^2$. This follows because $\nabla^2 = \frac{1}{r^2}\partial_r(r^2\partial_r) - \frac{1}{r^2}L^2$ and $\bar{L}$ does not act on the radial coordinate, see (24.2.8), and $\bar{L}$ commutes with L^2.

(ii) $\bar{\nabla}\cdot\bar{L}\psi = 0$ because, in Cartesian coordinates, $i\bar{\nabla}\cdot\bar{L} = \partial_i\epsilon_{ijk}x_j\partial_k = 0$ because $\partial_i\epsilon_{ijk}x_j\partial_k = \epsilon_{ijk}x_j\partial_i\partial_k + \epsilon_{ijk}\delta_{ij}\partial_k$ both terms of which are zero due to their symmetry under subscript interchange (whereas ϵ_{ijk} is antisymmetric).

(iii) $\bar{\mathrm{r}}\cdot\bar{L}\psi = 0$ because $i\bar{\mathrm{r}}\cdot\bar{L} = \epsilon_{ijk}x_i x_j\partial_k = 0$.

So, for example, if we assume $\bar{E} = \bar{L}\psi$ where ψ is of the form (24.2.16) then we have created a vector solution of the wave equation which, by the second bullet above, also obeys the divergence-free condition. However, the third bullet then shows that $\bar{\mathrm{r}}\cdot\bar{E} = 0$ so this would be a transverse electric (TE) mode. The magnetic field then follows from (24.2.3).

Conversely, if we assume $\bar{B} = \bar{L}\psi$ then we get a transverse magnetic (TM) mode and the electric field then follows from (24.2.3).

We will state without proof that the general solution is an arbitrary linear combination of TE and TM modes, so that it suffices to study these modes.

We note in passing that a solution for either field written as $\bar{L}\psi$ can be written in a different way, namely,

$$\bar{L}\psi = \mathrm{i}\bar{\nabla}\times(\psi\bar{r}) \tag{24.2.17}$$

This can readily be proved using the expression for $curl$ in spherical polars (see Appendix A2). This is worth noting because the above style of expression is used by other authors and might be thought to be a different solution.

24.2.1 There is No $l = 0$ Mode

Where one of the fields is given by $\bar{L}\psi$ it follows from (24.2.8) that if ψ is an $l = 0$ (spherically symmetric) scalar solution then the corresponding e/m solution is the trivial $\bar{E} = \bar{B} = 0$ everywhere. This suggests that there is no $l = 0$ mode in the e/m case, but would only be a proof if we could first prove that all e/m solutions were expressible as $\bar{L}\psi$.

An actual proof is topological. Consider a propagating solution. At large radii the propagation must be radial. It follows that $\bar{E}$ and $\bar{B}$ are both tangential to a large sphere of radius r. But the "hairy ball" theorem states that any tangential vector field which is continuous over the surface of a sphere must be zero somewhere on it. But if we also require independence of angle, then the fields must be zero everywhere on a large sphere – and hence everywhere, full stop. QED. This proof applies to propagation outwards or inwards and hence, by linear combinations thereof, also to standing wave solutions.

We now go on the consider specific cases, but first some important features of the spherical Bessel functions which will be required.

24.3 Properties of the Spherical Bessel Functions

Each of the spherical Bessel functions, $j_l(s)$ or $y_l(s)$, is of the form of a polynomial in $1/s$ times $\sin s$ plus a polynomial in $1/s$ times $\cos s$. All these functions, for integral l, can be found from,

$$j_l(s) = (-s)^l \left(\frac{1}{s}\frac{d}{ds}\right)^l \left(\frac{\sin s}{s}\right) \tag{24.3.1}$$

$$y_l(s) = -(-s)^l \left(\frac{1}{s}\frac{d}{ds}\right)^l \left(\frac{\cos s}{s}\right) \tag{24.3.2}$$

Here s should be understood as replaced by kr in the spherical wave solutions. The first three of each function are,

$$j_0(s) = \frac{\sin s}{s} \tag{24.3.3a}$$

$$j_1(s) = \frac{\sin s - s\cos s}{s^2} \tag{24.3.3b}$$

$$j_2(s) = \frac{(3-s^2)\sin s - 3s\cos s}{s^3} \tag{24.3.3c}$$

$$y_0(s) = -\frac{\cos s}{s} \tag{24.3.4a}$$

$$y_1(s) = -\frac{\cos s + s\sin s}{s^2} \tag{24.3.4b}$$

$$y_2(s) = -\frac{(3-s^2)\cos s + 3s\sin s}{s^3} \tag{24.3.4c}$$

For small values of their argument the functions are approximately as follows,

$$\text{Small } s: \quad j_l(s) \approx \frac{s^l}{(2l+1)!!} \tag{24.3.5a}$$

$$\text{Small } s: \quad y_l(s) \approx -\frac{(2l-1)!!}{s^{l+1}} \tag{24.3.5b}$$

where $n!! = n(n-2)(n-4) \dots 5 \times 3 \times 1$ for odd n.

Conversely, for sufficiently large arguments the asymptotic forms are as follows,

$$LIM_{s\to\infty} j_l(s) \to \frac{\sin\left(s - \frac{\pi}{2}l\right)}{s} \tag{24.3.6a}$$

$$LIM_{s\to\infty} y_l(s) \to -\frac{\cos\left(s - \frac{\pi}{2}l\right)}{s} \tag{24.3.6b}$$

The above asymptotic forms reveal that both the j_l and the y_l solutions, when combined with the time dependence assumed above, $e^{-i\omega t}$, are actually standing waves, not propagating waves. A propagating wave would have a radial dependence $e^{is} = e^{ikr}$. To reproduce this asymptotic behaviour we must use the following specific linear combination of the j_l and y_l functions,

$$g_l(s) = i^{l+1}\left(j_l(s) + iy_l(s)\right) \tag{24.3.7}$$

This is sometimes called a Hankel function, though the constant factor may differ in other sources. With this definition we have the asymptotic form,

$$LIM_{s\to\infty} g_l(s) \to \frac{e^{is}}{s} \tag{24.3.8}$$

Note that the variable phase $\frac{\pi}{2}l$ which appears in (24.3.6a,b) has disappeared from (24.3.8) because it has been killed by the factor of i^{l+1} in the definition of g_l.

Scalar spherical waves which are propagating outwards, and hence with asymptotic dependence $e^{is-i\omega t}$, will therefore be given, instead of by (24.2.16), by,

$$\psi = \sum_{l,m} C_{lm}^g g_l(kr)\, Y_l^m(\theta,\phi) e^{-i\omega t} \tag{24.3.9}$$

Note that replacing $g_l(kr)$ in (24.3.9) with the complex conjugate of $g_l(kr)$ which has the asymptotic dependence $\frac{e^{-is}}{s}$ would therefore be the general form of a scalar spherical wave propagating inwards, towards the origin.

24.4 Resonant Spherical Cavities

§23.6 considered the possible e/m field modes within a rectangular cavity with perfectly conducting boundaries. Here we look at the same problem for spherical cavities, $r = a$. These are the resonant modes, or equivalently the quantum states after reinterpretation of $\hbar k$ as the momentum and $\hbar\omega$ as the energy, where $\hbar = h/2\pi$ and h is Planck's constant.

We are interest here in standing waves, and solutions which are finite at the origin, assuming no sources are present. The latter condition rules out the y_l solutions. Consequently, the general solution involves linear combinations of the solutions where either the electric or the magnetic field is identified with $\bar{L}\psi$ where ψ is the scalar solution,

$$\psi = j_l(kr)Y_l^m(\theta,\phi)e^{-i\omega t} \tag{24.4.1}$$

Conducting boundary conditions require the normal component of the magnetic field to be zero at the boundary, $B_r(a) = 0$, and the tangential components of the electric field to be zero, $E_\theta(a) = E_\phi(a) = 0$.

24.4.1 TM Modes

This is the most obvious way of satisfying $B_r(a) = 0$ because B_r is zero everywhere. Since it is the magnetic field which is transverse the solution is,

$$\bar{B} = -i\bar{L}\psi \tag{24.4.2a}$$

$$\bar{E} = \frac{c^2}{\omega}\bar{\nabla}\times\bar{L}\psi \tag{24.4.2b}$$

(omitting a constant factor which carries the dimensions of B). Expressing the $curl$ in spherical polars (Appendix A2) and substituting (24.4.1) the boundary conditions $E_\theta(a) = E_\phi(a) = 0$ are both found to be ensured by the same eigenvalue condition on the wavenumber (or equivalently on the frequency, or energy in the quantum mechanics interpretation), namely,

$$\partial_r\big(rj_l(kr)\big)\big|_{r=a} = 0 \tag{24.4.3}$$

24.4.2 TE Modes

In this case it is the electric field which is everywhere transverse, i.e., having zero radial component. Hence the solution is,

$$\bar{E} = -i\bar{L}\psi \tag{24.4.4a}$$

$$\bar{B} = -\frac{1}{\omega}\bar{\nabla} \times \bar{L}\psi \tag{24.4.4b}$$

(omitting a constant factor which carries the dimensions of E). The electric field is proportional to $j_l(kr)$. To spell this out, using (24.2.8) the electric field is,

$$\bar{E} = j_l(kr)\left(\frac{\hat{\theta}}{\sin\theta}\partial_\phi - \hat{\phi}\partial_\theta\right)Y_l^m(\theta,\phi)e^{-i\omega t} \tag{24.4.5}$$

Hence, both boundary conditions, $E_\theta(a) = E_\phi(a) = 0$, require,

$$j_l(ka) = 0 \tag{24.4.6}$$

Using $curl$ in spherical polars (Appendix A2) the radial component of the magnetic field is,

$$B_r = \frac{ij_l(kr)}{\omega r\sin^2\theta}\left[\sin\theta\,\partial_\theta(\sin\theta\,\partial_\theta) + \partial_\phi^2\right]Y_l^m(\theta,\phi)e^{-i\omega t}\text{q}$$

$$= \frac{ij_l(kr)}{\omega r}l(l+1)Y_l^m(\theta,\phi)e^{-i\omega t} \tag{24.4.7}$$

So the boundary condition $B_r(a) = 0$ is also obeyed due to (24.4.6).

Table 24.1 gives the values of ka at which there are TM modes, (24.4.3), and TE modes, (24.4.6). For a given mode and order the TM modes have the smaller frequencies (or energies in quantum mechanics).

Table 24.1: The first three solutions for ka of (24.4.3) and (24.4.6), for each of the first three modes, $l = 1, 2, 3$ (there is no l=0 mode)

$l = 1$	$l = 2$	$l = 3$
	$\partial_r\left(rj_l(kr)\right)\big\rvert_{r=a} = 0$ (TM modes)	
2.7437	3.8702	4.9734
6.1168	7.4431	8.7218
9.3166	10.7130	12.0636
	$j_l(ka) = 0$ (TE modes)	
4.4934	5.7635	6.9879
7.7253	9.0950	10.4171
10.9041	12.3229	13.6980

24.5 Resonant Spherical Annular Cavities

We now consider the annular cavity between two concentric spheres of radii $r = a$ and $r = b$ with $a < b$. We cannot now rule out the y_l modes as the origin does not lie in the cavity. (24.4.1) is thus replaced by, for a given angular order,

$$\psi = \left(C_1 j_l(kr) + C_2 y_l(kr)\right) Y_l^m(\theta, \phi) e^{-i\omega t} \tag{24.5.1}$$

The boundary conditions are now $B_r(a) = 0$, $E_\theta(a) = E_\phi(a) = 0$ and $B_r(b) = 0, E_\theta(b) = E_\phi(b) = 0$.

TM Modes

This is the most obvious way of satisfying $B_r(a) = B_r(b) = 0$ because B_r is zero everywhere. The fields are then given by (24.4.2a,b) where ψ is now given by (24.5.1). The boundary conditions on the electric field become,

$$\partial_r\left(rC_1 j_l(kr) + rC_2 y_l(kr)\right)\Big|_{r=a} = 0 \tag{24.5.2a}$$

$$\partial_r\left(rC_1 j_l(kr) + rC_2 y_l(kr)\right)\Big|_{r=b} = 0 \tag{24.5.2b}$$

The two conditions (24.5.2a,b) are then used to find the two constants C_1 and C_2.

TE Modes

In this case it is the electric field which is everywhere transverse and the solution is given by (24.4.4a,b) but using ψ from (24.5.1). All the boundary conditions are ensured by,

$$C_1 j_l(ka) + C_2 y_l(ka) = 0 \tag{24.5.3a}$$

$$C_1 j_l(kb) + C_2 y_l(kb) = 0 \tag{24.5.3b}$$

The two conditions (24.5.3a,b) again allow both constants C_1 and C_2 to be found.

24.6 Propagating Spherical Waves

We now consider a point source at the origin creating outgoing spherical waves. As discuss in §24.3 for this purpose we need to use the g_l functions, so the general solution is to be found as a linear combination of the e/m fields corresponding to the scalar solution,

$$\psi = C g_l(kr) Y_l^m(\theta, \phi) \tag{24.6.1}$$

for some constant C. Because $g_l(s) = i^{l+1}\big(j_l(s) + iy_l(s)\big)$ we note that g_l is divergent at the origin, but that is to be expected now because we have placed a point source there. A question that arises is "what type of source?". The source might be an electric or magnetic dipole or quadrupole or one of the higher multipoles, as discussed in §14.1.2.3, or a combination thereof.

To generate radiation we will assume a source undergoing sinusoidal variations, hence we will then include a factor of $e^{-i\omega t}$ in (24.6.1).

24.6.1 TM Modes

We proceed as before so for TM modes,

$$\bar{B} = -i\bar{L}\psi \tag{24.6.2a}$$

$$\bar{E} = \frac{c^2}{\omega}\bar{\nabla} \times \bar{L}\psi = \frac{c}{k}\bar{\nabla} \times \bar{L}\psi \tag{24.6.2b}$$

where ψ is now given by (24.6.1). Using (24.3.5a,b) we see that near the source, for small s, we have, $g_l(s) \approx i^l \frac{(2l-1)!!}{s^{l+1}}$, so that, when static,

$$\psi \approx Ci^l \frac{(2l-1)!!}{(kr)^{l+1}} Y_l^m(\theta, \phi) \tag{24.6.3}$$

We now seek to simplify the expression, (24.6.2b) for the electric field when ψ is given by (24.6.3). Firstly we find,

$$(\bar{\nabla} \times i\bar{L})_i = \big(\bar{\nabla} \times (\bar{r} \times \bar{\nabla})\big)_i = \varepsilon_{ijk}\partial_j\varepsilon_{kmn}x_m\partial_n$$

$$= \varepsilon_{ijk}\varepsilon_{kmn}\big(\delta_{jm}\partial_n + x_m\partial_j\partial_n\big) = \big(\delta_{im}\delta_{jn} - \delta_{in}\delta_{jm}\big)\big(\delta_{jm}\partial_n + x_m\partial_j\partial_n\big)$$

$$= \delta_{ji}\partial_j + x_i\partial_j\partial_j - \delta_{jj}\partial_i - x_j\partial_j\partial_i = \partial_i + x_i\partial_j\partial_j - 3\partial_i - x_j\partial_j\partial_i$$

$$= x_i\partial_j\partial_j - 2\partial_i - \partial_i\big(x_j\partial_j\big) + \partial_i = x_i\partial_j\partial_j - \partial_i - \partial_i\big(x_j\partial_j\big)$$

$$= [\bar{r}\nabla^2 - \bar{\nabla}(1 + \bar{r} \cdot \bar{\nabla})]_i \tag{24.6.4}$$

Next we find, using $\nabla^2 = \left(\frac{1}{r^2}\partial_r(r^2\partial_r) - \frac{1}{r^2}L^2\right)$ from (24.2.11), and also $L^2Y_l^m = l(l+1)Y_l^m$ from (24.2.12,14), that, in the near field limit,

$$\nabla^2\left(\frac{Y_l^m}{r^{l+1}}\right) = \left(\frac{1}{r^2}\partial_r(r^2\partial_r) - \frac{1}{r^2}L^2\right)\left(\frac{Y_l^m}{r^{l+1}}\right)$$

$$= \left(\frac{1}{r^2}\partial_r(r^2\partial_r) - \frac{l(l+1)}{r^2}\right)\left(\frac{Y_l^m}{r^{l+1}}\right) = 0 \tag{24.6.5}$$

Also,

$$(1 + \bar{r} \cdot \bar{\nabla})\left(\frac{Y_l^m}{r^{l+1}}\right) = (1 + r\partial_r)\frac{Y_l^m}{r^{l+1}} = -l\frac{Y_l^m}{r^{l+1}} \tag{24.6.6}$$

Putting all that together we find, in the near field,

For small r: $$\bar{E} = \frac{c}{k}\bar{\nabla} \times \bar{L}\psi \approx -\frac{c}{k} C i^{l+1} \frac{(2l-1)!!}{k^{l+1}} l \bar{\nabla}\left(\frac{Y_l^m}{r^{l+1}}\right) \tag{24.6.7}$$

Compare this with the electric field derived from the scalar potential for a multipole field of order lm from (14.1.27c,d)

$$\bar{E} = -\frac{1}{4\pi\varepsilon_0}\sqrt{\frac{4\pi}{2l+1}} M_{lm} \bar{\nabla}\left(\frac{Y_l^m}{r^{l+1}}\right) \tag{24.6.8}$$

We see that the TM mode derived from (24.6.1) approximates in the near field to an electric multipole M_{lm} provided we define the constant C to be,

$$C = \frac{(-i)^{l+1}}{l(2l-1)!!\sqrt{4\pi(2l+1)}} \cdot \frac{k^{l+2}}{c\varepsilon_0} M_{lm} \tag{24.6.9}$$

The reader should check that C has the same dimensions as the B field, as required. This identification will remain true in the near field even when oscillation of the source is considered since, in the near field, the effect of retarded time is negligible.

> Hence, we conclude that the TM spherical wave mode given by (24.6.1,2) results from an electric multipole of the same lm order.

To examine the type of radiation produced by an oscillating source let us consider the large r asymptotic form of the fields using (24.3.8), that is the asymptotic limit,

For large r: $$\psi \to C Y_l^m(\theta,\phi)\frac{e^{-i(\omega t - kr)}}{kr} \tag{24.6.10}$$

Consequently, as $\bar{L}$ only acts on the angular variables, the magnetic field is,

For large r: $$\bar{B} = -iC\left(\bar{L}Y_l^m(\theta,\phi)\right)\frac{e^{-i(\omega t - kr)}}{kr} \tag{24.6.11a}$$

This has the expected form for a radiation field. The magnetic field is, of course, transverse ($B_r = 0$) as it is everywhere in a TM mode.

The electric field is found from $\bar{E} = \frac{ic}{k}\bar{\nabla} \times \bar{B}$. Using $curl$ in spherical polars, with $B_r = 0$, gives, (24.6.11b)

$$\bar{E} = \frac{ic}{k}\bar{\nabla} \times \bar{B} = \frac{ic}{k}\left\{\frac{\hat{r}}{r\sin\theta}\left(\partial_\theta(B_\phi \sin\theta) - \partial_\phi B_\theta\right) - \frac{\hat{\theta}}{r}\partial_r(rB_\phi) + \frac{\hat{\phi}}{r}\partial_r(rB_\theta)\right\}$$

All terms in (24.6.11b) are of order $1/r^2$ (and so can be neglected in the far field limit) except where ∂_r acts on the wave function $e^{-i(\omega t-kr)}$ which produces the dominant term in the far field of $\bar{E}$, proportional to $1/r$,

$$\bar{E} = iCc\left(\hat{r} \times \bar{L}Y_l^m(\theta,\phi)\right)\frac{e^{-i(\omega t-kr)}}{kr} \tag{24.6.11c}$$

Hence, the far field of $\bar{E}$ is also of radiation field form. Moreover, $\bar{E}$ is transverse in the far field limit, confirming that the Poynting vector is radial, i.e., the energy flow is radial, as expected.

(24.6.11a,c) give the angular variation of the magnetic and the electric fields in a radiation field due to an oscillating electric multipole of order lm, noting that this is not simply $Y_l^m(\theta,\phi)$ but involves its derivatives.

24.6.2 TE Modes

We proceed as before so for TE modes,

$$\bar{E} = -i\bar{L}\psi \tag{24.6.12a}$$

$$\bar{B} = -\frac{1}{ck}\bar{\nabla} \times \bar{L}\psi \tag{24.6.12b}$$

where ψ is given again by (24.6.1). The development closely follows §24.6.1 but with the electric and magnetic fields swapped around. We again seek the form of the solution in the near field using $g_l(s) \approx i^l \frac{(2l-1)!!}{s^{l+1}}$. Just as we derived (24.6.7) we now have for the magnetic field,

For small r: $$\bar{B} = \frac{1}{ck}\bar{\nabla} \times \bar{L}\psi \approx -\frac{1}{ck}Ci^{l+1}\frac{(2l-1)!!}{k^{l+1}}l\bar{\nabla}\left(\frac{Y_l^m}{r^{l+1}}\right) \tag{24.6.13}$$

Again this may be compared with the magnetostatic multipole of the same order, as it appears when derived via the magnetic scalar potential (see §14.1.2.5) when the magnetic field is expressed as a gradient,

$$\bar{B} = -\frac{\mu_0}{4\pi}\sqrt{\frac{4\pi}{2l+1}}M_{lm}^{mag}\bar{\nabla}\left(\frac{Y_l^m}{r^{l+1}}\right) \tag{24.6.14}$$

> Hence, we conclude that the TE spherical wave mode given by (24.6.1,12) results from a magnetic multipole of the same lm order.

In the far field the radiation fields can be found just as for the TM case, with the fields swapped. Both fields are again transverse in the far field limit, corresponding to a radial flow of power (radial Poynting vector).

25. The Motion of Charged Particles in EM Fields

In this chapter we consider a few examples of charged particles moving in a background e/m field. For this purpose it is convenient to work in the approximation that radiation is negligible, and hence that the charge does not experience a resistance to its motion that would result from the emission of radiation. This permits us to avoid the tricky issue of a consistent formulation of the force of radiation reaction, the problems with which we discussed briefly in §22.6. Much of this chapter is qualitative or descriptive. Such problems in realistic scenarios would, in practice, be subject to numerical computation rather than analytical solutions. However, simple cases are derived analytically, sufficient to motivate the qualitative extensions.

Contents

25.1 The Force Law and Energy

In §4.2.1 and (4.2.10) we defined the force acting on a point charge due to an electric and a magnetic field respectively. Their sum is,

$$\bar{f} = q(\bar{E} + \bar{v} \times \bar{B}) \tag{25.1.1}$$

In this chapter we consider some applications of this force law. Unless otherwise stated, non-relativistic conditions are assumed, i.e., speeds substantially less than light speed.

The rate at which the e/m field does work on the charge is,

$$\bar{v} \cdot \bar{f} = q\bar{v} \cdot (\bar{E} + \bar{v} \times \bar{B}) = q\bar{v} \cdot \bar{E} \tag{25.1.2}$$

Only the electric field does work on the particle. The magnetic field does no work because its force is always perpendicular to the charge's velocity. This should be borne in mind when considering the examples of the trajectories of charges in this chapter.

25.2 Limitations of This Chapter

In this chapter we treat only a simplified account of charged particle motion in which the radiation emitted by charges undergoing acceleration is ignored, and hence the resulting back-reaction, or resistance, to the motion of the charge is also ignored. In reality, if the charge emits radiation – hence emitting energy into the e/m field – it must loose energy itself. This manifests as a resistance to such motion, robbing the charge of some kinetic energy. This amounts to a damping force for motion in an e/m field. The matter is complicated because the e/m field – specifically the electric field – can also increase the particle's kinetic energy. In short, the presumption of a fixed background e/m field which is not influenced by the presence of a moving charge is a convenient fiction and can only be an approximation to reality, accurate enough under certain conditions.

25.3 The Mass Spectrometer

A mass spectrometer is a means of accurately measuring the charge-to-mass ratio of ions. Stripped down to its essentials, a mass spectrometer consists of,

- an ioniser which creates ions with charge q and mass m from the neutral atoms which are input and accelerates these ions over a potential difference of Φ;
- a uniform magnetic field, B, perpendicular to the velocity of the ions and extending over a distance D in the direction of their (initial) motion;

- a detector which measures the angular deflection, θ, of the ion beam due to the magnetic field.

The following is a simple analysis of the idealised model of the mass spectrometer.

Equating the kinetic energy to the loss of potential energy, the velocity attained by the ions is $v = \sqrt{\frac{2q\Phi}{m}}$. Whilst in the magnetic field the ion's trajectory is a circular arc whose radius, r, follows from equating the centripetal force, mv^2/r, to the magnetic force, qvB, from which the angle of deflection is obtained by geometry, thus,

$$r = \frac{mv}{qB} = \frac{1}{B}\sqrt{\frac{2m\Phi}{q}} \qquad \text{hence} \quad \sin\theta = \frac{D}{r} = DB\sqrt{\frac{q}{2m\Phi}} \tag{25.3.1}$$

It will be noted that the measured deflection only provides the ratio m/q, not the mass and charge separately.

25.4 Motion in a Uniform Magnetic Field

Suppose there is a uniform magnetic field $\bar{B} = B\hat{z}$ which persists over the whole region where the trajectory of a charge q is considered. Suppose the charge enters the region at an angle α to the field, such as,

$$\bar{v} = v(\hat{x}\sin\alpha + \hat{z}\cos\alpha) = v_T\hat{x} + v_z\hat{z} \tag{25.4.1}$$

where v_T is the initial component of the velocity in the x-direction and will thereafter be the speed in the x, y plane, transverse to the field. The magnetic force acts only in the x, y plane and projected onto that plane the charge executes circular motion with radius r. Equating the centripetal force, mv_T^2/r, to the magnetic force, qv_TB, the radius is $r = \frac{mv_T}{qB}$. Meanwhile the velocity in the z-direction, v_z, continues unchanged. Consequently, the charge is executing helical motion. The periodic time for one circuit in the x, y plane is $T = \frac{2\pi r}{v_T} = \frac{2\pi m}{qB}$.

The distance gone in the z-direction in one period is thus $\xi = Tv_z$. This is the pitch of the helical motion. Hence, the shape of the motion, given by the ratio of the pitch to the helical radius, is,

$$\frac{\xi}{r} = 2\pi\frac{v_z}{v_T} = 2\pi\cot\alpha \tag{25.4.2}$$

25.5 The Cyclotron

Cyclotrons were an early form of particle accelerator, subsequently largely superseded by synchrotrons. The device has a uniform vertical magnetic field. It also has two D-shaped electrodes with an alternating potential difference between them, with frequency f. Charged particles are injected near the centre of the device, which has a circular shape. The particles move in roughly circular horizontal orbits but whose radius gradually increases, i.e., actually a spiral motion. At a time when the tangential speed is v the radius is found as in the above examples to be $r = \frac{mv}{qB}$. Note that this corresponds to an angular speed of $\omega = \frac{qB}{m} = \frac{2\pi}{T}$. The angular speed is therefore independent of radius.

But the frequency of the alternating electric field is chosen to coincide with this "cyclotron resonance frequency", i.e., $f = \frac{1}{T} = \frac{qB}{2\pi m}$. Consequently, the particle is being accelerated during all parts of its circular orbits (noting that an electric field is necessary to do so as it means increasing the particle's energy). As the speed increases so does the radius of the orbit, according to $r = \frac{mv}{qB}$, hence the spiral path. When the particle reaches the boundary of the device the field is dropped and the particle exits the cyclotron down a tangential pipe or beamline.

25.6 Motion in a Magnetic Field with Non-Uniform Direction

Suppose a magnetic field has uniform magnitude but its direction gradually turns through an arc. What is the trajectory of a charged particle then? A heuristic argument runs as follows. Suppose the field is replaced by one which has a fixed direction for the duration of one revolution of the charge around it, i.e., for time T, so that the charge follows the usual helical path over this period. But, at the new location, the magnetic field abruptly changes to a slightly different direction and is fixed in that direction for another period, T. The particle will now execute its helical motion around the new direction as centre. And so on. This suggests that the charge will follow a curved helical path, as if a straight helix has been bent to follow the magnetic field.

In fact there is a theorem, Alfvén's theorem, which states that an ideal conducting fluid and a magnetic field are constrained to move together, in agreement with the above heuristic argument. In application, plasmas (clouds of charged particles) approximate ideal conductors and so respect Alfvén's theorem. The same will apply to individual particles of which they are composed. Alfvén's theorem holds only in the limit of large magnetic

Reynold's number, R_m, which is defined as $R_m = \frac{vL}{\eta}$ where v is the flow speed and L is a characteristic length scale of the flow. In fluid dynamics the ordinary Reynold's number is defined similarly when η is the fluid's kinematic viscosity. Here η is the magnetic field diffusion, defined as $1/\mu_0\sigma$ where σ is the electrical conductivity of the fluid. (The reader should confirm that R_m is therefore dimensionless).

Consider a magnetic field within a region $0 \leq z \leq Z/4$. The flux density is assumed to be constant but to vary spatially in the z-direction thus,

$$\bar{B} = B(\hat{x}\sin kz + \hat{z}\cos kz) \qquad k = 2\pi/Z \tag{25.5.1}$$

The force law, (25.1.1), with no electric field, implies,

$$m\dot{v}_x = qBv_y \cos kz \tag{25.5.2a}$$

$$m\dot{v}_y = qB(v_z \sin kz - v_x \cos kz) \tag{25.5.2b}$$

$$m\dot{v}_z = -qBv_y \sin kz \tag{25.5.2c}$$

These equations were solved numerically in the region $0 \leq z \leq Z/4$ assuming the charge started at the origin with velocity $v_0(\sin\alpha, 0, \cos\alpha)$ with $\alpha = 1$ and we have taken $\frac{mv_0}{qB} = 1$. The resulting trajectory is shown in Figure 25.1. A useful exercise for the reader is to carry out a similar numerical solution.

25.7 Aurora Borealis / Aurora Australis

The Sun ejects plasma from its surface, forming what is know as the solar wind. It consists mostly of energetic electrons and protons. In the vicinity of the Earth's outer atmosphere typical energies of these particles are 1 or 2 keV. (Hence they are far less energetic than cosmic ray particles, which are typically in the order of hundreds of MeV or around a GeV).

The Earth's magnetic field extends several Earth radii into space in a region known as the magnetosphere. Due to the effect discussed in §25.5, when the solar wind encounters the magnetosphere its trajectory bends to follow the field lines around the planet. The magnetosphere thus acts as a shield, keeping the solar wind away from the planet and its atmosphere. If it were not for the Earth's magnetic field the solar wind would erode the atmosphere and ultimately strip it completely, with obvious implications for life on Earth.

Figure 25.1: Numerical solution of equs.(25.5.2a-c). *The magnetic field is vertical on* $z = 0$ *(vertical axis) but turns parallel to* x *at large* z*. The helical path of the charge is seen to turn to follow the field lines.*

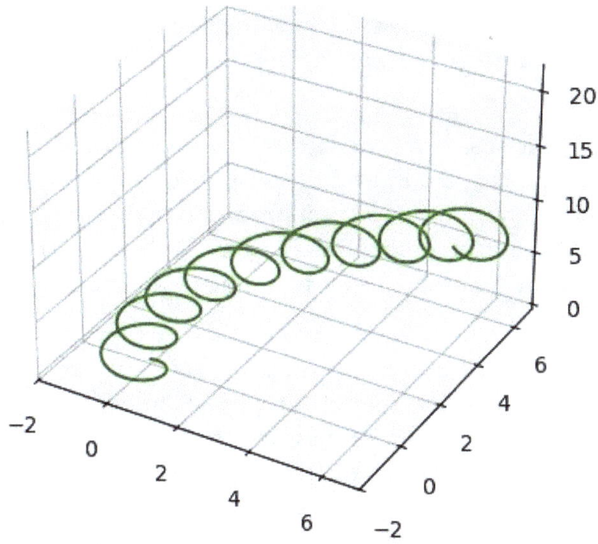

The magnetosphere similarly protects the Earth from much cosmic radiation.

However, some of the solar wind does penetrate the magnetosphere where it ends up trapped by the Alfvén effect, forming the van Allen radiation belt that surrounds the Earth in a toroidal shaped distribution.

A small proportion of the solar wind manages to penetrate nearer the Earth where the magnetic field directs the particles towards the poles. When reaching down far enough to encounter the upper atmosphere the nitrogen and oxygen there are ionised by the radiation, resulting in the northern lights (or southern lights).

25.8 Sunspots, Flares, Coronal Mass Ejections

The Sun also has powerful magnetic fields, unsurprisingly as it consists entirely of plasma undergoing violent convection – hence electrical currents, hence magnetic fields. Many transient phenomena occur from the Sun's surface (strictly, the photosphere) which involve both magnetic flux escaping from the interior of the Sun and also localised mass ejection. From the Alfvén theorem we expect those two things to go together.

Sunspots are such events. They are easily observed with a primitive solar scope (to avoid looking directly at the Sun). Sunspots appear as dark spots on the Sun's surface. They are actual regions of locally higher magnetic flux, which suppresses convention and so produces a cooler region (hence appearing 'black' in contrast). Their size ranges from tens of miles to hundreds of thousands of miles, and they last for between hours and months. The number of sunspots is a measure of solar activity, which varies on an 11 year cycle.

Sunspots are believed to be loops of magnetic flux that escape the Sun's surface. They often appear in pairs, the individual sunspots of a pair being at opposite ends of a flux loop. As we expect from Alfvén's theorem, sunspots involve plasma also escaping temporarily from the Sun's surface and moving along the magnetic flux tube, hence returning to the Sun.

In contrast, coronal mass ejections (CMEs) involve plasma permanently escaping the Sun and being emitted into interplanetary space. They are not rare. They occur between about once a week to several times per day, depending on the solar cycle. The amount of mass ejected in a single CME is around a Gigatonne. Just as in a sunspot the plasma follows the lines of flux. In a CME the same is true but best envisaged in reverse: the Sun's magnetic field gets dragged along with the ejected plasma. The result is that the Sun's magnetic flux can propagate into planet orbits. The speed of the ejected matter from a CME is such that typically they take several days to reach the Earth.

Most CMEs do not impact the Earth, but occasionally they do. When this happens the results can be dramatic. The Earth's magnetosphere can be severely disrupted, causing unusually extensive auroral displays. More seriously, in the modern world power grids may be disrupted. The CME in March 1989 caused a blackout in Canada's electricity grid lasting some 9 hours. The so-called Carrington event of 1859 was probably a severe CME. The world did not have large electrical grids then, but telegraph networks were plagued by sparking and fires.

Solar flares, which are massive outpourings of energy from the Sun, also cause disruptions of the Earth's magnetosphere. As I write there was a strong solar flare just a few days ago on 29 March 2026.

25.9 Magnetic Mirrors / Magnetic Confinement

Non-uniform magnetic fields can create conditions which act as a mirror for charged particles. Specifically, an increase in the magnetic flux density can constitute a magnetic mirror. Hot plasma (ionised gas) can be confined in the region between a pair of parallel mirrors. Such arrangements have stability problems but can work to a degree with suitable tweaks to the detailed design. Without delving into detailed magnetohydrodynamics, the basic reason for the magnetic mirror effect can be understood by the following argument.

Suppose the charged particle's velocities parallel to, and perpendicular to, the local $\bar{B}$ field are $v_{\parallel}$ and $v_{\perp}$. The kinetic energy of the particle is thus $K = \frac{m}{2}\left(v_{\parallel}^2 + v_{\perp}^2\right)$, assuming non-relativistic motion (though that is not an essential restriction). As the particle will be executing orbital motion around the field lines, as seen above, it's transverse motion will therefore constitute a near-circular current loop and hence possess a magnetic moment. Call this magnetic moment μ. Consideration of the need for the magnetic force to be centripetal leads to this magnetic moment being anti-aligned with $\bar{B}$ (see Figure 25.2) so that the magnetic moment has an energy contribution of μB (positive because anti-aligned, see (5.6.5)). Hence, assuming there is no electric field, the total energy is,

$$\xi = \frac{m}{2}\left(v_{\parallel}^2 + v_{\perp}^2\right) + \mu B \tag{25.8.1a}$$

$$\text{i.e., } \frac{m}{2}v_{\parallel}^2 = \xi - \frac{m}{2}v_{\perp}^2 - \mu B \geq 0 \tag{25.8.1b}$$

Hence the particle can only enter regions such that,

$$\mu B \leq \xi - \frac{m}{2}v_{\perp}^2 \tag{25.8.2}$$

If the magnetic field increases beyond this limit then the particle cannot enter such a region. Evidently, if the charge approaches such a region the particle's velocity, $v_{\parallel}$, must reverse in sign, i.e., the particle is reflected.

Note that the sense of the particle's circular motion is unchanged on reflection, and so the sign of its magnetic moment is unchanged also (an observation which is essential to make the above argument valid).

A simple analysis which shows why the parallel velocity component, $v_{\parallel}$, decreases in magnitude as the magnetic flux density increases is based on the assumption of adiabatic invariance of the magnetic moment. The magnetic moment is the current, I, times the area of the orbital loop, $\pi r^2 I$. The current is the charge divide by the periodic time, $T = 2\pi r/v_{\perp}$, hence,

$$\mu = \frac{1}{2}qrv_{\perp} \tag{25.8.3}$$

But the circular motion equation gives $r = \frac{mv_{\perp}}{qB}$, (25.3.1), hence,

$$\mu = \frac{mv_{\perp}^2}{2B} \tag{25.8.4}$$

The hypothesis that this is approximately a constant of the motion means that as B increases so $v_\perp$ must also increase. But the conservation of energy, (25.8.1a), then implies that $v_\parallel$ must decrease. A sufficiently large field will therefore ultimately result in $v_\parallel < 0$ (reflection).

The assumption of adiabatic invariance of the magnetic moment is an adequate approximation as long as,

- The field is quasi-static,
- The field does not vary much over the orbital size, r
- The particle motion does not become turbulent.

Figure 25.2: The anti-aligned orientation of the magnetic moment

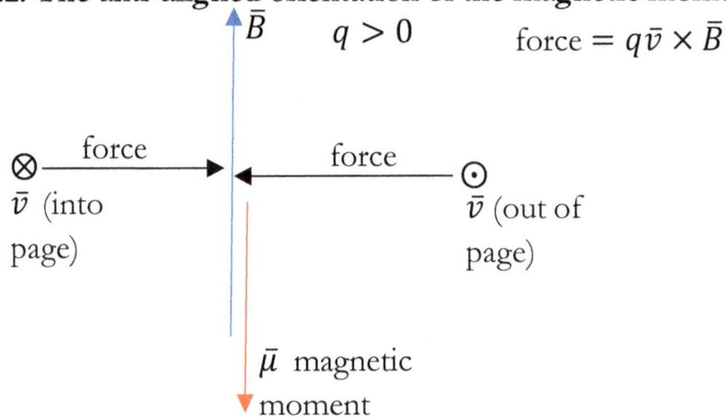

26. The Equivalence of Inertial Observers: Approaching Relativity

Here we approach the idea of combining space and time into a single four-dimensional manifold to create spacetime. In particular, we deduce what sort of spacetime results from considering all inertial observers to be equivalent. An "inertial observer" is an observer who is stationary or moving at constant velocity (but we shall duck the obvious question "with respect to what?"). It is shown that this motivates consideration of there being a finite maximum possible speed. The result is special relativity. The Lorentz transformation is here derived from its physical basis, and in an elementary fashion. Having done so, some of its key implications are explained and some famous paradoxes resolved. The Lorentz transformation of the e/m field is deferred to the following chapter.

Contents

26.1 How the Equivalence of Inertial Observers leads to the Lorentz Transformation

Inertial observers, or frames of reference, are defined as being in a constant state of motion or at rest. All inertial observers move at unvarying velocity, possibly zero, with respect to any other inertial observer. I shall pass over the issue of what the ultimate, or cosmic, frame of reference might be with respect to which all other inertial frames are defined.

It is easy to be beguiled by the algebraic elegance of Lie groups and so forth and to forget just what the Lorentz transformation means. There is nothing mysterious or surprising about transformations which mix space and time. Consider two observers in relative uniform motion in the x-direction. Suppose their x-origins coincide at time $t = 0$. Observer S' measures an object to be at a coordinate position x', whereas observer S places it at his coordinate x. If S' is moving with velocity u in the x-direction wrt S then, in pre-relativity physics, the transformation between their coordinates is simply $x' = x - ut$. So there we have it: the mixing of space and time in the transformation. Motion tends to do that, you know. Before relativity came along, time was regarded as universal. So our two observers would (we thought) have agreed on the time of any event, i.e., $t' = t$. The complete transformation, now known as a Galilean transformation, was thus,

$$\begin{pmatrix} x' \\ t' \end{pmatrix} = \begin{pmatrix} 1 & -u \\ 0 & 1 \end{pmatrix} \begin{pmatrix} x \\ t \end{pmatrix} \tag{26.1.1}$$

Prior to relativity emerging, (26.1.1) was regarded as so self-evident that it was not then dignified with so grand a sounding title as a "transformation", still less a "law of physics".

We now wish to explore the most general linear relationship between the (x, t) coordinates of two inertial observers constrained only by the principle that all inertial observers are equivalent. But firstly we address a couple of preliminaries.

Why confine attention to a linear transformation? The answer to this is that we can always confine attention to a small region of spacetime, so that we are really interested in the transformation of some small spacetime displacement (dx, dt). In this case, we could have an arbitrary non-linear transformation of coordinates, $x' = f(x, t)$ and $t' = g(x, t)$, but this then gives,

$$\begin{pmatrix} dx' \\ dt' \end{pmatrix} = \begin{pmatrix} \frac{\partial f}{\partial x} & \frac{\partial f}{\partial t} \\ \frac{\partial g}{\partial x} & \frac{\partial g}{\partial t} \end{pmatrix} \begin{pmatrix} dx \\ dt \end{pmatrix} \tag{26.1.2}$$

As long as we confine attention to a small region within which the gradients of f and g do not vary significantly, then this becomes simply,

$$\begin{pmatrix} dx' \\ dt' \end{pmatrix} = \begin{pmatrix} a & b \\ d & e \end{pmatrix} \begin{pmatrix} dx \\ dt \end{pmatrix} \tag{26.1.3}$$

i.e., a linear transformation. More technically, we may be dealing with a curved spacetime manifold – as in general relativity - but we are addressing the properties of the tangent space at a point. So confining attention to (26.1.3) with constant matrix elements does not represent a restriction as long as we are content with purely *local* transformations.

The second preliminary concerns what is implied regarding the matrix in (26.1.3) if the relative velocity of the observers is u. The relative velocity is the velocity of the origin of S' as observed by S. The origin of S' is at $dx' = 0$ so (26.1.3) gives $adx + bdt = 0$, which gives $\frac{dx}{dt} = u = -\frac{b}{a}$. Hence we can eliminate b and write (26.3) as,

$$\begin{pmatrix} dx' \\ dt' \end{pmatrix} = \begin{pmatrix} a & -au \\ d & e \end{pmatrix} \begin{pmatrix} dx \\ dt \end{pmatrix} \tag{26.1.4}$$

We are now ready to explore what further constraints on the form of the transformation in (26.1.4) follow from the principle that all inertial observers are equivalent. Note that the parameters a, d, e in (26.1.4) can depend upon the relative velocity of the two observers, u. The equivalence of S and S' means that we could equally write (26.1.4) with the S coordinates on the LHS and the S' coordinates on the RHS, so long as we change the relative velocity from u to $-u$. Displaying the functional velocity dependence of a, d, e explicitly we see that,

$$\begin{pmatrix} dx \\ dt \end{pmatrix} = \begin{pmatrix} a(-u) & a(-u) \cdot u \\ d(-u) & e(-u) \end{pmatrix} \begin{pmatrix} dx' \\ dt' \end{pmatrix} \tag{26.1.5}$$

However by inverting (26.1.4) we also have,

$$\begin{pmatrix} dx \\ dt \end{pmatrix} = \frac{1}{ae+adu} \begin{pmatrix} e & au \\ -d & a \end{pmatrix} \begin{pmatrix} dx' \\ dt' \end{pmatrix} \tag{26.1.6}$$

where it is understood in (26.1.6) that the parameters are all evaluated at u not $-u$. But the matrices in (26.1.5) and (26.1.6) must be equal since they both hold for arbitrary spacetime displacements. In particular, equating the 12 components gives,

$$a(-u) = \frac{a}{ae+adu} = \frac{1}{e+du} \tag{26.1.7}$$

where it is again understood in (26.1.7) that the parameters on the RHS are evaluated at u not at $-u$.

Now if we consider an interval in frame S with $dt = 0$ we have from (26.1.4) that $dx' = a(u) \cdot dx$. And if we consider points in frame S' with $dt' = 0$ we have from (26.1.5) that $dx = a(-u) \cdot dx'$. But the equivalence of observers means that these two situations should be equivalent, so that it must be that $a(-u) \equiv a(u)$, i.e., a is an even function of u. Using this in (26.1.7) gives,

$$a(e + du) = 1 \tag{26.1.8}$$

Equating the other components between (26.1.5) and (26.1.6) then gives,

$$e = a \quad \text{and} \qquad d(-u) = -d(u) \tag{26.1.9}$$

as well as $e = e(-u)$. Now consider a third observer, S'', moving with velocity u' wrt S'. We have,

$$\begin{pmatrix} dx'' \\ dt'' \end{pmatrix} = \begin{pmatrix} a' & -a'u' \\ d' & a' \end{pmatrix} \begin{pmatrix} dx' \\ dt' \end{pmatrix} \tag{26.1.10}$$

where the dashes on a' and d' refer to the same functions of velocity as in (26.4) but evaluated at u'. Substituting (26.1.4) into (26.1.10) gives,

$$\begin{pmatrix} dx'' \\ dt'' \end{pmatrix} = \begin{pmatrix} a'a - a'u'd & -a'a(u + u') \\ d'a + da' & -d'au + a'a \end{pmatrix} \begin{pmatrix} dx \\ dt \end{pmatrix} \tag{26.1.11}$$

But this must have the same form as a transformation direct from S'' to S, and so must equal,

$$\begin{pmatrix} dx'' \\ dt'' \end{pmatrix} = \begin{pmatrix} a'' & -a''u'' \\ d'' & a'' \end{pmatrix} \begin{pmatrix} dx \\ dt \end{pmatrix} \tag{26.1.12}$$

where the double dash represents quantities evaluated at the relative velocity of S'' wrt S, i.e., u''. We have made no assumption regarding how to find u'' in terms of u and u'. Indeed, avoiding assuming the usual non-relativistic law of relative velocities, namely that $u'' = u + u'$, is crucial because it turns out to be untrue.

Comparing (26.1.11) and (26.1.12) shows that the diagonal elements in (26.1.11) must be equal, so we require,

$$a'a - a'u'd = -d'au + a'a \tag{26.1.13}$$

Assuming $d \neq 0$ and $d' \neq 0$ this gives,

$$\frac{a'u'}{d'} = \frac{au}{d} = \kappa \tag{26.1.14}$$

(26.1.14) must hold for arbitrary velocities u and u' and hence this combination of the functions must be independent of velocity, i.e., a universal constant, κ. The transformation, (26.1.4), therefore must be,

$$\begin{pmatrix} dx' \\ dt' \end{pmatrix} = a \begin{pmatrix} 1 & -u \\ u/\kappa & 1 \end{pmatrix} \begin{pmatrix} dx \\ dt \end{pmatrix} \tag{26.1.15}$$

for some universal constant, κ, to be determined. Finally, (26.1.8) now becomes,

$$a\left(a + \frac{a}{\kappa}u^2\right) = 1 \tag{26.1.16}$$

Hence,

$$a = \frac{1}{\sqrt{1+\frac{u^2}{\kappa}}} \tag{26.1.17}$$

So finally, the most general transformation consistent with the principle of the equivalence of inertial observers is,

$$\begin{pmatrix} dx' \\ dt' \end{pmatrix} = \frac{1}{\sqrt{1+\frac{u^2}{\kappa}}} \begin{pmatrix} 1 & -u \\ u/\kappa & 1 \end{pmatrix} \begin{pmatrix} dx \\ dt \end{pmatrix} \tag{26.1.18}$$

for some universal constant, κ.

The Galilean transformation, (26.1.1), is regained if we take $\kappa \to \infty$.

More generally the meaning of the universal constant, κ, can be elucidated as follows. By dividing the spatial equation in (26.1.18) by the time equation, we get,

$$v' = \frac{v-u}{1+\frac{uv}{\kappa}} \tag{26.1.19}$$

where $v = \frac{dx}{dt}$ and $v' = \frac{dx'}{dt'}$ are the velocities of some particle as seen by the two observers. For convenience of exposition, replace u by $-u$ in (26.1.19), so that S' is moving in the negative x-direction wrt S. (Or, equivalently, S is moving in the positive x-direction wrt S'). Hence, S' sees a positive-going particle moving faster than does S. (26.1.19) becomes, with all positive numbers u, v, v',

$$v' = \frac{v+u}{1-\frac{uv}{\kappa}} \tag{26.1.20}$$

We next show that κ must be negative if finite. Assuming that $\kappa > 0$ and that κ is finite, (26.1.20) would imply that there are finite speeds u and v with $uv = \kappa$, for which S' sees the particle moving infinitely fast, $v' \to \infty$. This predisposes us to regard this as unphysical (because of the two observes with relative finite velocity, one will see a particle with finite energy whilst the other sees a particle with divergently large energy). A further argument against finite κ with $\kappa > 0$ is that $uv = \kappa$ establishes that infinite relative speeds can occur in this case, and hence we can consistently consider arbitrarily large u in the transform (26.1.18). But if we consider $u^2 >> \kappa$ then (26.1.18) implies $dx' \to -\sqrt{\kappa} \cdot dt$ and $dt' \to dx/\sqrt{\kappa}$, because the diagonal components become very small in this limit. But physically it makes no sense for a pure time interval to be transformed into a pure spatial interval, or vice-versa. Physically there is a preferred direction of time (future pointing). Such a transformation would induce a preferred spatial direction. Consequently, for finite κ we can rule out $\kappa > 0$.

We conclude that, if κ is finite, then $\kappa < 0$ and we write $\kappa = -c^2$, so that (26.1.20) becomes,

$$v' = \frac{v+u}{1+\frac{uv}{c^2}} \tag{26.1.21}$$

Now the opposite happens. No infinite speeds are possible. Instead, the form of (26.1.21) ensures that as $u \to c$ for a finite $v < c$ we get $v' \to c$. Thus, at whatever speed S sees a particle moving, S' will see it moving asymptotically to c if the relative velocity of the two observers is also asymptotic to c.

Similarly, as $v \to c$ at any $u < c$ we also get $v' \to c$. Hence a particle moving at the velocity c is seen to move at speed c wrt both observers, despite their relative motion.

Hence, in summary, the equivalence of inertial observers leads to the transformation (26.1.18) in terms of some universal constant, κ. The assumption that this universal constant is finite means that it must be negative, i.e., $\kappa = -c^2$ for some real c. The transformation law then shows that c acts as a maximum limiting speed that cannot be exceeded.

The most general transformation consistent with the principle of the equivalence of inertial observers and a finite κ is thus,

$$\begin{pmatrix} dx' \\ dt' \end{pmatrix} = \gamma \begin{pmatrix} 1 & -u \\ -u/c^2 & 1 \end{pmatrix} \begin{pmatrix} dx \\ dt \end{pmatrix} \tag{26.1.22a}$$

where,
$$\gamma = \frac{1}{\sqrt{1-\frac{u^2}{c^2}}} \tag{26.1.22b}$$

Note that in (26.1.22a,b) we have switched back to S' is moving with positive x-direction velocity wrt S (or, equivalently, S is moving in the negative x-direction wrt S'). The formula for relative velocity consistent with (26.1.22a,b) is,

$$v' = \frac{v-u}{1-\frac{uv}{c^2}} \tag{26.1.23}$$

which has the limiting case for $v \to c$ that $v' \to c$ whereas for $u \to c$ we have $v' \to -c$.

Equations (26.1.22a,b) are the familiar form of the Lorentz transformation. The important point is that the form of the Lorentz transformation has been deduced from the principle of the equivalence of inertial observers. This derivation makes coherent the notion of a limiting speed, via the natural emergence of the parameter κ, but it does not motivate identification of this limiting speed with light speed *in vacuo*.

Can the Galilean assumption, that the limiting speed is infinite, be ruled out by purely theoretical reasoning? No, ultimately experiment is the arbiter. The Michelson–Morley experiment showed directly that the speed of light does not vary for different motions of its source, thus having exactly the characteristics of a limiting speed in accord with (26.22a,b). However, the Michelson–Morley result did not automatically lead to relativity theory. Indeed, Einstein made no mention of the Michelson–Morley experiment in the epochal 1905 paper that introduced special relativity, [26.1]. However, Lorentz did in his paper from the year before, the paper that led to the transformation being named for Lorentz not Einstein, [26.2]. Indeed, as early as 1895, Lorentz had suggested that the results of the Michelson-Morley experiments could be explained on the basis of a change in length of the equipment according to its orientation with respect to the Earth's rotational motion, [26.3]. In fact, the Irish professor George Fitzgerald was the first to make this suggestion, in 1889, [26.4]. Hence the relativistic contraction of bodies along their direction of motion is known as the Lorentz-Fitzgerald contraction.

However, both Lorentz and Fitzgerald interpreted this contraction in terms of forces acting on bodies from the aether (which was then assumed to exist and to constitute the substance of electromagnetic fields). It took Einstein to recognise the significance of a limiting speed in terms of the nature of spacetime itself, and hence that the Lorentz-Fitzgerald contraction was essentially geometrical rather than to do with forces on ponderable bodies, and in particular did not require the existence of a luminiferous aether causing a force of contraction. The route Einstein took to demonstrating this was via the transformation properties of the electromagnetic field equations. Thus, the identification of the limiting speed with that of electromagnetic wave propagation through the vacuum was a theoretical insight.

26.2 The Lorentz-Fitzgerald Contraction

A rigid rod is moving at speed u in the direction parallel to its length. It has "rest length" L_0 wrt an observer who moves along with it, namely S'. What does the stationary observer, S, consider to be its length?

The length of the rod is the difference between the x coordinates, Δx, of its ends at the same time in the frame of reference in which the length is perceived. Inverting the Lorentz transformation, (26.1.22a), which involves merely changing the sign of u, we have,

$$\Delta x = \gamma(\Delta x' + u\Delta t') \qquad (26.2.1a)$$

$$\Delta t = \gamma(\Delta t' + u\Delta x'/c^2) \qquad (26.2.1b)$$

We wish to find Δx when $\Delta t = 0$, i.e., the length wrt S, given that the two spacetime points ("events") correspond to the ends of the rod and hence have $\Delta x' = L_0$ wrt observer S'. It follows from (26.2.1b) that $\Delta t'$ cannot be zero when $\Delta t = 0$ but rather $\Delta t' = -\frac{u\Delta x'}{c^2} = -\frac{uL_0}{c^2}$. Hence, the length wrt observer S is,

$$\Delta x = L = \gamma\left(L_0 - \frac{u^2}{c^2}L_0\right) = \frac{L_0}{\gamma} \qquad (26.2.2)$$

where we have used the relativistic γ factor from (26.1.22b). As $\gamma \geq 1$ it follows that the apparent length of a rod which is moving wrt an observer is reduced to a fraction $1/\gamma$ of its rest length.

By the same token, a rod at rest wrt S will be perceived by S' to have contracted. This is the great difference between Einstein's relativistic interpretation and the Fitzgerald/Lorentz view that the contraction had something to do with motion wrt a fixed ether. In that view, the contraction would not occur if the rod were stationary wrt the aether and only the observer was moving.

26.3 Time Dilation

Time dilation is analogous to Lorentz-Fitzgerald contraction but relates to a time interval at zero spatial interval, rather than the reverse. Suppose the time interval between two events wrt observer S' is $\Delta t'$ and that these events occur at the same place wrt S', namely at $\Delta x' = 0$. The Lorentz transformation, (26.1.22a), requires that observer S sees the same two events at different x coordinates, where $\Delta x = u\Delta t$ to be consistent with $\Delta x' = 0$. Hence we get,

$$\Delta t' = \gamma\left(\Delta t - \frac{u^2}{c^2}\Delta t\right) = \frac{\Delta t}{\gamma} \tag{26.3.1}$$

Hence the time interval perceived by the notionally stationary observer, Δt, is longer than the time interval perceived by an observer comoving with the two events (so that $\Delta x' = 0$), namely $\Delta t = \gamma\Delta t'$. In other words, a moving clock appears to be running slow.

As there is no assumed absolute state of rest, this is mutual. So whilst S perceives clocks comoving with S' to be running slow, so also does S' perceive clocks comoving with S to be running slow.

26.4 The Relativity of Simultaneity

When one first encounters relativity theory it takes some time to become comfortable with the notions of time dilation and Lorentz-Fitzgerald contraction. However, even more challenging is another immediate corollary of the Lorentz transform, namely the elimination of the idea that events at two different places can be unambiguously said to be simultaneous. The events may appear simultaneous to one observer, but occur at different times to another observer. Suppose observer S sees two events separated by a distance x but at the same time, so that their time difference according to S is $t = 0$. The Lorentz transform, (26.1.22a,b) or (26.2.1a,b), shows that

S' sees the two events occurring at times which differ by $t' = -\frac{u}{c^2}x' = -\gamma\frac{u}{c^2}x$. So S' does not agree that the events are simultaneous.

Furthermore, if we consider a third observer, S'', moving at velocity $-u$ wrt S then S'' sees the two events as occurring with a time gap of $t'' = \frac{u}{c^2}x'' = \gamma\frac{u}{c^2}x$. So S' and S'' will disagree on which of the two events came first. Temporal order itself is not invariant between observers but is relative to their state of motion – when considering events which occur at different spatial locations.

Temporal order, including simultaneity, ***is*** invariant between observers if we confine attention to events occurring at the same spatial location (wrt one or the other of the observers), though different observers will disagree on the duration of the time interval between the events (time dilation).

26.4.1 Impact on Causality

The fact that the temporal order of two events, A and B, can be changed according to our frame of reference initially seems at odds with causality. How can any event, A, ever be consistently claimed to be the cause of an event B if, on changing one's frame of reference one can contrive to make B precede A? We have already seen one way of avoiding this awkward conclusion, namely when events occur at the same location (according to either observer).

There is a more general class of cases for which causality is saved, i.e., temporal order is maintained invariant between observers. This is when the two events in question are "time-like or null separated". The term "time-like or null separated" means that the time and space interval between the two events as seen by S respects $\Delta t^2 - c^2\Delta x^2 \geq 0$. The Lorentz transformation immediately tells us that this is an invariant constraint, i.e., $c^2\Delta t'^2 - \Delta x'^2 = c^2\Delta t^2 - \Delta x^2 \geq 0$. Assuming without loss of generality that $\Delta x \geq 0$ and that the time interval $\Delta t > 0$ corresponds to what S perceives as the time between causal event A and caused event B then $c\Delta t \geq \Delta x$ for time-like or null separated events. In this case, the Lorentz transform $\Delta t' = \gamma(\Delta t - u\Delta x/c)$ implies that $\Delta t > 0$ and so the causative event order is preserved.

The lesson is that, in order for causality to be consistently attributable, the interval from cause (A) to effect (B) must be time-like or null, i.e., A and B must be "time-like or null separated".

The physical significance of this is that, as long as the causal influence on B propagates from A at a speed not exceeding that of light, then the causal temporal order is preserved. But since the Lorentz transform itself guarantees that no transluminal speed of any physical influence is possible, it follows that unambiguous causality still prevails provided that cause and effect are connected by a physical, propagating, intermediary. In relativistic quantum field theory this is manifest in the form of all forces being conveyed by the exchange of intermediary virtual particles, visualised in the form of Feynman diagrams.

26.5 The Ladder and Barn Paradox

Imagine a ladder (or long rod) which, when at rest, is longer than the length of a barn. Suppose the barn has two doors, an entrance and an exit on the opposite side. When at rest, the ladder cannot fit inside the barn with both doors shut – because it is too long. (No tilting allowed). But if the ladder is moving fast enough then Lorentz-Fitzgerald contraction can reduce its length sufficiently so that it can be contained within the barn – that is, there is a time when one could shut both doors simultaneously with the ladder inside.

In view of the previous section, the reader's concern should have been aroused by the use of the word "simultaneously" in that description. But let us analyse the situation.

From the perspective of the observer S' comoving with the ladder it is the barn which is contracted – so the ladder is even less able to be fully contained within the barn.

But, surely, one can either shut both barn doors at the same time with the ladder inside – or not. How can this be "relative"?

The answer is that it is indeed relative and the clue is the phrase "at the same time" in the above sentence. What is seen by S as both doors shut "at the same time" in seen by S' as the doors being shut at different times. S' never perceives a time when both doors are shut simultaneously.

Define "event 1" as being when the trailing end of the ladder is coincident with the front door of the barn, and define "event 2" as being when the front end of the ladder is coincident with the back door of the barn. S perceives event 1 as coming before event 2, so both doors can be shut "at the same time" – relative to S. In contrast, S' perceives event 1 as coming after event 2 so there is no time as gauged by S' at which both doors can be shut "at the same time" in his frame of reference. It is worth spelling this out in detail.

Take the origin of the x coordinate in frame S to be the front door of the barn, and the origin of x' in frame S' to be the trailing end of the ladder. These coincide at event 1 when we defined the time datums to be $t = 0$ and $t' = 0$. At this time in frame S the position of the ladder wrt to the barn doors is illustrated below, where the red lines indicated the two doors and the blue line is the ladder,

According to S the ladder is contracted but the barn is perceived with its resting length, w, where $w > L = L_0/\gamma$. Event 2 occurs wrt S at $x = w$ and at a time such that the ladder covers the distance $w - L$ at speed u, hence $t = (w - L)/u$ and so at a positive (later) time than event 1.

At $t' = 0$ (event 1) the perspective wrt observer S' is this,

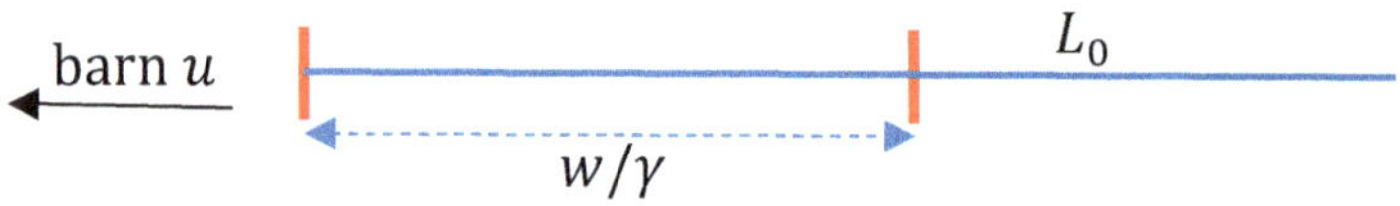

Event 2 has clearly already occurred. It did so sufficiently earlier for the barn (perceived to be moving wrt S') to cover the distance $L_0 - w/\gamma$, hence at $t' = -(L_0 - w/\gamma)/u$. Because the origin of x' is fixed at the trailing end of the ladder, event 2 occurs at $x' = L_0$. The reader should confirm that the Lorentz transform,

$$x' = \gamma(x - ut) \quad \text{and} \quad t' = \gamma(t - ux/c^2) \tag{26.5.1}$$

is obeyed by the coordinates derived above for event 2, i.e.,

$$x = w, t = (w - L)/u, x' = L_0, t' = -(L_0 - w/\gamma)/u \tag{26.5.2}$$

where $L = L_0/\gamma$. The paradox is resolved because $t > 0$ but $t' < 0$.

26.6 The Twin Paradox

Time dilation, §26.3, means that an observer on Earth will perceive an astronaut receding speedily from Earth as ageing more slowly than himself. Let's say the two are twins, and so originally the same age. The paradox arises from the fact that the astronaut perceives the observer back on Earth to be ageing more slowly than himself. The trouble inherent in this conflict of views is avoided provided the two never meet.

But what if we have the astronaut make a round trip so that he does return to Earth? What then? Are the two twins actually the same age after all?

No, they are not. The astronaut is younger than his less adventurous twin. In the limit, the astronaut may not have altered his youthful appearance perceptibly whilst his stationary twin has become an aged grey and bent old man. At an even greater extreme, many centuries may have passed on Earth whilst the astronaut is still in his prime. In effect the astronaut has travelled into the Earth's future.

How can this be understood from the Lorentz transform? It cannot, because the Lorentz transformation is based on the assumption that frames of reference are inertial – that is, free of any accelerations. The asymmetry of the problem comes from the fact that the astronaut must first accelerate to high speed and then undergo further deceleration/acceleration at some point in order to return to Earth. As is found in the general theory of relativity, acceleration/deceleration (or gravity) also has an influence on the flow of time. Indeed, it was considerations akin to this which led Einstein to the general theory. He had already established the effect of gravity on clocks by 1907, see [26.5]. The effect of gravitational fields on time has now been well established experimentally, so that, indirectly, the twin paradox is resolved in favour of the astronaut being younger than his stay-at-home twin.

26.7 The Closed Door Paradox

This is akin to the Twin Paradox but in the spatial domain rather than the time domain. For this we return to the Ladder & Barn scenario, §26.5, but with a twist: the exit door from the barn is shut – and we assume the barn and its door are infinitely strong and so will withstand the impact from the fast moving ladder. From the perspective of S, at the very instant of this collision with the back door the front door is also slammed shut (which S sees as possible). We assume this can be done extremely quickly. As the distance the door needs to close need not be great, this can be done without violating light speed.

Observer S says this is possible; observer S' says it is impossible because the rear end of the ladder has not yet entered the barn. The door either shuts or fouls on the ladder. It cannot be both. Which is it?

The answer is that the door shuts. But this appears to violate the equivalence of the observers. Indeed it does, because, once again the conditions under which the Lorentz transform is valid are violated because, when the front of the ladder impacts the back door, enormous decelerations come into play.

If we consider all parts of the ladder to decelerate to zero speed instantaneously in the frame of the barn, this will not be how it appears to the initially comoving observer S'. He perceives the deceleration occurring later near the rear of the ladder than near the front. Consequently, S' interprets the event as the ladder being compressed after the impact, and hence fitting into the barn.

Of course, immediately after the ladder is brought to rest, it is now longer than the barn as seen from the frame of the barn, and so – if the front door is very strong – the ladder will be obliged to compress or bend or snap after only fitting inside the barn for a fleeting period.

26.8 Lorentz Transformation in Three Spatial Dimensions

Including all three spatial dimensions is essentially trivial as regards the transformation of the coordinates. Assuming the relative velocity is parallel to the x axis then the other two Cartesian coordinates are invariant, that is $y' = y, z' = z$.

If the relative velocity is in some arbitrary direction, the simplest procedure is to first rotate the coordinate basis to bring the new x axis into alignment with the relative velocity and then to use the same Lorentz transformation equations in the new basis, (26.1.22a,b).

26.9 Boosts

Beginners may be confused by a terminology that is common in physics, namely that any combination of a rotation and a Lorentz transform due to relative velocity (or just a rotation alone) may be referred to as a Lorentz transformation (or an element of the homogeneous Lorentz group). To distinguish this usage from a transformation specifically referring to relative velocity, the latter is called a "boost". Thus, the most general Lorentz transformation is a combination of any boost in 3D with any rotation in 3D.

To add further to potential terminological confusion, sometimes a spacetime displacement is also included as a "Lorentz" transformation (or element of the inhomogeneous Lorentz group). I would recommend that adding displacements into the mix should always be referred to as a Poincaré transformation.

26.10 References

[26.1] Einstein, Albert (30 June 1905). *Zur Elektrodynamik bewegter Körper* [On the Electrodynamics of Moving Bodies]. Annalen der Physik 17 (10): 891–921. English translation available in *The Principle of Relativity: Original Papers by A. Einstein, H.A.Lorentz, H.Weyl and H. Minkowski*. Dover, 19226.

[26.2] Lorentz, H.A. (1904). *Electromagnetic Phenomena in a System Moving with Any Velocity Less Than That of Light*. English translation available in *The Principle of Relativity: Original Papers by A. Einstein, H.A.Lorentz, H.Weyl and H. Minkowski*. Dover, 19226.

[26.3] Lorentz, H.A. (1895). *Versuch einer Theorie der Elektrischen und Optischen Erscheinungen in bewegten Kopern*. English translation available in *The Principle of Relativity: Original Papers by A. Einstein, H.A.Lorentz, H.Weyl and H. Minkowski*. Dover, 19226.

[26.4] FitzGerald, George Francis (1889), *"The Ether and the Earth's Atmosphere"* , *Science*, **13** (328): 390.

[26.5] Schwartz, H.M. (1976). *Einstein's comprehensive 1907 essay on relativity* (in three parts). Originally published in American Journal of Physics vol.45, no.6 (1977) pp.512–517; no.9 (1977), pp.811–817; no.10 (1977), pp.899–902.

27. 4-Vectors and the E/M Field Tensor

Here the principle of the equivalence of inertial observers is extended to electromagnetism. This reveals that the electric and magnetic fields are not separate entities but actually components of a single six-component field entity, the electromagnetic field. The route to understanding this lies in the concept of 4-vectors and associated tensors. The electromagnetic field is then revealed as an antisymmetric second rank tensor. The electric and magnetic field "vectors", $\bar{E}, \bar{B}$, *are not really vectors at all when we move from space to spacetime. We will see why it was emphasised from chapter 1 onwards that we are dealing with geometrical objects not merely with sets of numbers. These geometrical objects do not live in 3D Euclidean space but in 4D Minkowski spacetime. The further key result which follows from this realisation is that the limiting speed identified in chapter 26 is the speed of light in vacuo. This is necessary if two inertial observers are to agree on the form of Maxwell's equations, i.e., that Maxwell's equations must be covariant – which is just another way of talking about the equivalence of inertial observers. Hence, the identification of the Lorentzian invariant speed with light speed follows from Maxwell's equations; it is a theoretical deduction.*

Contents

27.11 Summary of this Chapter

27.12 References

27.1 What does the Equivalence of Inertial Observers Imply for Electromagnetism?

The last chapter derived the implications of the equivalence of inertial observers, together with a maximum speed, for the observers' measures of space and time. But if inertial observers are truly equivalent this must apply to all physical phenomena, and to electromagnetism in particular. The Maxwell equations in vacuo (or in air, to a good enough approximation) are,

$$\bar{\nabla}\cdot\bar{E} = \rho/\varepsilon_0 \qquad \bar{\nabla}\cdot\bar{B} = 0 \tag{27.1.1a}$$

$$\bar{\nabla}\times\bar{B} = \mu_0\bar{J} + \frac{1}{c^2}\frac{\partial\bar{E}}{\partial t} \qquad \bar{\nabla}\times\bar{E} = -\frac{\partial\bar{B}}{\partial t} \tag{27.1.1b}$$

Let us assume these are the Maxwell equations as perceived by observer S and suppose an observer S' is moving at speed u in the positive x-direction. We have seen in chapter 26 how the observers' measures of spacetime differ according to the Lorentz boost, which we may write as,

$$\begin{pmatrix} x' \\ ct' \end{pmatrix} = \gamma\begin{pmatrix} 1 & -u/c \\ -u/c & 1 \end{pmatrix}\begin{pmatrix} x \\ ct \end{pmatrix} \tag{27.1.2a}$$

$$\begin{pmatrix} x \\ \mathrm{ct} \end{pmatrix} = \gamma\begin{pmatrix} 1 & u/c \\ u/c & 1 \end{pmatrix}\begin{pmatrix} x' \\ ct' \end{pmatrix} \tag{27.1.2b}$$

$$\text{where, } \gamma = \frac{1}{\sqrt{1-\frac{u^2}{c^2}}} \tag{27.1.2c}$$

So, observer S' will interpret spatial or temporal derivatives wrt his own coordinates, x', t'. Given that S' measures space and time differently from S, we must acknowledge that he may also measure the electric and magnetic fields differently. Let us denoted by $\bar{E}', \bar{B}'$ the fields observed by S'. If inertial observers truly are equivalent then we would require that S' will conclude the Maxwell equations to be,

$$\bar{\nabla}'\cdot\bar{E}' = \rho'/\varepsilon_0 \qquad \bar{\nabla}'\cdot\bar{B}' = 0 \tag{27.1.3a}$$

$$\bar{\nabla}'\times\bar{B}' = \mu_0\bar{J}' + \frac{1}{c^2}\frac{\partial\bar{E}'}{\partial t'} \qquad \bar{\nabla}'\times\bar{E}' = -\frac{\partial\bar{B}'}{\partial t'} \tag{27.1.3b}$$

Note that we have also acknowledged that the charge density and current density 'seen' by S' will differ from those observed by S. Clearly this must be so because the Lorentz-Fitzgerald contraction of volume will change the observed charge density. Also, a charge density seen as stationary by S will be seen as moving by S', and hence S' sees a current density whereas S does not.

We can write the derivative operators in the frame of reference of S' thus,

$$\frac{\partial}{\partial x'} = \frac{\partial t}{\partial x'}\frac{\partial}{\partial t} + \frac{\partial x}{\partial x'}\frac{\partial}{\partial x} = \gamma\left(\frac{u}{c^2}\frac{\partial}{\partial t} + \frac{\partial}{\partial x}\right)$$

$$\frac{\partial}{\partial t'} = \frac{\partial t}{\partial t'}\frac{\partial}{\partial t} + \frac{\partial x}{\partial t'}\frac{\partial}{\partial x} = \gamma\left(\frac{\partial}{\partial t} + u\frac{\partial}{\partial x}\right)$$

i.e., $$\begin{pmatrix} \dfrac{\partial}{\partial x'} \\ \dfrac{\partial}{c\partial t'} \end{pmatrix} = \gamma\begin{pmatrix} 1 & u/c \\ u/c & 1 \end{pmatrix}\begin{pmatrix} \dfrac{\partial}{\partial x} \\ \dfrac{\partial}{c\partial t} \end{pmatrix} \tag{27.1.4}$$

So $\begin{pmatrix} \dfrac{\partial}{\partial x'} \\ \dfrac{\partial}{c\partial t'} \end{pmatrix}$ transforms as the inverse matrix compared to $\begin{pmatrix} dx' \\ cdt' \end{pmatrix}$. But the Lorentz boost leaves coordinates $y' = y, z' = z$ unchanged. Hence we can write,

$$\bar{\nabla}' \cdot \bar{B}' = \gamma\left(\frac{u}{c^2}\frac{\partial}{\partial t} + \frac{\partial}{\partial x}\right)B'_x + \frac{\partial}{\partial y}B'_y + \frac{\partial}{\partial z}B'_z = 0 \tag{27.1.5}$$

If we supposed that the components of $\bar{B}$ transform only amongst themselves under a Lorentz boost, as they do under rotations, then (27.1.5) is problematic because it would involve the time derivatives of the magnetic field which can be replaced from $\bar{\nabla} \times \bar{E} = -\frac{\partial \bar{B}}{\partial t}$ with spatial derivatives of the electric field. This would give us an equation involving first order spatial derivatives only, but of both fields. But there is no such Maxwell equation. The inertial observers could not be equivalent.

We conclude that Lorentz boosts must 'mix up' the electric and magnetic field components, i.e., that what is seen as an electric field by S can be seen as including a magnetic component by S', and vice-versa. But what exactly is the transformation law?

27.2 4-Vectors

We leave dangling that key question about the transformation of the electric and magnetic field components while we develop a required preliminary concept: the 4-vector.

Throughout this book we have worked with vectors in 3D Euclidean space and emphasised that vectors are geometrical objects in their own right, not merely sets of numbers called coordinates. Whilst vectors can be expressed in terms of coordinates, these coordinates are always relative to a "coordinate system" or frame of reference. This is exemplified by (2.4.3), $\bar{r} = r_j\hat{e}_j = r'_k\hat{e}'_k$ (summation convention assumed). The set of coordinates r_j is meaningless without the coordinate system, i.e., the basis $\hat{e}_j$, to which they relate, being specified. We may choose an alternative basis (coordinate system or frame of reference), $\hat{e}'_k$, wrt which the vector's coordinates (or components) are r'_k. But both choices of coordinate system and components are equivalent because both describe the same geometrical object, $\bar{r}$.

Now we have entered 4D spacetime we must adopt the same philosophy. We now have spacetime coordinates which we will denote x^μ where the index runs over t, x, y, z rather than just x, y, z. The convention will be that Greek indices take spacetime values t, x, y, z whereas Latin indices take only spatial values x, y, z. The x^μ coordinates are defined wrt the basis $\hat{e}_\mu$, in which we can identify $\hat{e}_j$ with the usual spatial unit vectors, $\hat{x}, \hat{y}, \hat{z}$ but $\hat{e}_t$ (which we might sometimes write as $\hat{e}_0$) is a new concept: a unit vector pointing in the future time direction. The new geometrical object with which we are now dealing is the so-called "4-vector" $x^\mu \hat{e}_\mu$.

The reason for using both upper and lower indices (superscripts and subscripts) must now be addressed.

27.2.1 Covariant and Contravariant Components

The 4-vector $x^\mu \hat{e}_\mu$ can equivalently be specified as $x_\mu \hat{e}^\mu$ but the x^μ are not the same as the x_μ, and the $\hat{e}_\mu$ are not the same as the $\hat{e}^\mu$. Note the use of the word "equivalent", as opposed from "equal". Although $x^\mu \hat{e}_\mu$ and $x_\mu \hat{e}^\mu$ are in one-to-one correspondence with the geometrical object in question, they are not mathematically identifiable. (Strictly, as we will see in chapter 30, $x^\mu \hat{e}_\mu$ is the 4-vector whereas $x_\mu \hat{e}^\mu$ is the corresponding 1-form).

For $x^\mu \hat{e}_\mu$ to be a geometrical object it must be independent of the frame of reference. So, however the components x^μ transform the basis vectors $\hat{e}_\mu$ must transform as the inverse. This is the reason why their indices are written one with a superscript and the other with a subscript. This indicates which transformation matrix applies. So the x^μ and the alternative basis, $\hat{e}^\mu$, transform in the same way. And the x_μ transform in the same way as the basis $\hat{e}_\mu$. We now introduce a notation which will apply to any Lorentz transformation, which means a combination of any rotation and a boost in any direction. Quantities perceived by S' are indicated by dashes on the indices. Thus we define the transforms,

$$\hat{e}_{\mu'} = \hat{e}_\alpha \Lambda^\alpha{}_{\mu'} \quad \text{and} \quad \hat{e}_\mu = \hat{e}_{\alpha'} \Lambda^{\alpha'}{}_\mu \tag{27.2.1}$$

which demands that the two transforms are mutual inverses,

$$\Lambda^\alpha{}_{\mu'} \Lambda^{\mu'}{}_\beta = \delta^\alpha_\beta \qquad \Lambda^{\alpha'}{}_\mu \Lambda^\mu{}_{\beta'} = \delta^{\alpha'}_{\beta'} \tag{27.2.2}$$

Here we have written the usual Kronecker delta with upper and lower indices to be consistent with the positions of the indices on the LHS. Note that other authors use other conventions, but here we must understand that (say) ${\Lambda^t}_{x'}$ will not in general be the same as ${\Lambda^{t'}}_x$. These transforms are often written as matrices, which are convenient for computation, and in this case we use the bracket notation to indicate matrix convention. But we must distinguish between ${\Lambda^\alpha}_{\mu'}$ and ${\Lambda^{\alpha'}}_\mu$, for instance by writing,

$$(\Lambda)_{\alpha\mu} = {\Lambda^\alpha}_{\mu'} \quad \text{but} \quad (\Lambda^{-1})_{\alpha\mu} = {\Lambda^{\alpha'}}_\mu \tag{27.2.3}$$

The remaining quantities transform as,

$$x_{\mu'} = x_\alpha {\Lambda^\alpha}_{\mu'} \quad \text{and} \quad x_\mu = x_{\alpha'} {\Lambda^{\alpha'}}_\mu \tag{27.2.4a}$$

$$\hat{e}^{\mu'} = {\Lambda^{\mu'}}_\alpha \hat{e}^\alpha \quad \text{and} \quad \hat{e}^\mu = {\Lambda^\mu}_{\alpha'} \hat{e}^{\alpha'} \tag{27.2.4b}$$

$$x^{\mu'} = {\Lambda^{\mu'}}_\alpha \mathrm{x}^\alpha \quad \text{and} \quad x^\mu = {\Lambda^\mu}_{\alpha'} \mathrm{x}^{\alpha'} \tag{27.2.4c}$$

The coordinates x_μ transform in the same way as the vector basis, $\hat{e}_\mu$, and so are called "covariant components".

The coordinates x^μ transform as the inverse of the transform for the vector basis, $\hat{e}_\mu$, and so are called "contravariant components".

> The coordinates we have used thus far are identified with the contravariant components, x^μ. Thus, we write $\{x^\mu\} = \{ct, x, y, z\}$.

In matrix notation in which components are considered to be written as a 4 component column matrix these are,

<u>covariant (lower index) components</u>

$$x' = (\Lambda)^T x \quad \text{and} \quad x = (\Lambda^{-1})^T x' \tag{27.2.5a}$$

$$\hat{e}' = (\Lambda)^T \hat{e} \quad \text{and} \quad \hat{e} = (\Lambda^{-1})^T \hat{e}' \tag{27.2.5d}$$

<u>contravariant (upper index) components</u>

$$x' = (\Lambda^{-1}) x \quad \text{and} \quad x = (\Lambda) x' \tag{27.2.5b}$$

$$\hat{e}' = (\Lambda^{-1}) \hat{e} \quad \text{and} \quad \hat{e} = (\Lambda) \hat{e}' \tag{27.2.5c}$$

For pure boosts the matrices are symmetric so taking the transpose makes no difference, however rotations are not symmetric so the above expressions are needed to cover the general case.

These transformations, (27.2.4a-c) and (276.2.1), require little effort to remember because the simple rules are,

- The transforms, Λ, are always written with the superscript first and subscript second;
- Summation is over a pair of indices, one upper and one lower, and these are both dashed or both undashed;
- The free index on the LHS must match with the free index on the RHS in terms of upper/lower and in terms of dashed or undashed.

This makes each expression unique.

In the simple case of the transformed observer moving at speed u in the positive x-direction and no rotation the transformations are, in matrix form,

$$(\Lambda) = \begin{pmatrix} \gamma & \gamma\tilde{u} & 0 & 0 \\ \gamma\tilde{u} & \gamma & 0 & 0 \\ 0 & 0 & 1 & 0 \\ 0 & 0 & 0 & 1 \end{pmatrix} \quad (\Lambda^{-1}) = \begin{pmatrix} \gamma & -\gamma\tilde{u} & 0 & 0 \\ -\gamma\tilde{u} & \gamma & 0 & 0 \\ 0 & 0 & 1 & 0 \\ 0 & 0 & 0 & 1 \end{pmatrix} \tag{27.2.5}$$

where $\tilde{u} = u/c$ and (Λ) relates to $\Lambda^{\alpha}{}_{\mu'}$ whilst (Λ^{-1}) relates to $\Lambda^{\alpha'}{}_{\mu}$ and where the time coordinate is replaced by ct for dimensional consistency. Hence we see that this is consistent with (27.1.2a,b).

27.2.2 The Minkowski Metric

So far we have left the basis vectors, $\hat{e}_\mu$, as defined only intuitively as "little arrows" in the stated coordinate directions and with unit length. This vagueness becomes unacceptable when the intrinsically curved manifolds of general relativity must be addressed. The more rigorous approach identifies the $\hat{e}_\mu$ with the differential operators $\frac{\partial}{\partial x^\mu}$ whilst the $\hat{e}^\mu$ are identified with the differentials dx^μ. The transformations of $\frac{\partial}{\partial x^\mu}$ and dx^μ, e.g., (27.1.4), are consistent with those of $\hat{e}_\mu$ and $\hat{e}^\mu$ respectively.

So far we have not made any connection between the covariant and contravariant components of a vector, x_μ and x^μ. To do so we note from the transform (27.1.2a) that,

$$(ct')^2 - x'^2 = (ct)^2 - x^2 \tag{27.2.6a}$$

This remains true for a boost of any magnitude. It becomes true for a boost in any direction, including when combined with a rotation, if we write it in the 4D form,

$$(ct')^2 - r'^2 = (ct)^2 - r^2 \tag{27.2.6b}$$

where $r^2 = x^2 + y^2 + z^2$, as usual. Moreover, this can be reversed: any linear transformation in which $(ct)^2 - r^2$ remains invariant is a combination of a rotation and a boost in some direction (possibly with inversion of one or more coordinate axes). Hence the most general Lorentz transformation can be defined as any linear transformation respecting the invariance of $(ct)^2 - r^2$, excluding axis reversals. This is directly analogous to the definition of rotations in 3D as linear transformations leaving r^2 invariant.

This perspective on the meaning of Lorentz transformations was first clearly enunciated by Minkowski, [27.1], which is why the spacetime of special relativity is known as Minkowski space. It is defined by the invariant "length" measure, (27.2.6b).

But we also require that $x_\mu x^\mu$ is invariant under Lorentz transformations because the covariant and contravariant components transform as inverses. Since we have identified $\{x^\mu\} = \{ct, x, y, z\}$ it follows that $\{x_\mu\} = \{ct, -x, -y, -z\}$ so that,

$$x_\mu x^\mu = (ct)^2 - r^2 \tag{27.2.7}$$

This is how the concept of "length" is adopted into (flat) 4D spacetime, generalising the length measure in 3D Euclidean space $r^2 = x^2 + y^2 + z^2$. However, unlike the Euclidean length, which is real and positive definite for two non-identical points, the squared "length" (27.2.7) may be positive, negative or zero.

For any pair of points along the trajectory in spacetime of a material particle with non-zero rest mass the "distance" between them, the square root of (27.2.7), is real and positive. In the frame of reference comoving with the particle, since $r = 0$ in that frame, it follows that $c\tau = \sqrt{(ct)^2 - r^2}$ where τ is the time coordinate wrt that comoving frame, known as the particle's proper time. If the particle is unstable with a half life of $t_{1/2}$ then it is wrt the proper time, τ, that this half-life applies. In frames of reference which see the particle as moving its half-life will appear to exceed $t_{1/2}$, namely $\gamma t_{1/2}$, due to time dilation (see §26.3).

In contrast, a zero rest mass particle moves at light speed and so the squared "length", (27.2.7), is zero. Light, and any other zero rest mass particles, are said to move on null trajectories.

If two "points" (correctly "events") in spacetime seem to occur at different spatial locations but at the same time wrt frame S then wrt any other inertial frame the squared "length", (27.2.7), will be negative (referred to as "spacelike").

Note that some authors use the opposite sign convention, defining the squared "length" as $r^2 - (ct)^2$.

The connection between the covariant and contravariant components is written,

$$x_\mu = \eta_{\mu\nu} x^\nu \text{ where } \{\eta_{\mu\nu}\} = diag(1, -1, -1, -1) \tag{27.2.8a}$$

$\eta_{\mu\nu}$ is known as the Minkowski metric. We can equally write,

$$x^\mu = \eta^{\mu\nu} x_\nu \text{ where } \{\eta^{\mu\nu}\} = diag(1, -1, -1, -1) \tag{27.2.8b}$$

We mention in passing that, in general relativity in which flat spacetime becomes curved, this uniform metric is replaced by a metric $g_{\mu\nu}$ which varies with position, in general, and functions both as the defining feature of the particular spacetime geometry and is also such that $g_{\mu\nu} - \eta_{\mu\nu}$ is the generally covariant gravitational field.

27.2.3 Definition of a 4-Vector

So, finally we can define a 4-vector in general as…

> A 4-vector is a geometrical entity, $v^\mu \hat{e}_\mu$, in 4D Minkowski spacetime whose components Lorentz transform in the same way as (27.2.4) and whose covariant and contravariant components are related by $v_\mu = \eta_{\mu\alpha} v^\alpha$.

The square-magnitude of a 4-vector is defined as $v^\mu v_\mu$. It follows from the Lorentz transformations, (27.2.5a-d), that the square-magnitude is a Lorentz invariant. Space-like 4-vectors have a negative square-magnitude. Time-like 4-vectors have a positive square-magnitude. 4-vectors with zero square-magnitude as said to be "null" 4-vectors.

27.2.4 The 4-Velocity, 4-Momentum, 4-Force and Kinetic Energy

As an example of a 4-vector let's consider velocity. The ordinary 3-velocity is already defined. It is the ratio of the vectorial displacement to the increment of time, $\bar{v} = \frac{\delta \bar{r}}{\delta t}$. How do we form a 4-vector, which transforms as a 4-vector must but also captures the essence of what is meant by velocity? The 3-vector of displacement, $\delta \bar{r}$, is obviously generalised to the 4-vector of spacetime displacement, δx^μ. This is a good start because δx^μ is a 4-vector. We must convert it to a velocity by dividing by some time. But if we pick the time interval in an arbitrary frame of reference, δt, this wrecks its 4-vector nature, i.e., $\frac{\delta x^\mu}{\delta t}$ does not transform as a 4-vector because the denominator picks up a factor of γ when transformed between observers (where γ is given by (27.1.2c)). We need to divide by a measure of time interval which is unique not arbitrary. There is only one choice: the time interval as seen in the frame of reference comoving with the particle, i.e., the particle's rest frame. This is known as the proper time, denoted $\delta\tau$. Hence, 4-velocity is defined as,

$$v^\mu = \frac{dx^\mu}{d\tau} = \gamma(c, \bar{v}) \tag{27.2.9}$$

To make x^0 dimensionally consistent with the spatial components, x^i, we define $x^0 = ct$. The time component of v^μ results from time dilation, i.e., $v^0 = \frac{dx^0}{d\tau} = c\frac{dt}{d\tau} = c\frac{d\gamma\tau}{d\tau} = c\gamma$ (see §26.3). The spatial component similarly follows from "spatial part of v^μ" $= \frac{d\bar{r}}{d\tau} = \frac{d\bar{r}}{dt/\gamma} = \gamma\frac{d\bar{r}}{dt} = \gamma\bar{v}$, noting the potential for confusion between v^i understood as the contravariant components of a 4-vector and the components of the ordinary 3-vector, $\bar{v}$. The spatial components of v^μ also follow from the requirement that the 4-velocity must have a square-magnitude invariant under Lorentz transformation. This is satisfied by (27.2.9) for which,

$$v^\mu v_\mu = c^2\gamma^2\left(1 - \left(\frac{v}{c}\right)^2\right) = c^2 \tag{27.2.9b}$$

in all frames of reference. Note that whilst the ordinary speed has a finite upper bound, namely c, the components of the 4-velocity are unbounded because γ can be arbitrarily large. However, the square-magnitude of the 4-velocity remains bounded by c^2, and in fact always equal c^2. Note that this is true even for massive particles.

Once we have 4-velocity then the 4-momentum of a particle follows given its mass, m. The 4-momentum is simply,

$$p^\mu = mv^\mu = \gamma m(c, \bar{v}) \tag{27.2.10}$$

So that $p^\mu p_\mu = m^2c^2$. Here m is the rest mass. Hence $c^2 p^\mu p_\mu = (mc^2)^2$ which we recognise as the square of the energy-equivalent of the rest mass. But in the rest frame the LHS is $(cp_t)^2$ so this suggests that, in an arbitrary frame we might interpret cp_t as the energy of the particle (kinetic energy plus rest-mass equivalent energy). In other words we tentatively identify,

$$E = cp_t = \gamma mc^2 \qquad p^\mu = (E/c, \gamma m\bar{v}) \tag{27.2.11}$$

So we identify the total relativistic energy to be such that,

$$(mc^2)^2 = E^2 - (\gamma mcv)^2 \Rightarrow E = \sqrt{(mc^2)^2 + (\gamma mcv)^2} \quad (27.2.12)$$

Expanding the square-root using the binomial theorem in powers of v/c and keeping only the leading term the reader should be able to confirm that we get,

$$E \approx mc^2 + \frac{1}{2}mv^2 + O\left(mv^2\left(\frac{v}{c}\right)^2\right) \quad (27.2.13)$$

So we recover the usual non-relativistic equation for the kinetic energy, noting that E includes also the rest mass energy.

How are we to make relativistic the dynamical equation which links force to rate-of-change of momentum? "Relativistic" means Lorentz covariant, and Lorentz covariant means looking algebraically the same wrt any inertial frame (just as we require (27.1.3a,b) to be algebraically the same as (271.1a,b)). The general procedure for achieving this is, as we will shortly see, to ensure that all laws are expressed solely in tensorial form. For now one can interpret "tensorial" as meaning expressed as 4-vectors.

A 4-vector of rate-of-change of momentum involves dividing the 4-vector dp^μ by some time interval without undermining its 4-vector status. So the time interval must again be given by the proper time. Hence, the 4-vector force is,

$$f^\mu = \frac{dp^\mu}{d\tau} \quad (27.2.14)$$

In applications we shall, therefore, be looking for force laws formulated with a 4-vector force. Because $p^\mu = (E/c, \gamma m\bar{v})$ we see that cf^t is the rate of increase of the particle's energy which must equal the rate that the external force is doing work on the particle.

27.3 The Current Density 4-Vector

From the pre-relativistic perspective we have, on the one hand, charge density, ρ, a scalar, and on the other hand, current density, $\bar{J}$, a 3-vector. It is immediately obvious, without Einstein's insight, that a charge density which is at rest wrt observer S will be observed by S' as also constituting a current density. But does it make sense to regard $\{c\rho_0, \bar{0}\}$ as the components of a 4-vector as seen from the rest frame of the charge? If so then the current density seen by S' should follow from the transformation relations given in

§27.2. Assuming that $\{c\rho_0, \bar{0}\}$ are the contravariant components of some 4-vector J^μ as seen by S we would expect that S' would see,

$$J_x' = -\gamma u \rho_0 \quad \text{and} \quad c\rho' = \gamma c \rho_0 \tag{27.3.1a}$$

where ρ_0 is the charge density seen from the frame S where it is stationary. The difference in the charge density seen by S' compared with S is just what we would expect if charge itself is a scalar invariant but a volume element dV seen as stationary by S is reduced by Lorentz-Fitzgerald contraction to $dV' = dV/\gamma$ as seen by S', thus leading to an increase in the charge density seen by S' by a factor of γ.

Similarly, $J_x' = -\gamma u \rho_0$ is the current density we would expect to be seen by S' as he sees the charge density $\gamma\rho_0$ moving with velocity $-u\hat{x}$. So we conclude that the Lorentz transforms do indeed reproduce the physically expected result. Moreover, we conclude that the components of the 4-vector of current density in an arbitrary frame which sees the particle moving with velocity $\bar{v}$ must be, using (27.2.9),

$$\{J^\mu\} = \{\gamma c \rho_0, \gamma \rho_0 \bar{v}\} = \rho_0 v^\mu = \{c\rho, \rho\bar{v}\} = \{c\rho, \bar{J}\} \tag{27.3.1b}$$

where $\rho = \gamma\rho_0$ is the charge density seen from the frame in which it is moving and so has an associated current.

A key feature of charge is that it is conserved. This is expressed by the equation of continuity,

$$\frac{\partial \rho}{\partial t} + \bar{\nabla} \cdot \bar{J} = 0 \tag{3.8.12}$$

In terms of our hypothesised 4-vector of current density, $\{J^\mu\} = \{c\rho, \bar{J}\}$, and the 4-vector derivative operator, $\{\partial_\mu\} = \left\{\frac{\partial}{c\partial t}, \bar{\nabla}\right\}$, this can be written rather neatly as,

$$\partial_\mu J^\mu = 0 \tag{27.3.1}$$

If J^μ is indeed a 4-vector then it follows from the Lorentz transformations that'

$$\partial_{\mu'} J^{\mu'} = \partial_\mu J^\mu \tag{27.3.2}$$

This means that S' will see charge as conserved, just as does S, which, again, is what is expected physically.

Of relevance to the interpretation of the equation of continuity is the way in which volumes and times transform between observers. Suppose S sees charge dQ as stationary and occupying volume dV. Due to Lorentz-Fitzgerald contraction S' sees this same region as being of volume $dV' = \frac{dV}{\gamma}$. They agree on the quantity of charge enclosed because S' sees the charge density to be $\rho' = \gamma\rho$ so that $\rho' dV' = \rho dV$.

However, the two observers would disagree on the rate at which charge within a region was increasing. That is $\frac{\partial\rho}{\partial t} dV \neq \frac{\partial\rho'}{\partial t'} dV'$ because they disagree on the time interval. Due to time dilation, a time interval seen by S as dt is seen by S' to be $dt' = \gamma dt$. Hence, the 4-volume element is invariant,

$$dV'dt' = dVdt \tag{27.3.3}$$

It follows that $\partial_{\mu'} J^{\mu'} dV'dt' = \partial_\mu J^\mu dVdt$ is invariant. Integrating this means integrating over both volume and time in order for the two observers to agree. It is no longer sufficient to integrate only over volume and conclude from,

$$-\int_V \frac{\partial\rho}{\partial t} dV = \int_V \bar{\nabla} \cdot \bar{J}\, dV = \oiint_{\delta V} \bar{J} \cdot d\bar{S} \tag{27.3.4}$$

that the rate of decrease of charge within region V equals the rate at which charge is flowing out through its surface, δV. This remains true, of course, but the quantities in (27.3.4) refer to the specific frame of reference of S. However, S' will not see the same rates, when these are expressed as ***rates***. The two observers will agree only on the amount of charge lost from the region over corresponding, not equal, time intervals, linked by the Lorentz transformation. Thus, we have,

$$-\int_{\Delta t} \int_V \frac{\partial\rho}{\partial t} dVdt = \int_{\Delta t} \oiint_{\delta V} \bar{J} \cdot d\bar{S}\, dt = -\int_{\Delta t'} \int_{V'} \frac{\partial\rho'}{\partial t'} dV'dt' =$$
$$\int_{\Delta t}' \oiint_{\delta V'} \bar{J}' \cdot d\bar{S}'\, dt' \tag{27.3.5}$$

We conclude that the proposed 4-vector of charge density $\{J^\mu\} = \{c\rho, \bar{J}\}$ makes good sense, consistent with the spacetime transformations.

27.4 Tensors

The question we now ask is "what other sorts of geometrical object are there?". The question will not be answered in full generality, only sufficiently for our present needs. We start with this: consider two 4-vectors with components a_μ and b_μ wrt observer S. An entity defined by $a_\mu b_\nu \hat{e}^\mu \hat{e}^\nu$ is therefore a new type of geometrical object since it is clearly invariant. It has components with two indices, namely $T_{\mu\nu} = a_\mu b_\nu$.

The alert reader will notice that whilst the product of two numbers, $a_\mu b_\nu$, is perfectly well defined, we haven't defined what we mean by juxtaposing two basis vectors, $\hat{e}^\mu \hat{e}^\nu$. What we have strictly appealed to here is the tensor product, which would more correctly be written $\hat{e}^\mu \otimes \hat{e}^\nu$. This can be understood, heuristically at least, as little more than the ordered pair of $\hat{e}^\mu$ and $\hat{e}^\nu$, except with some structure added, namely that $\alpha\hat{e}^\mu \otimes \beta\hat{e}^\nu \equiv \alpha\beta(\hat{e}^\mu \otimes \hat{e}^\nu)$.

Slightly closer to respectable mathematics, the 16 pairs written $\hat{e}^\mu \otimes \hat{e}^\nu$ can be interpreted as the basis of a 16 dimensional vector space, the most general "vector" in such a space being $T_{\mu\nu}\hat{e}^\mu \otimes \hat{e}^\nu$ for arbitrary $T_{\mu\nu}$.

(The terminology is confusing as this "vector" in the tensor product vector space is actually a second rank tensor, i.e., a tensor with two indices. The confusion arises because "vector" is being used in two different ways. The mathematical concept of vector space is firmly embedded and not going away. The pedantic resolution is that what we have called a vector up until now should correctly have been called a rank 1 tensor – but how were we to know that in advance?).

$T_{\mu\nu}$ clearly has a broader scope than $T_{\mu\nu} = a_\mu b_\nu$ as the latter has only 4 + 4 = 8 degrees of freedom, whereas the former has 4 x 4 = 16 degrees of freedom. But $T_{\mu\nu}\hat{e}^\mu \otimes \hat{e}^\nu$ is still a geometrical object because it will still be invariant under Lorentz transforms provided that this generalised $T_{\mu\nu}$ transforms in the same way as $a_\mu b_\nu$. In terms of lower-index (covariant) components, $T_{\alpha\beta}$, the defining feature of a second rank tensor, i.e., one with two indices, is that its components transform under Lorentz transformations as indicated by (27.2.4a),

$$T_{\mu'\nu'} = T_{\alpha\beta}\Lambda^\alpha{}_{\mu'}\Lambda^\beta{}_{\nu'} \tag{27.4.1}$$

We raise indices in the same way as for vectors, i.e., from (27.2.8b) we see that the contravariant components of the tensor are,

$$T^{\mu\nu} = \eta^{\mu\alpha}\eta^{\nu\beta}T_{\alpha\beta} \tag{27.4.2}$$

We can also have components of mixed type, e.g.,

$$T^\mu{}_\nu = \eta^{\mu\alpha}T_{\alpha\nu} \tag{27.4.3}$$

And the transformations of these components follow from (27.2.4c), e.g.,

$$T^{\mu'\nu'} = \Lambda^{\mu'}{}_{\alpha}\Lambda^{\nu'}{}_{\beta}T^{\alpha\beta} \qquad (27.4.4)$$

There are many examples of tensors in physics, including rank three and rank four tensors with 3 and 4 indices respectively. An example of a second rank tensor, in 3D space, is the stress tensor. This has two indices because one index defines the element of area within the material over which the stress acts whilst the second index defines the force itself. Both these have (independent) directions. Another example of a second rank tensor would be the electric (or magnetic) susceptibility for an anisotropic material. An example of a fourth rank tensor in physics or engineering is provided by the elastic moduli which express the relationship between the stress tensor and the strain tensor for anisotropic materials.

There are some special cases of second rank tensor, namely symmetric and antisymmetric tensors. As the name implies these are such that $T^{\mu\nu} = T^{\nu\mu}$ and $T^{\mu\nu} = -T^{\nu\mu}$ respectively. These are coherent categories because the property of being symmetric or antisymmetric is invariant, i.e., if $T^{\mu\nu}$ is (anti)symmetric then so is $T^{\mu'\nu'}$. For example, in the antisymmetric case,

$$T^{\mu'\nu'} = \Lambda^{\mu'}{}_{\alpha}\Lambda^{\nu'}{}_{\beta}T^{\alpha\beta} = -\Lambda^{\mu'}{}_{\alpha}\Lambda^{\nu'}{}_{\beta}T^{\beta\alpha} = -\Lambda^{\mu'}{}_{\beta}\Lambda^{\nu'}{}_{\alpha}T^{\alpha\beta} =$$
$$= -\Lambda^{\nu'}{}_{\alpha}\Lambda^{\mu'}{}_{\beta}T^{\alpha\beta} = -T^{\nu'\mu'}$$

In 4D spacetime, a symmetric rank two tensor has 10 independent components whilst an antisymmetric rank two tensor has 6 independent components.

We concluded above that under Lorentz boosts the components of the $\bar{E}$ and $\bar{B}$ fields must become mixed. We are therefore looking for a geometrical entity with 6 independent components. One is led to surmise that the type of object required to describe the electromagnetic field may be an antisymmetric rank two tensor. But before turning our attention to that tensor we see how the hypothesis might be consolidated by considering how our understanding of the electric and magnetic potentials changes under the relativistic point of view.

27.5 The 4-Vector Potential

From the pre-relativistic perspective the electric field has a scalar potential, Φ, and the magnetic field has a vector potential, $\bar{A}$, with three spatial components. The fields are derived in the non-relativistic conditions considered so far by $\bar{E} = -\bar{\nabla}\Phi - \partial_t\bar{A}$ and $\bar{B} = \bar{\nabla} \times \bar{A}$. But we now suspect that Lorentz transformations will mix the components of $\bar{E}$ and $\bar{B}$, which implies that such transforms must mix Φ and $\bar{A}$ in some way. Noting that their dimensions can be made compatible if we replace Φ with Φ/c, we ask: does it make sense to consider $\{A^\mu\} = \{\Phi/c, \bar{A}\}$ to be a 4-vector? This would automatically provide the supposed mixing of Φ and $\bar{A}$ under Lorentz transformation. But does it make sense?

We know from chapter 7 that the usual potentials obey the wave equation with source terms, which in free space are,

$$\left(\frac{1}{c^2}\partial_t^2 - \nabla^2\right)\Phi = \frac{\rho}{\varepsilon_0} \tag{27.5.1a}$$

$$\left(\frac{1}{c^2}\partial_t^2 - \nabla^2\right)\bar{A} = \mu_0\bar{J} \tag{27.5.1b}$$

But we are now dealing with a 4-vector of charge density, $\{J^\mu\} = \{c\rho, \bar{J}\}$. We note that the wave operator itself can be written in terms of the 4-vector derivative, $\{\partial_\mu\} = \left\{\frac{\partial}{c\partial t}, \bar{\nabla}\right\}$, as,

$$\left(\frac{1}{c^2}\partial_t^2 - \nabla^2\right) = \partial^2 \equiv \partial^\mu\partial_\mu \tag{27.5.2}$$

where the minus sign results from lowering the index using (27.2.8a). This means that the wave operator is Lorentz invariant,

$$\partial^{\mu'}\partial_{\mu'} = \partial^\mu\partial_\mu \tag{27.5.3}$$

So (27.5.1a,b) is Lorentz covariant only if $\{A^\mu\} = \{\Phi/c, \bar{A}\}$ is indeed a 4-vector, and, using $\frac{1}{c^2\varepsilon_0} = \mu_0$, we can now rewrite (27.5.1a,b) succinctly as,

$$\partial^2 A^\alpha = \mu_0 J^\alpha \tag{27.5.4}$$

This is a considerable formal simplification. I reiterate the convention as regards the tensorial coordinate x^0,

$$x_0 = x^0 = ct \text{ and } \partial_0 = \frac{1}{c}\partial_t \tag{27.5.5}$$

We find that the electric and magnetic fields can be expressed very compactly in terms of the 4-vector potential thus,

$$F^{\alpha\beta} = \partial^\alpha A^\beta - \partial^\beta A^\alpha \tag{27.5.6}$$

This replaces both $\bar{E} = -\bar{\nabla}\Phi - \partial_t\bar{A}$ and $\bar{B} = \bar{\nabla}\times\bar{A}$. Recalling from (27.2.8a,b) that $\partial^0 = \partial_0$ but $\partial^i = -\partial_i$ we find,

$$F^{0i} = -F^{i0} = -E_i/c \tag{27.5.7a}$$

$$F^{ij} = -F^{ji} = -\varepsilon_{ijk}B_k, \text{ e.g., } F^{xy} = -B_z \tag{27.5.7b}$$

27.6 The Electromagnetic Field Tensor

At the close of §27.4 we surmised that the six components of the $\bar{E}$ and $\bar{B}$ fields might form an antisymmetric second rank tensor. From (27.5.6,7) we now know that that is indeed the case and we can write the electromagnetic field tensor explicitly in matrix form thus, with $\tilde{E}_i = E_i/c$,

$$(F^{\mu\nu}) = \begin{pmatrix} 0 & -\tilde{E}_x & -\tilde{E}_y & -\tilde{E}_z \\ \tilde{E}_x & 0 & -B_z & B_y \\ \tilde{E}_y & B_z & 0 & -B_x \\ \tilde{E}_z & -B_y & B_x & 0 \end{pmatrix} \tag{27.6.1a}$$

Using (27.2.8a,b) to convert contravariant to covariant indices this gives,

$$(F_{\mu\nu}) = \begin{pmatrix} 0 & \tilde{E}_x & \tilde{E}_y & \tilde{E}_z \\ -\tilde{E}_x & 0 & -B_z & B_y \\ -\tilde{E}_y & B_z & 0 & -B_x \\ -\tilde{E}_z & -B_y & B_x & 0 \end{pmatrix} \tag{27.6.1b}$$

We now return to where we started this chapter: if all this recasting into Lorentz covariant form is self-consistent we should find that (27.6.1a,b) together with the Lorentz transformations of §27.2 reproduce the Maxwell equations wrt the moving observer, S', as given by (27.1.3a,b). The elegant way of demonstrating this is to show that the Maxwell equations can be written in explicitly covariant form.

Taking the covariant divergence (∂_α) of (27.5.6) and using (27.5.4) together with the fact that the Lorentz gauge, $\partial_\alpha A^\alpha = 0$, used to derive the wave equations with source, (27.5.1a,b), we see that,

$$\partial_\alpha F^{\alpha\beta} = \mu_0 J^\beta \tag{27.6.2}$$

The reader should confirm that this becomes $\bar{\nabla} \times \bar{B} - \frac{1}{c^2}\frac{\partial \bar{E}}{\partial t} = \mu_0 \bar{J}$ when $\beta = i$ and when $\beta = 0$ it becomes $\bar{\nabla} \cdot \bar{E} = \rho/\varepsilon_0$, so (27.6.2) contains two of the Maxwell equations. The other two follow from (27.5.6) and are most neatly expressed in terms of the dual tensor defined by,

$$\mathcal{F}^{\alpha\beta} = \frac{1}{2}\varepsilon^{\alpha\beta\gamma\mu} F_{\gamma\mu} \tag{27.6.3}$$

Here $\varepsilon^{\alpha\beta\gamma\mu}$ is the "alternating tensor" which takes the value ± 1 according to whether $\alpha\beta\gamma\mu$ is an even or odd permutation of (t, x, y, z), and otherwise zero. Written explicitly in components the dual field tensor is,

$$(\mathcal{F}^{\mu\nu}) = \begin{pmatrix} 0 & -B_x & -B_y & -B_z \\ B_x & 0 & \tilde{E}_z & -\tilde{E}_y \\ B_y & -\tilde{E}_z & 0 & \tilde{E}_x \\ B_z & \tilde{E}_y & -\tilde{E}_x & 0 \end{pmatrix} \tag{27.6.4}$$

A further Lorentz covariant equation is,

$$\partial_\alpha \mathcal{F}^{\alpha\beta} = 0 \tag{27.6.5}$$

When $\beta = i$ this becomes $\overline{\nabla} \times \bar{E} + \frac{\partial \bar{B}}{\partial t} = 0$ and when $\beta = 0$ it is $\overline{\nabla} \cdot \bar{B} = 0$, so (27.6.5) contains the other two Maxwell equations. In summary, the complete Maxwell equations are expressed very compactly by,

$$\partial_\alpha F^{\alpha\beta} = \mu_0 J^\beta \quad \text{and} \quad \partial_\alpha \mathcal{F}^{\alpha\beta} = 0 \tag{27.6.6}$$

The contraction of indices (i.e., the summation over a pair of covariant and contravariant indices) provides an invariant quantity, as illustrated by (27.3.2) and (27.5.3). The same free index on both sides of an equation indicates the two quantities transform in the same way. In short, the form of these equations guarantees that they are Lorentz covariant. So we can immediately write,

$$\partial_{\alpha'} F^{\alpha'\beta'} = \mu_0 J^{\beta'} \quad \text{and} \quad \partial_{\propto'} \mathcal{F}^{\alpha'\beta'} = 0 \tag{27.6.7}$$

So we have demonstrated that the Maxwell equations in the moving frame, (27.1.3a,b), also hold, as well as (27.1.1a,b).

We can now find the explicit expressions for the transformation of the fields under a Lorentz boost in the x-direction by using the general transformation equation interpreted now for the tensorial e/m field, i.e., (27.4.4) gives,

$$F^{\mu'\nu'} = \Lambda^{\mu'}{}_\alpha \Lambda^{\nu'}{}_\beta F^{\alpha\beta} \tag{27.6.8}$$

The reader should confirm these are,

$$E_{x'} = E_x \quad E_{y'} = \gamma(E_y - uB_z) \quad E_{z'} = \gamma(E_z + uB_y) \tag{27.6.9a}$$

$$B_{x'} = B_x \quad B_{y'} = \gamma\left(B_y + \tfrac{u}{c^2}E_z\right) \quad B_{z'} = \gamma\left(B_z - \tfrac{u}{c^2}E_y\right) \tag{27.6.9b}$$

An exercise for the reader is to confirm directly that substitution of (27.6.9a,b) into (27.1.1a,b), given also (27.1.4), does indeed yield (27.1.3a,b).

27.7 The Relativistic Force Law

The non-relativistic force on a point charge q moving with velocity $\bar{v}$ in a combined electric and magnetic field is,

Non-relativistic: $\bar{f} = q(\bar{E} + \bar{v} \times \bar{B})$ (27.7.1)

For a distributed charge density, ρ, we can alternatively consider the force per unit volume to be,

Non-relativistic: $\bar{f} = \rho(\bar{E} + \bar{v} \times \bar{B}) = \rho\bar{E} + \bar{J} \times \bar{B}$ (27.7.2)

But from (27.2.14) we must look for a 4-vector of force if we wish to make this relativistic (i.e., Lorentz covariant). Moreover, the ingredients we now have to work with are the current density 4-vector, J^μ, (27.3.1b), and the field tensor, e.g., $F^{\alpha\beta}$, (27.6.1a). To make a 4-vector out of these ingredients we consider,

Relativistic: $f^\mu = F^{\mu\alpha} J_\alpha$ (27.7.3)

Recalling the rule on lowering indices, (27.2.8a), so $\{J_\alpha\} = \rho(c, -\bar{v})$ and using (27.6.1a) we see that, $\{f^\mu\} = \{f^t, \bar{f}\}$ where,

$$cf^t = \rho\bar{E} \cdot \bar{v} \qquad \bar{f} = \rho(\bar{E} + \bar{v} \times \bar{B}) \tag{27.7.4}$$

The spatial (vectorial) part, $\bar{f}$, is just the same as the non-relativistic expression for the force. For a point charge these therefore become,

$$f^\mu = \frac{dp^\mu}{d\tau} = qF^{\mu\alpha} v_\alpha \quad cf^t = q\bar{E} \cdot \bar{v} \quad \bar{f} = q(\bar{E} + \bar{v} \times \bar{B}) \tag{27.7.5}$$

The time component ($\times\, c$) is just the rate at which the e/m field does work on the charge and is thus consistent with the interpretation of the 4-vector of force as the rate-of-change of 4-momentum, the time component of which we saw following (27.2.14) must equal the rate of increase of the particle's energy, i.e., $cf^t = c\frac{dp^t}{d\tau}$.

The equation of motion which results from the relativistic formulation differs from that in the non-relativistic case, the former being,

$$\bar{f} = q(\bar{E} + \bar{v} \times \bar{B}) = \frac{d\bar{p}}{d\tau} = \frac{d}{d\tau}(\gamma m\bar{v}) = \gamma\frac{d}{dt}(\gamma m\bar{v}) \tag{27.7.6a}$$

i.e., $\gamma\frac{d}{dt}(\gamma m\bar{v}) = q(\bar{E} + \bar{v} \times \bar{B})$ (27.7.6b)

Hence, the relativistic equation of motion allows for continuously acting constant fields to continuously increase the energy of the charge but without violating the limiting speed. As $v \to c$ the LHS of (27.7.6b) can increase without limit due to the γ factors. If the LHS were replaced by the non-relativistic expression, i.e., $\frac{d}{dt}(m\bar{v})$, then light speed would be exceeded if the force acted long enough. A useful exercise is to consider an electron accelerated over some distance by a strong electric field and to find what difference the relativistic equation of motion makes to the final velocity.

As discussed in chapter 25 and §22.6, the above equation of motion does not allow for radiation reaction retarding the motion of the charge.

27.8 The Limiting Speed is the Speed of Light

There was a hidden assumption in the transformations carried out above, namely that the parameter c does not change between observers. This is second nature to us now, but before relativity would have been regarded as an error. In non-relativistic physics the fact that S' is moving with x-velocity u wrt S would imply that a light particle moving with x-velocity c wrt S would be seen by S' to have velocity $c - u$. Had S' been moving in the opposite direction he would see the light particle moving with velocity $c + u$ in non-relativistic physics.

But these are wrong. In both cases S' sees light speed being c irrespective of direction. We saw in chapter 26 that the law of relative velocities is different in relativity. In particular, the relativistic law of relative velocities is consistent with the limiting speed being invariant between observers. But chapter 26 was not able to identify the limiting speed as being light speed. We have now done so because Maxwell's equations together with the principle of the equivalence of inertial observers requires it. If c were not the limiting speed, and hence invariant between observers, then the Lorentz transformed Maxwell equations would not be correct as a representation of what S' experiences. To put that another way, the transformed Maxwell equations, (27.6.7), would be incorrect because they contain the same speed of light, c, as seen by S, not a different speed relative to S'. The Maxwell equations are consistent with the principle of the equivalence of inertial observers only because the speed of light is invariant between observers – and hence must also be the limiting speed.

27.9 The Vector Cross-Product Revisited

The reader may recall our earlier words of warning that, for a pair of ordinary Euclidean 3-vectors, $\bar{a}$ and $\bar{b}$, their cross-product $\bar{a} \times \bar{b}$ is not really a vector despite being treated as such in elementary accounts. Now we are dealing with 4-vectors is there a counterpart of the 3D cross-product?

Recall that in 3D Euclidean space we can interpret $\bar{a} \times \bar{b}$ as the vector area of the parallelogram defined by the two vectors, the direction of $\bar{a} \times \bar{b}$ being the normal to the plane of the parallelogram. Moving to 4D, and supposing this were a 4D Euclidean space (as opposed from Minkowskian) we might want the generalisation of $\bar{a} \times \bar{b}$ to again represent the area of the parallelogram, but what now is its direction? In 4D, a 2D element of area does not have a unique normal. It has two directions which are orthogonal to each other and both are orthogonal to the element of area. A quantity associated with two directions is a second rank tensor. This observation reveals the counterpart of $\bar{a} \times \bar{b}$ to be the antisymmetric second rank tensor $A_{\alpha\beta} = \varepsilon_{\alpha\beta\gamma\delta} a^{\gamma} b^{\delta}$. In chapter 30 we will see how this leads in turn to a more sophisticated formulation of these geometrical matters, and hence also of electromagnetism itself, namely exterior algebra, the algebra of differential forms, and the exterior derivative.

27.10 Other Geometrical Objects

Here we have concentrated upon vectors and rank two tensors (and the reader should have picked up on the fact that a vector is also a rank one tensor). The obvious generalisation is to tensors of arbitrary (positive integral) rank. We have already met the alternating tensor, $\varepsilon^{\alpha\beta\gamma\delta}$, which is rank four. The rank is, of course, just the number of indices. There are plenty of examples in physics of rank three and four tensors. An example is the elastic relationship between stress and strain in a crystalline material which can be written $\sigma_{ij} = C_{ijkl}\varepsilon_{kl}$, shown with all lower indices because in 3D Euclidean space we do not distinguish between covariant and contravariant indices. The rank four modulus tensor C_{ijkl} expresses the anisotropy that arises naturally due to the crystalline structure of the material in question.

However, we warned in §27.4 that tensors do not provide a complete description of all possible geometrical objects. I should clarify that I refer here to point objects only. If we allow extended objects then the scope becomes essentially inexhaustible. It may be contentious that there is any incontrovertible answer to the complete classification of geometrical point objects. However, a contender for such a classification might start with the symmetry group which encapsulates all the symmetries of the geometrical space of interest, e.g., Minkowski spacetime. The next step is to categorise the complete set of representations of that symmetry group, both as finite matrices and also as infinite dimensional representations (e.g., as differential operators). These representations may or may not be subject to restrictions such as unitarity, Hermiticity, etc., so ambiguity may arise as a result. The 'vectors' within the vector spaces upon which those representations act are then geometrical objects because their transformations are defined by operators (matrices) which isomorphically represent the symmetries of the parent geometrical space. Hence, the complete set of representations of the largest symmetry group of the geometrical space define (one might claim) the complete set of point geometrical objects that can arise within it. Some people might argue about that claim for completeness, but one can confidently assert that all entities constructed in such a manner are indeed geometrical objects.

This philosophical digression has been a way of introducing the most important types of geometrical objects which are not representable by tensors, of any integral rank. In Minkowski spacetime, these are the spinors. In fact, even in 3D Euclidean space there are spinors. Having alluded to their existence we will do no more here and the interested reader may consult standard texts on quantum mechanics, e.g., [27.2], or group representation theory.

27.11 Summary of this Chapter

We have covered a lot of formal and philosophical ground in this chapter. The following summarises the key results,

- The electric and magnetic potentials come together to form a 4-vector, $\{A^\mu\} = \{\Phi/c, \overline{A}\}$.
- The wave operator is a Lorentz invariant, $\partial^{\mu'}\partial_{\mu'} = \partial^\mu\partial_\mu$.
- The wave equation-with-sources expressed in terms of this 4-vector potential is Lorentz covariant, $\partial^2 A^\alpha = \mu_0 J^\alpha \Leftrightarrow \partial'^2 A^{\alpha'} = \mu_0 J^{\alpha'}$.
- The electric and magnetic field vectors come together to form an antisymmetric rank two tensor in 4D spacetime, $F^{\alpha\beta} = \partial^\alpha A^\beta - \partial^\beta A^\alpha$, which has a dual tensor $\mathcal{F}^{\alpha\beta} = \frac{1}{2}\varepsilon^{\alpha\beta\gamma\mu}F_{\gamma\mu}$.
- Maxwell's equations can be written in manifestly covariant form as $\partial_\alpha F^{\alpha\beta} = \mu_0 J^\beta$ and $\partial_\alpha \mathcal{F}^{\alpha\beta} = 0$.
- The components of the electric and magnetic fields transform under a boost in the x-direction as given by (27.6.9a,b).
- The limiting speed whose existence was motivated in chapter 26 has been identified as light speed *in vacuo* as the condition that the Maxwell equations are covariant between inertial observers.

- An element of area which appeared in 3D Euclidean space as the 'vector' cross product $\bar{a} \times \bar{b}$ we now understand in 4D space (e.g., Minkowski spacetime) to be an antisymmetric rank two tensor, $A_{\alpha\beta} = \varepsilon_{\alpha\beta\gamma\delta} a^{\gamma} b^{\delta}$.
- Tensors are not the only geometrical objects which can arise in a geometrical space. As a minimum there will (in cases of interest, at least) also be spinors, which we do not address here but which are crucial to the structure of matter.

In the final chapter we will see that antisymmetric rank two tensors such as $F^{\alpha\beta}, \mathcal{F}^{\alpha\beta}$ and $A_{\alpha\beta}$ are better regarded from the perspective of differential geometry as 2-forms.

27.12 References

[26.6] Minkowski, Hermann (21 September 1908). *Space and Time.* Address delivered at the 80th Assembly of German Natural Scientists and Physicians, Cologne. English translation available in *The Principle of Relativity: Original Papers by A. Einstein, H.A.Lorentz, H.Weyl and H. Minkowski*. Dover, 19226.

[26.7] Bradford, R.A.W. (2020). *The Unwierding: Not Just Another Quantum Mechanics Text*. Principia Publications Unlimited, May 2020.

28. The Lagrangian, the Energy-Momentum Tensor and Noether's Theorem

In this chapter we first make a discursion into non-relativistic particle dynamics, specifically Newtonian mechanics in Hamiltonian and Lagrangian formulations. This acts as the springboard for a generic presentation of the relativistic formulation of fields in Minkowski spacetime in terms of the Lagrange density. The Euler-Lagrange equations of field theory are derived from first principles in the general case. Noether's theorem is also derived in the general case, showing how a divergence-free second rank tensor, the energy-momentum tensor, arises from the Lagrange density if spatial and temporal homogeneity is assumed. This encapsulates the conservation of energy and momentum in fully relativistic (covariant) form. The reason why some authors refer to the energy-momentum tensor as the stress tensor is explained. In a similar manner the conservation of angular momentum follows from Noether's theorem due to the isotropy of space and is formulated in terms of a divergence-free third rank tensor. The intrinsic spin angular momentum of non-scalar fields is a by-product of the analysis. Candidate Lagrange densities for classical electromagnetism are discussed and the Maxwell equations shown to result from the corresponding Euler-Lagrange equations. The deficiency of Noether's prescription for the energy-momentum tensor in conjunction with standard Lagrange densities for the e/m field, namely its lack of symmetry, is exposed and the route to its "correction" described. Despite that shortcoming, it is elementary to derive the energy-momentum tensor consistent with the previously derived expressions for field energy density and momentum flux via tensorial reasoning, and the result is stated.

Contents

28.1 The Hamiltonian Formulation of Dynamics

Forget electromagnetism for a moment, and forget relativity. In 1833, William Rowan Hamilton formulated Newtonian mechanics in a form that is of great practical utility, [28.1]. Instead of being obliged to measure space in Cartesian coordinates, and have momentum also in Cartesian coordinates, Hamilton's formulation could deploy any curved coordinate system, provided that the corresponding momentum – the "conjugate" momentum – was suitably defined.

Hamilton's system starts with an algebraic expression for the total energy of the system in question, H, expressed in terms of a set of discrete positional variables (coordinates), q_i, and a corresponding discrete set of momentum variables, p_i. The subscript here covers both the three coordinates of a given particle and also the different particles. Thus two particles in 3D would have $i \in [1,6]$. Hamilton gave two sets of equations which define the dynamics. The first can be interpreted as an implicit definition of the conjugate momenta,

$$\frac{\partial H}{\partial p_i} = \frac{dq_i}{dt} \tag{28.1.1a}$$

For example, in the case that H depends on p_i only through a term $p_i p_i / 2m$ then this equation reduces to the usual definition of momentum in terms of velocity (rate of change of position), $\frac{p_i}{m} = \frac{dq_i}{dt}$.

The second of Hamilton's equations is the generalised form of Newton's law, essentially $F = ma$,

$$\frac{\partial H}{\partial q_i} = -\frac{dp_i}{dt} \tag{28.1.1b}$$

Considering one part of H to be a potential energy, V, dependent on position, and the remaining (kinetic energy) part to be independent of position, depending only on the momenta, then $-\frac{\partial H}{\partial q_i} = -\frac{\partial V}{\partial q_i}$ which is the definition of the corresponding components of force arising from the gradient of potential energy, and, as we would expect, this force is equated with the rate-of-change of the corresponding momenta.

The Hamiltonian formulation is non-relativistic. The Hamiltonian itself, H, being energy, is not a Lorentz invariant.

28.2 The Lagrangian Formulation of Dynamics

The Lagrangian formulation of discrete dynamics also deploys an algebraic expression with the dimensions of energy, the Lagrangian, L, which depends upon the generalised coordinates, q_i, and their time derivatives, $\dot{q}_i = \frac{dq_i}{dt}$. The corresponding momenta are defined by,

$$p_i = \frac{\partial L}{\partial \dot{q}_i} \tag{28.2.1}$$

This specifies the momenta in terms of q_i, and $\dot{q}_i$. It is assumed that this is uniquely invertible so that q_i, $\dot{q}_i$ can be converted to q_i, p_i and vice-versa. The Lagrangian can then be defined via the Hamiltonian as,

$$L = p_i\dot{q}_i - H = \frac{\partial L}{\partial \dot{q}_i}\dot{q}_i - H \tag{28.2.2}$$

(summation assumed). Interpreting $H = K + V$, the sum of the kinetic and potential energies, the term $p_i\dot{q}_i$ in Cartesian coordinates is twice the non-relativistic kinetic energy, i.e., $m_i v_i v_i$. Hence $L = K - V$, the difference between the kinetic and potential energies.

The above approach, deriving the Lagrangian from the Hamiltonian, is commonly used in modern texts. However, it belies the history in which the Lagrangian formulation preceded the Hamiltonian formulation by some 73 years, Joseph-Louis Lagrange having first enunciated his formulation in 1760, [28.2], and ultimately as part of his major survey of dynamics in 1788, [28.3]. The Lagrange approach starts with the definition $L = K - V$ and formulates dynamics in terms of a stationary action principle, Lagrange building on the earlier work of d'Alembert.

For any given postulated trajectory of the particles, $q_i(t)$, the action is defined as the time integral of the Lagrangian between specified time limits, thus,

$$S = \int_{t_1}^{t_2} L dt \tag{28.2.3}$$

All the dynamical equations may then be derived from a single variational principle, namely that the action is stationary wrt variations in the functions, $q_i(t)$. A terminology sometimes used is that S is a functional because it depends upon the functions, $q_i(t)$. The principle of stationary action can be written compactly as,

$$\delta S = \delta \int_{t_1}^{t_2} L dt = 0 \tag{28.2.4}$$

where δ denotes variation wrt any of the functions $q_i(t)$. The reader will see the principle referred to as the principle of least action. This is inaccurate. The action may be a minimum, or it may be a maximum or a saddle point wrt the variations in question.

The integral condition (28.2.4) can be converted into an equivalent set of dynamical equations which constrain the possible functions, $q_i(t)$. These equations are derived as follows. Consider the variation $\delta q_i(t)$ in one of the functions, and its associated $\delta \dot{q}_i(t) = \frac{d}{dt}\delta q_i(t)$, the dot notation denoting time derivative. Then,

$$\delta S = \delta \int_{t_1}^{t_2} L dt = \int_{t_1}^{t_2} \left(\frac{\partial L}{\partial q_i}\delta q_i + \frac{\partial L}{\partial \dot{q}_i}\delta \dot{q}_i \right) dt = 0 \tag{28.2.5}$$

But integrating by parts gives, (28.2.6)

$$\int_{t_1}^{t_2} \frac{\partial L}{\partial \dot{q}_i}\delta \dot{q}_i dt = \int_{t_1}^{t_2} \frac{\partial L}{\partial \dot{q}_i}\frac{d}{dt}\delta q_i(t) dt = \left[\frac{\partial L}{\partial \dot{q}_i}\delta q_i(t) \right]_{t_1}^{t_2} - \int_{t_1}^{t_2} \left(\frac{d}{dt}\frac{\partial L}{\partial \dot{q}_i} \right) \delta q_i(t) dt$$

We now introduce boundary conditions for the dynamics, namely that the coordinates q_i have specified values at the start and finish times, so that the variations δq_i must be zero at these times. The first term on the RHS of (28.2.6) is therefore zero. Substitution in (28.2.5) thus gives,

$$\delta S = \int_{t_1}^{t_2} \left[\frac{\partial L}{\partial q_i} - \left(\frac{d}{dt}\frac{\partial L}{\partial \dot{q}_i} \right) \right] \delta q_i dt = 0 \tag{28.2.7}$$

This must hold for arbitrary function variations δq_i and hence we require,

$$\left(\frac{d}{dt}\frac{\partial L}{\partial \dot{q}_i} \right) = \frac{\partial L}{\partial q_i} \tag{28.2.8}$$

These are the Euler-Lagrange equations which are the dynamical equations in the Lagrangian formulation.

We must now establish that (28.2.8) are equivalent to the Hamiltonian equations, (28.1.1a,b). This follows simply from, firstly by taking the derivative wrt p_i of (28.2.2), and using (28.2.1),

$$\frac{\partial L}{\partial p_i} = \frac{\partial L}{\partial \dot{q}_j}\frac{\partial \dot{q}_j}{\partial p_i} = p_j \frac{\partial \dot{q}_j}{\partial p_i} = \frac{\partial}{\partial p_i}\left(p_j \dot{q}_j - H \right) = \dot{q}_i + p_j \frac{\partial \dot{q}_j}{\partial p_i} - \frac{\partial H}{\partial p_i} \tag{28.2.9}$$

Cancelling the common term $p_j \frac{\partial \dot{q}_j}{\partial p_i}$ leaves $\dot{q}_i - \frac{\partial H}{\partial p_i} = 0$ which is (28.1.1a). Secondly, putting (28.2.1) and (28.2.2) into (28.2.8) gives,

$$\left(\frac{d}{dt} p_i \right) = \frac{\partial L}{\partial q_i} = \frac{\partial}{\partial q_i}\left(p_j \dot{q}_j - H \right) = -\frac{\partial H}{\partial q_i} \tag{28.2.10}$$

which is (28.1.1b). QED.

As formulated here, Lagrangian dynamics is also non-relativistic, because the action principle is expressed wrt a specific time coordinate and also because the Lagrangian, being energy, is not a Lorentz invariant.

28.3 Lagrangian Field Theory

Here we proceed to a formulation of the dynamics of fields in Minkowski spacetime with the intention of producing a relativistic (Lorentz covariant) formulation. The discrete set of dynamical variables, q_i, now becomes either a single field dependent on spacetime location, $x \equiv \{x^\mu\}$, e.g., $\Phi(x)$, or a set of such fields which we will denote $\varphi_i(x)$. Different i may refer to completely independent fields, or to the components of a non-scalar field, such as a vector or tensor field (or a spinor field). In general the subscript i can serve both purposes at once, thus covering, for example, two vector fields or a scalar field, a vector field and a tensor or spinor field, or any other combination.

Whereas in the case of discrete particles the formulation used both the dynamical variables q_i and their time derivatives, $\dot{q}_i$, we now use both the field variables themselves, φ_i, and their vectorial derivatives, $\varphi_{i,\mu} \equiv \partial_\mu \varphi_i$, where μ is understood to refer to the four spacetime coordinates. The other difference from the discrete case in which the Lagrangian, L, had dimensions of energy is that we will now deal with a Lagrange density, where 'density' signifies per spatial volume, i.e., per $d^3x = dV$. The Lagrange density, $\mathcal{L}$, thus has dimensions energy/volume. The action therefore becomes,

$$S = \int \mathcal{L} d^4x \tag{28.3.1}$$

The action therefore has the same dimensions as before, i.e., energy x time. (In the context of quantum field theory, a dimensionless action can thus be defined by dividing (28.3.1) by Planck's adjusted constant, $\hbar$, but this need not concern us in the classical theory). The principle of stationary action is essentially as before,

$$\delta S = \delta \int \mathcal{L} d^4x = 0 \tag{28.3.2}$$

where variations are in φ_i and consequently also in $\partial_\mu \varphi_i$. No limits have been specified in the integral of (28.3.1,2) but we may assume integration over all space and time unless otherwise specified. Moreover, if we consider only fields confined to a region of finite extent then the fields and their derivatives fall to zero at the limits of the integration.

The Hamiltonian density may be defined by the obvious generalisation of (28.2.2), i.e.,

$$\mathcal{H} = \frac{\partial \mathcal{L}}{\partial \varphi_{i,t}} \varphi_{i,t} - \mathcal{L} \tag{28.3.2b}$$

The Euler-Lagrange equations which result from (28.3.2) are derived just as before. For completeness the derivation is repeated explicitly for this case.

$$\delta S = \int \delta \mathcal{L} d^4x = \int \left(\frac{\partial \mathcal{L}}{\partial \varphi_i} \delta \varphi_i + \frac{\partial \mathcal{L}}{\partial \varphi_{i,\mu}} \delta \varphi_{i,\mu} \right) d^4x = 0 \tag{28.3.3}$$

But,

$$\int \frac{\partial \mathcal{L}}{\partial \varphi_{i,\mu}} \delta \varphi_{i,\mu} d^4x = \int \frac{\partial \mathcal{L}}{\partial \varphi_{i,t}} \left(\frac{d}{dt} \delta \varphi_i \right) d^4x + \int \frac{\partial \mathcal{L}}{\partial \varphi_{i,k}} \left(\frac{d}{dx^k} \delta \varphi_i \right) d^4x \tag{28.3.4}$$

where k is summed over the spatial coordinates. Considering the first term in (28.3.4) we carry out the time integral by parts,

$$\int \frac{\partial \mathcal{L}}{\partial \varphi_{i,t}} \left(\frac{d}{dt} \delta \varphi_i \right) d^4x = \int \left\{ \frac{\partial \mathcal{L}}{\partial \varphi_{i,t}} \delta \varphi_i - \int \delta \varphi_i \frac{d}{dt} \left(\frac{\partial \mathcal{L}}{\partial \varphi_{i,t}} \right) dt \right\} d^3x \tag{28.3.5}$$

The vanishing of the fields at the limits of integration (both temporal and spatial) means the first term is zero, so the first term in (28.3.4) reduces to,

$$\int \frac{\partial \mathcal{L}}{\partial \varphi_{i,t}} \left(\frac{d}{dt} \delta \varphi_i \right) d^4x = -\int \frac{d}{dt} \left(\frac{\partial \mathcal{L}}{\partial \varphi_{i,t}} \right) \delta \varphi_i d^4x \tag{28.3.6}$$

But the same approach can be taken with all the terms in (28.3.4), e.g.,

$$\int \frac{\partial \mathcal{L}}{\partial \varphi_{i,x}} \left(\frac{d}{dx} \delta \varphi_i \right) d^4x = -\int \frac{d}{dx} \left(\frac{\partial \mathcal{L}}{\partial \varphi_{i,x}} \right) \delta \varphi_i d^4x \tag{28.3.7}$$

So overall (28.3.3) gives us,

$$\delta S = \int \left(\frac{\partial \mathcal{L}}{\partial \varphi_i} - \partial_\mu \left(\frac{\partial \mathcal{L}}{\partial \varphi_{i,\mu}} \right) \right) \delta \varphi_i d^4x = 0 \tag{28.3.8}$$

And because this must hold for arbitrary variations of all the fields we find the Euler-Lagrange equations,

$$\partial_\mu \left(\frac{\partial \mathcal{L}}{\partial \varphi_{i,\mu}} \right) = \frac{\partial \mathcal{L}}{\partial \varphi_i} \tag{28.3.9}$$

The great advantage of the Lagrangian method in this field-theoretic form is that it results in a Lorentz covariant theory so long as the Lagrange density itself is Lorentz covariant (to be specific, the requirement is that $\mathcal{L}d^4x$, and hence the action, are Lorentz invariant). However, as an example from non-relativistic physics consider the stress analysis of elastic solids…

Example

Consider firstly the static case of a body in equilibrium. Stress is a symmetric rank two tensor, σ_{ij}, where i, j in this context run over the three spatial coordinates. Strain is also a symmetric rank two tensor and is given in terms of the displacement gradients as $\varepsilon_{ij} = (u_{i,j} + u_{j,i})/2$. Hooke's law for a generally anisotropic solid relates stress to strain, $\sigma_{ij} = C_{ijkl}\varepsilon_{kl} = C_{ijkl}u_{k,l}$

where the elastic modulus tensor has the symmetry properties $C_{ijkl} = C_{ijlk} = C_{jikl} = C_{klij}$. The potential energy per unit volume when such a material is loaded to a given stress is obtained by integrating $\sigma_{ij}\delta\varepsilon_{ij}$ and linearity then gives $\frac{1}{2}\sigma_{ij}\varepsilon_{ij} = \frac{1}{2}C_{ijkl}u_{i,j}u_{k,l}$. As there is no kinetic energy in this static case the Lagrange density in 3D is just $\mathcal{L} = -\frac{1}{2}C_{ijkl}u_{i,j}u_{k,l}$. There is no dependence on the displacements, u_i, only on their derivatives. Hence the Euler-Lagrange equations, (29.3.9), become,

$$\partial_j\left(\frac{\partial\mathcal{L}}{\partial u_{i,j}}\right) = -\partial_j\left(C_{ijkl}u_{k,l}\right) = -\partial_j\sigma_{ij} = -\sigma_{ij,j} = 0 \qquad (28.3.10)$$

which is the correct equation imposing equilibrium on every element of the body.

If we now allow the situation to be dynamic, with time-varying stresses and strains, the non-relativistic kinetic energy per unit volume is $\frac{1}{2}\rho u_{i,t}u_{i,t}$ where ρ is the density of the material. The Lagrange density thus becomes $\mathcal{L} = \frac{1}{2}\left(\rho u_{i,t}u_{i,t} - C_{ijkl}u_{i,j}u_{k,l}\right)$. There is still no dependence on the displacements, u_i, only on their derivatives, but these now include the time derivative. Hence the Euler-Lagrange equations, (29.3.9), become, for a density which is approximately unchanging,

$$\partial_\mu\left(\frac{\partial\mathcal{L}}{\partial u_{i,\mu}}\right) = \partial_t\left(\rho u_{i,t}\right) - \partial_j\left(C_{ijkl}u_{k,l}\right) = \rho u_{i,tt} - \sigma_{ij,j} = 0 \qquad (28.3.11)$$

In other words, the net force per unit volume, $\sigma_{ij,j}$, must equal the mass of that volume of material times its acceleration, $u_{i,tt}$, i.e., Newton's Law.

28.4 The Energy-Momentum Tensor

28.4.1 Shortcut to the Answer

We have already derived the energy density and momentum flux associated with an e/m field, which, in vacuum, (5.3.3), (5.6.11) and (7.5.1) give as,

$$u_{EM} = \frac{\varepsilon_0}{2}\left(E^2 + c^2B^2\right) \quad \text{and} \quad \bar{N}_{mf} = c\varepsilon_0\bar{E}\times\bar{B} \qquad (28.4.1)$$

We have seen in chapter 27 that the energy and momentum of particles can be combined to form a 4-vector? Can we do the same here? No, we cannot, because the transformation of the fields under a Lorentz boost, (27.6.9a,b), is inconsistent with interpreting $\left(u_{EM}, \bar{N}_{mf}\right)$ as a 4-vector.

We now assume something that will be explained and rationalised in §28.4.2, namely that the energy density and momentum flux are but parts of a second rank tensor. This tensor is often called simply the energy-momentum tensor, but sometimes the stress tensor, or, more completely, the stress-energy-momentum tensor. Moreover, based on (28.4.1), this tensor must be built from the electric and magnetic fields alone, and is quadratic in those fields. And finally, we now know that the proper relativistic expression for these fields is the antisymmetric second rank tensor $F^{\mu\nu}$, (27.6.1a). The most general expression that fits those criteria is,

$$T^{\alpha\beta} = \lambda_1 \eta^{\alpha\beta} F^{\mu\nu} F_{\mu\nu} + \lambda_2 \eta^{\alpha\mu} F_{\mu\nu} F^{\nu\beta} \tag{28.4.2}$$

where $\eta^{\alpha\beta}$ is the Minkowski metric, §27.2.2, and λ_1, λ_2 are some numerical constants. Here we see the great power of the principle of covariance; it constrains the form that tensorial expressions can take – often to within just one or two constant factors.

The reader should check that if we take $\lambda_1 \mu_0 = \frac{1}{4}$, $\lambda_2 \mu_0 = 1$ then the field tensors (27.6.1a,b) give,

$$T^{00} = u_{EM} \qquad \text{and} \qquad T^{0i} = c\left(\bar{N}_{mf}\right)_i \tag{28.4.3}$$

So, expressed as a matrix, the first row of $T^{\alpha\beta}$ is $\left(u_{EM}, \bar{N}_{mf}\right)$. In fact, $T^{\alpha\beta}$ is symmetrical, so its first column is the same (transposed). Proof of the symmetry of the matrix given by (28.4.2) is as follows. The first term is symmetric because $\eta^{\alpha\beta}$ is symmetric. The symmetry of the second term can be seen as follows, (28.4.4)

$$\eta^{\alpha\mu} F_{\mu\nu} F^{\nu\beta} = \eta^{\alpha\mu} \eta^{\nu\kappa} \eta^{\beta\lambda} F_{\mu\nu} F_{\kappa\lambda} = \eta^{\alpha\mu} \eta^{\nu\kappa} \eta^{\beta\lambda} F_{\nu\mu} F_{\lambda\kappa} = \eta^{\beta\lambda} F_{\lambda\kappa} F^{\kappa\alpha}$$

We now explain why, in field theory, the conservation of energy and momentum is expected to be formulated via a second rank tensor whose divergence is zero. That is, that the conservation of the energy and momentum of the field is expressed as $\partial_\alpha T^{\alpha\beta} = 0$.

28.4.2 Noether's Theorem

Noether's theorem (strictly Noether's First Theorem, [28.4]), states that to every continuous symmetry of the action there is a conserved quantity. This is the theorem that establishes the deep connection between symmetry and conservation laws in physics. Recall the continuity, or conservation, of charge, (3.8.12), in its Lorentz covariant form, (27.3.1), is manifest as a so-called "conserved current", that is, a current with zero 4-divergence, $\partial_\mu J^\mu = 0$. We therefore expect something of the sort to follow from Noether's theorem for every continuous symmetry. We shall see, however, that the form of the conservation law becomes a little different from $\partial_\mu J^\mu = 0$ when there are several continuous symmetries which are themselves connected by Lorentz transformations.

The case we consider here are the symmetries under space time translations. We expect the laws of physics to be the same on Mars – or in a distant galaxy – as they are on Earth. And we expect the laws of physics to be the same now as they were in the early universe and will continue to be into the far future. If those claims seem difficult to justify it suffices for a more limited perspective to assume at least that the laws of physics do not differ noticeably over more modest spatial and temporal intervals, such as between different laboratories on Earth or between the 21st and 19th centuries.

The upshot of this is that the Lagrange density of any theory does not have any explicit dependence on position or time. Of course it does depend upon position and time indirectly through the fields and their derivatives from which it is constructed. The distinction between these two types of dependence is summarised by saying that whilst the total derivative of $\mathcal{L}$ wrt x^μ is non-zero its partial derivative wrt x^μ is zero. Hence using the chain rule followed by (28.3.9)

$$\frac{d\mathcal{L}}{dx^\mu} = \frac{\partial\mathcal{L}}{\partial\varphi_i}\cdot\frac{d\varphi_i}{dx^\mu} + \frac{\partial\mathcal{L}}{\partial\varphi_{i,\nu}}\cdot\frac{d\varphi_{i,\nu}}{dx^\mu} = \partial_\nu\left(\frac{\partial\mathcal{L}}{\partial\varphi_{i,\nu}}\right)\cdot\frac{d\varphi_i}{dx^\mu} + \frac{\partial\mathcal{L}}{\partial\varphi_{i,\nu}}\cdot\frac{d\varphi_{i,\nu}}{dx^\mu}$$
$$= \partial_\nu\left(\frac{\partial\mathcal{L}}{\partial\varphi_{i,\nu}}\cdot\frac{d\varphi_i}{dx^\mu}\right) \tag{28.4.5}$$

Rearranging this gives,

$$\partial_\nu\left(\frac{\partial\mathcal{L}}{\partial\varphi_{i,\nu}}\cdot\frac{d\varphi_i}{dx^\mu} - \mathcal{L}\delta^\nu{}_\mu\right) = 0 \tag{28.4.6}$$

This demonstrates the existence of a second rank tensor constructed from the Lagrange density which has identically zero 4-divergence, i.e., defining,

$$T^\nu{}_\mu = \frac{\partial\mathcal{L}}{\partial\varphi_{i,\nu}}\cdot\frac{d\varphi_i}{dx^\mu} - \mathcal{L}\delta^\nu{}_\mu \qquad \text{then} \qquad \partial_\nu T^\nu{}_\mu = T^\nu{}_{\mu,\nu} = 0 \tag{28.4.7}$$

$T^\nu{}_\mu$ is the energy-momentum tensor. As noted previously, the fully descriptive name would be the stress-energy-momentum tensor. That all three terms, stress, energy and momentum, are valid physical interpretations is now addressed.

$\partial_\nu T^\nu{}_\mu = 0$ is reminiscent of the conserved current equation, $\partial_\mu J^\mu = 0$, except there are now four conserved quantities, one for each value of μ, corresponding to the translational symmetry in each of the four spacetime directions, x^μ. Moreover, these four conserved quantities are not scalar invariants like charge.

To identify what these conserved quantities are physically, consider,

$$T^t{}_t = \frac{\partial \mathcal{L}}{\partial \varphi_{i,t}} \cdot \frac{d\varphi_i}{dx^t} - \mathcal{L} \tag{28.4.8}$$

We see from (28.3.2b) that $T^t{}_t$ is the Hamiltonian density, $\mathcal{H}$, i.e., the energy density. This suggests that, in a given frame of reference, we can interpret

$$P_\mu = \int T^t{}_\mu \, d^3x \tag{28.4.9}$$

as the energy and momentum associated with the whole field. This would appear to leave P_μ with time dependence, as time is not integrated-out in (28.4.9), but $T^\nu{}_{\mu,\nu} = 0$ ensures that this is not so, because,

$$\int T^\nu{}_{\mu,\nu} d^3x = 0 = \partial_t \int T^t{}_\mu d^3x + \int T^i{}_{\mu,i} d^3x \tag{28.4.10}$$

The integrand in the second term on the RHS is a 3-divergence and hence can be replaced by the surface integral $\oint\!\!\oint T^i{}_\mu dS_i$ which will be zero on a remote surface as we are considering fields which are confined to a finite region. Hence (28.4.10) gives,

$$\partial_t P_\mu = \partial_t \int T^t{}_\mu \, d^3x = 0 \tag{28.4.11}$$

i.e., all four components of the 4-momentum are conserved. However, a note of caution is needed here because, despite appearances, P_μ is not a 4-vector because it is defined wrt a specific reference frame, as betrayed by the use of an explicit time component in the integrand of (28.4.9) and a specific volume measure, d^3x. Nevertheless, this does not detract from the interpretation of P_μ as the energy and momentum as seen by an observer stationary wrt the frame in which the fields are specified, nor does it undermine their conservation, (28.4.11). The issue of a fully covariant description of the total energy and momentum of the e/m field is inextricably linked to the issue of the self-energy of the charge, and hence the "structure" of the charge, or why it is stable against its own Coulomb repulsion. For further discussion see Jackson, [28.6], section 17.5.

As $T^t{}_t$ is the energy density and $T^\nu{}_{t,\nu} = 0$ it follows that $T^i{}_t$ can be interpreted as the flux of energy in direction i. Similarly, $T^i{}_j$ is the flux of the j-component of momentum in the direction i. This latter observation is what motivates the additional designation of $T^\alpha{}_\beta$, and specifically $T^i{}_j$, as the stress tensor. It is because the flux of the j-component of momentum, i.e., the amount of j-momentum through unit area in unit time, is the force in direction j per unit area. But the area in question is in the i-direction. Hence, $T^i{}_j$ is the stress σ_{ij}.

Readers familiar with stress analysis will know that the symmetry of the stress tensor is required for there to be no net turning moment on small elements of volume. Absent applied distributed torques, equilibrium requires $\sigma_{ij} = \sigma_{ji}$. This seems to require that the energy-momentum tensor is symmetric, at least its spatial part. We will see shortly that this conclusion can be made precise. For now we note that (28.4.2) is indeed symmetric.

28.5 The Conservation of Angular Momentum

In §28.4 the conservation of energy and momentum was derived from the translational symmetry of spacetime. But spacetime also has Lorentz symmetry, i.e., symmetry under 3D rotations and 3D Lorentz boosts. The same sort of derivation as in §28.4 can be used to find the corresponding conserved quantities, namely the three components of angular momentum. We include it here though it is of lesser significance in classical electromagnetism than in quantum field theory and elementary particle physics which must deal with fields with different spin.

The analysis must pay careful attention to the fact that not only do the spacetime coordinates change under the Lorentz transform but for fields other than scalar fields, e.g., vector, tensor or spinor fields, the components of the field are also transformed. The latter transformation depends upon the "spin" of the field, i.e., whether it is vectorial (spin 1), or a second rank tensor (spin 2), or a spin ½ spinor, etc. The matrix representation of the Lorentz group appropriate to the field, $S_{rs}^{\alpha\beta}$, therefore appears in the formulation. We outline the derivation here, details being available in [28.5] or many standard texts, e.g., on quantum field theory.

Under a (homogeneous) Lorentz transformation, without translation, the coordinates transform as $\delta x_\alpha = \varepsilon_{\alpha\beta} x^\beta$ where $\varepsilon_{\alpha\beta}$ is an antisymmetric matrix of rotation angles and boost parameters. We assume a single type of field but with arbitrary spin. The transformation of the field is,

$$\delta\varphi_r(x) = \frac{1}{2}\varepsilon_{\alpha\beta} S_{rs}^{\alpha\beta}\varphi_s(x) - \frac{\partial\varphi_r(x)}{\partial x_\alpha}\delta x_\alpha \quad (28.5.1)$$

The symmetry condition with the Euler-Lagrange equations gives, (28.5.2)

$$\partial_\mu\left(\frac{\partial\mathcal{L}}{\partial\varphi_{r,\mu}}\right)\delta\varphi_r + \frac{\partial\mathcal{L}}{\partial\varphi_{r,\alpha}}\delta\varphi_{r,\alpha} + \frac{\partial\mathcal{L}}{\partial x_\alpha}\delta x_\alpha = \partial_\mu\left(\frac{\partial\mathcal{L}}{\partial\varphi_{r,\mu}}\delta\varphi_r\right) + \frac{\partial\mathcal{L}}{\partial x_\alpha}\delta x_\alpha = 0$$

Substitution of (28.5.1) and $\delta x_\alpha = \varepsilon_{\alpha\beta} x^\beta$ gives,

$$\partial_\mu\left(\frac{\partial\mathcal{L}}{\partial\varphi_{r,\mu}}\left[\frac{1}{2}\varepsilon_{\alpha\beta}S_{rs}^{\alpha\beta}\varphi_s(x)-\frac{\partial\varphi_r(x)}{\partial x_\alpha}\varepsilon_{\alpha\beta}x^\beta\right]\right)+\frac{\partial\mathcal{L}}{\partial x_\alpha}\varepsilon_{\alpha\beta}x^\beta$$
$$=\partial_\mu\left(\frac{\partial\mathcal{L}}{\partial\varphi_{r,\mu}}\left[\frac{1}{2}S_{rs}^{\alpha\beta}\varphi_s(x)-\frac{\partial\varphi_r(x)}{\partial x_\alpha}x^\beta\right]+\eta^{\alpha\mu}\mathcal{L}x^\beta\right)\varepsilon_{\alpha\beta}=0 \qquad (28.5.3)$$

Note that the term x^β can be taken inside the derivative ∂_μ because $\partial_\mu(\eta^{\alpha\mu}x^\beta\varepsilon_{\alpha\beta})=\eta^{\alpha\mu}\varepsilon_{\alpha\beta}\delta_\mu{}^\beta=\eta^{\alpha\beta}\varepsilon_{\alpha\beta}=0$ by virtue of the anti-symmetry of $\varepsilon_{\alpha\beta}$. The second and third terms in (28.5.3) can be re-written in terms of the energy-momentum tensor using (28.4.7) as,

$$\partial_\mu\left(\frac{1}{2}\cdot\frac{\partial\mathcal{L}}{\partial\varphi_{r,\mu}}S_{rs}^{\alpha\beta}\varphi_s(x)-T^{\mu\alpha}x^\beta\right)\varepsilon_{\alpha\beta}=0 \qquad (28.5.4)$$

However we cannot conclude that the coefficient of $\varepsilon_{\alpha\beta}$ in (28.5.4) is zero because the anti-symmetry of $\varepsilon_{\alpha\beta}$ means that it is only the part of its factor which is antisymmetric in $\alpha\beta$ which must be zero. Hence we write,

$$\mathfrak{J}^{\mu\alpha\beta}=\frac{\partial\mathcal{L}}{\partial\varphi_{r,\mu}}S_{rs}^{\alpha\beta}\varphi_s(x)+\left(x^\alpha T^{\mu\beta}-x^\beta T^{\mu\alpha}\right) \qquad (28.5.5)$$

For a scalar field the term in $S_{rs}^{\alpha\beta}$ does not occur. Note that $S_{rs}^{\alpha\beta}=-S_{rs}^{\beta\alpha}$ and that $\mathfrak{J}^{\mu\alpha\beta}=-\mathfrak{J}^{\mu\beta\alpha}$, so $\partial_\mu\mathfrak{J}^{\mu\alpha\beta}\varepsilon_{\alpha\beta}=0$ now implies,

$$\partial_\mu\mathfrak{J}^{\mu\alpha\beta}=0 \qquad (28.5.6)$$

Consequently we have again found conserved quantities labelled by $\alpha\beta$. Since $\mathfrak{J}^{\mu\alpha\beta}=-\mathfrak{J}^{\mu\beta\alpha}$ there are six conserved quantities defined by,

$$M^{\alpha\beta}=\int\mathfrak{J}^{0\alpha\beta}\cdot d^3x \qquad (28.5.7)$$

As before, $\mathfrak{J}^{i\alpha\beta}$, for spatial indices i can be interpreted as the flux of these $M^{\alpha\beta}$ quantities. But what are they?

Here we will consider only the spatial components, M^{ij}. If we look firstly at the second term in (28.5.5), which would be the only term for a scalar field, we find,

$$M_{orbit}^{ij}=\int d^3x\,(x^iT^{0j}-x^jT^{0i}) \qquad (28.5.8)$$

But T^{0j} is momentum density. So $x^iT^{0j}d^3x$ is the ij component of angular momentum due to the field in a small volume element, i.e., due to the momentum in the j-direction multiplied by the distance in the i-direction. So if ijk is a permutation of x,y,z then $(x^iT^{0j}-x^jT^{0i})d^3x$ is the angular momentum in the k-direction due to the fields in this small region. So we can interpret M_{orbit}^{ij} as the angular momentum of a scalar field, or that part of the angular momentum of a field with spin degrees of freedom which does not depend upon the spin degrees of freedom. In other words, it is the orbital angular momentum.

The first term in (28.5.5) must have an interpretation compatible with the second term, otherwise it would make no sense to add them together. This allows us to interpret the first term in (28.5.5) as the angular momentum which arises from the spin degrees of freedom, rather than from products of linear momentum and perpendicular distance. This justifies the use of the word "spin" for the degrees of freedom associated with the rotational transformations under the representation $S_{rs}^{\alpha\beta}$ of the Lorentz group. The term "spin" encapsulates a form of angular momentum intrinsic to the field, rather than resulting from orbital angular momentum.

$$M_{spin}^{ij} = \int \frac{\partial L}{\partial \varphi_{r,0}} S_{rs}^{ij} \varphi_s(x) d^3x \qquad (28.5.9)$$

The field conjugate to φ_r, the field momentum, is defined in the same way as the momentum conjugate to the positional variable in (28.2.1), i.e.,

$$\pi_r = \frac{\partial L}{\partial \varphi_{r,0}} \qquad (28.5.10)$$

So (28.5.9) can also be written,

$$M_{spin}^{ij} = \int \pi_r S_{rs}^{ij} \varphi_s(x) d^3x \qquad (28.5.11)$$

The total angular momentum is thus,

$$M^{ij} = M_{orbit}^{ij} + M_{spin}^{ij} \qquad (28.5.12)$$

It is only this *total* angular momentum which is conserved as a consequence of (28.5.6), i.e.,

$$\frac{dM^{ij}}{dt} = 0 \qquad (28.5.13)$$

Where there is more than one field the M^{ij} must be summed over them all in order for (28.5.13) to apply, i.e., the different fields may transfer angular momenta between them; only the total is conserved.

The physical meaning of the remaining conserved quantities, M^{0j}, is a good question with which to plague undergraduates. It is discussed in [28.5].

28.6 Is the Energy-Momentum Tensor Symmetric?

The canonical derivation of the energy-momentum tensor from Noether's theorem, i.e., (28.4.7), does not ensure that it is symmetric. But earlier it was intimated that the energy-moment tensor might be required to be symmetric based on the spatial parts being interpreted as a stress tensor, which is symmetric if net turning moments are zero (equilibrium). As regards the whole energy-momentum tensor, a correct statement is that the energy-momentum tensor of (28.4.7) can be made symmetric without penalty.

Taking the derivative of $\mathfrak{J}^{\mu\alpha\beta}$ in (28.5.5), imposing (28.5.6) and assuming the energy-momentum tensor is indicative of conserved energy and momentum, i.e., $\partial_\mu T^{\mu\alpha} = 0$, we see that,

$$\partial_\mu \mathfrak{J}^{\mu\alpha\beta} = \partial_\mu \left(\frac{\partial \mathcal{L}}{\partial \varphi_{r,\mu}} S_{rs}^{\alpha\beta} \varphi_s(x) \right) + \left(x^\alpha \partial_\mu T^{\mu\beta} - x^\beta \partial_\mu T^{\mu\alpha} \right) + T^{\alpha\beta} - T^{\beta\alpha} =$$

$$\partial_\mu \left(\frac{\partial \mathcal{L}}{\partial \varphi_{r,\mu}} S_{rs}^{\alpha\beta} \varphi_s(x) \right) + T^{\alpha\beta} - T^{\beta\alpha} = 0 \qquad (28.6.1)$$

We already know that $\partial_\mu \mathfrak{J}^{\mu\alpha\beta}$ and $S_{rs}^{\alpha\beta}$ are antisymmetric in $\alpha \leftrightarrow \beta$, and so are consistent with $T^{\alpha\beta} - T^{\beta\alpha}$, the antisymmetric part of $T^{\alpha\beta}$ appearing in (28.6.1). This equation establishes that the antisymmetric part of $T^{\alpha\beta}$ is a 4-divergence. This permits us to redefine the energy-momentum tensor thus,

$$T^{\alpha\beta} \to \tilde{T}^{\alpha\beta} = T^{\alpha\beta} + \frac{1}{2} \partial_\mu \left(\frac{\partial \mathcal{L}}{\partial \varphi_{r,\mu}} S_{rs}^{\alpha\beta} \varphi_s(x) \right) \qquad (28.6.2)$$

With this replacement, (28.6.1) leads to $\tilde{T}^{\alpha\beta} = \tilde{T}^{\beta\alpha}$. It is valid to make this replacement only because the antisymmetric part of the original $T^{\alpha\beta}$ is a pure 4-divergence. This means that, when integrated over volume to form the total energy and momentum, (28.4.9), transformation to a surface integral shows the additional term to have zero contribution for localised fields.

The reader should confirm that the modified tensor still has zero divergence, $\partial_\alpha \tilde{T}^{\alpha\beta} = 0$ so that the modified tensor has the same conserved quantities as the original $T^{\alpha\beta}$.

28.7 One Option for the Electromagnetic Lagrangian

The form of the electromagnetic Lagrangian which is most often quoted is,

$$-\mathcal{L} = \frac{1}{4\mu_0} F_{\mu\nu} F^{\mu\nu} + J_\mu A^\mu \qquad (28.7.1)$$

where, for purposes of finding the stationary condition, it is understood that $F^{\alpha\beta}$ means $\partial^\alpha A^\beta - \partial^\beta A^\alpha$, as given by (27.5.6). The reader should confirm that both terms have units of energy density, as they should. The reader should confirm that the Euler-Lagrange equations, (28.3.9), become in this case simply,

$$\partial_\propto F^{\alpha\beta} = \mu_0 J^\beta \tag{28.7.2}$$

which is as it should be, i.e., (27.6.2), which is equivalent to the two Maxwell equations $\overline{\nabla} \times \bar{B} - \frac{1}{c^2}\frac{\partial \bar{E}}{\partial t} = \mu_0 \bar{J}$ and $\overline{\nabla} \cdot \bar{E} = \rho/\varepsilon_0$. The other two (homogeneous, source-free) Maxwell equations are, of course, implicit in the potential expressions for the field, i.e., that $F^{\alpha\beta} = \partial^\alpha A^\beta - \partial^\beta A^\alpha$. In terms of the dual field, $\mathcal{F}^{\alpha\beta} = \frac{1}{2}\varepsilon^{\alpha\beta\gamma\mu}F_{\gamma\mu}$, (27.6.3), we have immediately,

$$\partial_\propto \mathcal{F}^{\alpha\beta} = \frac{1}{2}\varepsilon^{\alpha\beta\gamma\mu}\partial_\propto F_{\gamma\mu} = \frac{1}{2}\varepsilon^{\alpha\beta\gamma\mu}\left(A_{\mu,\gamma\alpha} - A_{\gamma,\mu\alpha}\right) = 0 \tag{28.7.3}$$

Due to the full antisymmetry of $\varepsilon^{\alpha\beta\gamma\mu}$, i.e., as per (27.6.3) and which is equivalent to $\overline{\nabla} \times \bar{E} + \frac{\partial \bar{B}}{\partial t} = 0$ and $\overline{\nabla} \cdot \bar{B} = 0$.

However, despite the success of the Lagrangian (28.7.1) in reproducing the Maxwell equations via the principle of stationary action, when that Lagrangian is inserted into (28.4.7) to give an expression for the energy-momentum tensor the result, in regions with no current, is,

$$\mu_0 T^{\alpha\beta}_{Noether} = \eta^{\beta\gamma}F^{\alpha\lambda}A_{\lambda,\gamma} + \frac{1}{4}F^{\mu\nu}F_{\mu\nu}\eta^{\alpha\beta} \tag{28.7.4}$$

which is not a symmetrical tensor, contrary to what we need for a consistent physical interpretation.

At this point the "correction" defined in (28.6.2) is invoked. Specifically, in this case the final (symmetric) energy-momentum tensor, $T^{\alpha\beta}$, is obtained from the initial version $T^{\alpha\beta}_{Noether}$, (28.7.4), by,

$$T^{\alpha\beta} = T^{\alpha\beta}_{Noether} + \frac{1}{\mu_0}\partial_\mu\left(F^{\mu\alpha}A^\beta\right) \tag{28.7.5}$$

Note that the correction term is a divergence, as required to ensure it does not contribute to the total field energy and momentum, (28.4.9), and also it has zero divergence because $\partial_\alpha\partial_\mu\left(F^{\mu\alpha}A^\beta\right) = 0$ due to the antisymmetry of $F^{\mu\alpha}$ and hence continues to ensure energy and momentum are conserved. The result is, finally, consistent with §28.4.1,

$$T^{\alpha\beta} = \frac{1}{\mu_0}\left(\frac{1}{4}\eta^{\alpha\beta}F^{\mu\nu}F_{\mu\nu} + \eta^{\alpha\mu}F_{\mu\nu}F^{\nu\beta}\right) \tag{28.7.6}$$

Hence, the first row and column of $T^{\alpha\beta}$ are seen to be consistent with the expressions for the field energy density and momentum flux derived previously, (28.4.3).

Whilst this analysis does finally work out, the reader may agree with me that it is less than elegant that the energy-momentum tensor derived directly from Noether's theorem has such a serious shortcoming.

The reader might imagine that a better choice for the Lagrangian might permit a symmetric energy-momentum tensor to result directly from Noether, (28.4.7). All I can say is that I have not seen one.

Another problem with the Lagrangian (28.7.1) is that the time component of the conjugate field momentum, the field generalisation of (28.2.1), is identically zero and this causes problems in making the transition to quantum field theory because this is incompatible with the canonical commutation relations. The usual practice in that context is to employ the Lagrangian,

$$-\mathcal{L} = \frac{1}{2\mu_0} A_{\mu,\nu} A^{\mu,\nu} + J_\mu A^\mu \qquad (28.7.7)$$

This has the drawback that the Lorentz gauge has to be assumed in order to derive the Maxwell equations. Noether, (28.4.7), does then result in a symmetric energy-momentum tensor…but not the right one!

28.8 References

[28.1] Hamilton, W. R. (1833). *On a general method of expressing the paths of light, and of the planets, by the coefficients of a characteristic function*. Printed by P.D. Hardy, Dublin.

[28.2] Fraser, Craig. (1983). J. L. Lagrange's Early Contributions to the Principles and Methods of Mechanics. Archive for History of Exact Sciences, vol. 28, no. 3, 1983, pp. 197–241.

[28.3] Lagrange, J. L. *Mécanique analytique* (1788). See also Hobson, E.W., in *Mathematics, from the points of view of the Mathematician and of the Physicist* an address delivered to the Mathematical and Physical Society of University College London., 1912.

[28.4] Noether, E. (1918). Invariante Variationsprobleme. *Nachrichten von der Gesellschaft der Wissenschaften zu Göttingen*. Mathematisch-Physikalische Klasse, 1918: 235–257.

[28.5] Bradford, R.A.W. (20120. *Noethers Theorem and the Origin of Spin*. Blog entry on http://rickbradford.co.uk.

[28.6] J.D.Jackson, J.D. (1975). Classical Electrodynamics. John Wiley & Sons.

29. Maxwell's Equations in Quaternions

Here we show that a 4-vector can be expressed as an Hermitian biquaternion, p*, and a Lorentz transformation can then be expressed in a form which looks almost the same as the quaternionic expression for rotations, namely* $\mathrm{p} \to \mathrm{p}' = q\mathrm{p}q^{*\dagger}$*. If* q *is a quaternion this is a rotation. If* q *is an Hermitian biquaternion then it is a Lorentz boost. The 4-vector potential,* $A = \Phi + i\bar{A}$*, and the 4-vector current,* $J = \rho + i\bar{J}$*, are Hermitian biquaternions and hence 4-vectors, as required. The electric and magnetic fields can be written as a biquaternion* $F = \bar{B} + i\bar{E}$*, which is not Hermitian, and it is found that the complete set of Maxwell's equations can be expressed as just one equation,* $DF + J = 0$.

Contents

29.1 Quaternions

Hamilton had sought the quaternions for some time. His motivation was to find an algebraic means of implementing rotations in three-dimensional space. In the two-dimensional plane, a position can be represented by a complex number, and multiplying by the complex number $e^{i\theta}$ causes an anticlockwise rotation of that position about the origin by θ. Hamilton had been looking for a corresponding means of representing positions and rotations in three-dimensional space. The quaternions are the answer. Thus, a position in 3D space can be represented by the (pure) quaternion,

$$p = xI + yJ + zK \qquad (29.1.1)$$

where the entities I, J, K are three distinct square-roots of -1. Moreover, these entities, the basis quaternions, do not commute – they anticommute. Their defining properties, namely,

$$I^2 = J^2 = K^2 = IJK = -1 \qquad (29.1.2a)$$

were carved by William Rowan Hamilton into the stone of Brougham Bridge in Dublin on 16th October 1843, immediately he realised their significance. From these we can prove also that,

$$IJ = -JI = K \qquad JK = -KJ = I \qquad KI = -IK = J \qquad (29.1.2b)$$

$$IJ - JI = 2K \qquad JK - KJ = 2I \qquad KI - IK = 2J \qquad (29.1.2c)$$

$$IJI = J,\ IKI = K,\ JIJ = I,\ JKJ = K,\ KIK = I, KJK = J \qquad (29.1.2d)$$

The non-commutative property of quaternions is not incidental. It is essential for quantities to represent 3D rotations, as Hamilton required, because 3D rotations are themselves non-commutative. If the reader has never checked this for himself, I suggest you do so now. Take an object and perform two rotations about different axes in 3D space. Now carry out those rotations in the reverse order. In general, the final orientation of the object will be different.

Hamilton and his immediate followers thought of quaternions as a further extension of the number system. Just as the reals extend the rational numbers, and the complex numbers extend the reals, so the quaternions extend further to a type of 'number' with four components, the general quaternion being $w + xI + yJ + zK$, where the addition of the "real" or "scalar" part, w, is essential for the quaternions to be algebraically closed (because (29.1.2a) links their algebra to the reals).

In representing a position in 3D space as the pure quaternion (29.1.1) the basis quaternions I, J, K act as unit vectors. But that begs the question: what does the real part, w, of a general quaternion mean physically? Hamilton's answer – and the answer we would naturally adopt with our knowledge of relativity and the concept of spacetime – is that w is time. There is a snag, though, as we will see.

How do the quaternions represent rotations? The position represented by pure quaternion p (i.e., for $w = 0$) is rotated to p' by a rotation quaternion q where,

$$q = exp\{\theta(\hat{n}_x I + \hat{n}_y J + \hat{n}_z K)/2\} \quad \text{and} \quad p' = qpq^{-1} \tag{29.1.2}$$

Here q is a 3D rotation by angle θ about the axis defined by unit vector $\hat{n}$. If $p = v_x I + v_y J + v_z K$ such a rotation would normally be implemented using a 3 x 3 rotation matrix, (R), such that $\bar{v}' = (R)\bar{v}$. An exercise for the reader is to confirm that (29.1.2) reproduces the expected rotation about, say, the z-axis, for which the usual formulation is,

$$\begin{pmatrix} v_x' \\ v_y' \\ v_z' \end{pmatrix} = \begin{pmatrix} cos\theta & -sin\theta & 0 \\ sin\theta & cos\theta & 0 \\ 0 & 0 & 1 \end{pmatrix} \begin{pmatrix} v_x \\ v_y \\ v_z \end{pmatrix} \tag{29.1.3}$$

When I first came across the quaternionic expression of the rotation of a vector in 3D space, i.e., $p' = qpq^{-1}$, I was puzzled as to why this expression is quadratic in the "rotation operator", q. This was because the more familiar 3D matrix notation, $\bar{v}' = (R)\bar{v}$, is linear in the operator (R). A clue to the answer is to recall from chapter 27 that the number of factors of the transformation matrix equals the rank of the tensor. For example, a second rank tensor transforms as $T^{\mu'v'} = \Lambda^{\mu'}{}_{\alpha}\Lambda^{v'}{}_{\beta}T^{\alpha\beta}$ and a rank three tensor as $T^{\mu'v'\xi'} = \Lambda^{\mu'}{}_{\alpha}\Lambda^{v'}{}_{\beta}\Lambda^{\xi'}{}_{\gamma}T^{\alpha\beta\gamma}$, etc. The natural entities upon which quaternions operate are not vectors but spinors, those unexpected "square roots of geometry" (see §27.10 and [29.1]). From the perspective of spinors, a vector is a second rank tensor in spinor space. But we will say no more about the spinor perspective here.

The pure quaternions mimic vector algebra in an elegant fashion. If the real 3D vector $\bar{p}$ is represented by the quaternion $p_x I + p_y J + p_z K$ and similarly the real 3D vector $\bar{q}$ is represented by $q_x I + q_y J + q_z K$, then the quaternion product is, (29.1.4)

$$qp = -(q_x p_x + q_y p_y + q_z p_z) + (q_y p_z - q_z p_y)I + (q_z p_x - q_x p_z)J + (q_x p_y - q_y p_x)K$$

This shows that the algebra of pure quaternions (i.e., quaternions with no real part) is not closed, as the above product does include a real part. Denoting the real and pure (or "vector") parts by $\mathbb{S}$ and $\mathbb{V}$ we see that, in terms of the usual scalar and vector products of 3D vectors,

$$\mathbb{S}(qp) = \mathbb{S}(pq) = -\bar{p} \cdot \bar{q} \qquad (29.1.5a)$$

$$\mathbb{V}(qp) = -\mathbb{V}(pq) = \bar{q} \times \bar{p} = -\bar{p} \times \bar{q} \qquad (29.1.5b)$$

The anticommutation of the basis quaternions results in the familiar anticommutation of the vector cross product.

This is pretty neat and so it is not surprising that James Clerk Maxwell used quaternions in his famous equations of electromagnetism, including his 1873 Treatise on Electricity and Magnetism, [29.2]. However, he did not use quaternions to their full extent, generally only for the vector parts, and mixed the notation with full component expressions. The posthumous 1881 and 1892 second and third editions of Maxwell's Treatise, [29.2], broadly used the same notation as Maxwell's original. However, over this same period, 1980s-1990s, Gibbs and Heaviside began to use modern vector notation in other publications, and this had become the norm with physicists by the start of the 20th century. Tensor notation was introduced after Einstein's initial papers on special relativity in 1905, though vector notation continues to be used to this day as more suited to specific problems. The long history of changing notations for Maxwell's equations has been given by Hehl, [29.3].

Despite Hamilton's proselytising of quaternions, their use gave way to vector algebra under the influence of Grassmann, Gibbs, Heaviside and Lord Kelvin. The latter was the leading physicist of the mid- to late nineteenth century and was reputed to have a dim view of quaternions. Kelvin is quoted as opining, "Quaternions were invented by Hamilton after his truly important work had been completed. Although beautiful and of ingenious origin, they have been a curse on anyone who has come into contact with them in any way". Kelvin himself avoided both quaternions and vectors, so might have been something of an old fuddy-duddy. From Einstein's general relativity (1916) onwards, the use of tensors became *de rigueur* by mathematical physicists, and ultimately the language of exterior calculus or Clifford algebras would take over.

From the inclusion of time in the full quaternion, $q = t + xI + yJ + zK$, one might hope that (special) relativity, that is the Lorentz transform, would arise naturally. Lamentably, this is not so. The square modulus of the quaternion representing an 'event' (i.e., a point in spacetime) is Euclidean, not Minkowskian. That is,

$$q^\dagger q = (t - xI - yJ - zK)(t + xI + yJ + zK) = t^2 + x^2 + y^2 + z^2$$

where the dagger superscript denotes the quaternion-conjugate. In contrast, what we want to be invariant (thus defining the Lorentz transformations) is the indefinite (Minkowski) metric, $t^2 - x^2 - y^2 - z^2$.

This is terribly disappointing. But this flaw can be rectified by moving to biquaternions...

29.2 Biquaternions

I acknowledge the lovely review by Joachim Lambek, [29.4], as the inspiration for this section. As we have seen in the preceding section, the general quaternion is $w + xI + yJ + zK$ where w, x, y, z are real. In contrast, a biquaternion is given by $q = w + xI + yJ + zK$ where w, x, y, z may be complex. A general biquaternion therefore involves 8 real parameters.

The dagger notation denotes the quaternion-conjugate, such that,

$$q^\dagger = w - xI - yJ - zK \qquad (29.2.1)$$

noting that this leaves the (generally complex) coordinates, w, x, y, z, unchanged. In contrast, the complex-conjugate is, as normal,

$$q^* = w^* + x^*I + y^*J + z^*K \qquad (29.2.2)$$

Note that, for any two biquaternions, $(pq)^\dagger = q^\dagger p^\dagger$, which corresponds to the similar expression for the Hermitian conjugate of complex matrices. (An exercise for the reader is to show this). In contrast, $(pq)^* = p^*q^*$.

When dealing with biquaternions, the terms "real" and "imaginary" become ambiguous. We use these terms to relate solely to the complex context. For the quaternionic part we refer to w as the "scalar" or "temporal" part (whether it is real or complex), and the $xI + yJ + zK$ part as the "vector" or "spatial" part, whether the coefficients are real or complex. Hence, the term "the real part of q" would mean $\mathcal{R}(w) + \mathcal{R}(x)I + \mathcal{R}(y)J + \mathcal{R}(z)K$. In contrast the scalar part is denoted $\mathbb{S}(q) = w$ and the vector part is denoted $\mathbb{V}(q) = xI + yJ + zK$.

Whilst the quaternions form a division ring, as every non-zero quaternion has an inverse, the biquaternions do not form a division ring as not every non-zero element has an inverse (for example, $1 + iI$ has no inverse). For non-zero quaternions there is a positive definitive norm whose square is defined by $N(q) = qq^\dagger = w^2 + x^2 + y^2 + z^2$ so that $q^{-1} = q^\dagger / N(q)$. In contrast, for biquaternions, $qq^\dagger$ will not generally be real, or positive definite even when it is real, whilst $qq^{*\dagger}$ is not generally scalar. The lack of an inverse for the general biquaternion is essential for its utility in relativity since it relates directly to the Minkowski metric not being positive definite.

29.3 The Lorentz Transformation

Spacetime points (events) are denoted by special biquaternions, referred to as Hermitian biquaternions, defined as having a real scalar part and a purely imaginary vector part. Thus, if t is a real time coordinate and $\bar{r}$ is a real 3-vector, which in quaternion form is written $\bar{r} = xI + yJ + zK$, where x, y, z are real, then the event is represented by $q = t + i\bar{r}$. Note that, with $\bar{r}$ now interpreted as a pure quaternion, $\bar{r}\bar{r} = -r^2$.

Hence, Hermitian biquaternions are defined such that they equal their full conjugate, $q^{*\dagger} = q$ (equivalently $q^* = q^\dagger$). The squared-norm of an Hermitian biquaternion is thus,

$$N(q) = qq^\dagger = (t + i\bar{r})(t - i\bar{r}) = t^2 - [x^2 + y^2 + z^2] \qquad (29.3.1)$$

Noting that $\bar{r}$ is quaternionic not a real vector so the minus sign characteristic of the Minkowski metric occurs because $I^2 = J^2 = K^2 = -1$. This immediately links to the Lorentz transformation which can be defined as the most general transformation which preserves the norm of an Hermitian biquaternion.

It follows that the transformation of an arbitrary Hermitian biquaternion, p, defined by,

$$p \to p' = qpq^{*\dagger} \qquad (29.3.2)$$

where q is any biquaternion with unit norm (i.e., with $qq^\dagger = 1$) is a Lorentz transformation. This is because,

$$p'p'^{\dagger} = qpq^{*\dagger}(qpq^{*\dagger})^{\dagger} = qpq^{*\dagger}q^{*}p^{\dagger}q^{\dagger} = qpp^{\dagger}q^{\dagger} = pp^{\dagger}qq^{\dagger} = pp^{\dagger}$$

thus showing that the norm (which is the Lorentz scalar product for an Hermitian biquaternion) is invariant, as required. Note we have used the fact that $qq^{\dagger} = 1$ implies $q^{*\dagger}q^{*} = 1$ and that $pp^{\dagger}$ is a scalar.

If we write $q = u + iv$, where u and v are (real) quaternions, then the requirement for q to represent a Lorentz transformation, i.e., $qq^{\dagger} = 1$, becomes $uu^{\dagger} - vv^{\dagger} = 1$ and $uv^{\dagger} + vu^{\dagger} = 0$.

q is a rotation if it is a quaternion (i.e., real, hence $v = 0$). We already know that any rotation can be expressed in this way from §29.1. That a spacetime event is now expressed as an Hermitian biquaternion rather than a quaternion does not detract from this as $q(t + i\bar{r})q^{\dagger} = t + iq\bar{r}q^{\dagger}$, i.e., the spatial part is rotated but the temporal part is unchanged.

q is a boost if it is Hermitian, i.e., if u is purely scalar (temporal) and v is purely vector (spatial), i.e., $\mathbb{V}(u) = \bar{u} = 0$ and $\mathbb{S}(v) = v_0 = 0$. It is reasonable to expect that any boost, i.e., of any magnitude in any direction, can be represented in this way because there are three degrees of freedom remaining after the four variables u_0, v_x, v_y, v_z are subject to the constraint $qq^{\dagger} = 1$.

It is readily shown that any Lorentz transformation can be expressed as a rotation followed by a boost by explicit construction. Putting $\mu^2 = uu^{\dagger} = vv^{\dagger} + 1$ it follows that $\mu^2 \geq 1$ because $vv^{\dagger} \geq 0$ as a real quaternion. Hence μ can be taken as real, positive and non-zero. We can define the biquaternions $r = \mu^{-1}u$ and $s = \mu - i\mu^{-1}uv^{\dagger}$. Thus, r is a real quaternion and hence represents a rotation, whilst,

$$s^{*\dagger} = \mu + i\mu^{-1}vu^{\dagger} = \mu - i\mu^{-1}uv^{\dagger} = s$$

so that s is Hermitian and so represents a boost. Finally we have,

$$sr = (\mu - i\mu^{-1}uv^{\dagger})\mu^{-1}u = u - i\mu^{-2}uv^{\dagger}u = u + i\mu^{-2}vu^{\dagger}u = u + iv$$

So the rotation r followed by the boost s is equivalent to the initial combined Lorentz transformation.

I think you'll agree this is all very much neater than throwing 4 x 4 matrices around, or dealing with things of the form $exp\left\{-\frac{i}{2}wL_{\alpha\beta}\varepsilon_{\mu\nu}^{\alpha\beta}\right\}$.

Note that if an Hermitian biquaternion, b, transforms under q, i.e., $b \rightarrow b' = qbq^{*\dagger}$ then the complex conjugate Hermitian biquaternion, b^*, does not transform under q but under q^*, i.e.,

$$b^* \rightarrow b'^* = (qbq^{*\dagger})^* = q^*b^*q^\dagger \tag{29.3.3}$$

(Incidentally, this means that, for every Hermitian biquaternion, it is necessary to state which transformation applies. There are two different types of Hermitian biquaternion. This is analogous to the difference between covariant and contravariant tensor components).

The product of two Hermitian biquaternions is not Hermitian, just as it is not, in general, for complex matrices, because $(pq)^\dagger = q^\dagger p^\dagger = qp$ which will only be the same as pq if they commute, which generally they will not. Nevertheless, the product of two Hermitian biquaternions has a simple transformation as long as they transform oppositely, i.e., one under q and the other under q^*. To put that differently, if Hermitian biquaternions a and b both transform under q then,

$$ab^* \rightarrow a'b^{*\prime} = qaq^{*\dagger}q^*b^*q^\dagger = qab^*q^\dagger \tag{29.3.4}$$

This is the simple transformation rule for products ab^* where a and b are both Hermitian biquaternions transforming under q. Note that it is NOT the same as the transformation of an Hermitian biquaternion of either type.

If we write $a = a_0 + i\bar{a}$ and $b = b_0 + i\bar{b}$ the reader will note that the quaternion product $\bar{a}\bar{b} = -\bar{a} \cdot \bar{b} + \bar{a} \times \bar{b}$, where the terms on the RHS refer to the usual 3-vector notation. Hence we find,

$$ab^* = \left(a_0b_0 - \bar{a} \cdot \bar{b}\right) + \bar{a} \times \bar{b} + i\left(b_0\bar{a} - a_0\bar{b}\right) \tag{29.3.5}$$

The transformation $ab^* \rightarrow qab^*q^\dagger$ therefore tells us that the usual Lorentz scalar product, $a^\mu b_\mu = a_0b_0 - \bar{a} \cdot \bar{b}$, is in variant under any Lorentz transform, as it should be. The vector part of ab^*, however, is another matter as this changes under both rotations and boosts.

It follows from the complex conjugate of (29.2.2) that, if Hermitian biquaternions a and b both transform under q then,

$$a^*b \rightarrow a'^*b' = q^*a^*bq^{*\dagger} \tag{29.3.6}$$

29.4 Lorentz Covariant Biquaternionic Maxwell Equations

We shall consider here only the free space equations, though with source terms (charge density ρ, and current flux density $\bar{J}$). To avoid clutter in this section I shall use units in which $c = \varepsilon_0 = \mu_0 = 1$. (Do not be disconcerted by factors of 4π which you may find in other texts – this too is a matter of units).

Firstly, a recap. Chapter 27 has shown how the Maxwell equations, in standard vector notation, are invariant in form (i.e., covariant) between inertial observers. We saw that,

- the electric and magnetic potentials come together to form a 4-vector, $\{A^\mu\} = \{\Phi, \bar{A}\}$,
- the wave operator is a Lorentz invariant, $\partial^{\mu'}\partial_{\mu'} = \partial^\mu \partial_\mu$,
- the wave equation-with-sources expressed in terms of this 4-vector potential is Lorentz covariant, $\partial^2 A^\alpha = \mu_0 J^\alpha \Leftrightarrow \partial'^2 A^{\alpha'} = \mu_0 J^{\alpha'}$,
- the electric and magnetic field vectors come together to form an antisymmetric rank two tensor in 4D spacetime, $F^{\alpha\beta} = \partial^\alpha A^\beta - \partial^\beta A^\alpha$, which has a dual tensor $\mathcal{F}^{\alpha\beta} = \frac{1}{2}\varepsilon^{\alpha\beta\gamma\mu}F_{\gamma\mu}$,
- Maxwell's equations can be written in manifestly covariant form as $\partial_\alpha F^{\alpha\beta} = \mu_0 J^\beta$ and $\partial_\alpha \mathcal{F}^{\alpha\beta} = 0$,
- and the components of the electric and magnetic fields transform under a boost in the x-direction as given by (27.6.9a,b).

To express the Maxwell equations in biquaternion form, first define the biquaternion derivative operator by,

$$D = \partial_t - i\bar{\nabla} \tag{29.4.1}$$

Note that it is Hermitian.

The 4-vector potential $\{A_\mu\} = (\Phi, \bar{A})$ is written in biquaternion form as,

$$A = \Phi + i\bar{A} \tag{29.4.2}$$

Note that it too is Hermitian. In the above definitions it is, of course, to be understood that what we have previously written as vectors are now understood to be pure (vectorial) quaternions, e.g.,

$$\bar{\nabla} \equiv I\partial_x + J\partial_y + K\partial_z \quad \text{and} \quad \bar{A} \equiv IA_x + JA_y + KA_z \tag{29.4.3}$$

Because D and A are both Hermitian, and both transform under q, (29.3.6) gives the following Lorentz transformation,

$$D^*A \to D'^*A' = q^*D^*Aq^{*\dagger} \tag{29.4.4}$$

But,

$$D^*A = (\partial_t\phi + \bar{\nabla}\cdot\bar{A}) - \bar{\nabla}\times\bar{A} + i(\partial_t\bar{A} + \bar{\nabla}\phi) \tag{29.4.5}$$

We recognise $\bar{B} = \bar{\nabla}\times\bar{A}$ and $\bar{E} = -(\partial_t\bar{A} + \bar{\nabla}\phi)$. Moreover, we know from standard theory that we are free to set a gauge for the potentials, and the Lorentz invariant gauge is $\partial_t\phi + \bar{\nabla}\cdot A = 0$. Hence, if we define the biquaternion electromagnetic field by,

$$F = \bar{B} + i\bar{E} \tag{29.4.6}$$

then we get,

$$D^*A = -F \tag{29.4.7}$$

This is merely the biquaternion form of the fields in terms of derivatives of the potentials. However, (29.4.7) tells us that the biquaternion field F must also transform in the same was as D^*A, and so,

$$F \to F' = q^*Fq^{*\dagger} \tag{29.4.8}$$

This differs from the transformation law for either type of Hermitian biquaternion, because (29.4.6) shows that F is not an Hermitian biquaternion.

This is a great deal simpler than either vector or tensor formulations of the Lorentz transformations. In terms of the sources, which we can define as the Hermitian biquaternion,

$$J = \rho + i\bar{J} \tag{29.4.9}$$

all four of the Maxwell equations in vector form (or both of Maxwell's equations in tensor form) can be written in the highly compact form,

$$DF + J = 0 \tag{29.4.10}$$

I leave it as a simple exercise for the reader to check (29.4.10). It is a matter of expanding $DF + J = (\partial_t - i\bar{\nabla})(\bar{B} + i\bar{E}) + \rho + i\bar{J}$. Biquaternion equations imply four separate equations, according to the four combinations of real or imaginary parts, and scalar or vector parts. Each of these four equations corresponds to one of the Maxwell equations.

Whether the notational compactness provided by biquaternions has more substantive implications appears to have been answered in the negative by posterity, since it never caught on and was overtaken in mathematical circles by the coordinate-free language of differentiable manifolds, forms and the external derivative, etc., to which we turn next in the final chapter.

29.5 References

[29.1] Bradford, R.A.W. (2023). *Five Square Roots: The Remarkable Fecundity of Impossible Square Roots*. Principia Publications Unlimited, March 2023.

[29.2] Maxwell, James Clerk (1873) *A Treatise on Electricity and Magnetism*. Nature. 7 (182): 478–480. 24 April 1873. Maxwell was revising the Treatise for a second edition when he died in 1879. The revision was completed by William Davidson Niven for publication in 1881. A third edition was prepared by J. J. Thomson for publication in 1892.

[29.3] Hehl, F.W. (2010). *On the changing form of Maxwell's equations during the last 150 years*. UCL Department of Mathematics, Applied Mathematics Seminar, 22 February 2010.

[29.4] Lambek, J. (2013). *In Praise of Quaternions*. CR Math. Rep. Acad. Sci. Canada, 2013, math.mcgil29.ca.

30. The Exterior Derivative and the Language of Forms

This final chapter provides a brief introduction to the concept and notation of differential forms and the exterior algebra. This is a perspective which reaches its full utility when the spacetime manifold is curved, and hence in general relativity, but that should not mislead the reader into imagining that it has little to say about electromagnetism more generally, and specifically in flat Minkowski space. On the contrary, electromagnetism emerges from exterior algebra as being essentially the theory of a closed 2-form. Famously, the Cartan-Stokes theorem, which is naturally expressed in the language of forms, namely $\int_{\Omega} \boldsymbol{d\omega} = \int_{\delta\Omega} \boldsymbol{\omega}$, *is the overarching theorem of which Green's theorem, Stokes's theorem and Gauss's theorem are just specific examples. The ubiquity of Stokes's theorem and Gauss's theorem throughout electromagnetic theory illustrates the conceptual centrality of this theorem, based crucially upon differential forms. Readers will forgive the lack of mathematical rigour in this chapter, the purpose of which is only to motivate a deeper study for which specialist texts would be the next step.*

Contents

30.1 Visualising a 1-Form

This intuitive introduction assumes a 3D space.

One visualises vectors as little arrows. But there is a complementary means of visualisation. Consider wave motion, typically indicated by a term such as $e^{-i(\omega t-\bar{k}\cdot\bar{r})}$. The wavevector $\bar{k}$ indicates the direction of motion. But we may also talk of the wavefronts, i.e., surfaces of constant phase at a given time. For a plane wave such as given by $e^{-i(\omega t-\bar{k}\cdot\bar{r})}$ these wavefronts are parallel planes normal to $\bar{k}$ whose equations are $\bar{k}\cdot\bar{r}=C$ for some constant C. The parallel planes given by $C, C\pm 2\pi, C\pm 4\pi, \ldots$ all have the same phase. They are spaced apart by a distance $k\Delta r=2\pi$, i.e., by $\Delta r=\frac{2\pi}{k}=\lambda$, the wavelength.

So we have two complementary ways of visualising the wavevector. The first is as a vector which is defined by its direction and length. The second is to imagine a stack of parallel planes whose distance apart also specifies the magnitude of the vector, albeit by the inverse relationship $\frac{2\pi}{k}$. This is the 1-form description (in 3D).

We are accustomed to the scalar product of two vectors, $\bar{a}\cdot\bar{b}$. This is actually a set of linear maps from vectors to the reals, parametrised by $\bar{a}$. The idea of the 1-form provides a new means of understanding this. If we understand $\bar{a}$ to be a 1-form, and $\bar{b}$ to be a vector, then the scalar product, which we may write as $\langle\bar{a},\bar{b}\rangle$, can be visualised as the number of planes of $\bar{a}$ cut through by $\bar{b}$. Note that this increases in proportion to both the magnitude of the vector and the magnitude of the 1-form, in the latter case because doubling the magnitude of a 1-form means halving the spacing of the planes.

An immediate warning: in 4D spacetime the "planes" become 3-surfaces because the plane of constant phase sweeps forward in time.

All this is rather vague. Let's get algebraic.

30.2 Algebraic Description of Vectors and 1-Forms

A vector can be written in terms of a set of basis vectors, $\hat{e}_\mu$, as,

$$\bar{a}=a^\mu\hat{e}_\mu \tag{30.2.1}$$

Similarly, a 1-form, which we will distinguish by using bold font, can be written in terms of basis 1-forms, $\boldsymbol{\omega}^\mu$,

$$\boldsymbol{b}=b_\mu\boldsymbol{\omega}^\mu \tag{30.2.2}$$

A 1-form is a linear mapping from vectors to the reals. For a given vector basis, $\hat{e}_\mu$, we can choose an orthonormal 1-form basis, $\boldsymbol{\omega}^\mu$, such that,

$$\langle \boldsymbol{\omega}^\mu, \hat{e}_\sigma \rangle = \delta^\mu_\sigma \tag{30.2.3}$$

Intuitively this means if the $\hat{e}_\mu$ are of unit length then the planes of $\boldsymbol{\omega}^\mu$ have unit spacing, and are orthogonal to the corresponding $\hat{e}_\mu$.

(Note that a 1-form is defined here as a linear mapping to the reals only because we are considering real vectors and real 1-forms. If we considered complex vectors and 1-forms then the linear map would be to the complex numbers).

The reader will note that in tensorial language, the components of a vector are contravariant components whilst the components of a 1-form are covariant components.

30.3 *p*-Forms and the Wedge Product

p-forms are equivalent to completely antisymmetric tensors. However they are elegantly constructed within their own right, without recourse to tensor concepts. One means of doing so is the wedge product. For two 1-forms, $\boldsymbol{a}$ and $\boldsymbol{b}$, the wedge product creates a 2-form,

$$\boldsymbol{a} \wedge \boldsymbol{b} = \boldsymbol{a} \otimes \boldsymbol{b} - \boldsymbol{b} \otimes \boldsymbol{a} \tag{30.3.1}$$

Note that (30.3.1) is specific to 1-forms.

Here $\boldsymbol{a} \otimes \boldsymbol{b}$ is a tensor product. To the degree of rigour we are working in this chapter, the tensor product may be considered to be an ordered pair of symbols together with a linear structure such that, for complex numbers α, β we have, (30.3.2a)

$$\boldsymbol{a} \otimes (\alpha \boldsymbol{b} + \beta \boldsymbol{c}) = \alpha \boldsymbol{a} \otimes \boldsymbol{b} + \beta \boldsymbol{a} \otimes \boldsymbol{c} \qquad (\alpha \boldsymbol{b} + \beta \boldsymbol{c}) \otimes \boldsymbol{a} = \alpha \boldsymbol{b} \otimes \boldsymbol{a} + \beta \boldsymbol{c} \otimes \boldsymbol{a}$$

These imply, for example, that $\alpha \boldsymbol{a} \otimes \beta \boldsymbol{b} = \alpha\beta \boldsymbol{a} \otimes \boldsymbol{b}$.

From (30.3.2a) the linear and distributive properties of the wedge product itself follow, (30.3.2b)

$$\boldsymbol{a} \wedge (\alpha \boldsymbol{b} + \beta \boldsymbol{c}) = \alpha \boldsymbol{a} \wedge \boldsymbol{b} + \beta \boldsymbol{a} \wedge \boldsymbol{c} \qquad (\alpha \boldsymbol{b} + \beta \boldsymbol{c}) \wedge \boldsymbol{a} = \alpha \boldsymbol{b} \wedge \boldsymbol{a} + \beta \boldsymbol{c} \wedge \boldsymbol{a}$$

The important feature of (30.3.1) is that the wedge product is antisymmetric,

$$\boldsymbol{a} \wedge \boldsymbol{b} = -\boldsymbol{b} \wedge \boldsymbol{a} \tag{30.3.3}$$

A 3-form can be formed using the wedge product from three 1-forms by constructing the expression which is completely antisymmetric (i.e., antisymmetric wrt the interchange of any pair of 1-forms). Thus,

$$\boldsymbol{a}\wedge\boldsymbol{b}\wedge\boldsymbol{c} \equiv (\boldsymbol{a}\wedge\boldsymbol{b})\wedge\boldsymbol{c} \equiv \boldsymbol{a}\wedge(\boldsymbol{b}\wedge\boldsymbol{c}) \tag{30.3.4}$$
$$\equiv \boldsymbol{a}\otimes\boldsymbol{b}\otimes\boldsymbol{c} + \boldsymbol{b}\otimes\boldsymbol{c}\otimes\boldsymbol{a} + \boldsymbol{c}\otimes\boldsymbol{a}\otimes\boldsymbol{b} - \boldsymbol{c}\otimes\boldsymbol{b}\otimes\boldsymbol{a} - \boldsymbol{a}\otimes\boldsymbol{c}\otimes\boldsymbol{b} - \boldsymbol{b}\otimes\boldsymbol{a}\otimes\boldsymbol{c}$$

From which it follows that $\boldsymbol{a}\wedge\boldsymbol{b}\wedge\boldsymbol{b} = \boldsymbol{a}\wedge\boldsymbol{a}\wedge\boldsymbol{b} = \boldsymbol{a}\wedge\boldsymbol{b}\wedge\boldsymbol{a} = 0$.

Similarly,

$$\boldsymbol{a}\wedge\boldsymbol{b}\wedge\boldsymbol{c}\wedge\boldsymbol{d} \equiv (\boldsymbol{a}\wedge\boldsymbol{b}\wedge\boldsymbol{c})\wedge\boldsymbol{d} \equiv \boldsymbol{a}\wedge(\boldsymbol{b}\wedge\boldsymbol{c}\wedge\boldsymbol{d}) \tag{30.3.5}$$
$$\equiv \boldsymbol{a}\otimes\boldsymbol{b}\otimes\boldsymbol{c}\otimes\boldsymbol{d} + \boldsymbol{a}\otimes\boldsymbol{c}\otimes\boldsymbol{d}\otimes\boldsymbol{b} + \boldsymbol{a}\otimes\boldsymbol{d}\otimes\boldsymbol{b}\otimes\boldsymbol{c} + \boldsymbol{b}\otimes\boldsymbol{a}\otimes\boldsymbol{d}\otimes\boldsymbol{c} +$$
$$\boldsymbol{b}\otimes\boldsymbol{c}\otimes\boldsymbol{a}\otimes\boldsymbol{d} + \boldsymbol{b}\otimes\boldsymbol{d}\otimes\boldsymbol{c}\otimes\boldsymbol{a} + \boldsymbol{c}\otimes\boldsymbol{d}\otimes\boldsymbol{a}\otimes\boldsymbol{b} + \boldsymbol{c}\otimes\boldsymbol{a}\otimes\boldsymbol{b}\otimes\boldsymbol{d} +$$
$$\boldsymbol{c}\otimes\boldsymbol{b}\otimes\boldsymbol{d}\otimes\boldsymbol{a} + \boldsymbol{d}\otimes\boldsymbol{a}\otimes\boldsymbol{c}\otimes\boldsymbol{b} + \boldsymbol{d}\otimes\boldsymbol{b}\otimes\boldsymbol{a}\otimes\boldsymbol{c} + \boldsymbol{d}\otimes\boldsymbol{c}\otimes\boldsymbol{b}\otimes\boldsymbol{a} -$$
$$\boldsymbol{a}\otimes\boldsymbol{b}\otimes\boldsymbol{d}\otimes\boldsymbol{c} - \boldsymbol{a}\otimes\boldsymbol{c}\otimes\boldsymbol{b}\otimes\boldsymbol{d} - \boldsymbol{a}\otimes\boldsymbol{d}\otimes\boldsymbol{c}\otimes\boldsymbol{b} - \boldsymbol{b}\otimes\boldsymbol{a}\otimes\boldsymbol{c}\otimes\boldsymbol{d} -$$
$$\boldsymbol{b}\otimes\boldsymbol{c}\otimes\boldsymbol{d}\otimes\boldsymbol{a} - \boldsymbol{b}\otimes\boldsymbol{d}\otimes\boldsymbol{a}\otimes\boldsymbol{c} - \boldsymbol{c}\otimes\boldsymbol{d}\otimes\boldsymbol{b}\otimes\boldsymbol{a} - \boldsymbol{c}\otimes\boldsymbol{a}\otimes\boldsymbol{d}\otimes\boldsymbol{b} -$$
$$\boldsymbol{c}\otimes\boldsymbol{b}\otimes\boldsymbol{a}\otimes\boldsymbol{d} - \boldsymbol{d}\otimes\boldsymbol{a}\otimes\boldsymbol{b}\otimes\boldsymbol{c} - \boldsymbol{d}\otimes\boldsymbol{b}\otimes\boldsymbol{c}\otimes\boldsymbol{a} - \boldsymbol{d}\otimes\boldsymbol{c}\otimes\boldsymbol{a}\otimes\boldsymbol{b} \tag{30.3.6}$$

From this it follows that if any pair of $\boldsymbol{a}, \boldsymbol{b}, \boldsymbol{c}, \boldsymbol{d}$ are equal then $\boldsymbol{a}\wedge\boldsymbol{b}\wedge\boldsymbol{c}\wedge\boldsymbol{d}$ is zero.

If two 1-forms are colinear, i.e., $\boldsymbol{b} = \alpha\boldsymbol{a},$ then $\boldsymbol{a}\wedge\boldsymbol{b} = 0$ and vice-versa. Hence a zero wedge product between two 1-forms indicates collinearity.

If three 1-forms are coplanar, i.e., $\boldsymbol{c} = \alpha\boldsymbol{a} + \beta\boldsymbol{b}$, then $\boldsymbol{a}\wedge\boldsymbol{b}\wedge\boldsymbol{c} = 0$ and vice-versa. Note that this follows from,

$$\boldsymbol{a}\wedge\boldsymbol{b}\wedge\boldsymbol{c} = \boldsymbol{a}\wedge\boldsymbol{b}\wedge(\alpha\boldsymbol{a} + \beta\boldsymbol{b}) = \alpha\boldsymbol{a}\wedge\boldsymbol{b}\wedge\boldsymbol{a} + \beta\boldsymbol{a}\wedge\boldsymbol{b}\wedge\boldsymbol{b} = \boldsymbol{0} \tag{30.3.7}$$

So $\boldsymbol{a}\wedge\boldsymbol{b}\wedge\boldsymbol{c} = 0$ indicates coplanarity.

Similarly, $\boldsymbol{a}\wedge\boldsymbol{b}\wedge\boldsymbol{c}\wedge\boldsymbol{d} = 0$ iff the four 1-forms lie within the same 3D volume, i.e., we can write one of them as a linear combination of the other three.

From the above it follows that if we are working in a 2D space then all possible $\boldsymbol{a}\wedge\boldsymbol{b}\wedge\boldsymbol{c}$ are zero. If we are working in a 3D space then all possible $\boldsymbol{a}\wedge\boldsymbol{b}\wedge\boldsymbol{c}\wedge\boldsymbol{d}$ are zero.

If we are working in an N-dimensional space then an arbitrary p-form, where $p \le N$, can be written in terms of the basis 1-forms, $\boldsymbol{\omega}^\mu$, as, (30.3.8)

$$\boldsymbol{a} = \frac{1}{p!}a_{\mu_1\mu_2\dots\mu_p}\boldsymbol{\omega}^{\mu_1}\wedge\boldsymbol{\omega}^{\mu_2}\wedge\dots\wedge\boldsymbol{\omega}^{\mu_p} = a_{[\mu_1\mu_2\dots\mu_p]}\boldsymbol{\omega}^{\mu_1}\wedge\boldsymbol{\omega}^{\mu_2}\wedge\dots\wedge\boldsymbol{\omega}^{\mu_p}$$

(summation understood) and where $[\mu_1\mu_2 \dots \mu_p]$ restricts the sum to the order $\mu_1 < \mu_2 < \dots < \mu_p$. The equivalence of the two expressions in (30.3.8) is due to the wedge product of a given set of basis 1-forms being the same irrespective of order up to sign, and, moreover, that terms of the opposite sign are returned to the positive because the components $a_{\mu_1\mu_2\dots\mu_p}$ are always defined as being completely antisymmetric under interchange of any pair of indices, i.e., $a_{\mu_1\mu_2\dots\mu_p}$ is a completely antisymmetric tensor.

It follows from the above that, for any p-form $\boldsymbol{a}$ and any q-form $\boldsymbol{b}$ that,

$$\boldsymbol{a}\wedge\boldsymbol{b} = (-1)^{pq}\boldsymbol{b}\wedge\boldsymbol{a} \tag{30.3.9}$$

In particular if $\boldsymbol{a}$ is a 0-form (i.e., a scalar field, say f) then $f\wedge\boldsymbol{b} = \boldsymbol{b}\wedge f$. But what are we to understand by this wedge product involving a scalar? It is simplest to regard an additional axiom coming into play which sets,

$$f\wedge\boldsymbol{b} \equiv \boldsymbol{b}\wedge f \equiv f\boldsymbol{b} \tag{30.3.10}$$

30.4 The Coordinate Basis

If our space, or spacetime, has some coordinates x^μ then a set of basis 1-forms, called the coordinate basis, is written,

$$\boldsymbol{\omega}^\mu = \boldsymbol{d}x^\mu \tag{30.4.1}$$

The differential operator, $\boldsymbol{d}$, has been written in bold font because it is not the usual differential operator but is the exterior derivative. The properties of this object are specified in the next section. It creates a 1-form out of the coordinates.

(30.4.1) appears very alien when one has been brought up with the conventional perspective that has been used in this book up to now. It appears to align the basis 1-forms with infinitesimal quantities. The utility of this is that p-forms are constructed to be the natural entities over which p-dimensional integrals will be carried out.

In 3D space the general 1-form can thus be written,

$$\boldsymbol{a} = a_x\boldsymbol{d}x + a_y\boldsymbol{d}y + a_z\boldsymbol{d}z \tag{30.4.2}$$

The general 2-form in 3D is,

$$\boldsymbol{a} = a_{xy}\boldsymbol{d}x\wedge\boldsymbol{d}y + a_{yz}\boldsymbol{d}y\wedge\boldsymbol{d}z + a_{zx}\boldsymbol{d}z\wedge\boldsymbol{d}x \tag{30.4.3}$$

In 4D, say Minkowski spacetime, the general 2-form is,

$$\boldsymbol{a} = a_{xy}\boldsymbol{dx}\wedge\boldsymbol{dy} + a_{yz}\boldsymbol{dy}\wedge\boldsymbol{dz} + a_{zx}\boldsymbol{dz}\wedge\boldsymbol{dx}+a_{tx}\boldsymbol{dt}\wedge\boldsymbol{dx} + a_{ty}\boldsymbol{dt}\wedge\boldsymbol{dy} + a_{tz}\boldsymbol{dt}\wedge\boldsymbol{dz} \tag{30.4.4}$$

This has six independent components, corresponding to the number of independent components of an antisymmetric 4x4 matrix. As we will see, the electromagnetic field is this type of object, a 2-form in 4D spacetime.

In 3D the most general 3-form is simply a multiple of $\boldsymbol{dx}\wedge\boldsymbol{dy}\wedge\boldsymbol{dz}$. The physical interpretation of $\boldsymbol{dx}\wedge\boldsymbol{dy}\wedge\boldsymbol{dz}$ is an element of volume, but with the twist that, in the language of forms, volume is no longer a scalar but a 3-form. Hence it now has a sign in that $\boldsymbol{dy}\wedge\boldsymbol{dx}\wedge\boldsymbol{dz}$ reverses the sign of the volume.

In 4D spacetime the most general 3-form can be written,

$$\boldsymbol{a} = a_{xyz}\boldsymbol{dx}\wedge\boldsymbol{dy}\wedge\boldsymbol{dz} + a_{xyt}\boldsymbol{dx}\wedge\boldsymbol{dy}\wedge\boldsymbol{dt} + a_{tyz}\boldsymbol{dt}\wedge\boldsymbol{dy}\wedge\boldsymbol{dz} + a_{xtz}\boldsymbol{dx}\wedge\boldsymbol{dt}\wedge\boldsymbol{dz} \tag{30.4.4}$$

This is a hypersurface, only one component of which is a spatial 3-volume, the other three terms are spatial areas extended in the time direction. But in 4D spacetime there is only one independent 4-form, namely $\boldsymbol{dx}\wedge\boldsymbol{dy}\wedge\boldsymbol{dz}\wedge\boldsymbol{dt}$, the oriented 4D hypervolume.

A "simple p-form" is a p-form that can be written as the wedge product of p 1-forms. In general this will require a change of basis to bring about. For example, $\alpha\boldsymbol{dx}\wedge\boldsymbol{dz} + \beta\boldsymbol{dy}\wedge\boldsymbol{dz}$ is a simple 2-form as we only need to change to a basis including $x' = \alpha x + \beta y$ to find that $\alpha\boldsymbol{dx}\wedge\boldsymbol{dz} + \beta\boldsymbol{dy}\wedge\boldsymbol{dz} = \boldsymbol{dx}'\wedge\boldsymbol{dz}$.

In 4D Minkowski spacetime, all 0-forms, 1-forms, 3-forms and 4-forms are simple. But not all 2-forms are simple, though they can always be reduced to the form $\alpha\boldsymbol{d\xi}_1 \wedge \boldsymbol{d\xi}_2 + \beta\boldsymbol{d\xi}_3 \wedge \boldsymbol{d\xi}_4$ for some set of four 1-forms.

30.5 The Exterior Derivative

The exterior derivative creates a 1-form from the coordinates, e.g., $\boldsymbol{dx}$. This is a special case of the exterior derivative's creation of a 1-form from any scalar field, i.e., a function of the coordinates, thus,

$$\boldsymbol{df} = \frac{\partial f}{\partial x^\mu}\boldsymbol{dx}^\mu \tag{30.5.1}$$

But $\boldsymbol{\omega}^\mu = \boldsymbol{dx}^\mu$ form a basis for the 1-forms, analogous to unit vectors in vector analysis, and so we see that $\boldsymbol{df}$, whose components are $\frac{\partial f}{\partial x^\mu}$, is analogous to the $grad(f) = \overline{\nabla}f$, albeit in an arbitrary number of dimensions.

(30.5.1) defines the action of the exterior derivative on a 0-form, i.e., a scalar field. But more generally the exterior derivative may act on any p-form. It's properties, which for our purposes can be taken as axiomatic, are,

(i) If $\boldsymbol{a}$ is a p-form then $\boldsymbol{da}$ is a $(p+1)$-form;

(ii) $\boldsymbol{dd} = \boldsymbol{d}^2 = 0$;

(iii) For any two p-forms $\boldsymbol{d}$ acts linearly, $\boldsymbol{d}(\boldsymbol{a}+\boldsymbol{b}) = \boldsymbol{da}+\boldsymbol{db}$;

(iv) If $\boldsymbol{a}$ is a p-form and $\boldsymbol{b}$ is a q-form then,

$$\boldsymbol{d}(\boldsymbol{a}\wedge\boldsymbol{b}) = \boldsymbol{da}\wedge\boldsymbol{b} + (-1)^p\boldsymbol{a}\wedge\boldsymbol{db}$$

Together with (30.5.1), (iv) defines the operation of the exterior derivative on any p-form by induction.

For any p-form $\boldsymbol{a}$ and any scalar field f (i.e., a 0-form) then (iv) gives, using (30.3.10),

$$\boldsymbol{d}(f\wedge\boldsymbol{a}) = \boldsymbol{d}(f\boldsymbol{a}) = \boldsymbol{d}f\wedge\boldsymbol{a} + f\wedge\boldsymbol{da} = \boldsymbol{d}f\wedge\boldsymbol{a} + f\boldsymbol{da} \tag{30.5.2}$$

We shall need to appeal to this repeatedly.

30.5.1 3D Space

In 3D space, any 1-form can be written $\boldsymbol{a} = a_x\boldsymbol{dx} + a_y\boldsymbol{dy} + a_z\boldsymbol{dz}$. Hence we see, using (iii) then (30.5.2) then (ii), that,

$$\boldsymbol{da} = \boldsymbol{d}(a_x\boldsymbol{dx}) + \boldsymbol{d}(a_y\boldsymbol{dy}) + \boldsymbol{d}(a_z\boldsymbol{dz})$$

$$= (\boldsymbol{d}a_x)\wedge\boldsymbol{dx} + a_x\boldsymbol{d}^2x + (\boldsymbol{d}a_y)\wedge\boldsymbol{dy} + a_y\boldsymbol{d}^2y + (\boldsymbol{d}a_z)\wedge\boldsymbol{dz} + a_z\boldsymbol{d}^2z$$

$$= (\boldsymbol{d}a_x)\wedge\boldsymbol{dx} + (\boldsymbol{d}a_y)\wedge\boldsymbol{dy} + +(\boldsymbol{d}a_z)\wedge\boldsymbol{dz} \tag{30.5.3}$$

This simplifies further using $\boldsymbol{d}a_x = a_{x,i}\boldsymbol{dx}^i$, etc., thus,

$$\boldsymbol{da} = a_{x,i}\boldsymbol{dx}^i\wedge\boldsymbol{dx} + a_{y,i}\boldsymbol{dx}^i\wedge\boldsymbol{dy} + a_{z,i}\boldsymbol{dx}^i\wedge\boldsymbol{dz} \tag{30.5.4}$$

But, in 3D,

$$a_{x,i}\boldsymbol{dx}^i\wedge\boldsymbol{dx} = (a_{x,x}\boldsymbol{dx} + a_{x,y}\boldsymbol{dy} + a_{x,z}\boldsymbol{dz})\wedge\boldsymbol{dx}$$
$$= a_{x,y}\boldsymbol{dy}\wedge\boldsymbol{dx} + a_{x,z}\boldsymbol{dz}\wedge\boldsymbol{dx} \tag{30.5.5}$$

where we have used $\boldsymbol{dx}\wedge\boldsymbol{dx} = 0$ due to (30.3.1), and similarly for the other two terms. Hence we get, (30.5.6)

$$\boldsymbol{da} = a_{x,y}\boldsymbol{dy}\wedge\boldsymbol{dx} + a_{x,z}\boldsymbol{dz}\wedge\boldsymbol{dx} + a_{y,x}\boldsymbol{dx}\wedge\boldsymbol{dy} + a_{y,z}\boldsymbol{dz}\wedge\boldsymbol{dy} +$$
$$a_{z,x}\boldsymbol{dx}\wedge\boldsymbol{dz} + a_{z,y}\boldsymbol{dy}\wedge\boldsymbol{dz}$$
$$= (a_{z,y} - a_{y,z})\boldsymbol{dy}\wedge\boldsymbol{dz} + (a_{x,z} - a_{z,x})\boldsymbol{dz}\wedge\boldsymbol{dx} + (a_{y,x} - a_{x,y})\boldsymbol{dx}\wedge\boldsymbol{dy}$$

We recognise the components of the 2-form $\boldsymbol{da}$ to be the components of $curl(\bar{a}) = \bar{\nabla} \times \bar{a}$ if the 1-form $\boldsymbol{a}$ is understood instead to be a vector with the same (covariant) components. Consistent with all 2-forms, this can also be interpreted as an antisymmetric matrix (rank two tensor), $A_{ij} = a_{i,j} - a_{j,i}$. Unlike the vector expression $\bar{\nabla} \times \bar{a}$, the basis of $\boldsymbol{da}$ are the basis 2-forms rather than unit vectors, the correspondence being,

$$\hat{x} \rightarrow \boldsymbol{dy} \wedge \boldsymbol{dz} \quad \hat{y} \rightarrow \boldsymbol{dz} \wedge \boldsymbol{dx} \quad \hat{z} \rightarrow \boldsymbol{dx} \wedge \boldsymbol{dy} \tag{30.5.7}$$

and we recognise these as being elements of area whose normals are in the corresponding coordinate directions. Consequently $\boldsymbol{da}$ is not just the expression for $curl(\bar{a})$ in the language of forms but also includes the element of area. The correspondence is therefore best considered as being,

$$\boldsymbol{da} \leftrightarrow (\bar{\nabla} \times \bar{a}) \cdot d\bar{S} \tag{30.5.8}$$

As stated previously, forms are expressly constructed for the purposes of integration. (30.5.8) rather begs to be integrated over some finite surface S,

$$\int_S \boldsymbol{da} \leftrightarrow \int_S (\bar{\nabla} \times \bar{a}) \cdot d\bar{S} \tag{30.5.9}$$

But what about the 1-form we started with, $\boldsymbol{a}$? We see that it corresponds to the scalar (dot) product of a vector with the same components and a line element,

$$\boldsymbol{a} = a_i \boldsymbol{dx}^i = a_x \boldsymbol{dx} + a_y \boldsymbol{dy} + a_z \boldsymbol{dz} \leftrightarrow \bar{a} \cdot d\bar{s} \tag{30.5.10}$$

This also begs to be integrated over some curve, Γ,

$$\int_\Gamma \boldsymbol{a} \leftrightarrow \int_\Gamma \bar{a} \cdot d\bar{s} \tag{30.5.11}$$

Combining (30.5.9,11) we see that Stoke's theorem can be written in the language of forms as,

$$\int_S \boldsymbol{da} = \int_{\delta S} \boldsymbol{a} \tag{30.5.12}$$

where δS is the boundary curve of the surface S. The discomfort that this notation can cause to the neophyte is because the infinitesimal indicator, d, is missing from the RHS. But we see that this is actually just because it has been incorporated into the 1-form itself, (30.5.10). This is crucial to the construction of the forms because the d in question is now $\boldsymbol{d}$, the exterior derivative, and the algebraic properties of this operator, and the forms generally, are essential in deriving such things as (30.5.6) which links the 2-form $\boldsymbol{da}$ with both $curl(\bar{a})$ and the element of oriented area, $d\bar{S}$, in its 2-form manifestation, (30.5.8).

We note that Green's theorem follows also from (30.5.12) if we merely choose to consider a 2D space. But, as we shall see, (30.5.12) also holds,

- In any number of dimensions, and,
- For any p-form $\boldsymbol{a}$
- Provided the dimension of δS is p (dimension of S is $p+1$).

In this highly generalised form, (30.5.12) is known as the Cartan-Stokes theorem, though some sources just write "Stokes's theorem". It is less confusing to reserve the term "Stoke's theorem" for its original restricted form as relating surface and line integrals in 3D space.

Continuing to restrict attention to 3D, what if we now consider $\boldsymbol{a}$ in (30.5.12) to be a 2-form, so that $\boldsymbol{da}$ is a 3-form? Let's consider what the two sides of (30.5.12) become in that case. The general 2-form in 3D is,

$$\boldsymbol{a} = a_x \boldsymbol{dy} \wedge \boldsymbol{dz} + a_y \boldsymbol{dz} \wedge \boldsymbol{dx} + a_z \boldsymbol{dx} \wedge \boldsymbol{dy} \tag{30.5.13}$$

So that,

$$\boldsymbol{da} = \boldsymbol{d}\left(a_x \boldsymbol{dy} \wedge \boldsymbol{dz} + a_y \boldsymbol{dz} \wedge \boldsymbol{dx} + a_z \boldsymbol{dx} \wedge \boldsymbol{dy}\right) \tag{30.5.14}$$

But we note, for example, that,

$$\begin{aligned} \boldsymbol{d}(a_z \boldsymbol{dx} \wedge \boldsymbol{dy}) &= \boldsymbol{d}a_z \wedge \boldsymbol{dx} \wedge \boldsymbol{dy} \\ &= \left(a_{z,x}\boldsymbol{dx} + a_{z,y}\boldsymbol{dy} + a_{z,z}\boldsymbol{dz}\right) \wedge \boldsymbol{dx} \wedge \boldsymbol{dy} = a_{z,z}\boldsymbol{dz} \wedge \boldsymbol{dx} \wedge \boldsymbol{dy} \end{aligned} \tag{30.5.15}$$

where we have used $\boldsymbol{dx} \wedge \boldsymbol{dx} \wedge \boldsymbol{dy} = \boldsymbol{dy} \wedge \boldsymbol{dx} \wedge \boldsymbol{dy} = 0$ due to antisymmetry (see (30.3.4)). So (30.5.14) becomes,

$$\begin{aligned} \boldsymbol{da} &= a_{x,x}\boldsymbol{dx} \wedge \boldsymbol{dy} \wedge \boldsymbol{dz} + a_{y,y}\boldsymbol{dy} \wedge \boldsymbol{dz} \wedge \boldsymbol{dx} + a_{z,z}\boldsymbol{dz} \wedge \boldsymbol{dx} \wedge \boldsymbol{dy} \\ &= \left(a_{x,x} + a_{y,y} + a_{z,z}\right)\boldsymbol{dx} \wedge \boldsymbol{dy} \wedge \boldsymbol{dz} \end{aligned} \tag{30.5.16}$$

because,

$$\boldsymbol{dy} \wedge \boldsymbol{dz} \wedge \boldsymbol{dx} = -\boldsymbol{dy} \wedge \boldsymbol{dx} \wedge \boldsymbol{dz} = \boldsymbol{dx} \wedge \boldsymbol{dy} \wedge \boldsymbol{dz} \tag{30.5.17a}$$

$$\boldsymbol{dz} \wedge \boldsymbol{dx} \wedge \boldsymbol{dy} = -\boldsymbol{dx} \wedge \boldsymbol{dz} \wedge \boldsymbol{dy} = \boldsymbol{dx} \wedge \boldsymbol{dy} \wedge \boldsymbol{dz} \tag{30.5.17b}$$

We could have anticipated that for any 2-form $\boldsymbol{a}$ in 3D space $\boldsymbol{da}$ must be a scalar times $\boldsymbol{dx} \wedge \boldsymbol{dy} \wedge \boldsymbol{dz}$ because there is no other basis 3-form in 3D. However, the above derivation shows that $\boldsymbol{da}$ is proportional to the divergence of the vector $\bar{a}$ with the same covariant components as the 2-form $\boldsymbol{a}$. Moreover, the infinitesimal part of $\boldsymbol{da}$ is the oriented volume element, $\boldsymbol{dx} \wedge \boldsymbol{dy} \wedge \boldsymbol{dz}$. The correspondence with vector notation is thus,

$$\boldsymbol{da} \leftrightarrow (\bar{\nabla} \cdot \bar{a})dV \tag{30.5.18}$$

The versatility of the p-form notation together with the exterior derivative is illustrated by equs.(30.5.8,10,18) which show that these 1-form, 2-form and 3-form expressions can represent line, area and volume elements together with the dot product, the $curl$ and the div of the equivalent vector.

What does the Cartan-Stokes theorem give in 3D when $\boldsymbol{a}$ is a 2-form, i.e., (30.5.13). We see that it becomes,

$$\int_V \boldsymbol{da} = \int_{\delta V} \boldsymbol{a} \rightarrow \int_V (\bar{\nabla} \cdot \bar{a}) dV = \int_{\delta V} \bar{a} \cdot d\bar{S} \tag{30.5.19}$$

where we have again interpreted a 2-form being an area-element expression,

$$a_x \boldsymbol{dy} \wedge \boldsymbol{dz} + a_y \boldsymbol{dz} \wedge \boldsymbol{dx} + a_z \boldsymbol{dxdy} \leftrightarrow \bar{a} \cdot d\bar{S} \tag{30.5.20}$$

So, for a 2-form in 3D the Cartan-Stokes theorem is just Gauss's theorem (or the divergence theorem).

30.5.2 4D Spacetime and Electromagnetism

The general 1-form in 4D is,

$$\boldsymbol{A} = A_\mu \boldsymbol{dx}^\mu \tag{30.5.21}$$

Hence,

$$\boldsymbol{dA} = \boldsymbol{d}A_\mu \boldsymbol{dx}^\mu = A_{\mu,\eta} \boldsymbol{dx}^\eta \wedge \boldsymbol{dx}^\mu = \frac{1}{2}\left(A_{\mu,\eta} - A_{\eta,\mu}\right) \boldsymbol{dx}^\eta \wedge \boldsymbol{dx}^\mu \tag{30.5.22}$$

Put,

$$F_{\mu\eta} = A_{\mu,\eta} - A_{\eta,\mu} \tag{30.5.23}$$

Let us interpret A_μ as the electromagnetic 4-vector potential. The usual $\Phi, \bar{A}$ are the contravariant components (A^μ), so $\{A_\mu\} = \{\Phi/c, -\bar{A}\}$. Hence we see that, with $\tau = ct$, $\partial_\tau = \frac{\partial}{c\partial t}$,

$$F_{i\tau} = A_{i,\tau} - A_{\tau,i} = -\frac{(\partial_\tau (\bar{A})_i + \partial_i \Phi)}{c} = \frac{E_i}{c} = \tilde{E}_i \tag{30.5.24a}$$

$$F_{ij} = A_{i,j} - A_{j,i} = \varepsilon_{ijk} B_k \tag{30.5.24b}$$

$$\begin{aligned} \boldsymbol{dA} = & F_{x\tau} \boldsymbol{d\tau} \wedge \boldsymbol{dx} + F_{y\tau} \boldsymbol{d\tau} \wedge \boldsymbol{dy} + F_{z\tau} \boldsymbol{d\tau} \wedge \boldsymbol{dz} \\ & + F_{xy} \boldsymbol{dy} \wedge \boldsymbol{dx} + F_{yz} \boldsymbol{dz} \wedge \boldsymbol{dy} + F_{zx} \boldsymbol{dx} \wedge \boldsymbol{dz} \end{aligned} \tag{30.5.25a}$$

That is,

$$\begin{aligned} \boldsymbol{F} = \boldsymbol{dA} = & \tilde{E}_x \boldsymbol{d\tau} \wedge \boldsymbol{dx} + \tilde{E}_y \boldsymbol{d\tau} \wedge \boldsymbol{dy} + \tilde{E}_z \boldsymbol{d\tau} \wedge \boldsymbol{dz} \\ & + B_z \boldsymbol{dy} \wedge \boldsymbol{dx} + B_x \boldsymbol{dz} \wedge \boldsymbol{dy} + B_y \boldsymbol{dx} \wedge \boldsymbol{dz} \end{aligned} \tag{30.5.25b}$$

Hence, $\boldsymbol{F} = \boldsymbol{dA}$ is a 2-form whose components are $F_{\mu\eta}$ in agreement with (27.6.1b), which bears repeating,

$$(F_{\mu\nu}) = \begin{pmatrix} 0 & \tilde{E}_x & \tilde{E}_y & \tilde{E}_z \\ -\tilde{E}_x & 0 & -B_z & B_y \\ -\tilde{E}_y & B_z & 0 & -B_x \\ -\tilde{E}_z & -B_y & B_x & 0 \end{pmatrix} \tag{27.6.1b}$$

But from $\boldsymbol{F} = \boldsymbol{dA}$ we have immediately that,

$$\boldsymbol{dF} = \boldsymbol{d^2A} = 0 \tag{30.5.26}$$

What does this mean in the context of the Maxwell equations? Operate the exterior derivative on (30.5.25b) and use, for example,

$$\boldsymbol{d}\tilde{E}_x\boldsymbol{d\tau} \wedge \boldsymbol{dx} = \tilde{E}_{x,\mu}\boldsymbol{dx}^\mu \wedge \boldsymbol{d\tau} \wedge \boldsymbol{dx}$$

$$= \tilde{E}_{x,y}\boldsymbol{dy} \wedge \boldsymbol{d\tau} \wedge \boldsymbol{dx} + \tilde{E}_{x,z}\boldsymbol{dz} \wedge \boldsymbol{d\tau} \wedge \boldsymbol{dx} \tag{30.5.27}$$

and similarly for the other five terms. Collect together like terms, paying careful attention to the order of the 1-forms and the changes of sign required to bring them to a common order. One finds by equating all factors of the same basis 3-form that,

$$\boldsymbol{dF} = 0 \Rightarrow \bar{\nabla} \times \bar{E} = -\frac{\partial \bar{B}}{\partial t} \quad and \quad \bar{\nabla} \cdot \bar{B} = 0 \tag{30.5.28}$$

So $\boldsymbol{dF} = 0$ incorporates two of the four Maxwell equations in vector form.

Now consider the dual field tensor, in covariant form, which from (27.6.4) we find is,

$$(\mathcal{F}_{\mu\nu}) = \begin{pmatrix} 0 & B_x & B_y & B_z \\ -B_x & 0 & \tilde{E}_z & -\tilde{E}_y \\ -B_y & -\tilde{E}_z & 0 & \tilde{E}_x \\ -B_z & \tilde{E}_y & -\tilde{E}_x & 0 \end{pmatrix} \tag{30.5.29}$$

From this the dual field as a 2-form is,

$$\begin{aligned} \boldsymbol{\mathcal{F}} = {} & B_x\boldsymbol{d\tau} \wedge \boldsymbol{dx} + B_y\boldsymbol{d\tau} \wedge \boldsymbol{dy} + B_z\boldsymbol{d\tau} \wedge \boldsymbol{dz} \\ & -\tilde{E}_z\boldsymbol{dy} \wedge \boldsymbol{dx} - \tilde{E}_x\boldsymbol{dz} \wedge \boldsymbol{dy} - \tilde{E}_y\boldsymbol{dx} \wedge \boldsymbol{dz} \end{aligned} \tag{30.5.30}$$

Now calculate $\boldsymbol{d\mathcal{F}}$, using simplifications like (30.5.27) when 3-forms such as $\boldsymbol{dx} \wedge \boldsymbol{dt} \wedge \boldsymbol{dx} = 0$, hence we find,

$$\begin{aligned} \boldsymbol{d\mathcal{F}} = {} & \left(B_{x,y} - B_{y,x} + \tilde{E}_{z,\tau}\right)\boldsymbol{d\tau} \wedge \boldsymbol{dx} \wedge \boldsymbol{dy} \\ & +\left(B_{x,z} - B_{z,x} - \tilde{E}_{y,\tau}\right)\boldsymbol{d\tau} \wedge \boldsymbol{dx} \wedge \boldsymbol{dz} \end{aligned}$$

$$+\left(B_{y,z} - B_{z,y} + \tilde{E}_{x,\tau}\right)d\tau \wedge dy \wedge dz$$
$$+\left(\tilde{E}_{x,x} + \tilde{E}_{y,y} + \tilde{E}_{z,z}\right)dx \wedge dy \wedge dz \qquad (30.5.31a)$$

i.e.,

$$\boldsymbol{d\mathcal{F}} = -\left(\bar{\nabla} \times \bar{B} - \partial_\tau \bar{\tilde{E}}\right)_x d\tau \wedge dy \wedge dz + \left(\bar{\nabla} \times \bar{B} - \partial_\tau \bar{\tilde{E}}\right)_y d\tau \wedge dx \wedge dz$$
$$-\left(\bar{\nabla} \times \bar{B} - \partial_\tau \bar{\tilde{E}}\right)_z d\tau \wedge dx \wedge dy + \bar{\nabla} \cdot \bar{\tilde{E}} dx \wedge dy \wedge dz \qquad (30.5.31b)$$

We recognise the terms in this as relating to the Maxwell equations with source terms. But we cannot merely equate this to the 1-form made from the current density 4-vector because $\boldsymbol{d\mathcal{F}}$ is a 3-form, not a 1-form. So we consider the dual of the current density 1-form defined by,

$$\boldsymbol{\mathcal{J}} = J^\mu \varepsilon_{\mu\alpha\beta\gamma} dx^\alpha \wedge dx^\beta \wedge dx^\gamma \qquad (30.5.32a)$$

From this we find, (30.5.32b)

$$\boldsymbol{\mathcal{J}} = J^\tau dx \wedge dy \wedge dz - J^x d\tau \wedge dy \wedge dz + J^y d\tau \wedge dx \wedge dz - J^z d\tau \wedge dx \wedge dy$$

From (30.5.31b) and (30.5.32b) we see that $\boldsymbol{d\mathcal{F}} = \mu_0 \boldsymbol{\mathcal{J}}$ is equivalent to the remaining two Maxwell equations, recalling that $\{J^\mu\} = \{c\rho, \bar{J}\}$

$$\boldsymbol{d\mathcal{F}} = \mu_0 \boldsymbol{\mathcal{J}} \Leftrightarrow \bar{\nabla} \times \bar{B} = \frac{1}{c^2}\partial_t \bar{E} + \mu_0 \bar{J} \ \ and \ \ \bar{\nabla} \cdot \bar{E} = \rho/\varepsilon_0 \qquad (30.5.33)$$

In the latter we have used $c^2\mu_0 = 1/\varepsilon_0$.

In summary, in terms of the 2-form field $\boldsymbol{F}$ and its dual, $\boldsymbol{\mathcal{F}}$, the complete Maxwell equations are written simply,

$$\boldsymbol{dF} = 0 \quad \text{and} \quad \boldsymbol{d\mathcal{F}} = \mu_0 \boldsymbol{\mathcal{J}} \qquad (30.5.34)$$

where $\boldsymbol{\mathcal{J}}$ is the 4-current density written as a 3-form, (30.5.32a,b).

Note the formal similarity with the Maxwell equations in tensorial form,

$$\partial_\alpha F^{\alpha\beta} = \mu_0 J^\beta \quad \text{and} \quad \partial_\alpha \mathcal{F}^{\alpha\beta} = 0 \qquad (27.6.6)$$

Although the Maxwell equations in biquaternion form is even simpler,

$$DF + J = 0 \qquad (29.4.10)$$

But note something curious that I missed myself for 50 years. In $\boldsymbol{dF} = 0$ it is the field 2-form, derived from the field tensor, $F_{\mu\nu}$, (27.6.1b), which appears in the homogeneous equation, i.e., without the source terms. In contrast, in tensor notation the field tensor appears in the inhomogeneous equation, with the source terms, in $\partial_\alpha F^{\alpha\beta} = \mu_0 J^\beta$. And vice-versa, the dual 2-form field appears with the source term in $\boldsymbol{d\mathcal{F}} = \mu_0 \boldsymbol{\mathcal{J}}$ whilst the dual tensor appears in the source-free homogeneous equation $\partial_{\propto}\mathcal{F}^{\alpha\beta} = 0$. So the formal similarity between the tensor, (27.6.6), and 2-form, (30.5.34), equations is not quite what it initially appears.

Finally, note that the conservation of charge (the continuity equation) is now trivial because $\boldsymbol{d\mathcal{F}} = \mu_0\boldsymbol{\mathcal{J}}$ gives immediately $\boldsymbol{d^2\mathcal{F}} = \mu_0\boldsymbol{d\mathcal{J}} = 0$ and the reader should be able to show from (30.5.32b) that $\boldsymbol{d\mathcal{J}} = \frac{\partial\rho}{\partial t} + \overline{\nabla}\cdot\overline{J}$, hence reproducing the usual equation of continuity / conservation.

We close by noting that a proof of the Cartan-Stokes theorem is provide in Appendix A9, though purists will note a lack of attention to the topological varieties which may occur. This chapter has been only a cursory introduction to differential geometry and the exterior algebra, which subjects require specialist texts dedicated to them, of which there are many. But we will not dwell further on these matters here for this book is finished.

Appendix A1: Summary of Vector Derivative and Vector Integral Formulae (3D Euclidian)

A1.1 Basic Formulae

Area of parallelogram: $\bar{a} \times \bar{b}$ (A1.1)

Volume of parallelopiped: $(\bar{a} \times \bar{b}) \cdot \bar{c} = (\bar{b} \times \bar{c}) \cdot \bar{a} = (\bar{c} \times \bar{a}) \cdot \bar{b}$ (A1.2)

Triple product: $(\bar{a} \times \bar{b}) \times \bar{c} = (\bar{a} \cdot \bar{c})\bar{b} - (\bar{b} \cdot \bar{c})\bar{a}$ (A1.3a)

$$\bar{a} \times (\bar{b} \times \bar{c}) = (\bar{a} \cdot \bar{c})\bar{b} - (\bar{a} \cdot \bar{b})\bar{c} \qquad \text{(A1.3b)}$$

Solid angle of area element at position $\bar{r}$: $\frac{(\bar{a}\times\bar{b})\cdot\hat{r}}{r^2}$ (A1.4)

Conservation of a substance with density ρ and flux $\bar{J}$ which is being generated at a rate of Q per unit volume: $\frac{\partial \rho}{\partial t} + \bar{\nabla} \cdot \bar{J} = Q$. (A1.5)

A1.2 Derivative Formulae

$$\boldsymbol{curl}(\boldsymbol{grad}(f)) = \bar{\boldsymbol{\nabla}} \times (\bar{\boldsymbol{\nabla}} \boldsymbol{f}) \equiv \mathbf{0} \qquad \text{(A1.6)}$$

$$\boldsymbol{div}(\boldsymbol{curl}(\bar{v})) = \bar{\boldsymbol{\nabla}} \cdot (\bar{\boldsymbol{\nabla}} \times \bar{\mathbf{v}}) \equiv \mathbf{0} \qquad \text{(A1.7)}$$

$$\bar{\nabla}(fg) = f\bar{\nabla}g + g\bar{\nabla}f \qquad \text{(A1.8)}$$

$$\bar{\nabla} \cdot (f\bar{v}) = f\bar{\nabla} \cdot \bar{v} + \bar{v} \cdot \bar{\nabla}f \qquad \text{(A1.9)}$$

$$\bar{\nabla} \times (f\bar{v}) = f\bar{\nabla} \times \bar{v} - \bar{v} \times \bar{\nabla}f \qquad \text{(A1.10)}$$

$$\bar{\nabla} \cdot (\bar{v} \times \bar{w}) = \bar{w} \cdot \bar{\nabla} \times \bar{v} - \bar{v} \cdot \bar{\nabla} \times \bar{w} \qquad \text{(A1.11)}$$

$$\bar{\nabla}(\bar{v} \cdot \bar{w}) = \bar{v} \times (\bar{\nabla} \times \bar{w}) + \bar{w} \times (\bar{\nabla} \times \bar{v}) + (\bar{v} \cdot \bar{\nabla})\bar{w} + (\bar{w} \cdot \bar{\nabla})\bar{v} \qquad \text{(A1.12)}$$

$$\bar{\nabla} \times (\bar{v} \times \bar{w}) = \bar{v}(\bar{\nabla} \cdot \bar{w}) - \bar{w}(\bar{\nabla} \cdot \bar{v}) - (\bar{v} \cdot \bar{\nabla})\bar{w} + (\bar{w} \cdot \bar{\nabla})\bar{v} \qquad \text{(A1.13)}$$

For a constant vector $\bar{a}$ the last two identities become.

$$\bar{\nabla}(\bar{a} \cdot \bar{w}) = \bar{a} \times (\bar{\nabla} \times \bar{w}) + (\bar{a} \cdot \bar{\nabla})\bar{w} \qquad \text{(A1.14)}$$

$$\bar{\nabla} \times (\bar{a} \times \bar{w}) = \bar{a}(\bar{\nabla} \cdot \bar{w}) - (\bar{a} \cdot \bar{\nabla})\bar{w} \qquad \text{(A1.15)}$$

Finally, the second order scalar differential operator ∇^2, known as the "Laplacian", acts on a scalar field in the obvious manner, namely as $div(grad)$, i.e.,

$$\nabla^2 f = \bar{\nabla} \cdot (\bar{\nabla} f) \qquad \text{(A1.16)}$$

But on a vector field it acts thus,

$$\nabla^2 \bar{v} = \bar{\nabla}(\bar{\nabla} \cdot \bar{v}) - \bar{\nabla} \times (\bar{\nabla} \times \bar{v}) \tag{A1.17}$$

which follows from (A1.15) by setting $\bar{a} = \bar{\nabla}$ and rearranging. In Cartesian coordinates we have $\nabla^2 = \partial_i \partial_i = \partial_x^2 + \partial_y^2 + \partial_z^2$. For the Laplacian in other coordinate systems see Appendix A2.

A1.3 Integral Formulae

Stokes's Theorem: $$\int_\Gamma (\bar{\nabla} \times \bar{v}) \cdot d\bar{S} = \oint_{\delta\Gamma} \bar{v} \cdot d\bar{s} \tag{A1.18}$$

Divergence (or Gauss's) Theorem: $$\int_V \bar{\nabla} \cdot \bar{v}\, d\tau = \oiint_{\delta V} \bar{v} \cdot d\bar{S} \tag{A1.19}$$

$$\iint_\Gamma \bar{\nabla} f \times d\bar{S} = -\oint_{\delta\Gamma} f d\bar{s} \tag{A1.20}$$

$$\int_V (\bar{\nabla} \times \bar{v}) d\tau = -\oiint_{\delta V} \bar{v} \times d\bar{S} \tag{A1.21}$$

$$\int_V (\bar{\nabla} f) d\tau = \oiint_{\delta V} f d\bar{S} \tag{A1.22}$$

$$\oiint_{\delta V} d\bar{S} = 0 \tag{A1.23}$$

In the above, $d\tau, d\bar{S}, d\bar{s}$ are the element of volume, the vector element of area and the vector element of a line increment respectively.

A1.4 Solid Angle

Solid angle integrates over a closed surface thus,

- If the point wrt which the solid angle is defined lies within the closed surface δV then: $\oiint_{\delta V} d\Omega = 4\pi$.
- If the point wrt which the solid angle is defined lies outside the closed surface δV then: $\oiint_{\delta V} d\Omega = 0$.
- If the point wrt which the solid angle is defined lies on the closed surface δV then: $\oiint_{\delta V} d\Omega = 2\pi$.

Appendix A2: Vector Derivatives in Various Coordinates (3D Euclidean Space)

A2.1 The Alternating Tensor

Positive permutations of 1,2,3 or x, y, z: $\epsilon_{ijk} = 1$;
Negative permutations $\epsilon_{ijk} = -1$
Otherwise $\epsilon_{ijk} = 0$

A useful identity is $\epsilon_{ijk}\epsilon_{mnk} = \delta_{im}\delta_{jn} - \delta_{in}\delta_{jm}$ (sum over k understood)

A2.2 Generic Formulae

For any curvilinear coordinate system, $\boldsymbol{x_1}, \boldsymbol{x_2}, \boldsymbol{x_3}$, in which the local axes are everywhere mutually orthogonal, so that unit vectors can be defined wrt these axes to provide an orthonormal basis, the formulae for $\boldsymbol{grad}, \boldsymbol{div}, \boldsymbol{curl}$ can be written in terms of the metric functions defined as follows,

$$ds^2 = h_1^2 dx_1^2 + h_2^2 dx_2^2 + h_3^2 dx_3^2$$

where h_1, h_2, h_3 are functions of the coordinates x_1, x_2, x_3, sometimes known as the scale factors. Their squares are the diagonal components of the metric tensor, the off-diagonal components being zero due to the orthogonality of the coordinate system. In terms of the scale factors the generic formulae are,

$$grad(f) = \overline{\nabla}\mathrm{f} = \frac{1}{h_1}\frac{\partial f}{\partial x_1}\hat{x}_1 + \frac{1}{h_2}\frac{\partial f}{\partial x_2}\hat{x}_2 + \frac{1}{h_3}\frac{\partial f}{\partial x_3}\hat{x}_3$$

$$div(\bar{v}) = \overline{\nabla} \cdot \bar{v} = \frac{1}{h_1 h_2 h_3}\left\{\frac{\partial(h_2 h_3 v_1)}{\partial x_1} + \frac{\partial(h_1 h_3 v_2)}{\partial x_2} + \frac{\partial(h_1 h_2 v_3)}{\partial x_3}\right\}$$

$$curl(\bar{v}) = \overline{\nabla} \times \bar{v} = \frac{\hat{x}_1}{h_2 h_3}\left\{\frac{\partial(h_3 v_3)}{\partial x_2} - \frac{\partial(h_2 v_2)}{\partial x_3}\right\} + \frac{\hat{x}_2}{h_3 h_1}\left\{\frac{\partial(h_1 v_1)}{\partial x_3} - \frac{\partial(h_3 v_3)}{\partial x_1}\right\} + \frac{\hat{x}_3}{h_1 h_2}\left\{\frac{\partial(h_2 v_2)}{\partial x_1} - \frac{\partial(h_1 v_1)}{\partial x_2}\right\}$$

$$\nabla^2 f = \frac{1}{h_1 h_2 h_3}\left\{\frac{\partial}{\partial x_1}\left(\frac{h_2 h_3}{h_1}\frac{\partial f}{\partial x_1}\right) + \frac{\partial}{\partial x_2}\left(\frac{h_1 h_3}{h_2}\frac{\partial f}{\partial x_2}\right) + \frac{\partial}{\partial x_3}\left(\frac{h_2 h_1}{h_3}\frac{\partial f}{\partial x_3}\right)\right\}$$

A2.3 Cartesian

$ds^2 = dx^2 + dy^2 + dz^2$

$\bar{a} \cdot \bar{b} = a_i b_i = a_x b_x + a_y b_y + a_z b_z$

$\bar{a} \times \bar{b} = (a_y b_z - a_z b_y)\hat{x} + (a_z b_x - a_x b_z)\hat{y} + (a_x b_y - a_y b_x)\hat{z}$

$grad(f) = \bar{\nabla}\mathrm{f} = \hat{x}\frac{\partial f}{\partial x} + \hat{y}\frac{\partial f}{\partial y} + \hat{z}\frac{\partial f}{\partial z}$

$div(\bar{v}) = \bar{\nabla} \cdot \bar{v} = \frac{\partial v_x}{\partial x} + \frac{\partial v_y}{\partial y} + \frac{\partial v_z}{\partial z}$

From here on we use $\partial_x = \frac{\partial}{\partial x}$, etc.

$curl(\bar{v}) = \bar{\nabla} \times \bar{v} = (\partial_y v_z - \partial_z v_y)\hat{x} + (\partial_z v_x - \partial_x v_z)\hat{y} + (\partial_x v_y - \partial_y v_x)\hat{z}$

i.e., $grad(f) = \hat{x}_i \partial_i f \quad div(\bar{v}) = \partial_i v_i \quad curl(\bar{v}) = \epsilon_{ijk}\hat{x}_i \partial_j v_k$

$\nabla^2 f = \partial_i \partial_i f = \partial_x^2 f + \partial_y^2 f + \partial_z^2 f$

A2.4 Cylindrical Polars

$x = r\cos\theta \qquad y = r\sin\theta$

$ds^2 = dr^2 + r^2 d\theta^2 + dz^2$

Unit vectors in the directions of greatest increase of the coordinates are $\hat{r}, \hat{\theta}, \hat{z}$, such that $\hat{\theta} \cdot \hat{\theta} = 1$, etc.

$grad(f) = \hat{r}\partial_r f + \hat{\theta}\frac{1}{r}\partial_\theta f + \hat{z}\partial_z f$

$div(\bar{v}) = \frac{1}{r}\partial_r(r v_r) + \frac{1}{r}\partial_\theta(v_\theta) + \partial_z(v_z)$

$curl(\bar{v}) = \hat{r}\left(\frac{1}{r}\partial_\theta v_z - \partial_z v_\theta\right) + \hat{\theta}(\partial_z v_r - \partial_r v_z) + \hat{z}\frac{1}{r}(\partial_r(r v_\theta) - \partial_\theta v_r)$

$\nabla^2 f = \frac{1}{r}\partial_r(r\partial_r f) + \frac{1}{r^2}\partial_\theta^2 f + \partial_z^2 f$

A2.5 Spherical Polars

$$x = r\sin\theta\cos\phi \qquad y = r\sin\theta\sin\phi \qquad z = r\cos\theta$$

$$ds^2 = dr^2 + r^2 d\theta^2 + r^2 \sin^2\theta d\phi^2$$

Unit vectors in the directions of greatest increase of the coordinates are $\hat{r}, \hat{\theta}, \hat{\phi}$, such that $\hat{\theta}\cdot\hat{\theta} = 1, \hat{\phi}\cdot\hat{\phi} = 1$, etc.

$$grad(f) = \hat{r}\partial_r f + \frac{\hat{\theta}}{r}\partial_\theta f + \frac{\hat{\phi}}{r\sin\theta}\partial_\phi f$$

$$div(\bar{v}) = \frac{1}{r^2}\frac{\partial}{\partial r}(r^2 v_r) + \frac{1}{r\sin\theta}\frac{\partial}{\partial\theta}(v_\theta\sin\theta) + \frac{1}{r\sin\theta}\frac{\partial v_\phi}{\partial\phi}$$

$$curl(\bar{v}) = \frac{\hat{r}}{r\sin\theta}\left(\partial_\theta(v_\phi\sin\theta) - \partial_\phi v_\theta\right) + \frac{\hat{\theta}}{r}\left(\frac{1}{\sin\theta}\partial_\phi v_r - \partial_r(r v_\phi)\right) + \frac{\hat{\phi}}{r}\left(\partial_r(r v_\theta) - \partial_\theta v_r\right)$$

$$\nabla^2 f = \frac{1}{r^2}\partial_r(r^2\partial_r f) + \frac{1}{r^2\sin\theta}\partial_\theta(\sin\theta\partial_\theta f) + \frac{1}{r^2\sin^2\theta}\partial_\phi^2 f$$

A2.6 Toroidal Coordinates

$$x = \rho\cos\phi, \quad y = \rho\sin\phi,$$

$$\rho = \sqrt{x^2+y^2} = \frac{a\sinh\eta}{\cosh\eta - \cos\xi}, \qquad z = \frac{a\sin\xi}{\cosh\eta-\cos\xi}$$

See §20.2 and Figures 20.6,7 for further elucidation.

$$h_\xi = h_\eta = a,\ h_\phi = \rho = \frac{a\sinh\eta}{\cosh\eta-\cos\xi} \qquad ds^2 = a^2\left(d_\xi^2 + d_\eta^2\right) + \rho^2 d_\phi^2$$

$$grad(f) = \frac{1}{a}\left(\hat{\xi}\partial_\xi + \hat{\eta}\partial_\eta\right) + \frac{\cosh\eta-\cos\xi}{a\sinh\eta}\hat{\phi}\partial_\phi$$

$$div(\bar{v}) = \frac{\cosh\eta-\cos\xi}{a\sinh\eta}\left\{\frac{\partial}{\partial\xi}\left(\frac{\sinh\eta}{\cosh\eta-\cos\xi}v_\xi\right) + \frac{\partial}{\partial\eta}\left(\frac{\sinh\eta}{\cosh\eta-\cos\xi}v_\eta\right) + \frac{\partial v_\phi}{\partial\phi}\right\}$$

$$curl(\bar{v}) = \frac{\cosh\eta-\cos\xi}{a\sinh\eta}\left\{\frac{\partial}{\partial\eta}\left(\frac{\sinh\eta}{\cosh\eta-\cos\xi}v_\phi\right) - \frac{\partial v_\eta}{\partial\phi}\right\}\hat{\xi} + \frac{\cosh\eta-\cos\xi}{a\sinh\eta}\left\{\frac{\partial v_\xi}{\partial\phi} - \frac{\partial}{\partial\xi}\left(\frac{\sinh\eta}{\cosh\eta-\cos\xi}v_\phi\right)\right\}\hat{\eta} + \frac{1}{a}\left\{\frac{\partial v_\eta}{\partial\xi} - \frac{\partial v_\xi}{\partial\eta}\right\}\hat{\phi}$$

$$\nabla^2 f = \frac{\cosh\eta-\cos\xi}{a^2\sinh\eta}\left\{\frac{\partial}{\partial\xi}\left(\frac{\sinh\eta}{\cosh\eta-\cos\xi}\frac{\partial f}{\partial\xi}\right) + \frac{\partial}{\partial\eta}\left(\frac{\sinh\eta}{\cosh\eta-\cos\xi}\frac{\partial f}{\partial\eta}\right) + \frac{\partial}{\partial\phi}\left(\frac{\cosh\eta-\cos\xi}{\sinh\eta}\frac{\partial f}{\partial\phi}\right)\right\}$$

A2.7 Elliptical Coordinates

$x = c\cosh\xi\cos\eta = a\cos\eta = \tilde{a}\cosh\xi$

$y = c\sinh\xi\sin\eta = b\sin\eta = \tilde{b}\sinh\xi$

$x + iy = c\cosh(\xi + i\eta)$

$c^2 = a^2 - b^2\ (a > b > 0) \qquad \frac{x^2}{a^2} + \frac{y^2}{b^2} = 1 \qquad \frac{x^2}{\tilde{a}^2} - \frac{y^2}{\tilde{b}^2} = 1$

Curves of constant ξ are ellipses, semi-major axis a, semi-minor axis b.

Curves of constant η are hyperbolae, intersecting the x axis at $\pm\tilde{a}$ and asymptotic to the lines $y = \pm\frac{\tilde{b}}{\tilde{a}}x$.

Scale factors $h_\xi = h_\eta = \frac{c}{\sqrt{2}}\sqrt{\cosh 2\xi - \cos 2\eta}$, $h_z = 1$

$$ds^2 = \frac{c^2}{2}(\cosh 2\xi - \cos 2\eta)\left(d_\xi^2 + d_\eta^2\right) + d_z^2$$

$$grad(f) = \frac{\sqrt{2}}{c\sqrt{\cosh 2\xi - \cos 2\eta}}\left(\hat{\xi}\partial_\xi + \hat{\eta}\partial_\eta\right)f + \hat{z}\partial_z f$$

$$div(\bar{v}) = \frac{\sqrt{2}}{c(\cosh 2\xi - \cos 2\eta)}\left\{\partial_\xi\left(\sqrt{\cosh 2\xi - \cos 2\eta}\cdot v_\xi\right) + \partial_\eta\left(\sqrt{\cosh 2\xi - \cos 2\eta}\cdot v_\eta\right)\right\} + \partial_z v_z$$

$$curl(\bar{v}) = \left[\frac{\sqrt{2}}{c\sqrt{\cosh 2\xi - \cos 2\eta}}\partial_\eta v_z - \partial_z v_\eta\right]\hat{\xi} + \left[\partial_z v_\xi - \frac{\sqrt{2}}{c\sqrt{\cosh 2\xi - \cos 2\eta}}\partial_\xi v_z\right]\hat{\eta}$$
$$+ \frac{\sqrt{2}}{c(\cosh 2\xi - \cos 2\eta)}\left[\partial_\xi\left(v_\eta\sqrt{\cosh 2\xi - \cos 2\eta}\right) - \partial_\eta\left(v_\xi\sqrt{\cosh 2\xi - \cos 2\eta}\right)\right]\hat{z}$$

$$\nabla^2 f = \frac{2}{c^2(\cosh 2\xi - \cos 2\eta)}\left(\partial_\xi^2 + \partial_\eta^2\right) + \partial_z^2$$

Appendix A3: Proof of Green's Theorem

Green's Theorem says,

$$\oint_{\delta C}(gdx + hdy) = \iint_C \left(\frac{\partial h}{\partial x} - \frac{\partial g}{\partial y}\right) dxdy$$

providing the derivatives and integrals exist, and noting that the boundary (path) integral is conducted counter-clockwise. Consider the region C bounded by δC below,

Figure A3.1

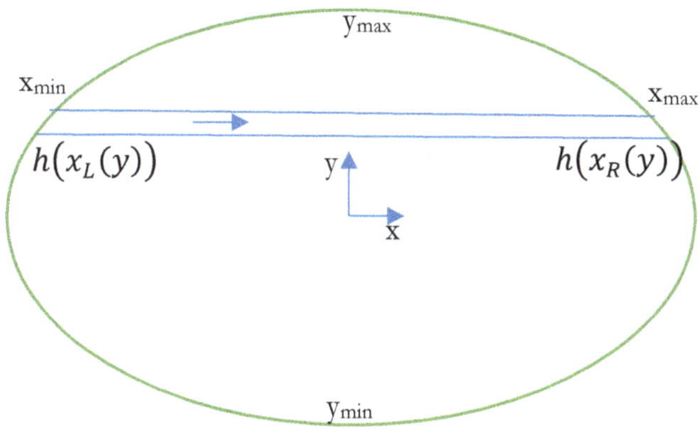

We seek first to establish the theorem when $g = 0$, i.e., that,

$$\oint_{\delta C} hdy = \iint_C \frac{\partial h}{\partial x} dxdy$$

As the diagram shows we may split the area integral on the RHS into narrow strips of constant y. The contribution to the area integral of this single strip can be found because the integral over x of $\frac{\partial h}{\partial x}dx$ is just Δh, which is $h(x_R(y)) - h(x_L(y))$, where those quantities are the values of the function h on the right-hand and left-hand boundaries at this y value. Note that the blue arrow indicates the sign of this integral, being from x_{min} to x_{max}. The area integral now requires us to carry out the y-integral, from y_{min} to y_{max}, giving,

$$RHS = \int_{y_{min}}^{y_{max}} [h(x_R(y)) - h(x_L(y))]dy$$

But this can equally be written,

$$RHS = \int_{y_{min}}^{y_{max}} h(x_R(y))dy + \int_{y_{max}}^{y_{min}} h(x_L(y))dy$$

whereupon we note that this is simply the LHS of the equation to be proved, since the first term is the counter-clockwise path integral on the right-hand part of the boundary and the second term is the counter-clockwise path integral on the left-hand part of the boundary.

The proof of the other half of Green's Theorem, i.e.,

$$-\oint_{\delta C} g dx = \iint_C \frac{\partial g}{\partial y} dxdy$$

proceeds in identical fashion, considering vertical y-integration strips at constant x, then integrating over x, thus,

$$RHS = \int_{x_{min}}^{x_{max}} [g(y_T(x)) - g(y_B(x))]dx$$

where y_T, y_B are the top and bottom of the vertical strips defining the top and bottom parts of the boundary respectively. This can be rewritten as,

$$RHS = -\int_{x_{max}}^{x_{min}} g(y_T(x))dx - \int_{x_{min}}^{x_{max}} g(y_B(x))dx$$

showing how an overall minus sign occurs, and establishing that this equals the LHS of the equation to be proved. Note that the sign is dictated by the requirement that the path integral is conducted counter-clockwise. Were we to reverse the limits of integration in the above expression, and thus to remove the minus signs, the resulting path integral would be clockwise. QED.

Appendix A4: $\bar{\nabla} \times \bar{v} \equiv 0 \Leftrightarrow \bar{v} \equiv \bar{\nabla}\varphi; \bar{\nabla} \cdot \bar{v} \equiv 0 \Leftrightarrow \bar{v} \equiv \bar{\nabla} \times \bar{a}$

That $\bar{v} \equiv \bar{\nabla}\varphi \Rightarrow \bar{\nabla} \times \bar{v} \equiv 0$ and $\bar{v} \equiv \bar{\nabla} \times \bar{a} \Rightarrow \bar{\nabla} \cdot \bar{v} \equiv 0$ follows immediately from the definitions, see §3.7.2. Here we tackle the proof of the reverse implications.

A4.1 Proof that a Vector Field with Identically Zero *Curl* can be Written as the *Grad* of Some Scalar Field

We require to prove that if $\bar{\nabla} \times \bar{v} \equiv 0$ a scalar field φ exists such $\bar{v} \equiv \bar{\nabla}\varphi$.

We can define the value of φ at any point $\bar{r}$, in terms of just one arbitrary constant, namely its value at the origin, $\varphi(\bar{0})$, by,

$$\varphi(\bar{r}) = \varphi(\bar{0}) + \int_0^{\bar{r}} \left(v_x dx + v_y dy + v_z dz\right)_C \qquad \text{(A2.1)}$$

where the subscript C denotes the curve in 3D space along which the integral is carried out. The obvious, and only, objection to defining a unique scalar field in this way is that, in general, the result of (A2.1) will depend upon the curve C chosen. If any two arbitrary curves from the origin to point $\bar{r}$, say C_1 and C_2, produce the same result for (A2.1), then defining a closed loop by $C_1 - C_2$ results in $\oint_{C_1 - C_2} \left(v_x dx + v_y dy + v_z dz\right) = 0$. Moreover, any closed loop through point $\bar{r}$ can be composed in this manner. But by Stokes's Theorem, (A1.18), $\oint_{\delta\Gamma} \left(v_x dx + v_y dy + v_z dz\right) \equiv \int_\Gamma (\bar{\nabla} \times \bar{v}) \cdot d\bar{A}$ which is identically zero by assumption. Hence, the only condition that is required to make (A2.1) a valid definition of a unique scalar field (up to an additive constant) is respected, so (A2.1) is the required scalar field. QED.

A4.2 Proof that a Vector Field with Identically Zero *Div* can be Written as the *Curl* of Some Vector Field

We require to prove that if $\bar{\nabla} \cdot \bar{v} \equiv 0$ a vector field $\bar{a}$ exists such $\bar{v} \equiv \bar{\nabla} \times \bar{a}$. The proof below assumes that the field $\bar{v}$ is zero at infinity.

We start with the Helmholtz decomposition, which is proved independently in Appendix A6. Hence there are scalar and vector fields such that $\bar{v} = \bar{\nabla} f + \bar{\nabla} \times \bar{a}$. But because *DivCurl* is identically zero, (see §3.7.2), we have $\bar{\nabla} \cdot \bar{v} = \nabla^2 f = 0$ and this is an identity and holds everywhere, including at infinity.

We now appeal to a uniqueness theorem for Laplace's equation, $\nabla^2 f = 0$, which states that if the scalar field, f, is specified at all points on some closed surface then there is a unique solution of $\nabla^2 f = 0$ everywhere within the surface. That theorem is proved in §13.2. But we have assumed that the field $\bar{v}$ disappears at infinity, and hence that $\bar{\nabla}f$ also disappears which implies that f tends to a constant. But this constant can be taken as zero without loss of generality as subtracting any constant from f leaves $\bar{v}$ unchanged everywhere.

But on any sufficiently large, closed surface (large in all directions) we must have $f = 0$ everywhere on it if $\bar{v}$ is zero at infinity. A solution of Laplace's equation consistent with this is clearly the trivial case $f = 0$ everywhere – and the uniqueness theorem tells us that this is the only possible solution. Hence the Helmholtz decomposition is left with just $\bar{v} = \bar{\nabla} \times \bar{a}$. QED.

Appendix A5: Proof of a Particular Representation of the Dirac Delta Function

Here we prove that we can write $\delta^3(\bar{r}-\bar{r}_1) = -\frac{1}{4\pi}\nabla^2\frac{1}{|\bar{r}-\bar{r}_1|}$

Without loss of generality we can shift the origin to $\bar{r}_1$. In spherical polar coordinates the Laplacian operator ∇^2 equals $\frac{1}{r^2}\partial_r(r^2\partial_r)$ plus terms which depend upon the derivatives wrt θ and ϕ. Hence,

$$\nabla^2\frac{1}{|\bar{r}-\bar{r}_1|} \to \nabla^2\frac{1}{\mathrm{r}} = \frac{1}{r^2}\partial_r(r^2\partial_r)\frac{1}{r} \tag{A5.1}$$

which is zero everywhere that $1/r$ is differentiable, i.e., everywhere except at $r=0$, or, more generally for $\delta^3(\bar{r}-\bar{r}_1)$ everywhere except $\bar{r}=\bar{r}_1$. Hence this is consistent with $\delta^3(\bar{r}-\bar{r}_1)$ which also vanishes when $\bar{r}\neq\bar{r}_1$. To complete the proof we need to show that the volume integral of $-\frac{1}{4\pi}\nabla^2\frac{1}{|\bar{r}-\bar{r}_1|}$ is 1 if the volume includes point $\bar{r}_1$. We have, by the divergence theorem, (A1.19),

$$-\frac{1}{4\pi}\int_V \nabla^2\frac{1}{|\bar{r}-\bar{r}_1|}d\tau = -\frac{1}{4\pi}\oint_{\delta V}\bar{\nabla}\frac{1}{|\bar{r}-\bar{r}_1|}\cdot d\bar{S} \tag{A5.2}$$

But $\bar{\nabla}$ in spherical polars is $\hat{r}\partial_r$ plus terms which depend upon the derivatives wrt θ and ϕ. Shifting the origin to $\bar{r}_1$ we get,

$$\bar{\nabla}\frac{1}{|\bar{r}-\bar{r}_1|} \to \hat{r}\partial_r\frac{1}{r} = -\frac{\hat{r}}{r^2} \quad \text{and} \quad -\frac{\hat{r}}{r^2}\cdot d\bar{S} = -d\Omega \tag{A5.3}$$

where $d\Omega$ is the solid angle subtended by the area element at the origin (which is the point initially specified by $\bar{r}_1$). Consequently, as long as the origin (i.e., the point initially specified by $\bar{r}_1$) lies within the integration surface then (A5.2) equals 1, as required (as the solid angle integrates to 4π). Note that if the origin (i.e., the point initially specified by $\bar{r}_1$) lies outside the integration surface, then the solid angle integrates to zero because every element of surface area is matched by another element which has the same magnitude of solid angle but of opposite sign.

Appendix A6: Proof of the Helmholtz Decomposition

We require to prove that "any" vector field can be written $\bar{a} = \bar{\nabla}\varphi + \bar{\nabla} \times \bar{A}$.

As always, sufficient smoothness is assumed so that derivatives up to the second exist. The proof starts with the representation of the Dirac delta function proved in Appendix A5, $\delta^3(\bar{r} - \bar{r}_1) = -\frac{1}{4\pi}\nabla^2\frac{1}{|\bar{r}-\bar{r}_1|}$. This allows any sufficiently smooth vector field, $\bar{a}$, to be represented as follows,

$$\bar{a}(\bar{r}) \equiv \int_V \bar{a}(\bar{r}_1)\delta^3(\bar{r} - \bar{r}_1)d\tau_1 = -\frac{1}{4\pi}\int_V \bar{a}(\bar{r}_1)\nabla^2\frac{1}{|\bar{r}-\bar{r}_1|}d\tau_1$$

where ∇^2 acts on $\bar{r}$ and $d\tau_1$ is the volume element wrt $\bar{r}_1$ and V is any volume containing the point $\bar{r}$. Because ∇^2 does not act on $\bar{r}_1$, we may commute $\bar{a}(\bar{r}_1)$ with ∇^2 and then use identity (A1.17) to re-express $\nabla^2\frac{\bar{a}(\bar{r}_1)}{|\bar{r}-\bar{r}_1|}$ as $\bar{\nabla}\left(\bar{\nabla}\cdot\frac{\bar{a}(\bar{r}_1)}{|\bar{r}-\bar{r}_1|}\right) - \bar{\nabla}\times\left(\bar{\nabla}\times\frac{\bar{a}(\bar{r}_1)}{|\bar{r}-\bar{r}_1|}\right)$ giving,

$$\bar{a}(\bar{r}) \equiv -\frac{1}{4\pi}\int_V \left[\bar{\nabla}\left(\bar{\nabla}\cdot\frac{\bar{a}(\bar{r}_1)}{|\bar{r}-\bar{r}_1|}\right) - \bar{\nabla}\times\left(\bar{\nabla}\times\frac{\bar{a}(\bar{r}_1)}{|\bar{r}-\bar{r}_1|}\right)\right]d\tau_1$$

In both the terms in the integrand we can commute $\bar{a}(\bar{r}_1)$ with $\bar{\nabla}$ and then change this operator to the del which acts on $\bar{r}_1$ instead, $\bar{\nabla}_1$, by simply changing the overall sign. Hence,

$$\bar{a}(\bar{r}) \equiv \frac{1}{4\pi}\int_V \left[\bar{\nabla}\left(\bar{a}(\bar{r}_1)\cdot\bar{\nabla}_1\frac{1}{|\bar{r}-\bar{r}_1|}\right) + \bar{\nabla}\times\left(\bar{a}(\bar{r}_1)\times\bar{\nabla}_1\frac{1}{|\bar{r}-\bar{r}_1|}\right)\right]d\tau_1$$

The first $\bar{\nabla}$ in each term can be taken out of the integral as it does not operate on the integration variable. Hence we see that this expression is in the form desired, where,

$$\varphi(\bar{r}) = \frac{1}{4\pi}\int_V \bar{a}(\bar{r}_1)\cdot\bar{\nabla}_1\frac{1}{|\bar{r}-\bar{r}_1|}d\tau_1 \qquad \text{(A3.1)}$$

and

$$\bar{A}(\bar{r}) \equiv \frac{1}{4\pi}\int_V \bar{a}(\bar{r}_1)\times\bar{\nabla}_1\frac{1}{|\bar{r}-\bar{r}_1|}d\tau_1 \qquad \text{(A3.2)}$$

QED, provided the integrals exist. Note, however, that the volume V is arbitrary provided it contains the point $\bar{r}$. The existence of the integrals requires $\bar{a}$ to be zero at infinity, so this proof fails if that is not the case. However, proofs are available even when $\bar{a}$ remains non-zero at infinity, so the Helmholtz decomposition is actually valid even in that case.

Appendix A7: Proof that Legendre's Equation Requires the Eigenvalue to be $l(l+1)$ for Integral $l \geq 0$

The most elegant derivation of the eigenvalues of the squared angular momentum operator is achieved on the basis of the commutation relations of the abstract angular momentum operators. Details can be found in most standard quantum mechanics texts, for example [A7.1], though such derivations conclude that the most general answer is $\frac{n}{2}\left(\frac{n}{2}+1\right)$ for $n = 0,1,2,3,\ldots$ which includes the half-integral spin cases. For strictly orbital angular momentum, where the representation of the squared angular momentum operator gives us Legendre's equation, the most general answer is $k = l(l+1)$ for $l = 0,1,2,3,\ldots$ So, here we prove that this is indeed the most general form of the eigenvalue, k, in Legendre's equation,

$$\left[\partial_\xi\left((1-\xi^2)\partial_\xi\right) - \frac{m^2}{(1-\xi^2)}\right]\Theta = -\mathrm{k}\Theta \tag{A7.1}$$

where m is a positive or negative integer or zero. The proof also shows that m must be constrained to be in the range $m \in [-l, +l]$.

A7.1 Proof

For $m = 0$ we can substitute a power series solution $\Theta = \sum_{r=0}^{\infty} A_r \xi^r$ so we get,

$$\sum_{r=0}^{\infty}(1-\xi^2)r(r-1)A_r\xi^{r-2} - 2\xi r A_r \xi^{r-1} + kA_r\xi^r = 0 \tag{A7.2}$$

Hence, (A7.3)

$$\sum_{r=0}^{\infty}(r+1)(r+2)A_{r+2}\xi^r - r(r-1)A_r\xi^r - 2rA_r\xi^r + kA_r\xi^r = 0$$

Equating coefficients of ξ^r to zero,

$$(r+1)(r+2)A_{r+2} - r(r+1)A_r + kA_r = 0 \tag{A7.4}$$

So,

$$\frac{A_{r+2}}{A_r} = \frac{r(r+1)-k}{(r+1)(r+2)} \tag{A7.5}$$

As $r \to \infty$ this gives $\frac{A_{r+2}}{A_r} \to 1$ unless the series of coefficients becomes zero at some point. But this would be a divergent series and so must be rules out. It follows that (A7.5) must be zero for some r. But that can only be the case if k takes the value $l(l+1)$ for some positive integer l (or zero). The power series then terminates at $r = l$.

The Legendre polynomials, $P_l(\xi)$, then result from the recurrence relation (A7.5) either by taking $A_1 = 0$ which gives a polynomial in even powers only, or by taking $A_0 = 0$ which gives a polynomial in odd powers only.

On physical grounds, though m may be non-zero for $l > 0$ the solution for the same total angular momentum, hence for the same k, must still exist when $m = 0$ so that we must still have $k = l(l+1)$.

Ratger more convincingly, we may substitute the following into Legendre's equation, (A7.1),

$$\Theta = (-1)^m (1-\xi^2)^{\frac{|m|}{2}} \left(\frac{d}{d\xi}\right)^m P_l(\xi) \tag{A7.6}$$

This shows that (A7.6) is indeed a solution given that $P_l(\xi)$ is the solution for $m = 0$ as long as $k = l(l+1)$.

A7.2 References

[A7.1] Bradford, R.A.W. (2020). *The Unweirding*. Principia Publications Unlimited, 15 May 2020 (see section 29.2).

Appendix A8: The Speed of Electrons in Wires

Within a current-carrying wire, it is the electrons whose motion conveys the current, albeit in the reverse direction (being negatively charged). Consequently, the issue of the speed of said electrons is perennially raised. This can be calculated in an elementary fashion, which is done here. This, however, does not constitute the "speed of electricity", which is another thing entirely.

Not all the electrons in a metal take part in the electric current, only those that are sufficiently mobile – which means the valence electrons, or some of the valence electrons. In copper, just one electron per atom is available as a conducting electron.

The main thing to emphasise as regards the speed of the electrons is the crucial distinction between the typical instantaneous speed of an individual electron and the collective drift velocity which constitutes the current (ignoring the direction). We address the former first.

If the conduction electron "gas" were subject to classical physics (it emphatically is not) then the average energy of an electron would be given by $\frac{3}{2}k_B T$ where k_B is Boltzmann's constant (1.38×10^{-23}J/K) and T is the absolute temperature, so about 293K at room temperature. Equating this energy (6.1×10^{-21}J) to $\frac{1}{2}m_e v^2$ and using the mass of the electron (9.1×10^{-31}kg) would therefore suggest an rms electron speed of 115,000 m/s (258,000 mph). However, this is wildly wrong – and an underestimate! This is not the place to derive quantum mechanical equations from first principles. However, the key issue is that electrons are fermions and hence no two electrons can occupy the same quantum state. Hence, whilst a few electrons can have very low energy, the bulk of the electrons are forced by the exclusion principle to have far higher energy. Ignoring thermal energy, the available quantum states are filled from "the bottom up" as regards their energy levels. The highest occupied energy is called the Fermi energy, E_F, and is given by,

$$E_F = \frac{\hbar^2 k_F^2}{2m_e}, \quad \text{where} \quad k_F = (3\pi^2 \rho_N)^{1/3} \qquad \text{(A8.1)}$$

where $\hbar = 1.054 \times 10^{-34}$Js is Planck's constant (divided by 2π) and ρ_N is the number of conduction electrons per unit volume. If we assume that the wire is copper, the number density of conduction electrons can be calculated from the atomic mass of copper (63.55), Avogadro's number (6.02×10^{23}) and the density of copper (8,960 kg/m^3), giving $8.48 \times 10^{28}\text{m}^{-3}$. (A8.1) then gives the Fermi energy to be $1.13 \times 10^{-18}\text{J} = 7.0\text{ eV}$. Even at absolute zero this is the most common electron energy and corresponds to a speed of 1,570,000 m/s (3,530,000 mph). Pretty nippy, then, but still only 0.5% of light speed. The average electron energy can be found from the density of states which varies with energy as $\sqrt{E}$, from which the reader should confirm that the average energy is $\frac{3}{5}E_F$ and so the rms speed is 1,220,000 m/s (2,730,000 mph).

However, none of this has any relevance at all to the issue of electric current!

The fast motion of the conduction electrons is in random orientations and hence does not in itself constitute an electric current. The electric current arises from a net drift velocity. This amounts, as we shall see, to only an extremely small bias in the electron motion in the direction of the applied potential gradient. This drift speed can be found very simply for a given current and a wire of a given cross-section.

A typical household ring main might be rated at 32A and have copper wires with a cross-section of 4 mm^2. At maximum rated current, this is a flow of electrons of $\frac{32}{e} = \frac{32}{1.6\times10^{-19}} = 2.0 \times 10^{20}$ /s, and hence a number flux of 5×10^{25} m^{-2}s^{-1}. But the number flux is the number density ($8.48 \times 10^{28}\text{m}^{-3}$) times the average drift velocity, so the latter is just 0.0006 m/s = 0.6 mm/s, which is extremely slow and some 9 orders of magnitude smaller than the rms speed of individual electrons.

But what is the speed of electricity? Well, chapters 19 and 20 establish that the energy conveyed by an electric current does not reside within the wires but travels through the air (or other surrounding medium) in the form of the e/m fields. This energy therefore travels at the speed of light, c (or whatever that speed is in the surrounding medium). The e/m fields that convey the energy are set up in part by charges which move to the surfaces of the wires (see chapters 19 and 20). So changes of emf, say, will not necessarily propagate at light speed but at a speed controlled by the time taken to create or change the required surface charges.

In short there are three or four different speeds involved in electricity flowing in wires, and these range from substantially less than 1 mm/s to c. And neither of these are remotely close to the rms speed of the mobile electrons.

Appendix A9: Proof of the Cartan-Stokes Theorem

The Cartan-Stokes theorem states that for any p-form, $\boldsymbol{\omega}$, and any $p+1$ dimensional region, V, whose p-dimensional boundary is ∂V, then,

$$\int_V \boldsymbol{d\omega} = \int_{\partial V} \boldsymbol{\omega} \tag{A9.1}$$

The theorem holds in a space of any dimension $N \geq p+1$.

Properly understood, the Cartan-Stokes theorem is an extension of the fundamental theorem of calculus into higher dimensions. The latter states,

$$\int_a^b \frac{\partial f(x)}{\partial x} dx = f(b) - f(a) \tag{A9.2}$$

This is the Cartan-Stokes theorem for $p = 0$ in which the region V is the line element $[a, b]$ and the boundary values of the 0-form, f, are just its values at the end points.

Appendix A3 has provided a proof of Green's theorem, which is the Cartan-Stokes theorem for $p = 1$ restricted to a 2D plane.

Stokes's original theorem is the Cartan-Stokes theorem for $p = 1$ restricted to a 3D space, though the generalised Cartan-Stokes theorem holds also for higher dimensional spaces.

Gauss's theorem, or the divergence theorem, is the Cartan-Stokes theorem for $p = 2$, originally conceived in 3D space but holds in the generalised form in higher dimensions.

Readers are likely to be familiar with the proof of Gauss's theorem and the proof of the general Cartan-Stokes theorem proceeds in essentially the same way, at least at the level of rigour we pursue here.

An arbitrary p-form, $\boldsymbol{\omega}$, in an N dimensional space can be written

$$\boldsymbol{\omega} = \sum f_{\alpha,\beta,\ldots\gamma}(x) \boldsymbol{dx}^\alpha \wedge \boldsymbol{dx}^\beta \wedge \ldots \wedge \boldsymbol{dx}^\gamma \tag{A9.3}$$

where there are p terms in $\boldsymbol{dx}^\alpha \wedge \boldsymbol{dx}^\beta \wedge \ldots \wedge \boldsymbol{dx}^\gamma$ and the sum is over every inequivalent such term (e.g., for $p = 2$ there would be 3 such terms in 3D and 6 such terms in 4D). Now construct,

$$\boldsymbol{d\omega} = \sum \frac{\partial f_{\alpha,\beta,\ldots\gamma}}{\partial x^\mu} \boldsymbol{dx}^\mu \wedge \boldsymbol{dx}^\alpha \wedge \boldsymbol{dx}^\beta \wedge \ldots \wedge \boldsymbol{dx}^\gamma \tag{A9.4}$$

where there are $p+1$ terms in $\boldsymbol{dx}^\mu \wedge \boldsymbol{dx}^\alpha \wedge \boldsymbol{dx}^\beta \wedge \ldots \wedge \boldsymbol{dx}^\gamma$ and the sum is over every inequivalent such term (e.g., for $p = 2$ there would be only one such term in 3D and 4 such terms in 4D). We now want to integrate over the $p+1$ dimensional region, V. Carrying out the 1D integral over x^μ first we get, (A9.5)

$$\int_{x^\mu\ only} \boldsymbol{d\omega} = \sum \left(f_{\alpha,\beta,\ldots\gamma}(b) - f_{\alpha,\beta,\ldots\gamma}(a) \right) \boldsymbol{dx^\alpha} \wedge \boldsymbol{dx^\beta} \wedge \ldots \wedge \boldsymbol{dx^\gamma}$$

Considering integration over small rectangular elements the a, b will be independent of the remaining coordinates. There will generally be a number of different x^μ which give rise to results like (A9.5). These correspond to different pairs of parallel faces on the hypercube.

The remaining integrations required to produce $\int_V \boldsymbol{d\omega}$ mean integrating over the hypersurfaces $\boldsymbol{dx^\alpha} \wedge \boldsymbol{dx^\beta} \wedge \ldots \wedge \boldsymbol{dx^\gamma}$. But the two terms that result from (A9.5) when this is done are just the contributions to $\int_{\partial V} \boldsymbol{\omega}$ of the pair of parallel faces.

A macroscopic region is addressed by considering it to be composed of many small rectangles. Contributions to $\int_V \boldsymbol{d\omega}$ are additive over the small rectangles. In contrast, adjoining 'faces' have contributions to $\int_{\partial V} \boldsymbol{\omega}$ which cancel, so that ultimately all that remains are the contributions to $\int_{\partial V} \boldsymbol{\omega}$ from the boundary beyond which there are no more rectangles. This completes the proof.

Obviously we have ridden rough-shod over niceties like topologically non-trivial regions. However, the theorem is rigorous and does not depend upon the embedding space being a metric space. The theorem is essentially just topological.

www.ingramcontent.com/pod-product-compliance
Lightning Source LLC
LaVergne TN
LVHW052213200926
848481LV00001B/98